T0328264

REMOTE SENSING OF AEROSOLS, CLOUDS, AND PRECIPITATION

REMOTE SENSING OF AEROSOLS, CLOUDS, AND PRECIPITATION

Edited by

TANVIR ISLAM
NASA Jet Propulsion Laboratory, Pasadena, CA, United States

YONGXIANG HU
NASA Langley Research Center, Hampton, VA, United States

ALEXANDER KOKHANOVSKY
EUMETSAT, Darmstadt, Germany

JUN WANG
The University of Iowa, Iowa City, IA, United States

Elsevier
Radarweg 29, PO Box 211, 1000 AE Amsterdam, Netherlands
The Boulevard, Langford Lane, Kidlington, Oxford OX5 1GB, United Kingdom
50 Hampshire Street, 5th Floor, Cambridge, MA 02139, United States

© 2018 Elsevier Inc. All rights reserved.

No part of this publication may be reproduced or transmitted in any form or by any means, electronic or mechanical, including photocopying, recording, or any information storage and retrieval system, without permission in writing from the publisher. Details on how to seek permission, further information about the Publisher's permissions policies and our arrangements with organizations such as the Copyright Clearance Center and the Copyright Licensing Agency, can be found at our website: www.elsevier.com/permissions.

This book and the individual contributions contained in it are protected under copyright by the Publisher (other than as may be noted herein).

Notices
Knowledge and best practice in this field are constantly changing. As new research and experience broaden our understanding, changes in research methods, professional practices, or medical treatment may become necessary.

Practitioners and researchers must always rely on their own experience and knowledge in evaluating and using any information, methods, compounds, or experiments described herein. In using such information or methods they should be mindful of their own safety and the safety of others, including parties for whom they have a professional responsibility.

To the fullest extent of the law, neither the Publisher nor the authors, contributors, or editors, assume any liability for any injury and/or damage to persons or property as a matter of products liability, negligence or otherwise, or from any use or operation of any methods, products, instructions, or ideas contained in the material herein.

Library of Congress Cataloging-in-Publication Data
A catalog record for this book is available from the Library of Congress

British Library Cataloguing-in-Publication Data
A catalogue record for this book is available from the British Library

ISBN: 978-0-12-810437-8

For information on all Elsevier publications visit our website at https://www.elsevier.com/books-and-journals

Publisher: Candice Janco
Acquisition Editor: Laura S. Kelleher
Editorial Project Manager: Hilary Carr
Production Project Manager: Anitha Sivaraj
Cover Designer: Victoria Pearson

Typeset by SPi Global, India

Dedication

Dr. Tanvir Islam
To my wife, Tabassum Rahman (Anila), for embracing the journey with me that all started from the Pink City.

Dr. Alexander Kokhanovsky
To my family.

Yongxiang Hu
To my professors and fellow researchers.

Prof. Jun Wang
To my parents, my parents-in-law, my wife Jing Zeng, and my children: Kerry Z. Wang, Cindy Z. Wang, and Justin Z. Wang.

Contents

7. Polarimetric Technique for Satellite Remote Sensing of Superthin Clouds

WENBO SUN, ROSEMARY R. BAIZE, GORDEN VIDEEN, YONGXIANG HU

8. Cloud Screening and Property Retrieval for Hyper-Spectral Thermal Infrared Sounders

YU SOMEYA, RYOICHI IMASU

9. Surface Remote Sensing of Liquid Water Cloud Properties

CHRISTINE KNIST, HERMAN RUSSCHENBERG

10. Measuring Precipitation From Space

FRANCISCO J. TAPIADOR

11. Measurement of Precipitation from Satellite Radiometers (Visible, Infrared, and Microwave): Physical Basis, Methods, and Limitations

ATUL K. VARMA

12. Development of a Rain/No-Rain Classification Method Over Land for the Microwave Sounder Algorithm

SATOSHI KIDA, TAKUJI KUBOTA, SHOICHI SHIGE, TOMOAKI MEGA

13. Remote Sensing of Precipitation from Airborne and Spaceborne Radar

STEPHEN J. MUNCHAK

14. Status of High-Resolution Multisatellite Precipitation Products Across India

SATYA PRAKASH, ASHIS K. MITRA, RAKESH M. GAIROLA, HAMID NOROUZI, DAMODARA S. PAI

15. Real-Time Wind Velocity Retrieval in the Precipitation System Using High-Resolution Operational Multi-radar Network

HAONAN CHEN, VENKATACHALAM CHANDRASEKAR

Contributors

Rosemary R. Baize NASA Langley Research Center, Hampton, VA, United States

Tirthankar Banerjee Institute of Environment and Sustainable Development, Banaras Hindu University, Varanasi, India

Venkatachalam Chandrasekar Colorado State University, Fort Collins, CO, United States

Haonan Chen Colorado State University, Fort Collins, CO, United States

Rakesh M. Gairola Space Applications Centre, Ahmedabad, India

Yongxiang Hu NASA Langley Research Center, Hampton, VA, United States

Ryoichi Imasu Atmosphere and Ocean Research Institute, The University of Tokyo, Chiba, Japan

Satoshi Kida Earth Observation Research Center, Japan Aerospace Exploration Agency, Tsukuba, Japan

Christine Knist Deutsche Wetterdienst, Offenbach am Main, Germany

Alexander Kokhanovsky EUMETSAT, Darmstadt, Germany; National Research Nuclear University, Moscow, Russia

Takuji Kubota Earth Observation Research Center, Japan Aerospace Exploration Agency, Tsukuba, Japan

Manish Kumar Institute of Environment and Sustainable Development, Banaras Hindu University, Varanasi, India

Kwon H. Lee Gangneung-Wonju National University, Gangneung, South Korea

Luca Lelli University of Bremen, Bremen, Germany

Tomoaki Mega Osaka University, Osaka, Japan

Alaa Mhawish Institute of Environment and Sustainable Development, Banaras Hindu University, Varanasi, India

Akhila K. Mishra Indian Institute of Remote Sensing, Dehradun, India

Ashis K. Mitra National Centre for Medium Range Weather Forecasting, Noida, India

Sonoyo Mukai The Kyoto College of Graduate Studies for Informatics (KCGI), Kyoto, Japan

Stephen J. Munchak NASA Goddard Space Flight Center, Greenbelt, MD, United States

Hamid Norouzi City University of New York, Brooklyn, NY, United States

Damodara S. Pai India Meteorological Department, Pune, India

Satya Prakash City University of New York, Brooklyn, NY, United States

Daniel Rosenfeld The Hebrew University of Jerusalem, Jerusalem, Israel

Herman Russchenberg Delft University of Technology, Delft, The Netherlands

Shoichi Shige Kyoto University, Kyoto, Japan

Yu Someya Atmosphere and Ocean Research Institute, The University of Tokyo, Chiba, Japan

Prashant K. Srivastava Institute of Environment and Sustainable Development, Banaras Hindu University, Varanasi, India

Wenbo Sun Science Systems and Applications Inc; NASA Langley Research Center, Hampton, VA, United States

Francisco J. Tapiador University of Castilla-La Mancha (UCLM), Toledo, Spain

Atul K. Varma Space Applications Centre, Ahmedabad, India

Gorden Videen Space Science Institute, Boulder, CO; US Army Research Laboratory, Adelphi, MD, United States

Marco Vountas University of Bremen, Bremen, Germany

Yi Wang The University of Iowa, Iowa City, IA, United States

Jun Wang The University of Iowa, Iowa City, IA, United States

Man S. Wong Hong Kong Polytechnic University, Kowloon, Hong Kong

Xiaoguang Xu The University of Iowa, Iowa City, IA, United States

Author Biographies

Tanvir Islam is presently with the NASA Jet Propulsion Laboratory, who specializes in remote sensing observations. Currently, he is engaged with the development of advanced microwave calibration and retrieval algorithms for NASA's Earth observing missions.

Prior to joining NASA/JPL in 2015, he was with the NOAA/NESDIS/STAR, and worked on the development of satellite remote sensing algorithms, with an emphasis on microwave variational inversion techniques (2013–15). He also held visiting scientist positions at the University of Tokyo as part of the NASA/JAXA precipitation measurement missions (PMM) algorithm development team in 2012, and at the University of Calgary in 2015. He received his PhD in remote sensing from the University of Bristol, United Kingdom, in 2012.

Dr. Islam received the Faculty of Engineering Commendation from the University of Bristol (nominated for a University Prize for his outstanding PhD thesis) in 2012, the JAXA visiting fellowship award in 2012, the CIRA postdoctoral fellowship award in 2013, the Calgary visiting fellowship award in 2015, and the Caltech postdoctoral scholar award in 2015. He has served as a lead guest editor for a special issue on "Microwave Remote Sensing" for *Physics and Chemistry of the Earth* (Elsevier), and currently serving on the editorial board of *Atmospheric Measurement Techniques* (EGU) and *Scientific Reports* (Nature). He has published 4 books and more than 60 peer-reviewed papers in leading international journals. His primary research interests include microwave remote sensing, radiometer calibration, retrieval algorithms, radiative transfer theory, data assimilation, mesoscale modeling, cloud and precipitation system, and artificial intelligence in geosciences.

Jun Wang is a professor in the University of Iowa, with joint appointments in the Department of Chemical and Biochemical Engineering and Iowa Informatics Initiative, and secondary affiliation in Center for Global and Regional Environmental Studies and the Department of Civil and Environmental Engineering. Prior to joining the University of Iowa, he worked at the University of Nebraska, Lincoln, first as an assistant professor, and then as an associate professor. His current research focuses on the integration of satellite remote sensing and chemistry transport model to study air quality, wildfires, aerosol-cloud-radiation interaction, and land-atmospheric interaction. He has authored/co-authored 100+ peer-reviewed articles, and has been a science team member of several NASA satellite missions, focusing on the development and improvement of aerosol retrieval algorithms. He is the journal section editor for the "New Directions" column in Atmospheric Environment, and a guest editor for Remote Sensing. In 2005, Jun Wang received his PhD in atmospheric sciences from the University of Alabama, Huntsville. In 2005–07, he was a postdoctoral researcher at Harvard University. More about his research can be found at: http://arroma.uiowa.edu.

Yongxiang Hu, Atmospheric Sciences Division, NASA Langley Research Center. Dr. Hu earned his PhD from the University of Alaska, Fairbanks. Since 1995, Dr. Hu has been a research scientist/senior research scientist at NASA's

Langley Research Center. Dr. Hu began his career working on radiative transfer and climate modeling in his PhD study. He worked on the ERBE and CERES projects, and then joined the CALIPSO team studying lidar remote sensing. Dr. Hu is currently working on developing innovative remote sensing concepts, such as photon orbital angular momentum measurements and studying subdiffraction limit telescopes. Dr. Hu's primary scientific accomplishments include theoretical radiative transfer studies for active and passive remote sensing; discovery of the relation between lidar depolarization and multiple scattering for water cloud droplets; development of highly accurate global cloud phase product using CALIPSO observations; high spatial resolution global ocean surface wind speed retrieval technique and data product using CALIPSO lidar measurements; innovative lidar remote sensing techniques, such as using space-based lidar for studying ocean primary productivity and carbon cycle, as well as deriving value added vegetation canopy, snow and sea ice product from CALIPSO, and theoretical and engineering studies of differential absorption radar concept for measurements of ocean/land surface atmospheric pressure.

Dr. Hu author/co-authored more than 150 peer-reviewed scientific journal articles with an SCI index of 47 on Google Scholar (https://scholar.google.com/citations?user=YySlI2oAAAAJ&hl=en) and 39 on ResearcherID (http://www.researcherid.com/rid/K-4426-2012).

Alexander A. Kokhanovsky received his MS in theoretical physics from the Belarussian State University, Minsk, Belarus, in 1983, and his PhD in optical physics from the B. I. Stepanov Institute of Physics, National Academy of Sciences of Belarus, Minsk, in 1991. His PhD work focused on modeling light scattering properties of aerosol media, clouds, and foams.

He is the editor of the Springer Series in Light Scattering and the Wiley Series in Atmospheric Physics and Remote Sensing. He is the author of the books *Light Scattering Media Optics: Problems and Solutions* (Springer-Praxis, 1999, 2001, 2004), *Polarization Optics of Random Media* (Springer-Praxis, 2003), *Cloud Optics* (Springer, 2006), and *Aerosol Optics* (Springer-Praxis, 2008). He has published more than 200 papers in the field of environmental optics, radiative transfer, remote sensing, and light scattering. His research is directed toward the solution of various forward and inverse problems of atmospheric optics.

Dr. Kokhanovsky is a member of the European Geophysical Union.

Preface

REMOTE SENSING OF AEROSOLS, CLOUDS, AND PRECIPITATION

Atmospheric parameters, including aerosols, clouds, and precipitation, play a very important role in the Earth's climate system. The factors influencing global climate variability are directly or indirectly affected by aerosols, clouds, and precipitation, and thus, modulate the Earth's radiation and surface energy balance. Therefore, global observations of aerosols, clouds, and precipitation are crucial not only to better understand the processes involved in the climate system but also to predict future climate change.

Satellite remote sensing can be a very valuable tool for acquiring such global observations of aerosols, clouds, and precipitation for monitoring Earth's climate system. Some regions of the Earth, in particular, are not easily accessible for obtaining in situ data, and while in those regions, space-borne instruments can act as complementary data sources. Also, large-scale data mapping and monitoring are only feasible with space-based observations.

In recent years, good progress has been made in dealing with the remote sensing of aerosols, clouds, and precipitation. Notably, the remote-sensing community has seen the launches of satellite instruments covering a broader range of frequencies in the electromagnetic spectrum involving advanced, active remote-sensing instruments (e.g., radars and lidars), as well as passive remote-sensing instruments with optical and microwave imagers. The advent of modern remote-sensing instruments brings new opportunities and challenges to this domain allowing for the exploration of novel methods and applications, and to potentially further progress. However, while combining different measurements from a broad suite of payloads (e.g., A-train) would certainly improve our understanding on different aspects of aerosols, clouds, and precipitation research, newer techniques and retrieval approaches are necessary for synergistic advancements.

The atmospheric remote sensing of aerosols, clouds, and precipitation is a growing research field, and it is still evolving towards the further potential of future developments. This book compiles some of the recent advancements made within the area of aerosols, clouds, and precipitation remote sensing. The book covers studies from a wide range of measurements including microwave (both active and passive), visible, and infrared portions of the spectrum. These contributions are gathered from state-of-the-art research in atmospheric remote sensing using space-borne, airborne, and ground-based datasets, but with the focus on supporting earth-observation satellite missions for aerosols, clouds, and precipitation studies.

The book is expected to provide the readers with a critical, all-inclusive overview of aerosols, clouds, and precipitation remote sensing originating from studies using earth-observation satellite datasets. We hope that the book will serve as a reference for scientists from a wide range of subject domains, including remote sensing, earth science, electromagnetics, climate physics, and space engineering. Operational forecasters, meteorologists, geospatial experts, modelers, policy makers, and environmental experts should also find the book very valuable.

Tanvir Islam
Jet Propulsion Laboratory, California Institute of Technology, Pasadena, CA, United States

Yongxiang Hu
NASA Langley Research Center, Hampton, VA, United States

Alexander Kokhanovsky
EUMETSAT, Darmstadt, Germany

Jun Wang
The University of Iowa, Iowa City, IA, United States

1

Passive Remote Sensing of Aerosol Height

Xiaoguang Xu, Jun Wang*, Yi Wang*, Alexander Kokhanovsky*[†,‡]

*The University of Iowa, Iowa City, IA, United States [†]EUMETSAT, Darmstadt, Germany
[‡]National Research Nuclear University, Moscow, Russia

List of Acronyms and Abbreviations

3MI	multiview multichannel multipolarization imager
AATSR	advanced along-track scanning radiometer
ADEOS-I/II	advanced earth observing satellite-I/II
AIRS	atmospheric infrared sounder
AOD	aerosol optial depth
APS	aerosol polarimetry sensor
APT	automatic picture transmission (a camera on NASA's Nimbus II satellite)
ASTER	advanced spaceborne thernal emission and reflection
ATSR	along-track scanning radiometer
BT	brightness temperature
CALIOP	cloud-aerosol LIDAR with orthogonal polarization
CALIPSO	cloud-aerosol LIDAR and infrared pathfinder satellite observation
CarbonSpec	carbon dioxide spectrometer
CPL	cloud physics LIDAR
CrIS	cross-track infrared sounder
DSCOVR	Deep Space Climate ObserVatoRy
ENVISAT	ENVIronment SATellite
EPIC	Earth polychromatic imaging camera
ERS-1/2	European remote sensing-1/2
FOV	field of view
FWHM	full width at half maximum
GOES	geostationary operational environmental satellite
GOES-CAPE	GEOstationary coastal and air pollution events
GOME	global ozone monitoring experiment
GOSAT	Greenhouse gases Observing SATellite
IASI	infrared atmospheric sounding interferometer
IDL	interative data language
IR	infrared
Lidar (LIDAR)	LIght Detection and Ranging
MAIA	multiangle imager for aerosols
MERIS	medium resolution imaging spectrometer
METOP-SG	METeorological OPerational satellite-Second Generation
MINX	MISR INteractive eXplorer
MISR	multiangle imaging spectroradiometer
NASA	National Aeronautics and Space Administration
OCI	ocean color instrument for the PACE mission
OCO-2	orbiting carbon observatory-2
PACE	the plankton, aerosol, cloud, ocean ecosystem (NASA mission)
PARASOL	polarization and anisotropy of reflectances for atmospheric sciences coupled with observations from a LIDAR
PM	particulate matter
PODEX	polarimeter definition experiment
POLDER	polarization and directionality of the Earth's reflectance

Remote Sensing of Aerosols, Clouds, and Precipitation
https://doi.org/10.1016/B978-0-12-810437-8.00001-3

© 2018 Elsevier Inc. All rights reserved.

RSP research scanning polarimeter
SCIAMACHY SCanning Imaging Absorption spectroMeter for Atmospheric CHartographY
SEAC4RS studies of emissions, atmospheric composition, clouds and climate coupling by regional surveys
SEVIRI spinning enhanced visible and infrared imager
SRON Netherlands Institute for Space Research
TanSat Tan (Carbon) satellite
TANSO-FTS thermal and near infrared sensor for carbon observation—Fourier transform spectrometer
TOA top of the atmosphere
TROPOMI TROPOspheric Monitoring Instrument
UV ultraviolet

1 INTRODUCTION

Atmospheric aerosols from both natural and anthropogenic sources have diverse impacts on earth's climate and environment. They affect earth's energy budget directly by scattering and absorbing solar and terrestrial radiation and indirectly through altering the lifetime and radiative properties of clouds. The average of the aerosol radiative forcing across the globe was estimated to range from -0.1 to $-1.9\ \mathrm{W\,m^{-2}}$ with the best estimate of $-0.9\ \mathrm{W\,m^{-2}}$ (Boucher et al., 2013), indicating that the cooling effects of aerosols might counteract the warming effects caused by the increase in carbon dioxide ($1.82 \pm 0.19\ \mathrm{W\,m^{-2}}$) since the industrial revolution (Myhre et al., 2013). The magnitudes, and even the sign, of the aerosol radiative effects are strongly influenced by the vertical distribution of aerosols (Peters et al., 2011; Wilcox, 2012; Zhang et al., 2013). Particularly, absorption of solar radiation by smoke and dust aerosols can change the air temperature profile and modify atmospheric stability in the boundary layer and free troposphere (Wendisch et al., 2008; Babu et al., 2011). The resulting effect on the atmospheric stability depends on the altitude of aerosol layers. Furthermore, the altitude of absorbing aerosols relative to a cloud is a determining factor in the effect of aerosols on cloud cover and lifetime (Koch and Del Genio, 2010). However, aerosol profiles simulated by current climate models can differ by up to an order of magnitude (Koffi et al., 2012; Kipling et al., 2016), and the resulting estimated aerosol radiative effects are therefore subjective to considerable uncertainties. In addition, elevated dense aerosol plumes such as airborne volcanic ash, which is invisible to aircraft radar, can pose significant hazards to aviation safety (Sears et al., 2013).

Knowledge of aerosol vertical distribution is also important to the remote sensing and climate modeling studies in many aspects. For instance, the selection of appropriate aerosol vertical distributions is essential in retrieving aerosol optical depth (AOD) in the UV channels (Torres et al., 1998). It could also affect the retrieval accuracy of aerosol microphysical properties from photopolarimetric measurements (Waquet et al., 2009; Chowdhary et al., 2005), the atmospheric correction for ocean color remote sensing (Duforêt et al., 2007), and the retrieval of the thermal state of the atmosphere from IR sounders (Maddy et al., 2012). In addition, an aerosol profile is needed for remote sensing of the surface concentration of particulate matters (PM) from space (Wang and Christopher, 2003). The plume height information of smoke, volcanic, and dust aerosols can provide initial conditions for the modeling of their emissions and the physics of the plume formation (such as the injection height of the plume) (Kahn et al., 2007, 2008; Yang et al., 2013b). Consequently, there is a critical need to observe the vertical distribution of aerosols on a global scale for validating transport models (e.g., Yu et al., 2010) and for better understanding the impact of aerosols on earth's climate and environment.

The only way to obtain the spatial and temporal variations of aerosol profiles across the globe is through satellite remote sensing. A detailed profile of aerosol backscattering can be accurately probed by active remote sensing techniques using LIDAR, such as CALIOP equipped on the CALIPSO platform (Winker et al., 2009). However, the spatial coverage of this type of measurement suffers from its narrow swath. For example, CALIOP scans a given site once every 16 days. The gaps between its adjacent suborbital "curtains" can be as wide as 2200 km.

By contrast, passive remote sensing techniques have been mainly used to retrieve total column quantities in cloud-free scenes with less information on the aerosol vertical distribution. Nevertheless, recent efforts have developed various techniques for retrieving aerosol layer height or profile from passive sensing measurements in the UV, visible, and thermal IR bands. While not being able to provide the same level of accuracy as a LIDAR, these passive techniques can add an important extension due to better spatial coverage and more frequent measurements. Existing passive techniques include stereo photogrammetry (e.g., Muller et al., 2007; Fisher et al., 2014; Zakšek et al., 2013), UV-visible polarization approach (e.g., Wu et al., 2016), spectroscopy in the oxygen (O_2) absorption bands (e.g., Corradini and Cervino, 2006; Dubuisson et al., 2009; Kokhanovsky and Rozanov, 2010; Sanders et al., 2015; Ding et al., 2016) and the O_2-O_2 band (Park et al., 2016; Chimot et al., 2017), and the IR retrieval technique (e.g., Pierangelo et al., 2004; Vandenbussche et al., 2013). These techniques make use of measurements from a number of passive sensors listed in Tables 1–4. This chapter is devoted to reviewing progress in the recent developments of these passive techniques and to discussing their associated physical principles as well as retrieval challenges. In Section 2, we briefly describe the characteristics of typical aerosol vertical profiles and discuss various representations of vertical profiles used in remote sensing algorithms. Section 3 is devoted to the description of retrieval techniques for the passive remote sensing of an aerosol profile. Lastly, Section 4 offers conclusions and outlooks on the passive remote sensing of aerosol vertical distribution.

TABLE 1 The Characteristics of Satellite Instruments With Dual- or Multiview Capability

Instrument/Satellite, Available for	Spectral Channels	Swatch/Spatial Resolution	Remarks
ATSR-1/ERS-1, 07/1991–06/1996	1.6, 3.7, 11, and 12 μm	500 km/1 × 1 km^2	Dual-view conical scan with forward VZA of 55 degrees and at nadir of close 0 degrees
ATSR-2/ERS-2, 04/1995–09/2011	0.55, 0.66, 0.87, 1.6, 3.7, 11, and 12 μm	500 km/1 × 1 km^2	Followup of ARSR-1
AATSR/ENVISAT, 05/2002–04/2012	Same to ATSR-2	500 km/1 × 1 km^2	Followup of ATSR-2
MISR/Terra, 12/1999–present	446, 558, 672, and 866 nm	380 km/275 × 250 m^2	Nine 14-bit push-broom cameras at 9 VZAs of 0, 26.1, 45.6, 60.0, and 70.5 degrees for both forward and afterward viewing
POLDER, 3MI, and MAIA	–	–	See Table 2

TABLE 2 The Characteristics of Satellite Instruments With Polarization Measurements in the Near-UV Spectral Region

Instrument/Satellite, Available for	Spectral Channels[a]	Swath/Spatial Resolution	Remarks
POLDER-1/ADEOS-I, 11/1996–07/1997 POLDER-2/ ADEOS-II, 04/2003–10/2003	**443**, 490, 565, **670**, 763, 765, **865**, 910 nm	2200 km/6 × 7 km^2	Up to 16 viewing angles
POLDER-3/PARASOL, 12/2004–12/2013	443, **490**, 565, **670**, 763, 765, **865**, 910, 1020 nm	2200 km/6 × 7 km^2	Up to 16 viewing angles
APS/Glory, Failed launch in 2011	**410, 443, 555, 672, 865, 910, 1378, 1610, 2250** nm	5.6 km/5.6 km nadir scan	Along-track scanner seeing the same scene from about 250 views
MAIA/unknown, Around 2022	367, 386, **445**, 543, **645**, 751, 763, **862**, 945, **1620**, 1888, **2185** nm	300 × 300 km^2 target area/<300 m	5 or 7 viewing angles
3MI/METOP-SG, Planned 2019	**410, 443, 490, 555, 670**, 763, 765, **865**, 910, 1370, **1650, 2130** nm	2200 km/4 × 4 km^2	14 different viewing angles

[a] *Polarization channels are indicated in bold.*

TABLE 3 The Characteristics of Satellite Instruments that Perform Measurements in O_2 Absorption Bands

Instrument/Satellite, Available for	Spectral Channels[a]	Spectral Resolution[b]	Swatch/Spatial Resolution[c]	Remarks
MERIS/ENVISAT, 05/2002–04/2012	753.75 nm; 761.75 nm	7.5 nm; 3.75 nm	1150 km/ocean: 1040 × 1200 m^2; Land: 260 × 260 m^2	
POLDER-series	763 nm; 765 nm	10 nm; 40 nm		See Table 2
EPIC/DSCOVR 06/2015–present	680 nm; 688 nm; 764 nm	3 ± 0.6 nm; 0.8 ± 0.2 nm; 1 ± 0.2 nm; 2 ± 0.4 nm	Sunlit area of the earth/12 × 12 km^2	DSCOVR spacecraft is located at the Earth-Sun Largrange-1 point
3MI/METOP-SG, Planned 2019	754 nm; 763 nm	10 nm; 10 nm	2200 km/4 × 4 km^2	See Table 2
GOME-2/MET OP-A/B/C METOP-A 10/2006— METOP-B 09/2012— METOP-C planned 2018	240–790 nm	0.2–0.5 nm	1920 km/80 × 40 km^2	
SCIAMACHY/ ENVISAT, 2002–12	240–2380 nm	0.48 nm in O_2 A band	960 km/30 × 60 km^2	
TANSO-FTS/GOSAT, 2009–present	758–775 nm	0.5 cm^{-1}	±35-degree cross-track/10.5 × 10.5 km^2	Two-axis pointing mechanism performs 3-point cross-track scan

TABLE 3 The Characteristics of Satellite Instruments that Perform Measurements in O_2 Absorption Bands—cont'd

Instrument/Satellite, Available for	Spectral Channels[a]	Spectral Resolution[b]	Swatch/Spatial Resolution[c]	Remarks
OCO-2	757–775 nm	0.042 nm	10.3 km/ 1.3×2.25 km^2	OCO-2 instrument is a grating spectrometer
CarbonSpec/TanSat, 12/2016-present	758–778 nm	0.044 nm	20 km/2×2 km^2	Also called CarbonSat as Tan means "carbon" in Chinese A mission similar to the GOSAT and OCO-2
TROPOMI/Sentinel-5 Precursor, Planned 09/2017	675–775 nm every 0.14 nm	0.38 nm	2600 km/up to 3.5×7 km^2	7×7 km^2 for 675–725, and 3.5×7 km^2 for 725–775
OCI/PACE around 2022–23	350–890 nm	5 nm	60-degree view angle/1 km^2	2-day global coverage

[a] *Listed here are only the channels including O_2 bands.*
[b] *Spectral resolution is in terms of the spectral full width at half maximum (FWHM).*
[c] *Spatial resolution is for a nadir field of view (FOV).*

TABLE 4 The Characteristics of Satellite Infrared Spectrometers (Sounders)

Instrument/Satellite, Available for	Spectral Range	Spectral Resolution	Swatch/Spatial Resolution (km)	Remarks
AIRS/Aqua 2002–present	2169–2674 cm^{-1} 1265–1629 cm^{-1} 649 1136 cm^{-1}	0.5–2.0 cm^{-1}	1650/50	Each 50-km fields of regard (FOR) contains 3×3 array of 13.5-km field-of-views (FOVs)
IASI/MetOp-A, 2006–present IASI/MetOp-B, 2012–present	2000–2760 cm^{-1} 210–2000 cm^{-1} 645–1210 cm^{-1}	0.25 cm^{-1}	2200/50	Each 50-km FOR contains a 2×2 array of 12-km FOVs
CrIS/Suomi-NPP, 2011–present	2155–2550 cm^{-1} 1210–1750 cm^{-1} 650–1095 cm^{-1}	0.6 cm^{-1}	2200/50	Each 50-km FOR contains a 3×3 array of 13-km FOVs

2 AEROSOL VERTICAL DISTRIBUTION

2.1 What Controls the Vertical Distribution of Aerosol?

Aerosol particles from industrial emissions are often well-mixed in the planetary boundary layer and decay rapidly with altitude in the free troposphere. However, in some cases the concentration of aerosol reaches a maximum at a certain height above the ground. For example, soil particles from dust outbreaks can be easily injected into the free troposphere by large-scale vertical advection. Forest and agriculture fires can lift smoke plumes into the upper level of the troposphere by strong thermal radiative energy. Ash aerosol plumes from an eruptive volcano often break through the tropopause and could remain in the stratosphere for several months. Therefore, the vertical distribution of atmospheric aerosols is governed by a complex interplay among the emission and deposition processes, aerosol microphysical properties (e.g., size and composition), and meteorological conditions (including wind, atmospheric stability, planetary boundary layer evolution, and precipitation).

Recently, Kipling et al. (2016) investigated a range of controlling factors that influence the aerosol vertical distribution through model sensitivity tests. They found that the processes having the greatest impact vary between different aerosol components and over the particle size distribution. For instance, the initial injection height of a biomass-burning emission is important for carbonaceous aerosols. The role of dry deposition dominates for large-sized particles like mineral dust, while in-cloud scavenging is important for all other aerosol types. Convective transport is found to be very important in controlling the vertical profile of all aerosol components. Studies based on CALIOP measurements (e.g., Thomas et al., 2013; Yu et al., 2010) have indicated that more than 70% of aerosols are located below 1 km in the planetary boundary layer, which mostly corresponds to anthropogenic aerosols in industrial pollution source regions and sea salts over oceans. Elevated aerosol layers are found over the dust belt and source regions of biomass-burning particles (e.g., southern Africa and South America).

2.2 Representing Aerosol Profile and Height

The development of a height retrieval algorithm requires a representation of the aerosol vertical profile. Fundamental radiative transfer defines the AOD τ_λ as the integrated extinction of light along its path through the atmosphere:

$$\tau_\lambda(z) = \int_z^{z_{TOA}} \alpha(\lambda, z')dz' \qquad (1)$$

where z is the physical path in meters, λ is the spectral wavelength of light, and α is the extinction coefficient of aerosol in m^{-1}. The above equation gives $\tau_\lambda = 0$ at the top of the atmosphere (TOA) and $\tau_\lambda = \tau^*$ is columnar AOD for $z = 0$ at the surface. Therefore, the vertical distribution of aerosol particles is commonly characterized by remote sensing techniques as the profile of aerosol extinction coefficient at the given spectral wavelength. From active LIDAR measurements, the extinction profile can be determined from measured backscatter with a prescribed extinction-to-backscatter ratio. The extinction coefficient is linearly related to the aerosol mass (or volume) concentration by mass (or volume) extinction coefficient, an optical parameter that depends on aerosol size and chemical composition. This extinction-to-mass conversion is often necessary when a satellite-derived aerosol profile is compared with in situ measurements or numerical model simulations.

Different algorithms have employed a variety of definitions on the aerosol height and profile thus far. A common assumption is that aerosol is homogeneously distributed within a layer of given depth (e.g., Pierangelo et al., 2004; Kokhanovsky and Rozanov, 2010). This layer can extend from the surface or be an aloft layer in the atmosphere. Under this assumption, the retrieval target is usually the top altitude of the aerosol layer (e.g., Kokhanovsky and Rozanov, 2010) or the central altitude of the aerosol layer (e.g., Pierangelo et al., 2004). Alternatively, some studies assume that the profile of the aerosol extinction coefficient follows a certain type of distribution function. For instance, an exponential-decay profile characterized by a scale height is often used in atmospheric correction algorithms (Gordon, 1997). Some studies have used Gaussian function (e.g., Dubovik et al., 2011; Ding et al., 2016; Wu et al., 2016; Xu et al., 2017) and lognormal distribution function (e.g., Hollstein and Fischer, 2014; Sanghavi et al., 2012) that are characterized by a peak height (the altitude with peak aerosol extinction) and a half-width parameter. Another way to represent the aerosol profile is to explicitly allocate aerosol concentration (or extinction coefficient) on each discrete atmospheric layer (e.g., Koppers and Murtagh, 1997; Vandenbussche et al., 2013).

In addition, profile representations also differ in the choice of pressure or height for the vertical coordinate. The retrieval techniques using the O_2 spectroscopy and the near-UV polarization can benefit from the pressure coordinate because the optical depths of O_2 absorption and Rayleigh scattering are directly contingent on the atmospheric pressure level. On the other hand, the stereoscopic technique favors the height coordinate for a pure geometric derivation. The conversion between pressure and height can be made through atmospheric sounding or modeling data.

It is difficult to judge whether one type of representation is superior to another. Ideally, a good representation should be able to sufficiently describe the realistic aerosol layering, and, at the same time, ensure that retrieved profile parameters are resolvable from the satellite measurements. If the number of independent profile parameters exceeds the independent pieces of profile information of the measurements, the inverse problem will become ill posed and can lead to unrealistic solutions. In this regard, sensitivity and information content studies are necessary to provide general guidance for algorithm development (e.g., Frankenberg et al., 2012; Ding et al., 2016).

However, there remains a lack of standard metric for aerosol vertical distribution in the remote sensing and modeling community. One needs to introduce an "effective height" when it comes to comparisons between the satellite retrievals and models or the validation of passive remote sensing of aerosol height against the active LIDAR measurements. From the perspective of passive remote sensing, the definition of "effective height" depends on how the profile is represented. It could be any of these parameters: the scale height of an exponential-decay profile, the center height of a homogeneously aerosol layer, or the peak height of a Gaussian or lognormal profile. In contrast, from the model or LIDAR perspective where the profile is discretely represented, the "effective height" has been defined as aerosol layer-mass-weighted mean height or aerosol layer-extinction-weighted mean height (Koffi et al., 2012; Wu et al., 2016). In addition, Yu et al. (2010) used an aerosol scale height to compare the aerosol layer height between CALIOP measurements and GOCART simulations, which is defined as the above-ground-level altitude below which the aerosol extinction accounts for 63% (equals to $1-e^{-1}$) of total columnar aerosol extinction.

3 PASSIVE REMOTE SENSING TECHNIQUES FOR RETRIEVAL OF AEROSOL LAYER HEIGHT

Over the last two decades, literally dozens of retrieval algorithms have been developed for deriving aerosol profile information from passive satellite sensors. Those algorithms were mostly developed based on heritages of the remote sensing of cloud altitude. However, the retrieval of aerosol height is much more challenging because aerosols are in general less optically thick and have more complex optical properties that depend on a wider range of aerosol size spectrum and chemical composition. This complexity is compounded by the heterogeneity of the underlying surface. Those passive algorithms can be grouped into two main categories: (1) the stereoscopic technique that relies on the principle of parallax; and (2) the spectroscopic techniques that are based on radiative transfer methods. The spectroscopic approach includes three major techniques: (1) the polarimetric analysis of reflected sunlight at the near-UV spectrum, (2) the analysis of scattered radiances in the oxygen absorption bands, and (3) the thermal IR spectral radiance technique. These techniques differ with respect to the underlying physics and applied spectral range. Aerosol height retrieval algorithms developed based upon these techniques have been utilized to process satellite data from many existing passive instruments. Tables 1–4 list those existing and future instruments corresponding to each technique. In this section, we discuss the progress of recent studies of these techniques as well as their physical principles, advantages, and challenges.

3.1 Stereo Photogrammetry

Stereo photogrammetry is also called stereography, stereoscopy, or stereo matching technique. The height estimate based on this technique is often called stereo height. The stereoscopic technique relies on the principle of parallax. As illustrated in Fig. 1, an elevated plume is seen from two different viewing directions with forward and afterward viewing zenith angles of θ_F and θ_A, respectively. The plume projects a displacement (d) on the background images from these two views. The height of the plume (h) is proportionate to the displacement by $h = d/(\tan\theta_F + \tan\theta_A)$ if neglecting the curvature of the earth's surface and the refraction of light in the atmosphere. Therefore, the major task of a stereo height algorithm is to detect aerosol contrast features and estimate parallax shift d through image matching procedures.

Satellite stereoscopy was originally introduced as a way to determine cloud-top height by Ondrejka and Conover (1966), who showed its possibility by nephanalysis of the images taken by the Nimbus II APT camera. As one of NASA's second-generation polar-orbiting meteorological satellites, Nimbus II was the first satellite that could provide images well suited to stereography. Ondrejka and Conover (1966)

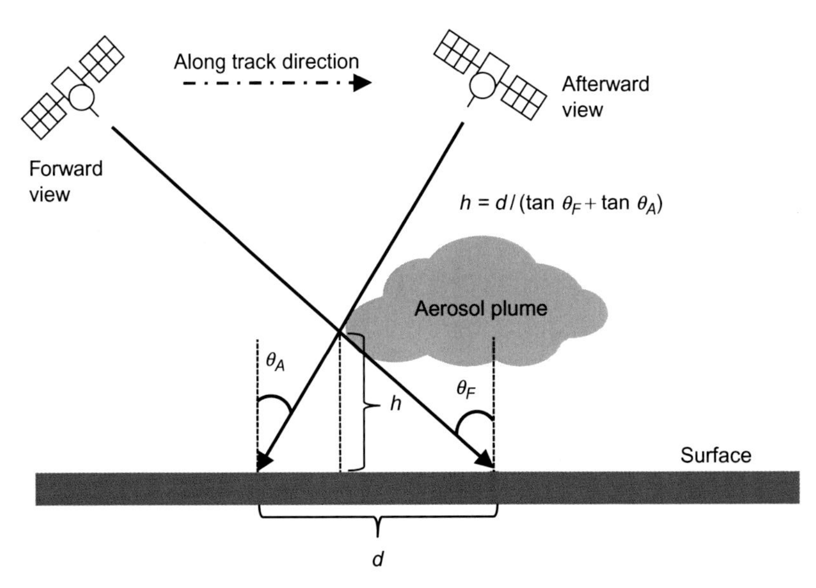

FIG. 1 A simplified diagram illustrating how aerosol height is observed by stereoscopic technique. The aerosol plume is observed by a sensor (or two sensors) from two directions, for which the forward and afterward viewing zenith angles are θ_F and θ_A, respectively. The altitude of the plume edge (h) is proportional to the separation (parallax, d) of plume position at ground between these two observations.

pointed out that cloud heights can be obtained above the overlapping area of two successive APT images taken with a 3-min difference between exposures. Their analysis suggests a practical accuracy of about 2 km for the height determination. Two years later, Kikuchi and Kasai (1968) investigated several Nimbus II APT images through manual image matching, for which they could distinguish the areas of sea ice from the areas of cloud and classify clouds into three grades of height. Adachi and Kasai (1970) further improved the estimation of cloud height by Kikuchi and Kasai (1968) through correcting the parallax due to the earth's spherical surface and the inclination of the optical axis, providing five grades of cloud-top height from surface to 12.5 km with 2.5 km intervals. In the early 1980s, Hasler (1981) and Hasler et al. (1983) estimated stereo heights of cloud tops using images acquired from two geostationary satellites (GOES west and east) with horizontal resolution of 1 km and vertical resolution of about 0.5 km. In the last two decades, stereo photogrammetry has been employed in operational retrievals of cloud-top height from several multiangle satellite instruments such as ATSR (Prata and Turner, 1997), ASTER (Seiz et al., 2006), MISR (Moroney et al., 2002; Muller et al., 2002), ATSR-2 (Muller et al., 2007), and AATSR (Fisher et al., 2016). The retrieval accuracy of some of these operational algorithms can reach up to 0.3 km due to improvements in both the algorithm and instrument performance.

With the heritage of cloud stereo height, several stereo photogrammetric algorithms have been developed for deriving the height of volcanic ash, wildfire smoke, and dust plumes. Based upon MISR's standard stereoscopic algorithm, the MISR team at the Jet Propulsion Laboratory developed the MINX software for the MISR Plume Height Project (Diner et al., 2008; Nelson et al., 2008, 2013). MINX is an IDL-based manual digitalization toolkit for retrieving the heights of individual pixels of smoke, dust, and volcano plumes with uncertainty less than 0.5 km. It uses seven inner cameras of MISR to determine wind-corrected stereo height from the pixel shifts relative to surface between six camera pairs. The horizontal and vertical resolution for MINX-retrieved plume heights are 1.1 km and 275 m, respectively. MINX has been used to retrieve stereo heights of wildfire plumes from multiyear MISR data, which improves the injection height estimate of biomass-burning emissions for chemistry transport models (Chen et al., 2009; Val Martin et al., 2010). MINX was also used to retrieve the height and optical depth of volcano-erupted ash clouds from Mount Etna in 2001 and 2002 (Scollo et al., 2012) and Eyjafjallajökull in 2010 (Kahn and Limbacher, 2012).

The stereoscopic technique has also been employed by other satellite sensors for smoke and ash plume retrievals. Virtanen et al. (2014) investigated the stereo height of ash plumes also from the Eyjafjallajökull 2010 eruption by using the AATSR/ENVISAT imagery. Their algorithm relies on matching the 10.85-μm brightness temperature (BT) of AATSR's nadir and 55-degree forward views, which also includes automatic detection of volcanic ash pixels using the BT difference between 11 and 12 μm channels (negative for ash pixels and positive for others). Their stereo heights estimated from AATSR were found to be about 0.5 km lower than those retrieved from MISR imagery with MINX. The ash plume height of the same volcano eruption was also studied by Zakšek et al. (2013) based on imagery parallax between a geostationary (SEVIRI) and a polar-orbiting (MODIS) satellite. To minimize the wind-induced displacement, Zakšek et al. (2013) used two sequential SEVIRI images, one before and one after the MODIS overpass, and interpolated the ash position from SEVIRI data to the time of MODIS overpass. They also compared this with the MINX-retrieved plume heights showing their retrieved heights are about 1 km higher.

The advantage of the stereo height technique, as pointed out by many studies (e.g. Hasler, 1981; Scollo et al., 2012; Fisher et al., 2016; Zakšek et al., 2013), is that it is a pure geometric approach. Consequently, the height estimates do not rely on radiometric calibration uncertainties or the emissivity of the plume material. For the same reason, it neither requires any radiative transfer computation nor needs the datasets of aerosol optical property or air temperature profile. The only ancillary data requirement is an accurate knowledge of satellite viewing geometry and geographic registration of scanned pixels for stereo reconstruction and for the conversion of the parallax in the imagery into actual observation.

Meanwhile, this technique has some limitations. First, it requires that the aerosol plume has texturally rich features. If the aerosol plume is texturally homogeneous, the imagery matching process will fail and the plume heights could be derived incorrectly. Therefore, this technique only favors the plume-type aerosols that are heterogeneous enough to enable the height retrieval to succeed. Second, the texture of the targeting aerosol plume must be discernable from the surface background. The quality of height retrievals can be substantially degraded for low-optical-depth aerosols, under which conditions the dominant visible textures are more likely attributed to the surface terrain. This problem could be partially remedied by including more channels for cross-correlation, such as the multiband matcher used in MINX. Third, wind-induced parallax displacement, if neglected or falsely corrected, can cause an error in the retrieved heights, as the paired images used for stereo matching are usually taken with a time difference from seconds to tens of minutes. Although MINX can efficiently detect the wind speed of an aerosol plume for the cross-track motion, it has difficulty detecting the motion when the wind is in the along-track direction.

3.2 Polarization at the Near-UV Spectrum

The application of polarization measurements for height estimation also started with cloud retrievals, which was first introduced by Buriez et al. (1997) for deriving cloud properties from POLDER-1 onboard the ADEOS-I satellite. The polarization of scattered light at the TOA in the near-UV spectrum is mainly contributed by the Rayleigh scattering above an opaque cloud layer. Buriez et al. (1997) derived cloud-top pressure from the molecular optical thickness above a cloud, which was calculated from polarized reflectance at 443 nm following a single-scattering approximation. For a more accurate retrieval, they also corrected the polarized reflectance of a cloud with polarization measurements at 865 nm where molecular scattering is negligible. This technique of Buriez et al. (1997) was later improved and applied to POLDER-2 onboard the ADEOS-II satellite by He et al. (2009), who used a vector radiative transfer model to address the multiple scattering between air molecules and cloud particles. He et al. (2009) also found the polarized reflectance at the TOA has negligible sensitivity to cloud albedo and aerosol scattering above a cloud. Their retrievals of cloud layer height were overall 0.8 km lower than the MODIS operational cloud-top heights determined by the CO_2-slicing technique (Menzel et al., 2008).

In a cloud-free atmosphere, by contrast, the polarization state of the scattered light at the TOA also depends on the aerosol scattering and molecular scattering underneath the aerosol layer, in addition to the molecular scattering above the aerosol layer. The aerosol height information is mainly due to the distinct polarizing effects between air molecules and aerosol particles. Air molecules are purely 100% linearly polarizing at the scattering angle of 90 degrees, while aerosol particles are partially polarizing with a maximum degree of linear polarization at scattering angles larger than 90 degrees (Zeng et al., 2008; Kalashnikova et al., 2011). Therefore, an elevated aerosol layer partially blocks the polarization signal of Rayleigh scattering by air molecules below the aerosol layer. An increase of the aerosol layer altitude thus causes a decrease of polarized reflectance around the 90-degree scattering angle (Fig. 2). As shown in several recent studies (e.g., Chowdhary et al., 2005; Waquet et al., 2009; Kalashnikova et al., 2011; Wu et al., 2016), the sensitivity of polarization to aerosol height is strong at near-UV and visible-blue spectra, around 410–470 nm, and decreases with the spectral wavelength. These studies also indicated that an incorrect assumption of aerosol vertical distribution leads to an inaccurate forward simulation of polarized reflectance at near-UV and, therefore, biases in the polarimetric retrieval of AOD and microphysical properties when AOD is above 0.3. It should be noted that the TOA reflectance in UV and near-UV channels is also sensitive to aerosol layer height for absorbing aerosol (Fig. 2A) (Torres et al., 1998). Only with additional constraints, such as aerosol single-scattering albedo and AOD from independent sources, can reflectance measurements be used to derive the aerosol height of a highly elevated absorbing aerosol layer (Satheesh et al., 2009). By contrast, the sensitivity of near-UV polarization to aerosol height is much less influenced by aerosol absorptivity.

Practically, it is difficult to independently extract aerosol height information from the polarization measurements because the polarization is also equally or even more sensitive to aerosol microphysical properties (e.g., Mishchenko and Travis, 1997; Cairns et al., 1997; Xu and Wang, 2015). Therefore, the algorithm developers seek to retrieve aerosol vertical information simultaneously with the particle size distribution and refractive index (Dubovik et al., 2011; Wu et al., 2016; c.f. Kokhanovsky et al., 2015). And the polarimetric measurements have to be taken from multiple directions at multiple spectra to ensure sufficient aerosol

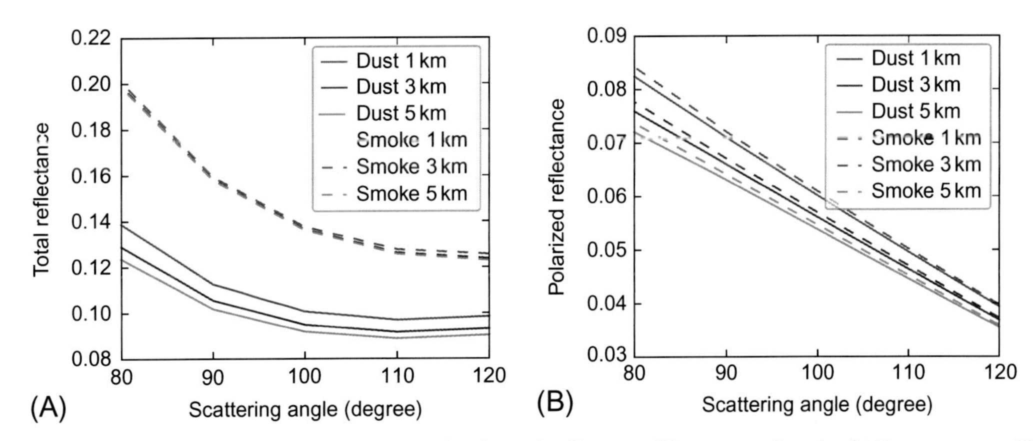

FIG. 2 Sensitivity of total TOA reflectance (A) and polarized reflectance (B) at a wavelength of 443 nm to aerosol height (1 km: *red line*, 3 km: *blue line*, and 5 km: *green line*). Aerosol height is defined as the altitude of peak aerosol concentration of a Gaussian profile. Dust aerosols are represented by *solid lines*, and smoke aerosols by a *dashed line*. Radiative transfer simulation is performed with the UNL-VRTM model (Wang et al., 2014) with AOD of 0.5. Lognormal particle size distribution is used with geometric median radius and standard deviation of 0.12 μm and 1.7 for smoke, and 1.94 μm and 1.8 for dust. Aerosol particles are assumed spherical. Complex reflectance index is assumed $1.45 - 0.033i$ for smoke and $1.50 - 0.004i$ for dust, with corresponding single-scattering albedo of 0.832 and 0.734, respectively.

information for robust retrievals. Dubovik et al. (2011) developed an algorithm that utilizes statistical optimization principles to retrieve a variety of aerosol properties from POLDER onboard the PARASOL satellite. The POLDER family have been the only satellite instruments with multiangle, multispectral polarimetric capability. The POLDER/PARASOL measures reflectance at 9 spectral bands (3 of them are polarization bands) at up to 16 view directions (Table 2). The total number of observations in the atmospheric window channels over each pixel exceeds a hundred, from which Dubovik et al. (2011) retrieve several tens of aerosol parameters including columnar volume concentration, volume size distribution represented by a dozen size bins, the complex refractive index, and vertical distribution. Their algorithm assumes the aerosol concentration profile follows a Gaussian function characterized by a mean aerosol altitude and a width factor. The width factor is fixed at 0.75 km, leaving the mean aerosol altitude as the retrieval parameter for the aerosol profile.

Most recently, Wu et al. (2016) presented an algorithm for a simultaneous retrieval of aerosol microphysical properties and layer height from airborne research scanning polarimeter (RSP) measurements. As a prototype of the APS instrument (Mishchenko et al., 2007), RSP measures Stokes parameters I, Q, and U at 152 viewing angles in nine spectral bands between 410 nm and 2250 nm over each observed pixel. Their algorithm utilizes measurements of six RSP bands with minimum gaseous absorption (at 410, 470, 550, 670, 865, and 1590 nm) and nine even viewing angles (between −60 and 40 degrees). Their algorithm was based on the Netherlands Institute for Space Research (SRON) aerosol retrieval algorithm (Hasekamp et al., 2011), which was developed to retrieve effective particle size and complex refractive index associated with a bimodal lognormal size distribution from POLDER/PARASOL measurements. Wu et al. (2016) added aerosol layer height as an additional retrieval parameter to the SRON algorithm. The aerosol layer height is defined as the mean altitude of a Gaussian-shaped profile with a fixed width of 2 km. For aerosol layer height, three retrievals are performed with different a priori (1, 3, and 5 km) and the retrieval that best fits with RSP data is selected. Wu et al. (2016) performed retrievals with RSP measurements in the PODEX and SEAC4RS campaigns. The validation against measurements from a Cloud Physics LIDAR (CPL) showed good agreement with absolute height difference less than 1 km.

The UV-polarization-based retrieval algorithms (Dubovik et al., 2011; Wu et al., 2016) are superior to the stereo technique in that they can retrieve aerosol height for a variety of aerosol types rather than only plume-type aerosols for the stereo technique. Meanwhile, the developments of the polarimetric algorithms are often more challenging, which demand advanced nonlinear optimized inversion techniques and extensive computations of aerosol scattering and rigorous vector radiative transfer (Kokhanovsky et al., 2015).

3.3 Oxygen (O_2) Absorption Spectroscopy

In contrast to the polarization approach that uses measurements at wavelengths with negligible gas absorption to probe the profile of aerosols in the context of air molecular scattering, another approach is to use measurements in absorption bands of trace gases, such as the A and B bands of molecular oxygen O_2 located in the spectral ranges of 755–775 nm and 685–695 nm, respectively. The physical principle behind the aerosol altitude retrieval in the O_2 bands is that the aerosol layer can scatter photons back to space and reduce the possibility of photons being captured by the underneath O_2 molecules. A higher scattering layer provides more chances for photons to be scattered back to space, and therefore enhances the reflectivity in the O_2 A band as detected by a satellite. Furthermore, the large variability of absorption optical depth of O_2 in the A and B bands enables sunlight from the TOA to penetrate the atmosphere at different depths (Fig. 3). For instance, the penetration altitude is as high as 30 km in A band and 15 km in B band. Given that O_2 is a well-mixed gas in the atmosphere with a well-defined vertical structure, the absorption profile of molecular O_2 as well as Rayleigh scattering can be accurately obtained. The spectral dynamics of the apparent reflectance in the O_2 A band as measured by a space-borne spectrometer can tell how aerosol particles interact with O_2 absorption through multiple scattering in different altitudes.

This idea dated back to Hanel (1961) and Yamamoto and Wark (1961), who were among the first to suggest cloud-top pressure retrievals based on the amount of absorption by CO_2 and O_2 above the cloud layer. Following their pioneering work, passive remote sensing in the O_2 A band has been subject to intensive studies (see brief reviews by Kokhanovsky et al. (2015) and Ding et al. (2016)). Currently, a number of operational cloud retrieval frameworks have been developed to retrieve cloud altitude based on spectral measurements in the O_2 A band

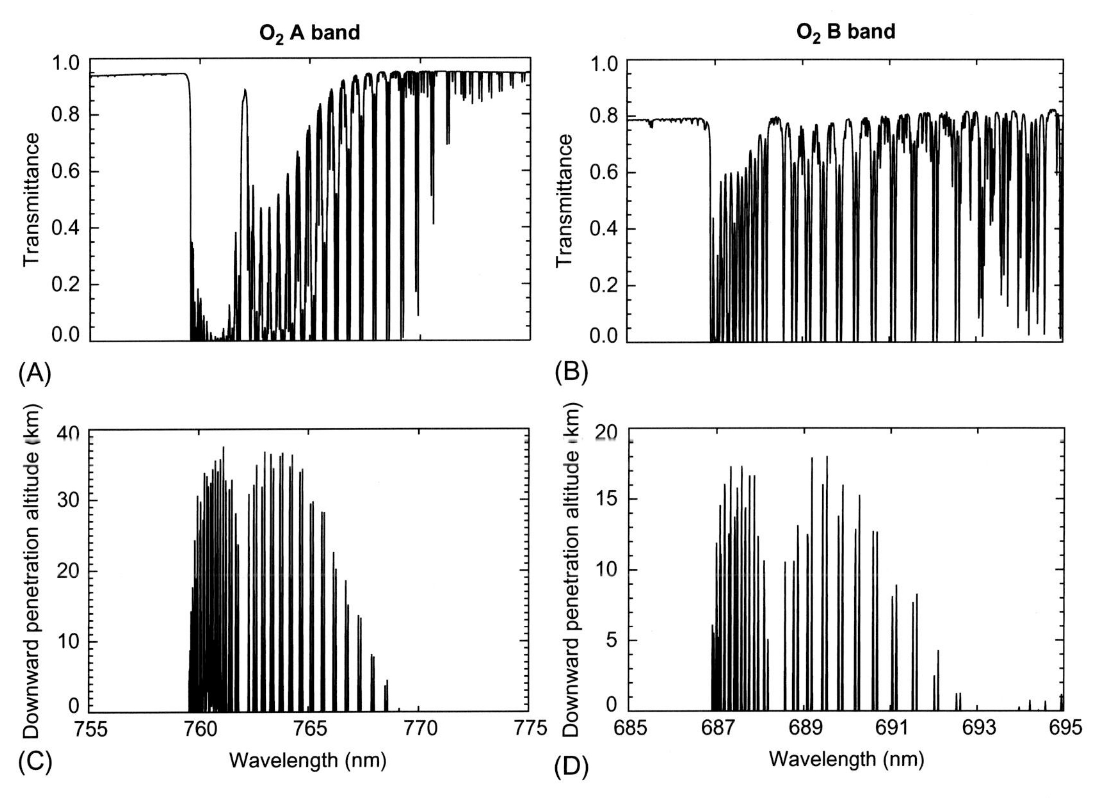

FIG. 3 (A) Atmospheric one-way transmittance in the O_2 A band. (B) Same as in (A) but in the O_2 B band. (C) Penetration altitude for various wavelengths in the O_2 A band. (D) Same as (C) but for the O_2 B band. The spectral interval and resolution (FWHM of a Gaussian spectral response) of 0.01 nm is used for a mid-latitude-summer atmospheric profile. The penetration altitude is defined as the altitude at which the spectral intensity of the downward solar radiation (for solar zenith angle of 0 degree) is attenuated by a factor of *e* as compared to its value at the TOA. In atmospheric window channels the penetration altitude is set to 0 (i.e., at the surface). Simulations were performed by UNL-VRTM (Wang et al., 2014).

from GOME, GOME-2, and SCIAMACHY (e.g., Kokhanovsky and Rozanov, 2004; Wang et al., 2008; Loyola et al., 2007; Yang et al., 2013a). With the heritage of cloud retrieval in the O_2 A band, recent years have seen many attempts to use O_2 absorption bands for the retrieval of aerosol vertical distribution, though aerosol retrieval is more challenging due to its smaller optical thickness, larger variability of aerosol microphysical properties, and unknown surface reflectance.

The retrieval algorithms for aerosol vertical distribution using O_2 spectroscopy can be grouped into two categories from the measurement perspective. The first category directly applies the principle that apparent reflectivity in the O_2 absorption band increases as the aerosol scattering layer rises. Algorithms in this category derive aerosol height based on the reflectance ratio of channels inside and outside the absorption band (e.g., Duforêt et al., 2007; Dubuisson et al., 2009; Xu et al., 2017). This reflectance ratio approach is well suited for narrow-band measurements in the O_2 absorption spectrum, such as those from MERIS, POLDER, and EPIC (Table 3). However, this approach can only yield a single piece of aerosol height information. An accurate height retrieval relies on the appropriate assumption of aerosol optical properties and surface reflectance. Duforêt et al. (2007) showed that the altitude of a single aerosol layer over a dark surface can be retrieved from POLDER measurements with an error of less than 1 km when AOD is larger than 0.2. Dubuisson et al. (2009) suggested the retrieval accuracy could reach 0.5 km for POLDER and 0.2 km for MERIS when AOD is over 0.3 and the surface albedo is below 0.06. Xu et al. (2017) derived dust-plume height over the ocean from DSCOVR/EPIC measured radiances in the O_2 A and B bands, and found that the plume height retrievals were in good agreement with aerosol extinction profiles probed by CALIOP. The error of retrieved aerosol height increases rapidly for brighter surfaces because the surface reflection dominates the apparent reflectance at the TOA. So, it is challenging to retrieve aerosol height over land in the O_2 A band, for which the reflectance of both the vegetation and soil surface is often larger than 0.3.

To overcome the retrieval issue caused by brighter surfaces, adding polarization measurements and/or measurements in the O_2 B band have been suggested. In the O_2 absorption band, the sensitivity of the degree of polarization to aerosol height is negligibly influenced by the change of surface reflectance (Boesche et al., 2009; Wang et al., 2014; Ding et al., 2016). Observations in the O_2 B band contain almost the same amount of aerosol height information as those in the O_2 A band (Pflug and Ruppert, 1993), but the reflectance of a vegetation surface is much lower in the O_2 B band. Therefore, a combined use of spectrally differentiated reflectance and polarization in both the O_2 A and B bands would enable aerosol height retrieval over both land and ocean surfaces (Ding et al., 2016). Such enhancement was shown primarily through theoretical studies but with limited analysis of real data and validation (Sanghavi et al., 2012).

Algorithms of the second category are based on the spectral fitting of reflectance measured across the O_2 absorption bands. These algorithms take advantage of reflectance measured by spectrometers at a moderate to high spectral resolution. Recent case studies include Koppers and Murtagh (1997) for GOME data, Kokhanovsky and Rozanov (2010) and Sanghavi et al. (2012) for SCIAMACHY data, and Sanders et al. (2015) for GOME-2 and TROPOMI data. Koppers and Murtagh (1997) retrieved AOD in five atmospheric layers and a Lambertian surface albedo from GOME spectral measurements in the O_2 A band. Kokhanovsky and Rozanov (2010) fitted spectral reflectance in the O_2 A band from SCIAMACHY data to retrieve the top height of a homogeneous-extinction layer extending from the surface, and performed case studies for Saharan dust outflows over the Atlantic Ocean and validated retrieved AOD with AERONET measurements. With the combination of O_2

A and B bands of SCIAMACHY measurements, Sanghavi et al. (2012) performed a year of aerosol profile (peak height and width of a lognormal distribution) retrievals over the Indo-Gangetic Plain and validated AOD against AERONET measurements at the Kanpur site. Sanders et al. (2015) developed an operational aerosol height retrieval algorithm targeted for TROPOMI instruments and applied the algorithm to GOME-2 data. Sanders et al. (2015) characterized aerosol profile as a scatter layer with a fixed pressure thickness of 50 hPa, and they evaluated the height retrievals with LIDAR measurements.

There is no doubt that the spectrally resolved measurements contain more pieces of information than a single reflectance ratio. As indicated by Corradini and Cervino (2006), SCIAMACHY measurements (of spectral resolution about 0.4 nm in the O_2 A band) could provide the retrieval of a three-layer profile in the troposphere (i.e., a degree of freedom of 3), though the retrieval depends strongly on the aerosol optical properties and the surface reflectivity. Higher spectral resolution measurements of the O_2 A band, such as those taken by GOSAT, OCO-2, and TanSat (spectral resolution as high as 0.04 nm), can considerably improve the aerosol height information content (reaching a degree of freedom of 4–5) and thereby the profile retrieval (Hollstein and Fischer, 2014; Geddes and Bösch, 2015; Colosimo et al., 2016). A degree of freedom of about 7 can be reached if multiangular measurements of the O_2 A band are used (Frankenberg et al., 2012). Moreover, the influence of aerosol optical properties and surface albedo to the retrieval accuracy can be significantly reduced for such an enhanced spectral resolution (Colosimo et al., 2016). However, these aforementioned studies also indicated that measurements in the O_2 A band, even at high spectral resolution, contain inadequate information for aerosols near the surface. This could lead to large biases if the surface albedo is simultaneously retrieved (Sanders et al., 2015). It therefore was suggested to either prescribe an accurate surface reflectivity or to use alternative profile parameterizations that better account for low-level boundary layer aerosols. In addition, the O_2 A band is influenced by the chlorophyll fluorescence from terrestrial vegetation. Studies showed that the fluorescence emission, if ignored, could introduce significant biases in retrieved aerosol parameters (Frankenberg et al., 2011). To overcome this issue, Sanders and de Haan (2013) modeled the contribution of fluorescence emission to the upwelling radiance field at the surface, and they retrieved a fluorescence emission simultaneously with aerosol properties.

It should be noted that there is a tradeoff between retrieval information and the spatial coverage of the instruments (see Table 3). Satellite instruments operated for spectral fitting techniques often have an either significantly coarser spatial resolution (e.g., GOME, GOME-2, and SCIAMCHY) or more restricted spatial coverage (e.g., GOSAT, OCO-2, and TanSat). By contrast, while the reflectance ratio approach provides a less-detailed aerosol profile than the spectral fitting algorithms, the corresponding instruments are often built with better spatial resolution and more extensive spatial coverage (e.g., POLDER, MERIS, and EPIC).

3.4 IR Technique

Observations of outgoing thermal IR radiance from space can provide information on mineral aerosols together with discernable information on the thermal state of the atmosphere and surface (DeSouza-Machado et al., 2006). Early studies have employed narrow-band satellite observations for the detection of dust aerosols (Ackerman, 1997; Miller, 2003). Recently, algorithms have been developed for inferring dust radiative altitude by taking advantage of rich dust information contained in the IR radiance data (Pierangelo et al., 2004, 2005; DeSouza-Machado et al., 2010; Peyridieu et al., 2010, 2013; Vandenbussche et al., 2013; Klüser et al., 2015).

These algorithms rely on the spectral signature of dust in thermal-infrared radiance. As shown by Sokolik (2002), the presence of dust aerosol over the ocean decreases the BT at the TOA across the IR atmospheric window at a spectral region of 8–12 μm. It also produces a unique "negative slope" of BT (along the wavenumber dimension from 820 to 1000 cm^{-1}) around 11 μm, widening the typical ozone-absorption-caused V-shape (Fig. 4A). As shown in Fig. 4B, this "negative slope" feature is sensitive to both dust optical depth and height as well as particle size distribution but with a distinct spectral variability. Moreover, spectral BT measured in the window channels around 4 μm is more sensitive to dust optical depth (Pierangelo et al., 2004). Together, high-resolution spectral radiance data from these two window regions (around 4 and 11 μm) allow for simultaneous retrieval of dust AOD and layer height.

Based upon the above principle, Pierangelo et al. (2004) used a look-up-table (LUT) approach to simultaneously retrieve dust altitude and 10-μm dust AOD over ocean from BT measured in eight selected AIRS channels. Their channel selection considers the best dust sensitivity and least sensitivity to water vapor and ozone. The LUT contains BT computed for five dust AOD values and four altitudes of dust layer under a variety of atmospheric thermodynamic situations. Their retrieval of dust AOD was found to be in good correlation with MODIS AOD in the visible band; the height retrievals also compared favorably with CALIPSO measurements. Pierangelo et al. (2005) further added one retrieval step to estimate the coarse-mode particle effective radius using one additional channel at 9.32 μm. Their algorithm has been applied by Peyridieu et al. (2010, 2013) for more than 9 years of AIRS observations and 4 years of IASI observations to characterize global distribution of dust loading, effective size, and height.

Most recently, Vandenbussche et al. (2013) used an optimal estimation approach to retrieve a five-layer profile of dust concentration from IASI observations with the assumption of known particle microphysical properties, surface temperature and emissivity, and atmospheric state. The algorithm of Vandenbussche et al. (2013) follows two steps to ensure a statistically stable solution. First, it retrieves an average dust profile by fitting a spatially averaged

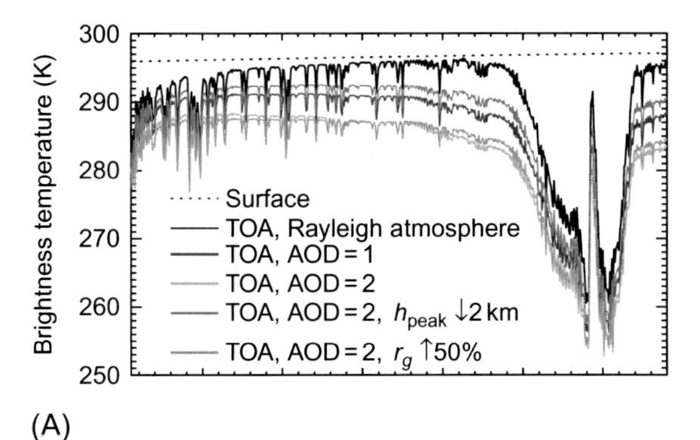

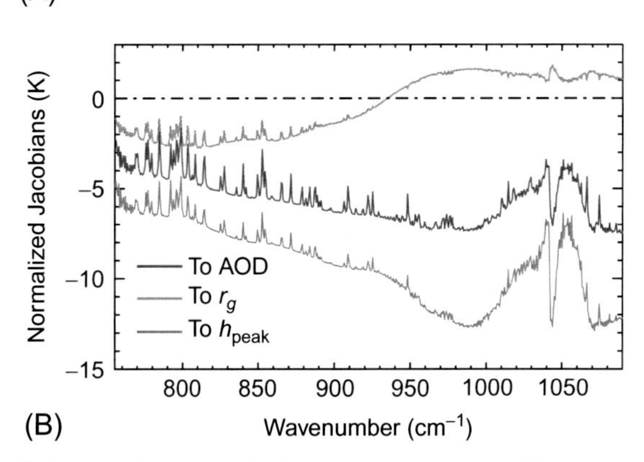

FIG. 4 (A) Simulated brightness temperature (BT) in thermal IR spectral region for various atmospheric conditions. (B) Corresponding normalized Jacobians of BT with respect to dust AOD (τ), dust height (h_{peak}), and geometric median radius (r_g), i.e., $\tau\frac{\partial BT}{\partial \tau}$, $h_{peak}\frac{\partial BT}{\partial h_{peak}}$, $r_g\frac{\partial BT}{\partial r_g}$, respectively. Dust particles are assumed to follow a lognormal size distribution with r_g of 0.5 μm and geometric standard deviation of 2.0. Dust refractive indice are adapted from Di Biagio et al. (2014). Unless labeled otherwise, h_{peak} = 3.0 km, AOD = 2.0 at 0.55 μm. Aerosol height h_{peak} is defined as the altitude of peak aerosol centration for a Gaussian vertical profile. Simulations are conducted by the UNL-VRTM model (Wang et al., 2014). Spectral resolution is assumed as 0.625 cm^{-1}, which is same to the CrIS instrument.

IASI spectra for field-of-views (FOVs) contained in a 100-km range. Second, the averaged profile is used as a priori with a 10% variance to retrieve the dust profile of concentration from the IASI spectra observed at each FOV.

The advantage of the IR technique is that the thermal IR radiance is sensitive to dust aerosols only, making it well suited for extracting dust information from other aerosol types. This is particularly useful for studies of dust radiative forcing, transport, and deposition. Another advantage is that IR observations are available for both the daytime and nighttime from AIRS, IASI, and CrIS (Table 4), which provide observations with a moderate spatial resolution and large spatial coverage. On the other hand, IR retrievals of aerosol height are dependent on the knowledge of surface emissivity, ambient temperature, lapse rate, and the spectrally dependent dust refractive index. So, the challenge is to select appropriate data regarding the aerosol, surface, and atmospheric thermal state. In addition, the heterogeneity of land surface cover and complexity of land surface emissivity make the dust altitude retrieval over land even more challenging.

4 CONCLUSIONS AND OUTLOOK

Aerosol vertical distribution is one of the important factors that contribute to the uncertainty in understanding aerosol climate effects. Accurate observational constraints of aerosol height information are critically demanding. While space-borne LIDAR can probe a detailed aerosol backscattering profile, its spatial coverage suffers from capturing only a "curtain" along its suborbital track. Passive remote sensing of aerosol height, while less accurate, can add a valuable extension due to its better spatial coverage. In the present survey, we have discussed the progress in the recent developments of four passive remote sensing techniques for aerosol vertical distribution: stereo photogrammetry, polarimetric approach in the near-UV bands, the oxygen absorption spectroscopy, and the thermal IR technique. We also discussed the associated physical principles as well as the challenges of each technique.

In general, different retrieval techniques can provide different aspects of aerosol height information and, at the same time, face different challenges. The stereoscopic technique, while not being affected by complex aerosol microphysical properties and radiative transfer computation, can only be applied for plume-type aerosols with discernable textures. The polarimetric technique in the near-UV spectrum demands multispectral, multiangle measurements and entails the development of an optimized inversion algorithm for retrieving aerosol height with complete aerosol microphysical properties. The O_2 absorption spectroscopy can enable the retrieval of aerosol concentration on several layers based on averaging kernel sensitivity. The retrieval accuracy from the O_2 absorption spectroscopy relies on appropriate a priori information of aerosol optical properties and surface reflectivity. While the IR radiance technique is well suited for retrieval of the dust aerosol profile, its accuracy is critically dependent on the assumptions of atmospheric thermodynamic status and dust optical properties.

To increase the information content one needs to take advantage of multiangular polarimetric measurements and to carefully select spectral bands. For instance, using multiangular observations in the O_2 absorption band may give more clues on the aerosol vertical distribution. As Duforêt et al. (2007) and Frankenberg et al. (2012) suggested, multiview angles enable a variety of air mass and therefore more variability of atmospheric weighting functions for aerosol scattering changes, particularly for elevated aerosols. Combining multiangular spectral differential absorption in O_2 absorption bands and multiangular multispectral polarimetric observations in the near-UV spectrum is an interesting perspective, which would lead to more complete information on aerosol vertical structure and, therefore, more accurate retrieval of aerosol microphysical properties. This would particularly improve our understanding of the impact of aerosols on environmental and climate changes by providing improved observational constraints for numerical climate models.

In the near future, hyperspectral measurements of the O_2 A and B bands will become available from TROPOMI instruments on the Sentinel-5 Precursor satellite, the spatial resolution of which is significantly improved over the existing instruments like SCIAMACHY and GOME-2. A retrieval algorithm for operational aerosol height from TROPOMI has been developed and tested with GOME-2 data (Sanders et al., 2015). Operational retrieval aerosol heights have also been planned for 3MI instrument onboard the METOP-SG satellite (Manolis et al., 2013), and for PACE (http://pace.gsfc.nasa.gov) and GEO-CAPE satellite missions (Fishman et al., 2012). These new instruments with better radiometric characteristics and spatial resolution and coverage will enable further promising applications of those passive techniques.

Acknowledgments

The authors thank Olga Kalashnikova and David Diner for the discussions about satellite remote sensing of aerosol height. This research is supported by NASA under grant (NNX17AB05G) issued through the DSCOVR Earth Science Algorithms Program, and is also supported in part by the NASA GEO-CAPE program, and in part by the Office of Naval Research under award number N00014-16-1-2040. Alexander Kokhanovsky acknowledges the support of the Center for Applied Mathematics and Theoretical Physics within the *Moscow Engineering Physics Institute's* Academic Excellence Project (contract No. 02.a03.21.0005, 27.08.2013).

References

Ackerman, S.A., 1997. Remote sensing aerosols using satellite infrared observations. J. Geophys. Res. Atmos. 102 (D14), 17069–17079.

Adachi, T., Kasai, T., 1970. Stereoscopic analysis of photographs taken by NIMBUS II APT System (II): an improvement in the method of the stereoscopic analysis. J. Meteorol. Soc. Jpn. Ser. II 48 (3), 234–242.

Babu, S.S., Moorthy, K.K., Manchanda, R.K., Sinha, P.R., Satheesh, S.K., Vajja, D.P., Kumar, V.H.A., 2011. Free tropospheric black carbon aerosol measurements using high altitude balloon: do BC layers build "their own homes" up in the atmosphere? Geophys. Res. Lett. 38 (8), L08803.

Boesche, E., Stammes, P., Bennartz, R., 2009. Aerosol influence on polarization and intensity in near-infrared O_2 and CO_2 absorption bands observed from space. J. Quant. Spectrosc. Radiat. Transf. 110 (3), 223–239.

Boucher, O., Randall, D., Artaxo, P., Bretherton, C., Feingold, G., Forster, P., Kerminen, V.-M., Kondo, Y., Liao, H., Lohmann, U., Rasch, P., Satheesh, S.K., Sherwood, S., Stevens, B., Zhang, X.Y., 2013. Clouds and aerosols. In: Stocker, T.F., Qin, D., Plattner, G.-K., Tignor, M., Allen, S.K., Boschung, J., Nauels, A., Xia, Y., Bex, V., Midgley, P.M. (Eds.), Climate Change 2013: The Physical Science Basis. Contribution of Working Group I to the Fifth Assessment Report of the Intergovernmental Panel on Climate Change. Cambridge University Press, Cambridge/New York, NY.

Buriez, J.C., Vanbauce, C., Parol, F., Goloub, P., Herman, M., Bonnel, B., Seze, G., 1997. Cloud detection and derivation of cloud properties from POLDER. Int. J. Remote Sens. 18 (13), 2785–2813.

Cairns, B., Carlson, B.E., Lacis, A.A., Russell, E.E., 1997. An analysis of ground-based polarimetric sky radiance measurements. In: Chipman, R.A., Goldstein, D.H. (Eds.), Polarization: Measurement, Analysis, and Remote Sensing, Proceedings of SPIE.

Chen, Y., Li, Q., Randerson, J.T., Lyons, E.A., Kahn, R.A., Nelson, D.L., Diner, D.J., 2009. The sensitivity of CO and aerosol transport to the temporal and vertical distribution of North American boreal fire emissions. Atmos. Chem. Phys. 9 (17), 6559–6580.

Chimot, J., Veefkind, J.P., Vlemmix, T., de Haan, J.F., Amiridis, V., Proestakis, E., Marinou, E., Levelt, P.F., 2017. An exploratory study on the aerosol height retrieval from OMI measurements of the 477 nm O_2-O_2 spectral band using a neural network approach. Atmos. Meas. Tech. 10 (3), 783–809.

Chowdhary, J., Cairns, B., Mishchenko, M.I., Hobbs, P.V., Cota, G.F., Redemann, J., Russell, E., 2005. Retrieval of aerosol scattering and absorption properties from photopolarimetric observations over the ocean during the CLAMS experiment. J. Atmos. Sci. 62 (4), 1093–1117.

Colosimo, S.F., Natraj, V., Sander, S.P., Stutz, J., 2016. A sensitivity study on the retrieval of aerosol vertical profiles using the oxygen A-band. Atmos. Meas. Tech. 9 (4), 1889–1905.

Corradini, S., Cervino, M., 2006. Aerosol extinction coefficient profile retrieval in the oxygen A-band considering multiple scattering atmosphere. Test case: SCIAMACHY

nadir simulated measurements. J. Quant. Spectrosc. Radiat. Transf. 97 (3), 354–380.
DeSouza-Machado, S.G., Strow, L.L., Hannon, S.E., Motteler, H.E., 2006. Infrared dust spectral signatures from AIRS. Geophys. Res. Lett. 33 (3), L03801.
DeSouza-Machado, S.G., Strow, L.L., Imbiriba, B., McCann, K., Hoff, R.M., Hannon, S.E., Torres, O., 2010. Infrared retrievals of dust using AIRS: comparisons of optical depths and heights derived for a North African dust storm to other collocated EOS A-Train and surface observations. J. Geophys. Res. 115 (D15), D15201.
Di Biagio, C., Boucher, H., Caquineau, S., Chevaillier, S., Cuesta, J., Formenti, P., 2014. Variability of the infrared complex refractive index of African mineral dust: experimental estimation and implications for radiative transfer and satellite remote sensing. Atmos. Chem. Phys. 14 (20), 11093–11116.
Diner, D.J., Nelson, D.L., Chen, Y., Kahn, R.A., Logan, J., Leung, F.-Y., Val Martin, M., 2008. Quantitative studies of wildfire smoke injection heights with the Terra Multi-Angle Imaging SpectroRadiometer. In: Hao, W.M. (Ed.), Proceedings of SPIE Remote Sensing of Fire: Science and Application, vol. 7089.
Ding, S., Wang, J., Xu, X., 2016. Polarimetric remote sensing in oxygen A and B bands: sensitivity study and information content analysis for vertical profile of aerosols. Atmos. Meas. Tech. 9 (5), 2077–2092.
Dubovik, O., Herman, M., Holdak, A., Lapyonok, T., Tanré, D., Deuzé, J.L., Lopatin, A., 2011. Statistically optimized inversion algorithm for enhanced retrieval of aerosol properties from spectral multi-angle polarimetric satellite observations. Atmos. Meas. Tech. 4 (5), 975–1018.
Dubuisson, P., Frouin, R., Dessailly, D., Duforêt, L., Léon, J.-F., Voss, K., Antoine, D., 2009. Estimating the altitude of aerosol plumes over the ocean from reflectance ratio measurements in the O_2 A-band. Remote Sens. Environ. 113 (9), 1899–1911.
Duforêt, L., Frouin, R., Dubuisson, P., 2007. Importance and estimation of aerosol vertical structure in satellite ocean-color remote sensing. Appl. Opt. 46 (7), 1107–1119.
Fisher, D., Muller, J.-P., Yershov, V.N., 2014. Automated stereo retrieval of smoke plume injection heights and retrieval of smoke plume masks from AATSR and their assessment with CALIPSO and MISR. IEEE Trans. Geosci. Remote Sens. 52 (2), 1249–1258.
Fisher, D., Poulsen, C.A., Thomas, G.E., Muller, J.-P., 2016. Synergy of stereo cloud top height and ORAC optimal estimation cloud retrieval: evaluation and application to AATSR. Atmos. Meas. Tech. 9 (3), 909–928.
Fishman, J., Iraci, L.T., Al-Saadi, J., Chance, K., Chavez, F., Chin, M., Wang, M., 2012. The United States' next generation of atmospheric composition and coastal ecosystem measurements: NASA's geostationary coastal and air pollution events (GEO-CAPE) mission. Bull. Am. Meteorol. Soc. 93 (10), 1547–1566.
Frankenberg, C., Butz, A., Toon, G.C., 2011. Disentangling chlorophyll fluorescence from atmospheric scattering effects in O_2A-band spectra of reflected sun-light. Geophys. Res. Lett. 38 (3), L03801.
Frankenberg, C., Hasekamp, O., O'Dell, C., Sanghavi, S., Butz, A., Worden, J., 2012. Aerosol information content analysis of multi-angle high spectral resolution measurements and its benefit for high accuracy greenhouse gas retrievals. Atmos. Meas. Tech. 5 (7), 1809–1821.
Geddes, A., Bösch, H., 2015. Tropospheric aerosol profile information from high-resolution oxygen A-band measurements from space. Atmos. Meas. Tech. 8 (2), 859–874.
Gordon, H.R., 1997. Atmospheric correction of ocean color imagery in the Earth Observing System era. J. Geophys. Res. Atmos. 102 (D14), 17081–17106.
Hanel, R.A., 1961. Determination of cloud altitude from a satellite. J. Geophys. Res. 66 (4), 1300.
Hasekamp, O.P., Litvinov, P., Butz, A., 2011. Aerosol properties over the ocean from PARASOL multiangle photopolarimetric measurements. J. Geophys. Res. Atmos. 116 (D14), D14204.
Hasler, A.F., 1981. Stereographic observations from geosynchronous satellites: an important new tool for the atmospheric sciences. Bull. Am. Meteorol. Soc. 62 (2), 194–212.
Hasler, A.F., Mack, R., Negri, A., 1983. Stereoscopic observations from meteorological satellites. Adv. Space Res. 2 (6), 105–113.
He, X., Bai, Y., Pan, D., Zhu, Q., Gong, F., 2009. Cloud top height retrieval using polarizing remote sensing data of POLDER. Terr. Atmos. Ocean. Sci. Lett. 2 (2), 73–78.
Hollstein, A., Fischer, J., 2014. Retrieving aerosol height from the oxygen A band: a fast forward operator and sensitivity study concerning spectral resolution, instrumental noise, and surface inhomogeneity. Atmos. Meas. Tech. 7 (5), 1429–1441.
Kahn, R.A., Limbacher, J., 2012. Eyjafjallajökull volcano plume particle-type characterization from space-based multi-angle imaging. Atmos. Chem. Phys. 12 (20), 9459–9477.
Kahn, R.A., Li, W.H., Moroney, C., Diner, D.J., Martonchik, J.V., Fishbein, E., 2007. Aerosol source plume physical characteristics from space-based multi-angle imaging. J. Geophys. Res. 112, D11205.
Kahn, R.A., Chen, Y., Nelson, D.L., Leung, F.-Y., Li, Q., Diner, D.J., Logan, J.A., 2008. Wildfire smoke injection heights: two perspectives from space. Geophys. Res. Lett. 35 (4), L04809.
Kalashnikova, O.V., Garay, M.J., Davis, A.B., Diner, D.J., Martonchik, J.V., 2011. Sensitivity of multi-angle photo-polarimetry to vertical layering and mixing of absorbing aerosols: quantifying measurement uncertainties. J. Quant. Spectrosc. Radiat. Transf. 112 (13), 2149–2163.
Kikuchi, K., Kasai, T., 1968. Stereoscopic analysis of photographs taken by NIMBUS II APT system. J. Meteorol. Soc. Jpn. Ser. II 46 (1), 60–67.

Kipling, Z., Stier, P., Johnson, C.E., Mann, G.W., Bellouin, N., Bauer, S.E., Zhang, K., 2016. What controls the vertical distribution of aerosol? Relationships between process sensitivity in HadGEM3–UKCA and inter-model variation from AeroCom Phase II. Atmos. Chem. Phys. 16 (4), 2221–2241.

Klüser, L., Banks, J.R., Martynenko, D., Bergemann, C., Brindley, H.E., Holzer-Popp, T., 2015. Information content of space-borne hyperspectral infrared observations with respect to mineral dust properties. Remote Sens. Environ. 156, 294–309.

Koch, D., Del Genio, A.D., 2010. Black carbon semi-direct effects on cloud cover: review and synthesis. Atmos. Chem. Phys. 10, 7685–7696.

Koffi, B., Schulz, M., Bréon, F.-M., Griesfeller, J., Winker, D., Balkanski, Y., Takemura, T., 2012. Application of the CALIOP layer product to evaluate the vertical distribution of aerosols estimated by global models: AeroCom phase I results. J. Geophys. Res. Atmos. 117 (D10), D10201.

Kokhanovsky, A.A., Rozanov, V.V., 2004. The physical parameterization of the top-of-atmosphere reflection function for a cloudy atmosphere—underlying surface system: the oxygen A-band case study. J. Quant. Spectrosc. Radiat. Transf. 85 (1), 35–55.

Kokhanovsky, A.A., Rozanov, V.V., 2010. The determination of dust cloud altitudes from a satellite using hyperspectral measurements in the gaseous absorption band. Int. J. Remote Sens. 31 (10), 2729–2744.

Kokhanovsky, A.A., Davis, A.B., Cairns, B., Dubovik, O., Hasekamp, O.P., Sano, I., Munro, R., 2015. Space-based-remote sensing of atmospheric aerosols: the multi-angle spectro-polarimetric frontier. Earth Sci. Rev. 145, 85–116.

Koppers, G.A.A., Murtagh, D.P., 1997. Retrieval of height resolved aerosol optical thickness in the atmospheric band. In: Koppers, G. (Ed.), Radiative Transfer in the Absorption Bands of Oxygen: Studies of Their Significance in Ozone Chemistry and Potential for Aerosol Remote Sensing. Stockholm University, Stockholm, pp. 1–24 (Chapter 5).

Loyola, D.G., Thomas, W., Livschitz, Y., Ruppert, T., Albert, P., Hollmann, R., 2007. Cloud properties derived from GOME/ERS-2 backscatter data for trace gas retrieval. IEEE Trans. Geosci. Remote Sens. 45 (9), 2747–2758.

Maddy, E.S., DeSouza-Machado, S.G., Nalli, N.R., Barnet, C.D., Strow, L., Wolf, W.W., Schou, P., 2012. On the effect of dust aerosols on AIRS and IASI operational level 2 products. Geophys. Res. Lett. 39 (10), L10809.

Manolis, I., Grabarnik, S., Caron, J., Bézy, J.-L., Loiselet, M., Betto, M., Meynart, R., 2013. The MetOp second generation 3MI instrument. In: Proceedings of SPIE, 8889, Sensors, Systems, and Next-Generation Satellites XVII, 88890J, Desden, Germany, October 16.

Menzel, W.P., Frey, R.A., Zhang, H., Wylie, D.P., Moeller, C.C., Holz, R.E., Gumley, L.E., 2008. MODIS global cloud-top pressure and amount estimation: algorithm description and results. J. Appl. Meteorol. Climatol. 47 (4), 1175–1198.

Miller, S.D., 2003. A consolidated technique for enhancing desert dust storms with MODIS. Geophys. Res. Lett. 30 (20), 2071.

Mishchenko, M.I., Travis, L.D., 1997. Satellite retrieval of aerosol properties over the ocean using polarization as well as intensity of reflected sunlight. J. Geophys. Res. 102 (D14), 16989–17013.

Mishchenko, M.I., Cairns, B., Hansen, J.E., Travis, L.D., Kopp, G., Schueler, C.F., Itchkawich, T., 2007. Accurate monitoring of terrestrial aerosols and total solar irradiance: introducing the glory mission. Bull. Am. Meteorol. Soc. 88 (5), 677–691.

Moroney, C., Davies, R., Muller, J.P., 2002. Operational retrieval of cloud-top heights using MISR data. IEEE Trans. Geosci. Remote Sens. 40 (7), 1532–1540.

Muller, J.P., Mandanayake, A., Moroney, C., Davies, R., Diner, D.J., Paradise, S., 2002. MISR stereoscopic image matchers: techniques and results. IEEE Trans. Geosci. Remote Sens. 40 (7), 1547–1559.

Muller, J.P., Denis, M.A., Dundas, R.D., Mitchell, K.L., Naud, C., Mannstein, H., 2007. Stereo cloud-top heights and cloud fraction retrieval from ATSR-2. Int. J. Remote Sens. 28 (9), 1921–1938.

Myhre, G., Shindell, D., Bréon, F.-M., Collins, W., Fuglestvedt, J., Huang, J., Koch, D., Lamarque, J.-F., Lee, D., Mendoza, B., Nakajima, T., Robock, A., Stephens, G., Takemura, T., Zhang, H., 2013. Anthropogenic and natural radiative forcing. In: Stocker, T.F., Qin, D., Plattner, G.-K., Tignor, M., Allen, S.K., Boschung, J., Nauels, A., Xia, Y., Bex, V., Midgley, P.M. (Eds.), Climate Change 2013: The Physical Science Basis. Contribution of Working Group I to the Fifth Assessment Report of the Intergovernmental Panel on Climate Change. Cambridge University Press, Cambridge, United Kingdom and New York, NY, USA, pp. 659–740.

Nelson, D.L., Chen, Y., Diner, D., Kahn, R.A., Mazzoni, D., 2008. Example applications of the MISR INteractive eXplorer (MINX) software tool to wildfire smoke plume applications. Proc. SPIE 7089, 11.

Nelson, D., Garay, M., Kahn, R., Dunst, B., 2013. Stereoscopic height and wind retrievals for Aerosol Plumes with the MISR INteractive eXplorer (MINX). Remote Sens. 5 (9), 4593–4628.

Ondrejka, R.J., Conover, J.H., 1966. Note on the stereo interpretation of nimbus ii apt photography. Mon. Weather Rev. 94 (10), 611–614.

Park, S.S., Kim, J., Lee, H., Torres, O., Lee, K.-M., Lee, S.D., 2016. Utilization of O_4 slant column density to derive aerosol layer height from a space-borne UV–visible hyperspectral sensor: sensitivity and case study. Atmos. Chem. Phys. 16, 1987–2006.

Peters, K., Quaas, J., Bellouin, N., 2011. Effects of absorbing aerosols in cloudy skies: a satellite study over the Atlantic Ocean. Atmos. Chem. Phys. 11 (4), 1393–1404.

Peyridieu, S., Chédin, A., Tanré, D., Capelle, V., Pierangelo, C., Lamquin, N., Armante, R., 2010. Saharan dust infrared optical depth and altitude retrieved from AIRS: a focus over North Atlantic—comparison to MODIS and CALIPSO. Atmos. Chem. Phys. 10 (4), 1953–1967.

Peyridieu, S., Chédin, A., Capelle, V., Tsamalis, C., Pierangelo, C., Armante, R., Scott, N.A., 2013. Characterisation of dust aerosols in the infrared from IASI and comparison with PARASOL, MODIS, MISR, CALIOP, and AERONET observations. Atmos. Chem. Phys. 13 (12), 6065–6082.

Pflug, B.M., Ruppert, T., 1993. Information content of measurements in the O_2A- and O_2B-bands for monitoring of aerosols from space. In: SPIE Vol. 1968 Atmospheric Propagation and Remote Sensing II, pp. 533–544.

Pierangelo, C., Chédin, A., Heilliette, S., Jacquinet-Husson, N., Armante, R., 2004. Dust altitude and infrared optical depth from AIRS. Atmos. Chem. Phys. 4 (7), 1813–1822.

Pierangelo, C., Mishchenko, M.I., Balkanski, Y., Chedin, A., 2005. Retrieving the effective radius of Saharan dust coarse mode from AIRS. Geophys. Res. Lett. 32 (20), L20813.

Prata, A.J., Turner, P.J., 1997. Cloud-top height determination using ATSR data. Remote Sens. Environ. 59 (1), 1–13.

Sanders, A.F.J., de Haan, J.F., 2013. Retrieval of aerosol parameters from the oxygen A band in the presence of chlorophyll fluorescence. Atmos. Meas. Tech. 6 (10), 2725–2740.

Sanders, A.F.J., de Haan, J.F., Sneep, M., Apituley, A., Stammes, P., Vieitez, M.O., Veefkind, J.P., 2015. Evaluation of the operational Aerosol Layer Height retrieval algorithm for Sentinel-5 Precursor: application to O_2 A band observations from GOME-2A. Atmos. Meas. Tech. 8 (11), 4947–4977.

Sanghavi, S., Martonchik, J.V., Landgraf, J., Platt, U., 2012. Retrieval of aerosol optical depth and vertical distribution using O_2 A- and B-band SCIAMACHY observations over Kanpur: a case study. Atmos. Meas. Tech. 5 (5), 1099–1119.

Satheesh, S.K., Torres, O., Remer, L.A., Babu, S.S., Vinoj, V., Eck, T.F., Holben, B.N., 2009. Improved assessment of aerosol absorption using OMI-MODIS joint retrieval. J. Geophys. Res. 114 (D5), D05209.

Scollo, S., Kahn, R.A., Nelson, D.L., Coltelli, M., Diner, D.J., Garay, M.J., Realmuto, V.J., 2012. MISR observations of Etna volcanic plumes. J. Geophys. Res. Atmos. 117 (D6), D06210.

Sears, T.M., Thomas, G.E., Carboni, E., Smith, A.J.A., Grainger, R.G., 2013. SO_2 as a possible proxy for volcanic ash in aviation hazard avoidance. J. Geophys. Res. Atmos. 118, 5698–5709.

Seiz, G., Davies, R., Grün, A., 2006. Stereo cloud-top height retrieval with ASTER and MISR. Int. J. Remote Sens. 27 (9), 1839–1853. https://doi.org/10.1080/01431160500 380703.

Sokolik, I.N., 2002. The spectral radiative signature of wind-blown mineral dust: implications for remote sensing in the thermal IR region. Geophys. Res. Lett. 29 (24), 2154.

Thomas, M.A., Devasthale, A., Kahnert, M., 2013. Exploiting the favourable alignment of CALIPSO's descending orbital tracks over Sweden to study aerosol characteristics. Tellus B 65, 21155.

Torres, O., Bhartia, P.K., Herman, J.R., Ahmad, Z., Gleason, J., 1998. Derivation of aerosol properties from satellite measurements of backscattered ultraviolet radiation: theoretical basis. J. Geophys. Res. 103 (D14), 17099–17110.

Val Martin, M., Logan, J.A., Kahn, R.A., Leung, F.Y., Nelson, D.L., Diner, D.J., 2010. Smoke injection heights from fires in North America: analysis of 5 years of satellite observations. Atmos. Chem. Phys. 10 (4), 1491–1510.

Vandenbussche, S., Kochenova, S., Vandaele, A.C., Kumps, N., De Mazière, M., 2013. Retrieval of desert dust aerosol vertical profiles from IASI measurements in the TIR atmospheric window. Atmos. Meas. Tech. 6 (10), 2577–2591.

Virtanen, T.H., Kolmonen, P., Rodríguez, E., Sogacheva, L., Sundström, A.M., de Leeuw, G., 2014. Ash plume top height estimation using AATSR. Atmos. Meas. Tech. 7 (8), 2437–2456.

Wang, J., Christopher, S.A., 2003. Intercomparison between satellite-derived aerosol optical thickness and PM2.5 mass: implications for air quality studies. Geophys. Res. Lett. 30, 2095.

Wang, P., Stammes, P., van der, A., Pinardi, G., van Roozendael, M., 2008. FRESCO+: an improved O_2 A-band cloud retrieval algorithm for tropospheric trace gas retrievals. Atmos. Chem. Phys. 8 (21), 6565–6576.

Wang, J., Xu, X., Ding, S., Zeng, J., Spurr, R., Liu, X., Mishchenko, M., 2014. A numerical testbed for remote sensing of aerosols, and its demonstration for evaluating retrieval synergy from a geostationary satellite constellation of GEO-CAPE and GOES-R. J. Quant. Spectrosc. Radiat. Transf. 146, 510–528.

Waquet, F., Riedi, J., Labonnote, L.C., Goloub, P., Cairns, B., Deuzé, J.L., Tanré, D., 2009. Aerosol remote sensing over clouds using A-train observations. J. Atmos. Sci. 66 (8), 2468–2480.

Wendisch, M., Hellmuth, O., Ansmann, A., Heintzenberg, J., Engelmann, R., Althausen, D., Mao, J., 2008. Radiative and dynamic effects of absorbing aerosol particles over the Pearl River Delta, China. Atmos. Environ. 42 (25), 6405–6416.

Wilcox, E.M., 2012. Direct and semi-direct radiative forcing of smoke aerosols over clouds. Atmos. Chem. Phys. 12 (1), 139–149.

Winker, D.M., Vaughan, M.A., Omar, A., Hu, Y., Powell, K.A., Liu, Z., Young, S.A., 2009. Overview of the CALIPSO mission and CALIOP data processing algorithms. J. Atmos. Ocean. Technol. 26 (11), 2310–2323.

Wu, L., Hasekamp, O., van Diedenhoven, B., Cairns, B., Yorks, J.E., Chowdhary, J., 2016. Passive remote sensing of aerosol layer height using near-UV multiangle polarization measurements. Geophys. Res. Lett. 43 (16), 8783–8790.

Xu, X., Wang, J., 2015. Retrieval of aerosol microphysical properties from AERONET photopolarimetric measurements: 1. Information content analysis. J. Geophys. Res. Atmos. 120, 7059–7078.

Xu, X., Wang, J., Wang, Y., Zeng, J., Omar, T., Yang, Y., Marshak, A., Reid, J., Miller, S., 2017. Passive remote sensing of altitude and optical depth of dust plumes using the oxygen A and B bands: first results from EPIC/DSCOVR at Lagrange-1 point. Geophys. Res. Lett. 44. https://doi.org/10.1002/2017GL073939.

Yamamoto, G., Wark, D.Q., 1961. Discussion of the letter by R. A. Hanel, "Determination of cloud altitude from a satellite". J. Geophys. Res. 66 (10), 3596.

Yang, Y., Marshak, A., Mao, J., Lyapustin, A., Herman, J., 2013a. A method of retrieving cloud top height and cloud geometrical thickness with oxygen A and B bands for the Deep Space Climate Observatory (DSCOVR) mission: radiative transfer simulations. J. Quant. Spectrosc. Radiat. Transf. 122, 141–149.

Yang, Z., Wang, J., Ichoku, C., Hyer, E., Zeng, J., 2013b. Mesoscale modeling and satellite observation of transport and mixing of smoke and dust particles over northern sub-Saharan African region. J. Geophys. Res. Atmos. 118 (21), 12139–12157.

Yu, H., Chin, M., Winker, D.M., Omar, A.H., Liu, Z., Kittaka, C., Diehl, T., 2010. Global view of aerosol vertical distributions from CALIPSO LIDAR measurements and GOCART simulations: regional and seasonal variations. J. Geophys. Res. 115, D00H30.

Zakšek, K., Hort, M., Zaletelj, J., Langmann, B., 2013. Monitoring volcanic ash cloud top height through simultaneous retrieval of optical data from polar orbiting and geostationary satellites. Atmos. Chem. Phys. 13 (5), 2589–2606.

Zeng, J., Han, Q., Wang, J., 2008. High-spectral resolution simulation of polarization of skylight: sensitivity to aerosol vertical profile. Geophys. Res. Lett. 35 (20), L20801.

Zhang, L., Li, Q.B., Gu, Y., Liou, K.N., Meland, B., 2013. Dust vertical profile impact on global radiative forcing estimation using a coupled chemical-transport-radiative-transfer model. Atmos. Chem. Phys. 13 (14), 7097–7114.

Further Reading

Boesche, E., Stammes, P., Preusker, R., Bennartz, R., Knap, W., Fischer, J., 2008. Polarization of skylight in the O_2A band: effects of aerosol properties. Appl. Opt. 47 (19), 3467–3480.

Diner, D.J., Beckert, J.C., Reilly, T.H., Bruegge, C.J., Conel, J.E., Kahn, R.A., Verstraete, M.M., 1998. Multi-angle Imaging SpectroRadiometer (MISR) instrument description and experiment overview. IEEE Trans. Geosci. Remote Sens. 36 (4), 1072–1087.

Jeong, M.-J., Hsu, N.C., 2008. Retrievals of aerosol single-scattering albedo and effective aerosol layer height for biomass-burning smoke: synergy derived from "A-Train" sensors. Geophys. Res. Lett. 35 (24), L24801.

Lee, J., Hsu, N.C., Bettenhausen, C., Sayer, A.M., Seftor, C.J., Jeong, M.-J., 2015. Retrieving the height of smoke and dust aerosols by synergistic use of VIIRS, OMPS, and CALIOP observations. J. Geophys. Res. Atmos. 120 (16), 8372–8388.

Sreekanth, V., 2014. Dust aerosol height estimation: a synergetic approach using passive remote sensing and modelling. Atmos. Environ. 90, 16–22.

CHAPTER

2

Vertical Profiling of Aerosol Optical Properties From LIDAR Remote Sensing, Surface Visibility, and Columnar Extinction Measurements

Kwon H. Lee, Man S. Wong†*

*Gangneung-Wonju National University, Gangneung, South Korea

†Hong Kong Polytechnic University, Kowloon, Hong Kong

1 INTRODUCTION

1.1 Aerosol Sources and Vertical Distribution

Atmospheric aerosols, including inorganic, water-soluble, carbon, mineral, heavy metals, or organic/inorganic acids, have significant implications for air quality, human health, and the ecosystem (Myhre et al., 2013). It is well known that aerosols can affect the Earth's radiation budget and contribute to climate change (WHO, 2013). In general, most aerosols are generated from the Earth's surface, although some arise from aviation emissions or meteorite debris. Pollution particles such as carbonaceous and sulfate particles originate from human activities, while dust and sea-salt particles originate from nature's dynamic interaction between the atmosphere and earth's surface. The height of aloft aerosol layers is a critical determinant of global aerosol transport and dispersion. Moreover, the vertical distribution of aerosols varies depending on the weather conditions and their dynamic processes.

An aerosol vertical profile from remote-sensing techniques is often expressed as the extinction coefficient (σ) at a given altitude, which is the fraction of light lost due to scattering and absorption by aerosol particles. The vertical aerosol extinction coefficients are important parameters because the aerosol extinction coefficient at surface level is used for the derivation of visibility (or "visual range"). In addition, a columnar integrated extinction coefficient over a vertical atmosphere corresponds to aerosol optical thickness (AOT, τ_a), which estimates aerosol amounts for the whole atmospheric column as a unitless measure. Obviously, aerosol

https://doi.org/10.1016/B978-0-12-810437-8.00002-5
© 2018 Elsevier Inc. All rights reserved.

vertical profile is an important parameter for understanding the radiative effects of aerosols and for generating more accurate aerosol models.

1.2 Remote Sensing of Aerosol Vertical Distribution

Measuring the aerosol vertical profile is an important issue in remote sensing. The past three decades have seen a turnaround in our capacity to study atmospheric aerosols as a global environmental issue through the use of satellite remote-sensing technology (Lee et al., 2009). Various optical sensors mounted onboard the remote-sensing satellites scan vast areas of the Earth's atmosphere during both day and night. With their continuous development and improvement, remote-sensing techniques have been increasingly utilized for studying the atmospheric environment and climate change. LIght Detection And Ranging (LIDAR) has endeavored to build a competent infrastructure in the field of ground-based remote sensing technology and its applications for local to continental network measurements. This technology is now being extensively utilized for sensor development and technological uplift of the national air quality and climate change.

Although ground-based sensors or satellite observations are well developed, they are only used to gather two-dimensional or horizontal information on aerosols, while vertical profiles are required for "in situ" (e.g., from aircraft, balloons, and unmanned aviation vehicles) or active remote sensing (e.g., from LIDAR and RADAR) measurements (see Fig. 1). Satellites have also been used as a surrogate for measuring aerosol

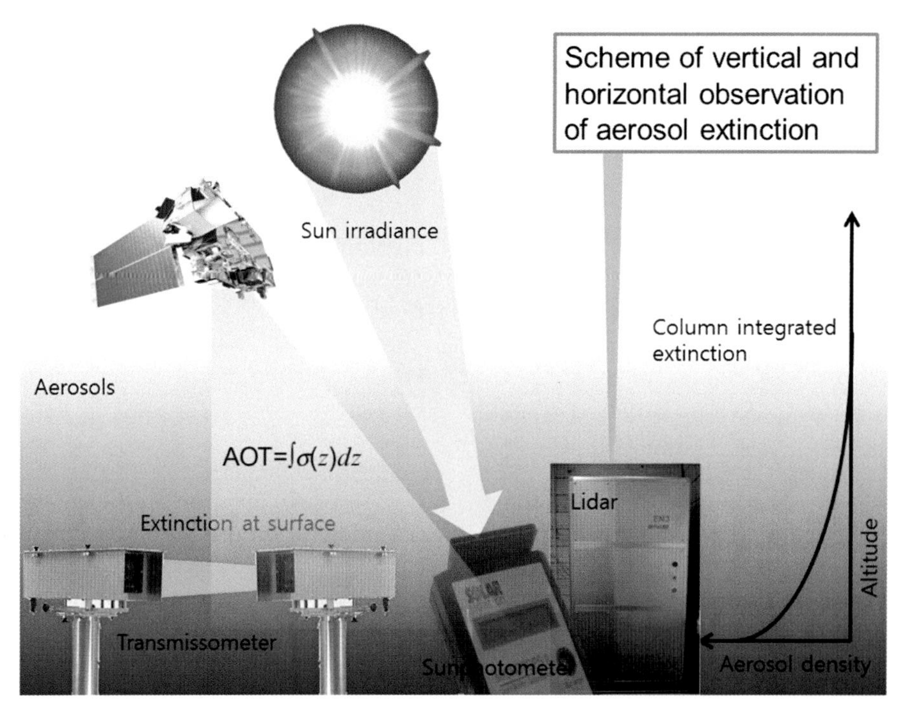

FIG. 1 Illustration of the vertical and horizontal aerosol observation concept. Passive remote-sensing techniques such as a radiometer or spectrometer are measuring path-integrated extinction. The LIDAR technique provides discrete information through the atmosphere while also making it possible to determine the altitudes of the observed extinction.

properties synoptically. Satellite remote sensing offers an opportunity to derive particulate matter through the estimation of AOT. τ_a is a relative measure of aerosol loading in the atmosphere. A higher AOT value indicates higher aerosol loading and hence lower visibility.

The main objectives of this chapter are: (i) to introduce the active and passive remote-sensing techniques for aerosol vertical distribution; and (ii) to explore the applications in the visibility estimation and radiative processes of the aerosols in relation to the vertical distribution of aerosol layers. This chapter is structured as follows: the instruments, theory, experimental methods employed, and results using the active remote-sensing techniques are introduced in Section 2. Section 3 presents the use of passive remote-sensing techniques. Section 4 explores the applications regarding visibility estimation and radiative impacts of aerosol vertical distribution. The work is concluded in Section 5, and further suggestions are also presented.

2 CHARACTERIZATION OF VERTICAL VARIABLES WITH ACTIVE REMOTE-SENSING TECHNIQUES

2.1 Remote Sensing From the Ground

LIDAR is an active remote-sensing technique using a laser beam for measuring the vertical distribution of atmospheric aerosols at a high temporal resolution. It is one of the active remote-sensing techniques that measures light scattering by atmospheric molecules and aerosols. From an observed signal, aerosol backscatter and extinction profiles can be determined using an analytical solution called the "LIDAR equation," (Klett, 1981; Fernald, 1984) with an initial estimation. A Typical LIDAR equation for two distinct scatters (Rayleigh and aerosol scattering) is expressed as,

$$P(z) = E \cdot C \cdot z^{-2} \cdot \{\beta_1(z) + \beta_2(z)\} \cdot T_1^2(z) \cdot T_2^2(z) \tag{1}$$

where, z = height, $P(z)$ = LIDAR receiving signal, E = output photon energy, C = the calibration constant for the instrument, $\beta_1(z), \beta_2(z)$ = backscattering cross-sections of aerosol and atmospheric molecules and $T_1(z)$, $T_2(z)$ = the aerosol and molecular transmittance, respectively. The solution to this equation for the aerosol backscattering cross sections becomes,

$$\beta_1(z) = \frac{P(z)z^2 \exp\left[-2(S_1 - S_2)\int_0^z \beta_2(z)dz\right]}{\mathrm{CE} - 2S_1\int_0^z P(z)z^2 \exp\left[-2(S_1 - S_2)\int_0^z \beta_2(z')dz'\right]dz} - \beta_2(z) \tag{2}$$

where, $S_1 = \sigma_2(z)/\beta_1(z)$; extinction to backscattering ratio of aerosols, $S_2 = \sigma_2(z)/\beta_2(z) = 8\pi/3$; extinction to backscattering ratio for molecular scatters, $\sigma_1(z)$; extinction cross-sections of aerosols, $\sigma_2(z)$; extinction cross-sections of molecules. To solve this equation, the reference height (z_0) where $\beta_2(z_0) \gg \beta_1(z_0)$ should be considered. Also, an extinction-to-backscatter ratio is assumed in determining the aerosol extinction from the backscattering cross-section.

For example, LIDAR measurements of the aerosol extinction profiles were performed during the 2001 Asian Pacific Regional Aerosol Characterization Experiment (ACE-Asia) (Huebert et al., 2003) campaign conducted at Gosan, South Korea (33°17′N, 126°10′E) (Hong et al., 2004). Fig. 2 shows the location of the Gosan super site for this observation. In Fig. 3, timely aerosol extinction profiles for the Asian dust events on March 22, 2001, April 13, 2001, and April 26, 2001 were plotted. On April 13, multiple layers observed at 1, 4, and 7 km altitudes are very strong Asian dust aerosols. After passing the Asian dust storm, a strong aerosol layer under 2 km altitude in the marine boundary layer was observed. A weak aerosol signal

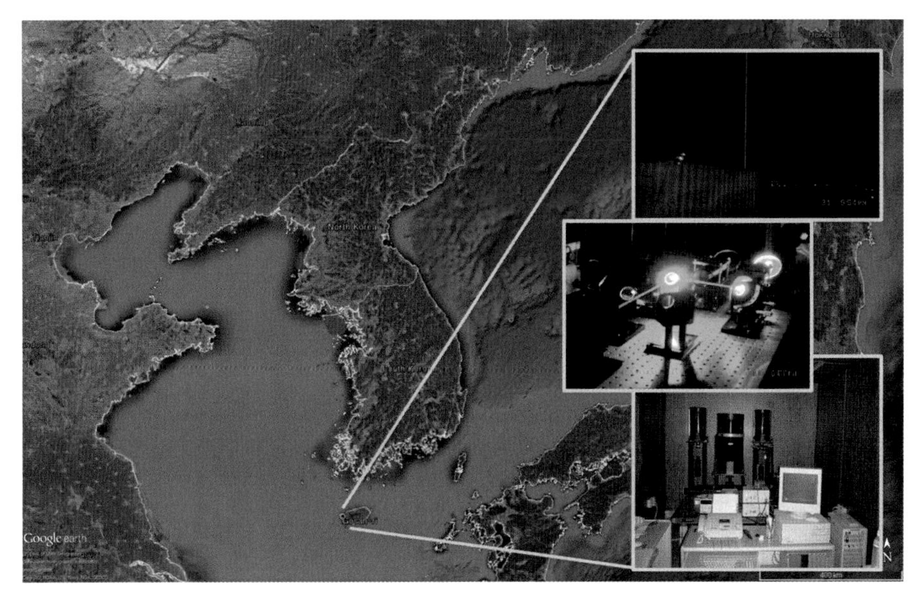

FIG. 2 Geographical location of the Gosan super site during the ACE-Asia 2001 campaign and photos of the LIDAR system used.

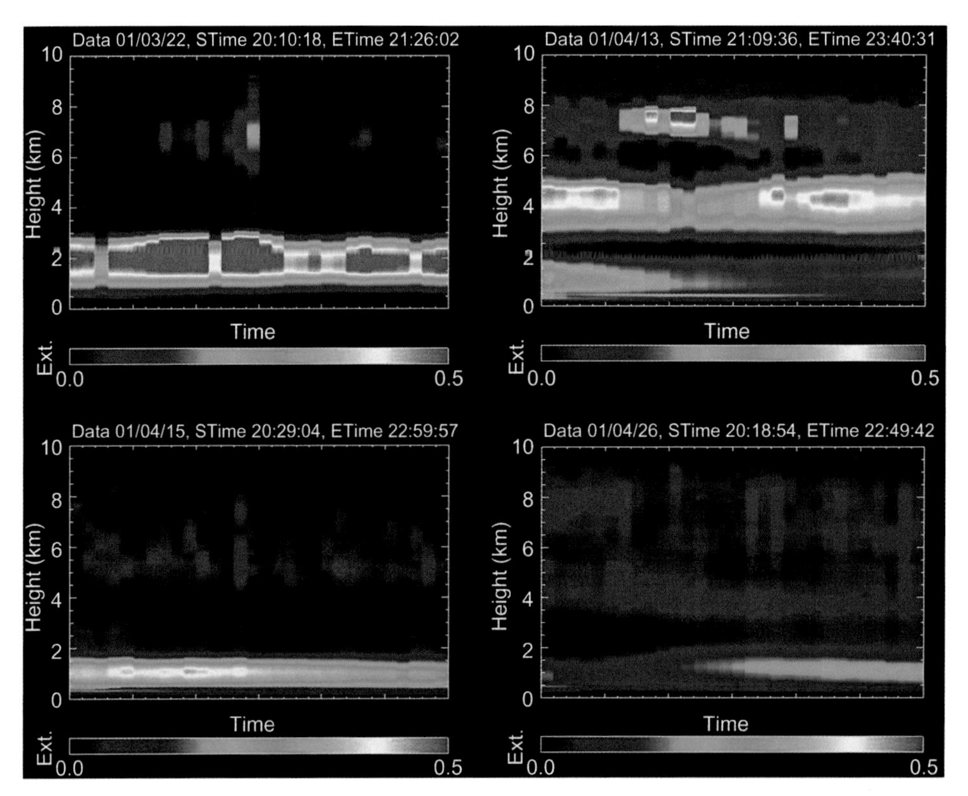

FIG. 3 Aerosol extinction profiles observed during the ACE-Asia 2001 campaign. *Reproduced with permission Springer, Hong, C.S., Lee, K.H., Kim, Y.J., Iwasaka, Y., 2004. Lidar measurements of the vertical aerosol profile and optical depth during the ACE-Asia 2001 IOP at Gosan, Jeju Island, Korea. Environ. Monit. Assess. 92, 43–57.*

was measured in the lower troposphere on April 26 under clear skies.

However, due to the use of the assumed LIDAR ratio (extinction-to-backscatter ratios, hereafter LR) in solving Eq. (2), large errors may arise in aerosol extinction retrieval from the backscatter signal. To overcome this, a LIDAR technique using the Raman spectroscopy (i.e., Raman LIDAR) was developed; it measures more reliable aerosol extinction profiles (Ansmann et al., 1991). Raman spectroscopy is commonly used to identify a fingerprint of molecules by measuring inelastic scattering (or Raman scattering) of monochromatic light, usually from a laser (Gardiner, 1989). The Raman scattering signal is useful in estimating aerosol optical properties and identifying aerosol type because it is able to measure the backscattering and extinction coefficient simultaneously, without the need for an assumption of LR (Mattis et al., 2003). The only disadvantage is the weak Raman scattering signals are used compared to the elastic scattering signals (i.e., Rayleigh scattering, aerosol scattering), but this limitation can be overcome by the use of high-power lasers and/or long integration times.

For the sake of multipurpose (i.e., deriving the mixing state, shape, size, absorption, etc.), the LIDAR technique uses more than two wavelengths (i.e., multiwavelength LIDAR). For example, 355 and 532 nm are used for derivation of backscatter coefficient, and the nitrogen vibration Raman cannels at 387 and 607 nm are implemented for the LIDAR inverion technique (Ansmann et al., 1990). The LIDAR equation used for the Raman backscatter signal is expressed as:

$$P_{\lambda R}(z)= K_{\lambda R}\frac{O(z)}{z^2}N_R(z)\frac{d\sigma_{\lambda R}(\pi)}{d\Omega} \times \exp\left\{-\int_0^z\left[\alpha_{\lambda 0}^{\text{aer}}(\xi)+\alpha_{\lambda 0}^{\text{mol}}(\xi)+\alpha_{\lambda R}^{\text{aer}}(\xi)+\alpha_{\lambda R}^{\text{aer}}(\xi)\right]\right\} \tag{3}$$

where $P_{\lambda R}(z)$ = the return signals from distance z at the Raman wavelength of λ_R; $O(z)$ = the overlap correction factor; $K_{\lambda R}$ = the depth-independent system parameters; $N_R(z)$ = the molecule number density of the Raman-active gas; and $d\sigma_{\lambda r}(\pi)/d\Omega$ = the range-independent differential Raman cross-section for the backward direction. Using the reference signal, the particle extinction coefficient (α_{aer} or α_p) can be determined as:

$$\alpha_{\text{aer}}(z)=\frac{\frac{d}{dz}\left\{\ln\left[\frac{N_{\text{ref}}(z)}{P_{\text{ref}}(z)\cdot z^2}\right]\right\}-\alpha_{\lambda}^{\text{mol}}(z)-\alpha_{\lambda_{\text{ref}}}^{\text{mol}}(z)}{1+\left(\frac{\lambda}{\lambda_{\text{ref}}}\right)^k} \tag{4}$$

where $N_{\text{ref}}(z)$ = the molecular number density of the reference gas, and $k=1$, except for ice clouds. The backscattering coefficient (β_{aer} or β_p) can be derived as:

$$\beta_{\text{aer}}(z)=[\beta_{\text{aer}}(z_0)+\beta_{\text{mol}}(z_0)] \times\frac{P_{\lambda_{\text{ref}}}(z_0)P_\lambda(z)}{P_\lambda(z_0)P_{\lambda_{\text{ref}}}(z)}\frac{N_{\text{ref}}(z)}{N_{\text{ref}}(z_0)} \times\frac{\exp\left\{-\int_{z_0}^{z}\left[\alpha_{\lambda_{\text{ref}}}^{\text{aer}}(\xi)+\alpha_{\lambda_{\text{ref}}}^{\text{mol}}(\xi)\right]\right\}}{\exp\left\{-\int_{z_0}^{z}\left[\alpha_{\lambda}^{\text{aer}}(\xi)+\alpha_{\lambda}^{\text{mol}}(\xi)\right]\right\}} -\beta_{\text{mol}}(z) \tag{5}$$

where z_0 is reference height with $\beta_{\text{mol}}(z_0)\gg\beta_{\text{aer}}(z_0)$. Using α_p and β_p determined by Eqs. (4), (5), LR is then determined as:

$$\text{LR}(z)=\frac{\alpha_{\text{aer}}(z)}{\beta_{\text{aer}}(z)} \tag{6}$$

The particle depolarization ratio (PDR, δ_p) using polarized light signals is defined as:

$$\delta_p=\frac{\delta(z)R(z)-\delta_m}{R(z)-1} \tag{7}$$

where δ_m denotes the molecular depolarization ratio; and $R(z)$ is the scattering ratio.

The dust-mixing ratio (DMR, R_D) is now defined as the ratio between dust backscattering to total backscattering coefficient, and can be mathematically expressed by δ_p for dust particles (δ_D) and for nondust particles (δ_{ND}) in previous studies (Noh et al., 2012).

$$R_D = \frac{(\delta_P - \delta_{ND})(1+\delta_D)}{(\delta_D - \delta_{ND})(1+\delta_P)} \quad (8)$$

Thus aerosol profiles for the various aerosol optical properties can be resolved, enabling a vertical snapshot for the precise identification of aerosol distribution.

Fig. 4 shows the aerosol extinction coefficient and the volume depolarization ratio derived from continuous LIDAR measurements between May 4 and 9, 2009, at the Gwangju Institute of Science and Technology (GIST), Gwangju, South Korea (35.10°N, 126.53°E). The extinction coefficients provide qualitative information on the vertical distribution of aerosols. In Fig. 4A, relatively high extinction coefficients at low altitudes indicate that most aerosols existed under the planetary boundary layer (1–2 km altitude). Meanwhile, depolarization ratio is a reliable indicator of particle shape; it is also useful in distinguishing between nonspherical and spherical particles (Shimizu et al., 2004). A shape-based aerosol type observed during this period (Fig. 4B) and depolarization ratios greater than 10% suggest that the aerosols contain irregularly shaped dust particles. The higher depolarization ratios detected from May 4 and 9, 2009, indicate a higher amount of nonspherical particles than on other observation days.

The multiwavelength Raman LIDAR (MRL) introduced here uses an Nd:YAG laser as the light source, operating at three wavelengths ($\lambda = 355, 532$, and 1064 nm) with a fixed repetition rate of 10 Hz. The pulse energy at the three wavelengths is 140, 154, and 640 mJ, respectively (Lee and Noh, 2015). Fig. 5 illustrates the

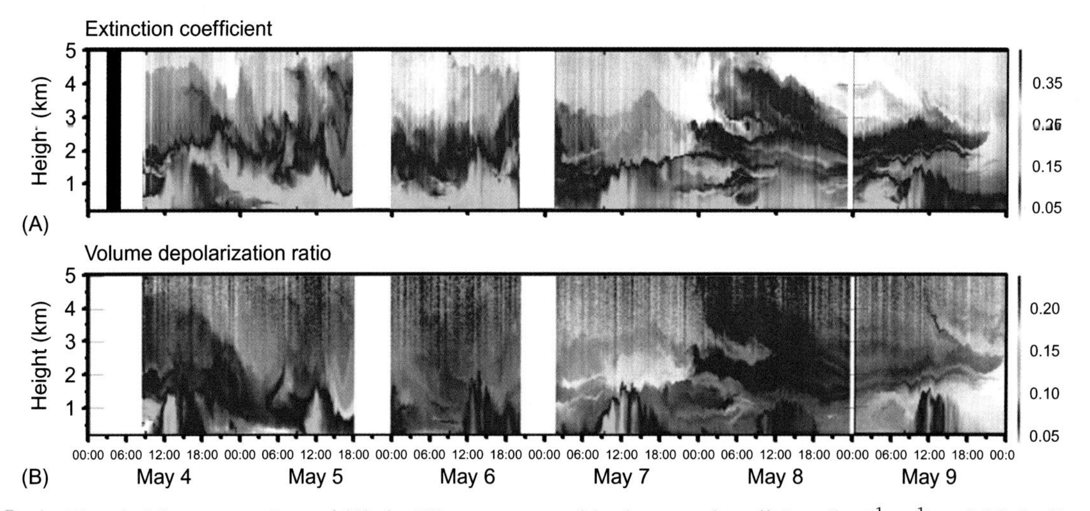

FIG. 4 Time-height cross-sections of (A) the 532 nm attenuated backscattered coefficient ($km^{-1} sr^{-1}$) and (B) the linear volume depolarization ratios (%) observed by the depolarization ratio LIDAR system between April 5 (0000 LT) and 12 (2400 LT) 2009. *Reproduced with permission from Copernicus publications, Noh, Y.M., Lee, H., Mueller, D., Lee, K., Shin, D., Shin, S., Choi, T.J., Choi, Y.J., Kim, K.R. 2003. Investigation of the diurnal pattern of the vertical distribution of pollen in the lower troposphere using LIDAR. Atmos. Chem. Phys. 13, 7619-7629. https://doi:10.5194/acp-13-7619-2013.*

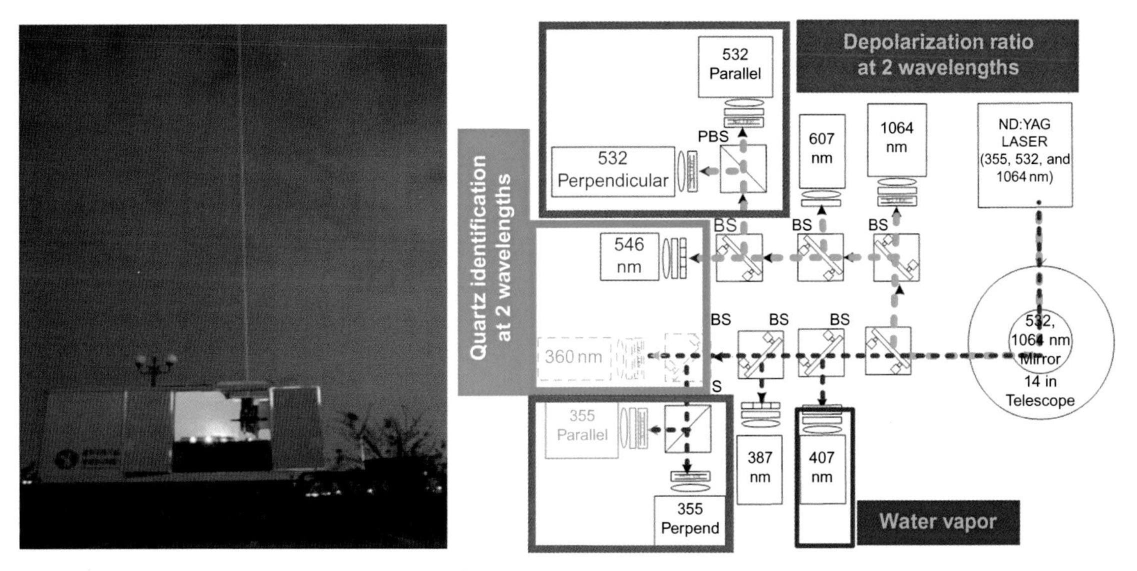

FIG. 5 Systematic diagram of the multiwavelength Raman LIDAR (MRL). MRL receives the light signals at 355, 532, and 1064 nm (elastic backscattering) and 361, 387, 407, 546, and 607 nm (molecular Raman backscattering signals). With those signals, particle backscatter coefficients at 355, 532, and 1064 nm; particle extinction coefficients at 355 and 532 nm; quartz-backscatter coefficients at 361 and 546 nm; and the water vapor mixing ratio can be derived.

systematic characteristics of MRL. A 14-in. Schmidt-Cassegrain telescope (focal length $f = 3.91$ m, field of view $= 0.4$–0.5 mrad) is used to collect and receive the backscattered signal. The MRL measures three elastic backscatter signals at 355, 532, and 1064 nm; two nitrogen vibration Raman signals at 387 and 607 nm; polarized signals at 532 nm; particle extinction coefficients at 355 and 532 nm; quartz-backscatter coefficients at 361 and 546 nm; and the water vapor mixing ratio. The linear PDR at 532 nm follows from the signals measured.

MRL has been measuring aerosol backscattering, extinction coefficients, the depolarization ratio, and LR of aerosol at local atmosphere. Fig. 6 shows an example of aerosol extinction coefficient profiles measured from May 27 to 30, 2005. In this figure, two different extinction coefficients derived using the Raman (*green line*) and Klett (*red*

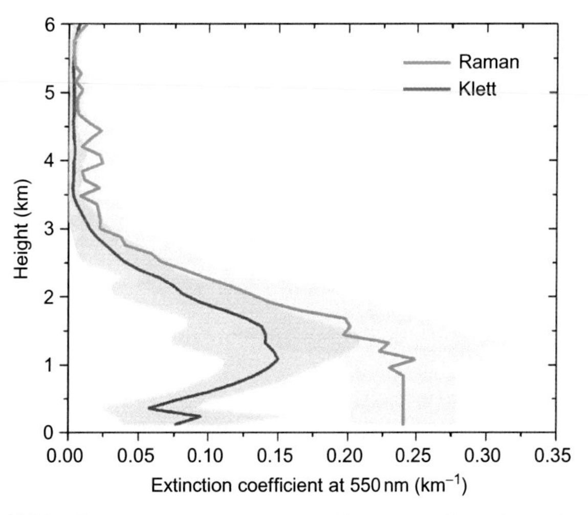

FIG. 6 Aerosol extinction coefficient profiles derived using the Raman method *(green line)* and the Klett method *(red line)* during May 27–30, 2005. *Shaded area* represents $\pm 1\sigma$ during a given period.

line) methods are plotted to compare the retrieval results. Each plot shows large discrepancies for the stronger aerosol loadings under 3 km height and the mean bias is about $0.04 \pm 0.06\ \text{km}^{-1}$. Since the Klett method uses a fixed LR value of 50, it may contain certain errors during retrieval of aerosol particles in the atmosphere. Meanwhile, aerosol extinction using the Raman method with simultaneous acquisition of LR is known to be more accurate (Ansmann et al., 1992). Fig. 7 shows the LR values retrieved using the Raman method simultaneously. In this plot, we can observe that the direct measured LR values vary with altitudes. LR values range from 45 to 90 sr^{-1} on May 27, 2005 and 75 to 120 sr^{-1} on May 30, 2005. Obviously these values are different from the value of the Klett method's assumption. These higher values are mainly caused by light-absorbing particles in the atmosphere.

Thus, MRL observes multispectral aerosol optical properties that include the backscattering coefficient, extinction coefficient, LIDAR ratio, depolarization ratio, ångström exponents, and the mineral-quartz concentration. Fig. 8 shows an example of the MRL-derived aerosol optical properties on March 20 and November 11, 2010, during the Asian dust storm periods. The backscatter and extinction coefficients show an optically thick aerosol layer up to 1 km in height. The mean values of the backscatter coefficients are 50–55 Mm sr^{-1} at 355 and 532 nm in the altitude range of 0.24–1.02 km. The extinction coefficients are 2695 mm^{-1} at 532 nm and 2226 mm^{-1} at 355 nm. The values at 532 nm correspond to a meteorological visibility of approximately 1.4 km. The optical depth in the altitude range from 0.24 to 1.02 km was 2.1 at 532 nm and 1.75 at 355 nm. If we assume that the extinction coefficients are similar throughout the dust plume, the total optical depth will exceed 3.5 at 532 nm and 3 at 355 nm, respectively.

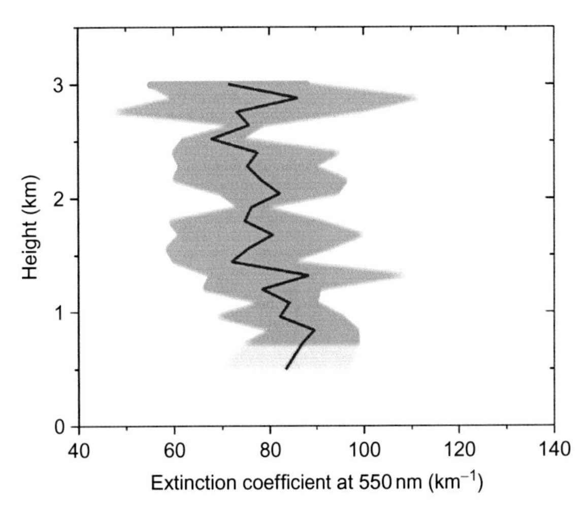

FIG. 7 LIDAR ratio profile measured by the MRL observation during May 27–30, 2005. *Shaded area* represents $\pm 1\sigma$ during a given period.

2.2 Remote Sensing From Space

A fixed ground-based LIDAR has challenges in providing synoptic measurements over a space. Thus, space-borne LIDAR was developed and it can provide the anatomy of aerosols with a geospatial extent of the vertical structure of aerosols. It bridges a critical gap between the point measurements of ground-based LIDAR and the horizontal measurements for optical remote sensors. The first experiment with a three-wavelength (1064, 532, and 256 nm) backscatter LIDAR, the LIDAR In-space Technology Experiment (LITE), was carried out from the space shuttle Discovery for 10 days in September 1994 (Winker et al., 1996). This mission confirmed that space-borne LIDAR is effective in detecting the major features of aerosols, including Saharan dust and African/South American biomass burning. This mission has continued with the development of new space LIDARs such as the Geoscience Laser Altimeter System (GLAS) (Spinhirne et al., 2005) and the Cloud-Aerosol LIDAR and Infrared Pathfinder Satellite Observations (CALIPSO) (Vaughan et al., 2004) satellites, which are currently operating and deriving aerosol products. The CALIPSO data are also useful for global estimation of AOT with regard to the aerosol radiative effect (ARE).

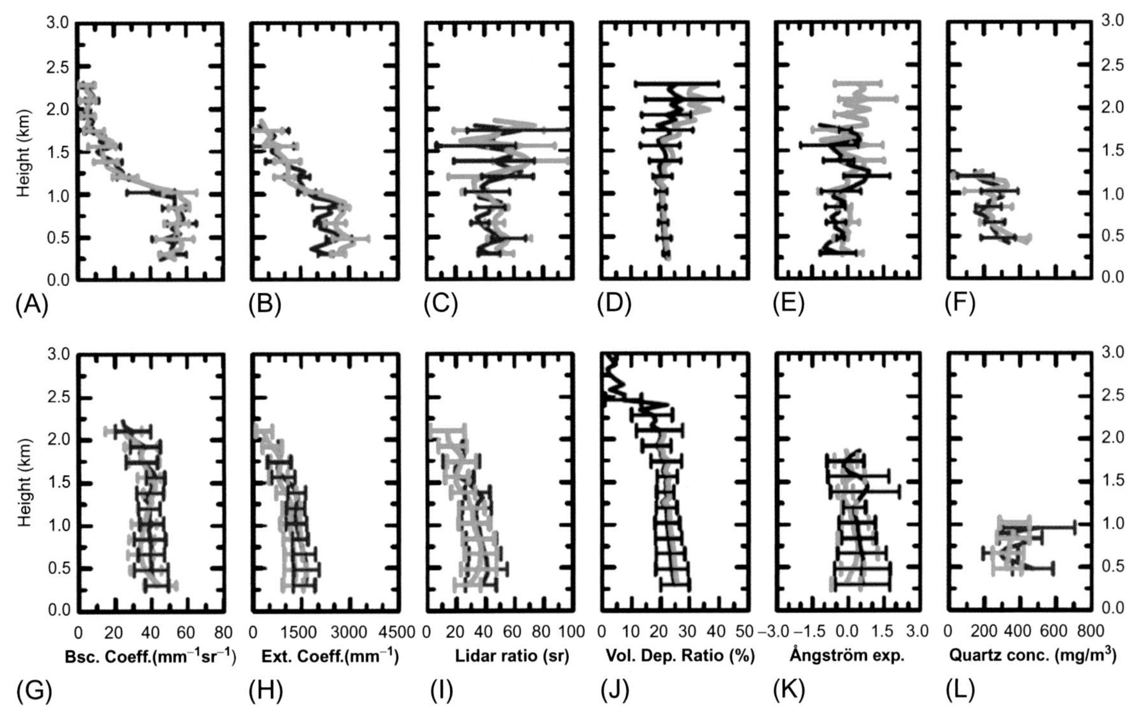

FIG. 8 (A)–(F) Dust optical properties measured from 1115 to 1315 UTC on March 20, 2010 and (G)–(L) measured from 1615 to 1815 UTC on November 11, 2010. (A and G) the particle backscatter coefficient at 532 nm *(green)* and 355 nm *(blue)*; (B and H) the particle extinction coefficient at 532 nm *(green)* and 355 nm *(blue)*; (C and I) the LIDAR ratio at 532 nm *(green)* and 355 nm *(blue)*; (D and J) the linear volume (molecules + particles) depolarization ratio at 532 nm *(black)* and the particle depolarization ratio *(green)*; (E and K) the extinction-related ångström exponent 355/532 nm *(black)* and the backscatter-related ångström exponent 355/532 nm *(green)*; and (F and L) the mineral-quartz concentration at 546 nm *(green)* and 360 nm *(blue)*. Error bars denote one-standard deviation statistical noise. *Reproduced with permission from John Wiley and Sons, Tatarov, B., Müller, D., Noh, Y.-M., Lee, K.-H., Shin, D.-H., Shin, S.-K., Sugimoto, N., Seifert, P., Kim, Y.-J., 2012. Record heavy mineral dust outbreaks over Korea in 2010: two cases observed with multiwavelength aerosol/depolarization/Raman-quartz lidar. Geophys. Res. Lett. 39, L14801. https://doi:10.1029/2012GL051972.*

Since its launch on April 28, 2006, CALIPSO has been widely used to measure three-dimensional data as well as for the new aerosol-cloud interaction study and understanding the aerosol direct/indirect effect. These observations help in understanding the overall processes of aerosols from source/origin to transport. The primary instrument aboard the CALIPSO satellite, a Cloud-Aerosol LIDAR with Orthogonal Polarization (CALIOP), has two wavelengths, meaning polarization-sensitive LIDAR (see Table 1) designed to quantify aerosols as well as thin clouds. The CALIOP uses a diode-pumped Nd:YAG laser producing the linearly polarized pulses of light at 1064 and 532 nm. The return signal is collected by a 1-m telescope measuring the backscattered intensity at 1064 nm and the two orthogonal polarization components at 532 nm (parallel and perpendicular to the polarization plane of the transmitted beam). The Nd:YAG lasers are Q-switched and frequency-doubled to produce simultaneous pulses at 1064 and 532 nm. Each laser produces 110 mJ of energy at each of the two wavelengths at a pulse repetition rate of 20.2 Hz. The sampling resolution of the return signal is 30 m

TABLE 1 Characteristics of the Cloud-Aerosol LIDAR With Orthogonal Polarization (CALIOP) Instrument Onboard the CALIPSO Satellite

Index	Properties
Laser	Nd: YAG, diode-pumped, Q-switched, frequency doubled
Wavelengths	532 nm, 1064 nm
Pulse energy	110 mJ/channel
Repetition rate	20.25 Hz
Receiver telescope	1.0 m diameter
Polarization	532 nm
Footprint/FOV	100 m/130 μrad
Vertical resolution	30–60 m
Horizontal resolution	333 m
Linear dynamic range	22 bits
Data rate	316 kbps

vertical and 333 m horizontal determined by the laser pulse repetition rate. Backscatter data will be acquired from the surface to 40 km with 30 m vertical resolution. After aerosol and cloud layers are detected using a threshold technique, a discrimination algorithm (Liu et al., 2009) is applied to separate cloud and aerosol. Aerosol extinction profiles are only retrieved for the detected aerosol layers in this process.

CALIOP Level 2 algorithms provide distinct identification of aerosol and cloud layers, classify aerosols into six major types (i.e., clean continental, clean marine, dust, polluted dust, polluted continental, smoke, and others) with the LR values from 20 to 70 sr (Omar et al., 2009). CALIOP Level 2 products contain retrieved profile data along the CALIPSO overpass track, which include aerosol backscatter, extinction, depolarization, and results of the aerosol type classification.

Fig. 9 shows an example of aerosol extinction and aerosol-type products observed on March 31, 2007. Two panels on the right illustrate the observations from CALIOP of aerosol loadings and types. Dust layers are observed near 2–3 km height and 35–42 degrees north, over the Yellow Sea. In the southern region (near 20–32 degrees north), an uplifted smoke layer originating from the East Asian continent can be seen in the CALIOP level 2 vertical feature mask (VFM) product.

Aerosol extinction profiles with aerosol types observed from the space-borne LIDAR provide spatial distribution in three-dimensional context as well as temporal variations for atmospheric aerosols. Fig. 10 shows the continuous observations for aerosol profiles analyzed during 2012 using a CALIOP. The statistical analysis on the extinction coefficient profiles for each aerosol type was conducted according to time and space in order to estimate the variation of optical properties of aerosols over Northeast Asia (E110°–140°, N20°–50°). The most frequent altitudes of aerosols are clearly identified and type-dependent aerosol profiles vary with the season. Since relatively high extinction values for dust and polluted dust are found during all seasons, it is considered that the nonspherical dusty aerosols mixed with pollution mainly exist over the region. The result also shows that the smoke aerosol profiles provide quantitative characterization of elevated aerosol layers in the summer.

3 AEROSOL VERTICAL PROFILE DERIVED FROM PASSIVE REMOTE SENSING

3.1 Remote Sensing From the Ground Based Sunphotometry

Active remote sensing techniques such as LIDAR provide height-dependent extinction values with a vertically high resolution. Therefore, their complex installation and high instrument cost prohibit use. Although aerosol profiles are not directly measured from most

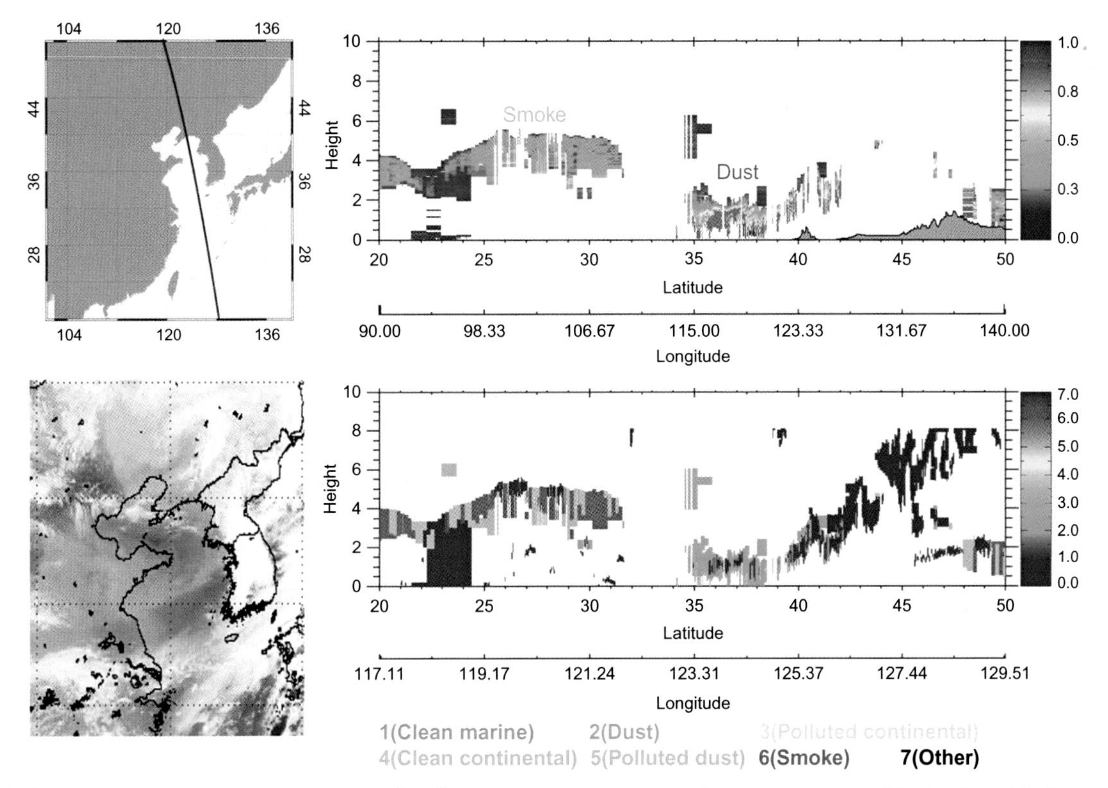

FIG. 9 Aerosol extinction coefficient (km^{-1} sr^{-1}) at 532 nm and the vertical feature mask derived by the CALIOP observation on March 31, 2007. The CALIPSO overpass orbit on the same day is shown as a *gray solid line* in the *upper right map*. AQUA/MODIS color composite images show a strong Asian dust plume.

of the ground-based passive remote sensing measurements, recently there have been some studies that attempt to overcome this limitation (Qiu, 2003; Wong et al., 2008, 2009). They demonstrated that a multichannel sunphotometry can derive the extinction profile based on the theoretical computation of aerosol scaling height from visibility at the surface.

In theory, the aerosol extinction coefficient at the surface level can be derived from visibility, which is an important factor in determining aerosol loadings near the surface. By assuming an exponential decay of aerosol loading, aerosol extinction coefficients with altitudes can be derived. This method is effective for a normal aerosol profile under nondust and cloud-free atmospheric conditions. Thus, by knowing the AOT, the derivation of extinction coefficients for altitudes, and the surface extinction visibility, the scaling height can be derived numerically.

A network of sun/sky radiometers, NASA's Aerosol Robotic NETwork (Holben et al., 1998), monitors AOT and aerosol optical properties globally. In this study, AOT at 550 nm at the Hong Kong AERONET site (http://aeronet.gsfc.nasa.gov) and visibility acquired from the Hong Kong Observatory (HKO) are used for the subsequent calculation of the extinction profile and aerosol scaling height. Visibility data acquired from the HKO, which is only 500 m from the AERONET site, showed that the paired AOT and visibility readings were within 30 min. Aerosol extinction ($\sigma_a(z)$) at altitude z is defined

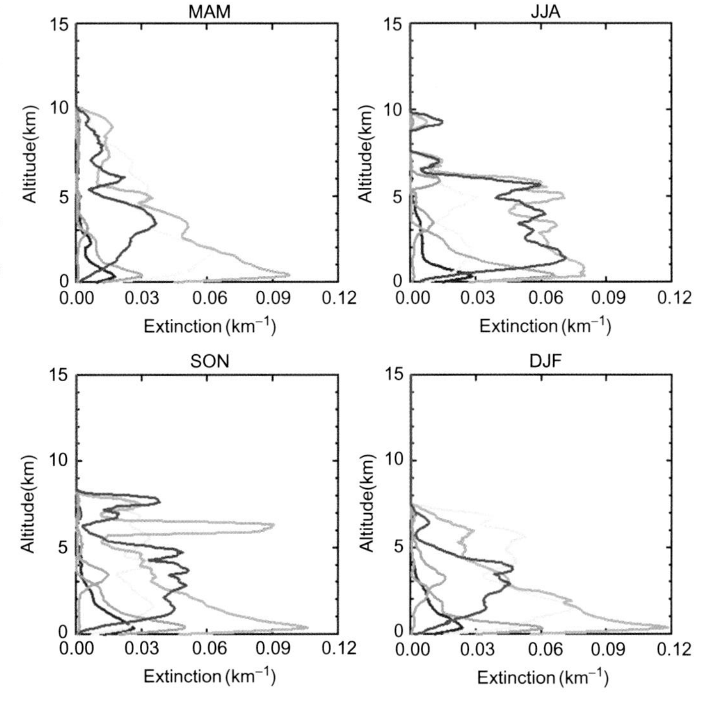

FIG. 10 Seasonal mean daytime extinction (at 532 nm) profile for different aerosol types. Colors represents clean maritime *(blue)*, dust *(yellow)*, polluted continental *(gray)*, clean continental *(sky blue)*, polluted dust *(orange)*, and smoke *(red)*, respectively. *Reproduced with permission from the Korean Journal of Remote Sensing, Lee, K.H., 2014. 3-D perspectives of atmospheric aerosol optical properties over northeast Asia using LIDAR on-board the CALIPSO satellite. Korean J. Remote Sens. 30 (5), 559–570. https://doi:10.7780/kjrs.2014.30.5.2.*

as an exponential function of aerosol scaling height (z_a), which is defined as;

$$\sigma_a(z) = \sigma_a(z=0)\exp\left[-\frac{z}{z_a}\right] \tag{9}$$

Integration of the aerosol extinction with heights, AOT (τ_a), is expressed as;

$$\begin{aligned}\tau_a &= \int_0^{\mathrm{TOA}} \sigma_a(z)dz \\ &= \sigma_a(z=0)\cdot z_a[1-\exp(-z_{\mathrm{TOA}}/z_a)]\end{aligned} \tag{10}$$

Simplifying Eq. (10) is possible by ignoring the exponential term because z_a values are much less than z_{TOA}. Also, $\sigma_a(z=0)$ is converted from the visibility (Vis), and can be estimated by the well-known Koschmieder equation (Koschmieder, 1924).

$$z_a = \frac{\tau_a}{\sigma_a} = \frac{\tau_a}{3.912/\mathrm{Vis} - \sigma_m(z=0)} \tag{11}$$

where σ_m is the surface-level molecular extinction coefficient. The error of extinction coefficients calculated from the aerosol scale height is reported as a very small value 0.0023 km^{-1} in the case of background aerosols (Qiu, 2003). It is worth noting that relative humidity (RH) can change an aerosol's chemical, physical, and optical properties due to the increase of water content. This changing factor by RH shows in the aerosol size distribution, major type, and the vertical structure of aerosol extinction. The RH effects can be partially overcome in the AOT-visibility relationship because both parameters are ambient optical quantities affected by the same RH effect on particle extinction (Kessner et al., 2013).

To visualize the inverted extinction profiles, the AOT values at different atmospheric levels ($\Delta\tau$) at different elevations are derived and linked with the ArcScene and ArcObject script (lsgi, n.d.) for a nearly real-time analysis.

$$\Delta\tau = \Delta\sigma_z \cdot \Delta z \tag{12}$$

The scripts include the predefinition of the aerosol layers and polygons at the vertical intervals of 75 m as well as their color assignment with six transparent shades representing different AOT levels at different altitudes. It is worth noting that a multilayered aerosol condition such as the dust storm or stratospheric volcanic aerosols could not be accurately retrieved using exponential profile assumption.

Fig. 11 demonstrates the visual output from the selected days in February 2007. In Fig. 11B, on February 1, 2007, relatively high AOT values ranging from 0.56 to 0.71 are observed, which corresponds to the high urban pollutants. For 1300 on this day, the aerosol vertical profile shows that aerosols are mainly concentrated below 100 m. This suggests a local source such as vehicle emissions are the major sources of pollution. Interestingly, evidence shows that better visibility and air quality on higher building floors gives the option of ventilating an indoor environment by opening windows. However, for a weekend on February 3, 2007, higher wind speeds and moderate pollution levels (AOT = 0.3–0.5) were shown in Fig. 11D. The vertical profile image at midday (Fig. 11C) shows relatively low surface level concentrations; they are very low near the mountaintops. The AOT values, visibilities, and scaling heights for both days are listed in Table 2.

3.2 Remote Sensing From Space

The spatially inhomogeneous aerosol loadings over a region/city are difficult to depict in both horizontal and vertical contexts using discrete point observations. To overcome this limitation, satellite observation is an alternative for monitoring air quality data by estimating the retrieved τ_a. Two MODIS instruments on Terra and Aqua platforms are passive remote-sensing sensors used worldwide. These satellites cross a 2330-km wide swath, providing daily complete global coverage with equatorial crossing times at 10:30 UTC for Terra and 13:30 UTC for Aqua. The MODIS level 1B (L1B) calibrated radiance data (code names: MOD02 and MYD02) are available from the NASA Goddard Earth Science Distributed Active Archive Center (DAAC) (ladsweb, n.d.). MODIS L1B data contain the calibrated and geolocated radiances in $W/m^2/\mu m/sr$ for 36 bands generated from the MODIS Level 1A data (code names: MOD01 and MYD01).

Based on the radiative transfer theory, aerosol's contribution to satellite receiving radiance can be determined by the subtraction of the atmospheric molecular scattering (Rayleigh reflectance, ρ_{Ray}) and surface reflection (ρ_{Surf}) from satellite-observed reflectance (ρ_{TOA}) defined as:

$$\rho_{TOA}(\lambda) = \frac{\pi \cdot L(\lambda)}{F_0 \cdot \cos\theta_0} \tag{13}$$

$$\rho_{Aer}(\theta_0, \theta_S, \varphi) = \rho_{TOA}(\theta_0, \theta_S, \varphi) - \rho_{Ray}(\theta_0, \theta_S, \varphi) - \frac{T_{Tot}(\theta_0) \cdot T_{Tot}(\theta_S) \cdot \rho_{Surf}(\theta_0, \theta_S)}{1 - \rho_{Surf}(\theta_0, \theta_S) \cdot r_{Hem}(\tau_{Tot}, g)} \tag{14}$$

where $L(\lambda)$ = satellite observed radiance, $F_0(\lambda)$ = the extraterrestrial irradiance, θ_0 = sun zenith angle, θ_S = satellite viewing angle, φ = relative azimuth angle between sun and satellite, τ_{Tot} = total optical thickness, T_{Tot} = the total atmospheric transmittance, g = the asymmetry parameter, and r_{Hem} = the hemispheric reflectance. In Eq. (14), ρ_{Ray} is determined using the Rayleigh scattering cross-section equation (Bucholtz, 1995). Two atmospheric environment terms, T_{Tot} and r_{Hem}, are parameterized as a function of air mass and atmospheric transmission (Gerilowski et al., 2007). The surface term, ρ_{Surf}, is then estimated using a liner mixing model of Eq. (15), which is a dynamic combination of vegetation and bare-soil spectra tuned by the empirical weighting factor (w).

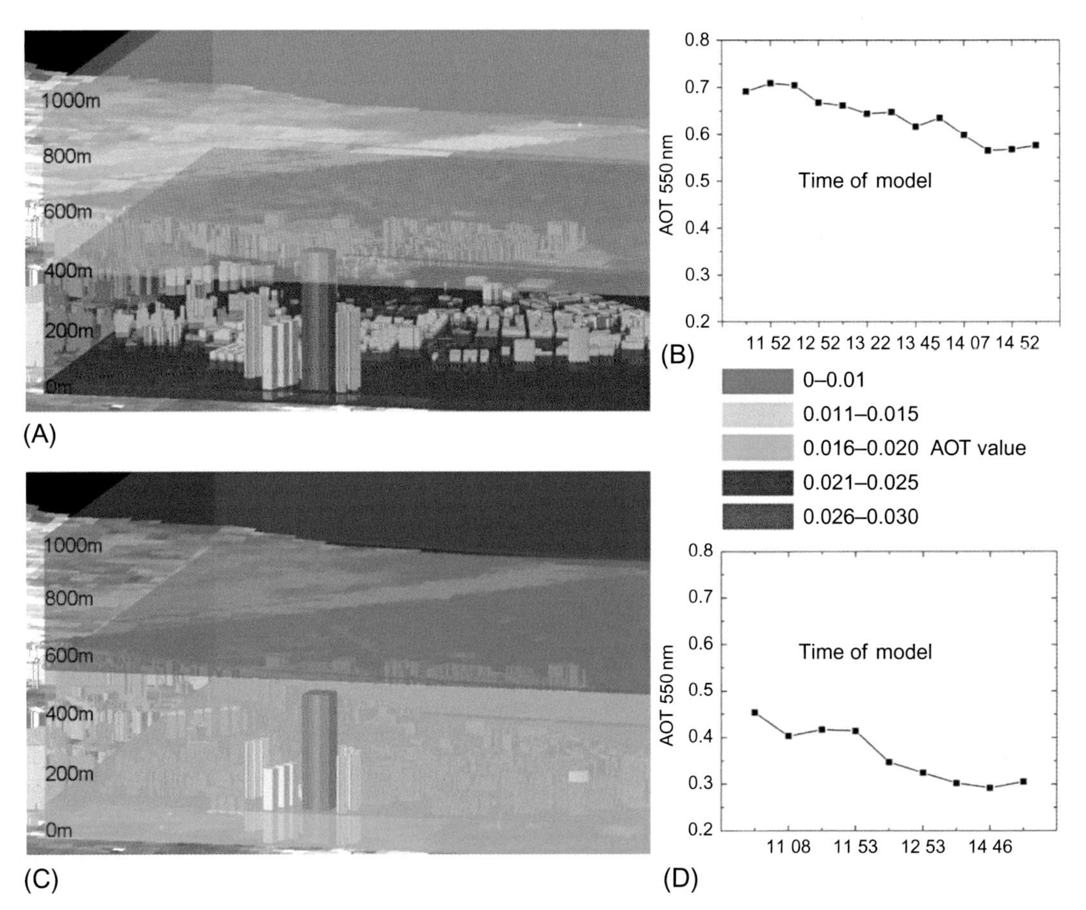

FIG. 11 Derived AOT for different atmospheric layers, three dimensional view from across Victoria harbor to the high-rise buildings on Kowloon peninsula on (A) February 1, 2007 (1322 LT), (C) February 3, 2007 (1153 LT). The graphs (B and D) represent processed level 1.5 AERONET AOT data for 550 nm collected over the course of the day. *Reproduced with permission from the MDPI, Wong, M.S., Nichol, J., Lee, K.H. 2009. Modeling of arosol vertical profiles using GIS and remote sensing. Sensors 9 (6), 4380–4389. https://doi:10.3390/s90604380.*

TABLE 2 Summary of AOT Value, Visibility, and Scaling Height

Date	Time	AOT at 550 nm	Visibility (km)	Scaling Height, z_a (km)
February 1, 2007	1152	0.71	10	1.81
February 1, 2007	1252	0.67	10	1.71
February 1, 2007	1322	0.64	9	1.48
February 1, 2007	1507	0.58	10	1.47
February 3, 2007	1153	0.41	9	0.95

Reproduced with permission from the MDPI, Wong, M.S., Nichol, J., Lee, K.H., 2009. Modeling of arosol vertical profiles using GIS and remote sensing. Sensors 9 (6), 4380–4389. https://doi:10.3390/s90604380.

$$\rho_{\text{Surf}}(\lambda) = w\left[\text{NDVI}_{\text{SW}} \cdot \rho_{\text{Veg}}(\lambda) + (1 - \text{NDVI}_{\text{SW}}) \cdot \rho_{\text{Soil}}(\lambda)\right] \tag{15}$$

$$w = \frac{\left(\left(\rho_{\text{TOA}}(0.66\,\mu\text{m}) - \rho_{\text{ray}}(0.66\,\mu\text{m})\right)\text{slope} + 0.00025\Theta_{\text{sc}} + 0.033\right)}{\text{NDVI}_{\text{SW}} \cdot \rho_{\text{Veg}}(0.66\,\mu\text{m}) + (1 - \text{NDVI}_{\text{SW}}) \cdot \rho_{\text{Soil}}(0.66\,\mu\text{m})} \tag{16}$$

$$\text{Slope} = sl_{\text{NDVI}_{\text{sw}}} + 0.0002 \times \Theta_{\text{sc}} - 0.27 \tag{17}$$

where λ = the wavelength, $\rho_{\text{Veg}}(\lambda)$ and $\rho_{\text{Soil}}(\lambda)$ = the spectral reflectance of "green vegetation" and "bare soil," $sl_{\text{NDVI}_{\text{sw}}}$ = determined by the vegetation index (NDVI_{sw}), Θ_{sc} = scattering angle.

To derive AOT from the MODIS L1b data, the MODIS satellite aerosol retrieval (MSTAR) method (Lee and Kim, 2010) was used. This method uses the look-up table (LUT) technique. LUT is a series of precalculated aerosol reflectances (ρ^*_{Aer}) for a given AOT condition. With the best matching pair of MODIS-observed aerosol reflectances (ρ_{Aer}) and ρ^*_{Aer}, AOT is inversely determined for a ρ^*_{Aer} value. The error of the MSTAR τ_a was reported to be <10% for the retrieval of the following periods: May 19–30, 2007, July 6–16, 2007, September 29–October 8, 2007, December 15–24, 2007, July 4–9, 2008, October 17–24, 2008, May 11–19, 2009, and November 16–22, 2009.

In addition, ground-based aerosol measurements including particulate mass concentrations and optical properties were also conducted at an urban site in Seoul, South Korea, from January 2007 to June 2009. Fig. 12 shows the location of the air quality measurement station at Yonsei

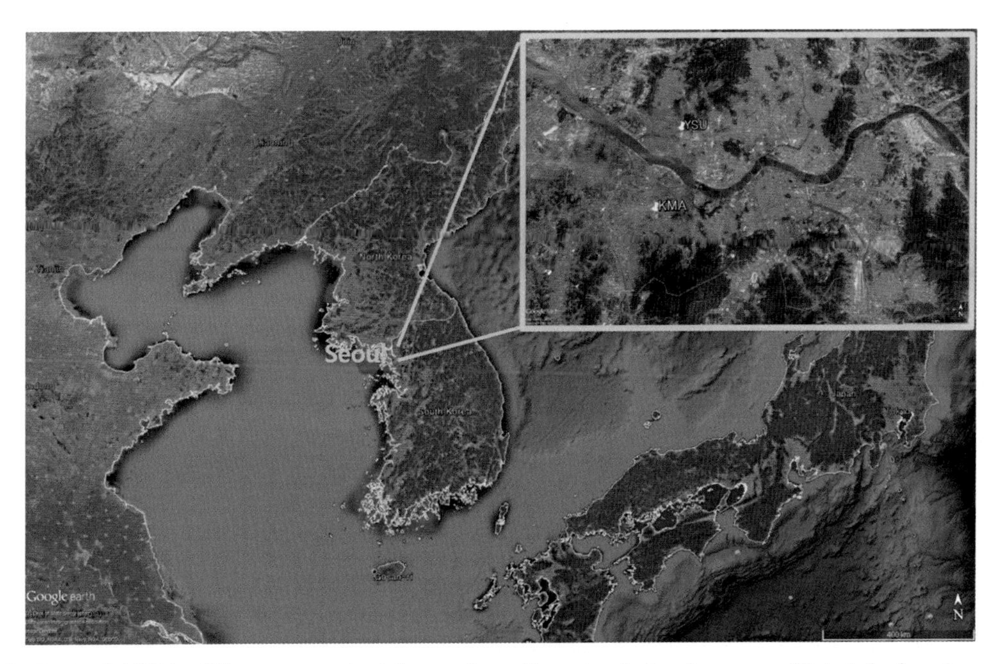

FIG. 12 Derived AOT for different atmospheric layers, three-dimensional view from across Victoria harbor. Area of study and geographical location of the air quality monitoring site at Yonsei University (YSU) (37.564°N, 126.935°E) and the meteorology observation site at Korea Meteorological Administration (KMA) (37.491°N, 126.921°E). Distance between YSU and KMA is 7.9 km.

University (YSU) (37.564°N, 126.935°E) and the meteorological observatory at the Korea Meteorological Administration (KMA) (37.491°N, 126.921°E). Both sites are at the convergence of commercial, residential, and educational zones without apparent industrial pollution sources. The air quality at these locations is considered to be typical for an urban environment.

As explained in Chapter 3.1, visibility is inversely proportional to light extinction, which is a measure of the attenuation of light passing through the atmosphere caused by scattering and absorption by aerosol particles. The visibility can also be derived from satellite observations using an aerosol optical analytical model. Computation of the near real-time vertical profiles is conducted by a linkage between MODIS L1b satellite data processing and the Google Earth platform. A customized script written in KML in Google Earth links the visibility from MODIS AOT data, then calculates the scaling height, extinction vertical profiles, and visibility based on the aerosol optical analytical model explained below.

The MODIS AOT and ground-based visibility data were used to estimate aerosol extinction at different vertical heights. For a conversion of visibility into extinction coefficient during the daytime, Koschmieder's equation is applied. However, Koschmieder's relationship has some assumptions and limitations such as: (1) sky brightness is similar for both observer and observed object; (2) pollutants are homogeneously distributed; (3) the viewing angle is horizontal; (4) earth curvatures are ignored; (5) objects are considered to be large and black; and (6) the threshold contrast is 0.02. Therefore, the visibility and extinction for the local atmospheric environment was evaluated first. A new equation was developed to derive visibility from the extinction coefficient σ_{ext}, as shown in Fig. 13A. The visibility values by the Koschmieder equation are also shown as a *dashed line* in this figure. Obviously, the visibility predicted by Koschmieder's equation differs considerably from the measured visibility, since Koschmieder's relationship is valid only under the limited atmospheric conditions discussed above. In

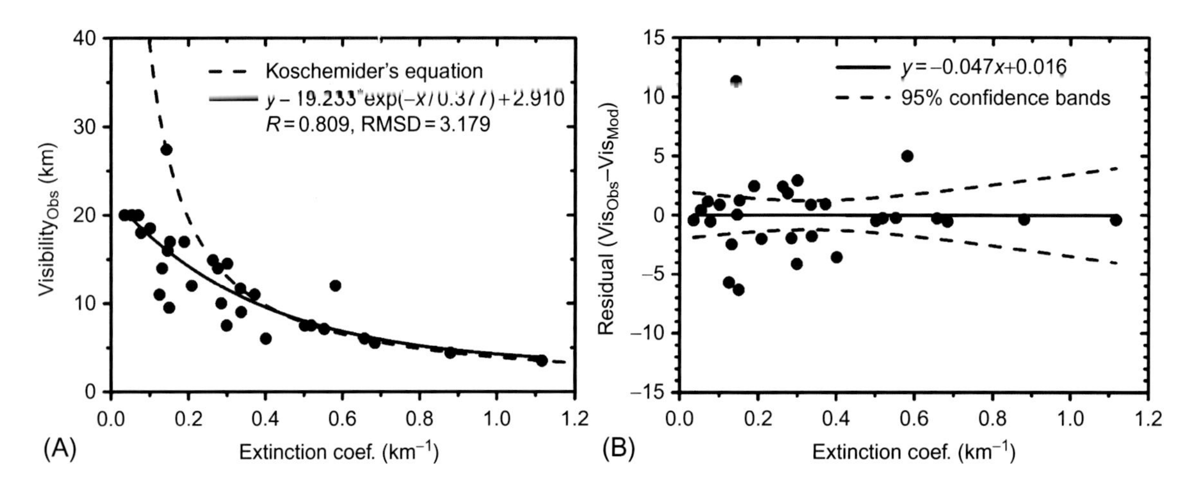

FIG. 13 (A) Scatter plots of observed visibility versus extinction coefficient. Exponential decay *fitting line* and visibility calculated by Koschemeider's equation are plotted as *solid* and *dashed lines.* (B) Scatter plots of residual versus extinction coefficient. *Reproduced with permission from the Elsevier, Lee, K.H., Wong, M.S., Kim, K.W., Park, S.S., 2014. Analytical approach to estimating aerosol extinction and visibility from satellite observations. Atmos. Environ. 91, 127–136. https://doi:10.1016/j.atmosenv.2014.03.050.*

terms of the fitting analysis, the exponential fitting equation obtained here is expressed as:

$$\text{Vis}\,(\text{km}) = 19.23\exp\left(-\sigma_{\text{ext}}/0.38\right) + 2.91 \quad (18)$$

Fig. 13B gives the residual between observed visibility (Vis_{obs}) and modeled visibility (Vis_{Mod}) using Eq. (18) as the function of extinction coefficient. Vis_{Mod} derived by Eq. (14) outperforms because the linear regression line is close to zero and a mean absolute residual is 3.178 km. The error in the estimated visibility with a new conversion equation is within $\Delta\text{Vis} = \pm 2.359$, $\sigma_{\text{ext}} \pm 0.986$.

Also, a relation between visibility and satellite-derived τ_a, based on Eq. (18) and the fitting equation obtained in Fig. 13A, was established. The modified linear regression equation can be expressed as:

$$\text{Vis}\,(\text{km}) \asymp 19.233 \cdot \exp\left(-\frac{\tau_a}{0.377 \cdot z_a}\right) + 2.910 \quad (19)$$

Eq. (19) was used to derive visibility from MODIS τ_a measurements in Seoul, South Korea. Fig. 14 compares the visibility derived from MODIS ($\text{Vis}_{\text{MODIS}}$) with the Vis_{obs} data. Good agreements are observed between the visibility derived from our methodology and the observed values, with a linear regression slope of 0.88 and correlation coefficient of 0.88, which is higher than the Vis_{KOS} (linear regression slope = 0.67, $R = 0.44$). Mean absolute residuals are 2.173 km for this study and 8.831 km for Vis_{KOS}.

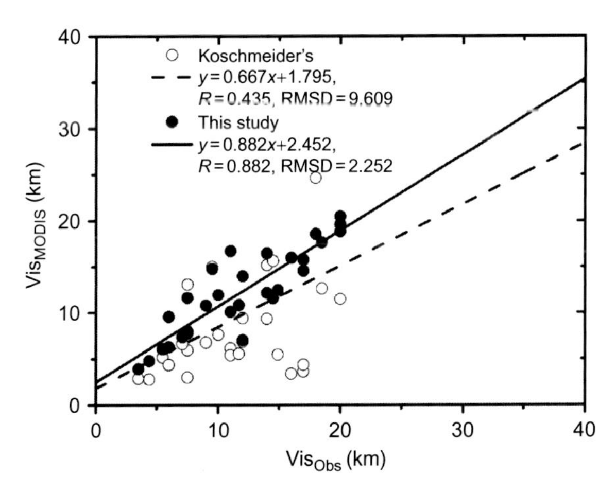

FIG. 14 Scatter plots of ground-observed versus MODIS-derived visibility. *Reproduced with permission from the Elsevier, Lee, K.H., Wong, M.S., Kim, K.W., Park, S.S., 2014. Analytical approach to estimating aerosol extinction and visibility from satellite observations. Atmos. Environ. 91, 127–136. https://doi.org/10.1016/j.atmosenv.2014.03.050.*

The MODIS retrieved τ_a data were used to study major episodes of heavy aerosol conditions such as those from dust storms and haze in the study area. Increased aerosol loadings during the hazy and dusty sky conditions represented the relatively high τ_a (~1.2). MODIS-derived τ_a and visibility for clear, dusty, and hazy sky conditions are summarized in Table 3. MODIS τ_a were retrieved as 0.09 for a clear sky, 0.72 for dusty conditions, and 1.40 for a hazy sky. The differences between observed visibility and MODIS-derived visibility for clear, dusty, and hazy sky days are 0.4, 0.1, and 0.8 km, respectively, which represents a good agreement.

The results indicate that this approach is able to retrieve visibility over a large urban coverage. In addition, it is noteworthy that residual extinction or visibility may result in systematic errors, due to: (1) uncertainty of MODIS-retrieved τ_a values over urban areas; (2) spatial resolution of MODIS compared to air quality single-point measurements; and (3) the time lag between satellite and ground observations. However, in view of the highly variable nature of the measurements and uncertainties in the proposed method, the basic agreement between modeled and observed data is still promising.

4 APPLICATIONS TO THE RADIATIVE IMPACTS OF AEROSOLS

It is well known that aerosols affect the earth's climate system by reflecting, absorbing, and scattering electromagnetic radiation. Depending on

TABLE 3 Summary of Air Quality in Seoul for the Clear, Dusty, and Hazy Sky Conditions

Sky Condition	RH (%)	Obs. Vis. (km)	ASH (km)	MODIS AOT	MODIS Vis. (km)
Clear	37.5	20	1.66	0.09	19.58
Dusty	52	10	2.24	0.72	10.08
Hazy	49	4	1.59	1.40	4.77

their chemical composition and spatial distribution, aerosols cause imbalance in the radiation budget and affect global atmospheric/hydrologic circulation. The presence of black aerosols such as soot particles absorbs more radiation and contributes to atmospheric warming. In contrast, white aerosols such as sulfate particles scatter more radiation and contribute to atmospheric cooling or modifying cloud brightness and cloud cover. More detailed information can be found in the recent IPCC report (IPCC, n.d.). Despite its importance in the geolocational characteristics of aerosols, aerosol vertical information on radiative impacts of aerosols is largely unknown. In this chapter, the AREs due to aerosol vertical profiles are examined and discussed. These effects are estimated using integrated data from ground-based measurements, satellite observations, and the radiative transfer model (RTM) calculation.

The incoming and outgoing radiation can be derived from the radiative transfer simulation using aerosol optical properties such as aerosol extinction, asymmetry parameters, and single scattering albedo (SSA). This radiative transfer simulation using aerosol's optical profiles acquired from remote-sensing techniques is explained in this chapter and Chapter 3. For example, the MRL observation data, including the extinction coefficient, refractive index, *reff*, and SSA, were used to derive incoming and outgoing shortwave radiation. Radiative transfer simulation with these parameters provides the downward ($F_{\downarrow}$) and upward flux ($F_{\uparrow}$) at a given height. These results are used to define the aerosol radiative forcing (ARF), indicating the difference in the net flux with and without aerosols, which is a measure of the impact of aerosols on the climate system as:

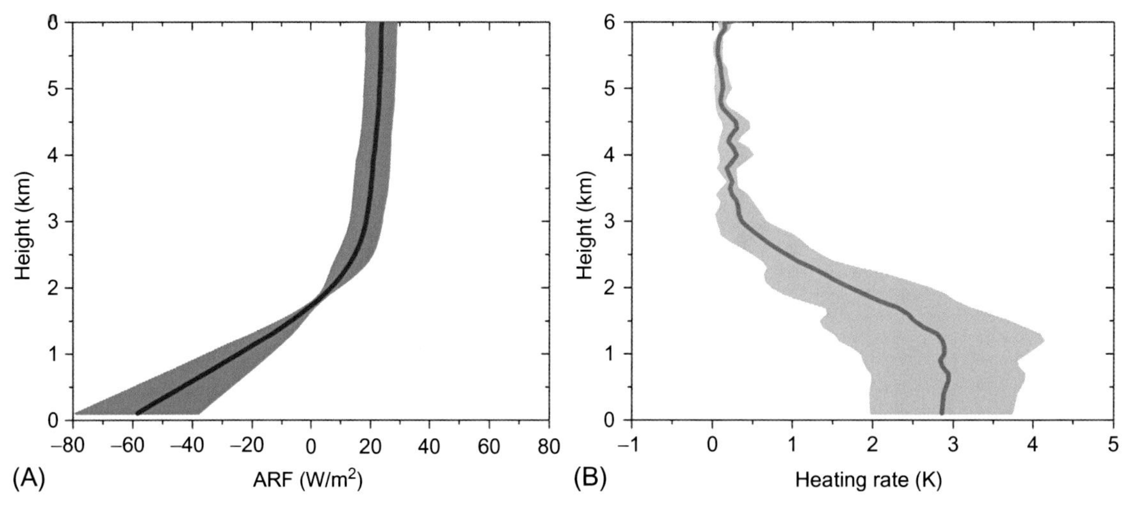

FIG. 15 (A) Aerosol radiative forcing (ARF) *(blue line)* and (B) heating rate *(red line)* by vertically resolved aerosol properties derived from the MRL.

$$\text{ARF} = (F_{\downarrow} - F_{\uparrow})_{\text{Aerosol}} - (F_{\downarrow} - F_{\uparrow})_{\text{Clean}} \quad (20)$$

Fig. 15 shows the instantaneous ARF estimated for the selected case studies. In these results, the magnitude of surface level ARF is very large at $80\ \text{W/m}^2$ and $-30\ \text{W/m}^2$. Because these aerosols absorb more radiation, the atmospheric heating rates are increased. As shown in Fig. 15 the largest heating rates are found to be 3.9 and 1.7 K. These results imply that the strongest cooling occurs at the surface, and that the strong warming occurs in the atmosphere because of absorbed aerosol loading. In addition, a large amount of incoming solar radiation is trapped inside the aerosol layer, and this is a significant source of heating to the atmosphere. It is well known that ARF can substantially alter atmospheric stability and influence the dynamic system (Li et al., 2010).

5 CONCLUSION

Estimating the impact of the atmospheric aerosols on climate change in an accurate manner is still challenging due to the spatial/temporal variability of the atmospheric aerosols. The vertical distribution of aerosols is particularly important for understanding their transport patterns and radiative effects in atmosphere. In addition, variations of radiative effects-including ARF and heating rate due to inhomogeneous aerosol loadings and their vertical distributions-are major unknown issue during the long-range transportations. In addition, there is an urgent need for more accurate vertical information of aerosol properties in radiative transfer simulations for climatic study.

Both ground-based and space-borne multichannel LIDAR systems are able to estimate inhomogeneous aerosol loadings vertically with their optical and microphysical properties. To overcome the limitations of the narrow-swath of satellite-borne LIDAR observations, passive remote-sensing techniques such as the ground-based sunphotometry or satellite-borne optical sensing methods can be supplemented using analytical models to derive vertical aerosol distribution. These methods provide unique and important observations of aerosols. The integrated dataset from both active and passive remote sensing techniques will provide a comprehensive observation of aerosols and radiation that are urgently required for air quality and climatic study.

A developed radiative property calculation coupled with a multiwavelength inversion technique is able to deliver ARF and the atmospheric heating rate profile. From the RTM simulation with that observational data, the aerosol layers with different vertical profiles are characterized with accompanying increases in the extinction coefficients but at different absorbing levels. The decline in ARF associated with increased aerosol loadings is explained by the fact that aerosol particles can result in the reflection of incoming solar radiation and atmospheric heating by light absorption.

Finally, the multidimensional aerosol loading information can be linked to the near real-time three-dimensional visualization. This may help policy decisions by environmental authorities, the health-hazard warning system, and the energy management system by supplying both near real-time vertical aerosols and visibility data. Realizing this potential will require careful coordination with supporting modeling and correlative measurement efforts.

Acknowledgments

This work was funded by the Korea Meteorological Administration Research and Development Program under grant KMIPA2015-2012. Dr. Man Sing Wong was supported in part by a grant from General Research Fund (project id: 15205515) from the Research Grants Council of Hong Kong; the grant PolyU 1-ZVAJ from the Faculty of Construction and Environment, the Hong Kong Polytechnic University; and the grants PolyU 1-ZVBP, PolyU 1-ZVBR from the Research Institute for Sustainable Urban Development, the Hong Kong Polytechnic University. The authors thank GSFC/NASA for providing the MODIS, CALIPSO, and AERONET data used.

References

Ansmann, A., Riebesell, M., Weitkamp, C., 1990. Measurement of atmospheric aerosol extinction profiles with a Raman Lidar. Opt. Lett. 15, 746–748.

Ansmann, A., Riebesell, M., Wandinger, U., Weitkamp, C., Michaelis, W., 1991. In optical remote sensing of the atmosphere. In: Optical Society of America, Summaries of Papers Presented at the Optical Remote Sensing of the Atmosphere Topical Meeting, Williamsburg, VA, pp. 206–208.

Ansmann, A., Wandinger, U., Riebesell, M., Weitkamp, C., Michaelis, W., 1992. Independent measurement of extinction and backscatter profiles in cirrus clouds by using a combined Raman elastic-backscatter lidar. Appl. Opt. 31, 7113–7131.

Bucholtz, A., 1995. Rayleigh-scattering calculations for the terrestrial atmosphere. Appl. Opt. 34, 2765–2773.

Fernald, F.G., 1984. Analysis of atmospheric lidar observations: some comments. Appl. Opt. 23, 652–653.

Gardiner, D.J., 1989. Practical Raman Spectroscopy. Springer-Verlag, Berlin, Germany, ISBN: 978-0-387-50254-0.

Gerilowski, K., Burrows, J.P., Latter, B., Siddans, R., Kerridge, B.J., 2007. Validation of SCIAMACHY top-of-atmosphere reflectance for aerosol remote sensing using MERIS L1 data. Atmos. Phys. Chem. 7, 97–106.

Holben, B.N., Eck, T.F., Slutsker, I., Tanré, D., Buis, J.P., Setzer, A., Vermote, E., Reagan, J.A., Kaufman, Y., Nakajima, T., Lavenu, F., Jankowiak, I., Smirnov, A., 1998. AERONET—a federated instrument network and data archive for aerosol characterization. Remote Sens. Environ. 66, 1–16.

Hong, C.S., Lee, K.H., Kim, Y.J., Iwasaka, Y., 2004. Lidar measurements of the vertical aerosol profile and optical depth during the ACE-Asia 2001 IOP at Gosan, Jeju Island, Korea. Environ. Monit. Assess. 92, 43–57.

http://www.lsgi.polyu.edu.hk/rsrg/resources/pj/UHI/3d_coding.txt.

https://ladsweb.nascom.nasa.gov.

https://www.ipcc.ch/report/ar5/.

Huebert, B.J., Bates, T., Russell, P.B., Shi, G., Kim, Y.J., Kawamura, K., Carmichael, G., Nakajima, T., 2003. An overview of ACE-Asia: strategies for quantifying the relationships between Asian aerosols and their climatic impacts. J. Geophys. Res. 108 (D23), 8633. https://doi.org/10.1029/2003JD003550.

Kessner, A.L., Wang, J., Levy, R.C., Colarco, P.R., 2013. Remote sensing of surface visibility from space: a look at the United States East Coast. Atmos. Environ. 81, 136–147.

Klett, J.D., 1981. Stable analytical inversion solution for processing lidar returns. Appl. Opt. 20 (2), 211–220.

Koschmieder, H., 1924. Theorie der horizontalen Sichtweite. Beitr. Phys. Freien Atmos. 12, 33–53.

Lee, K.H., Kim, Y.J., 2010. Satellite remote sensing of Asian aerosols: a case study of clean, polluted and dust storm days. Atmos. Meas. Tech. 3, 1771–1784. https://doi.org/10.5194/amt-3-1771-2010.

Lee, K.H., Noh, Y.M., 2015. Multi-wavelength Raman Lidar for use in determining the microphysical, optical, and radiative properties of mixed aerosols. Asian J. Atmos. Environ. 9 (1), 91–99.

Lee, K.H., Li, Z., Kim, Y.J., Kokhanovsky, A., 2009. Aerosol monitoring from satellite observations: a history of three decades. In: Kim, Y.J., Platt, U., Gu, M.B., Iwahashi, H. (Eds.), Atmospheric and Biological Environmental Monitoring. Springer, Netherlands, pp. 13–38. https://doi.org/10.1007/978-1-4020-9674-7_2.

Li, Z., Lee, K.-H., Wang, Y., Xin, J., Hao, W.-M., 2010. First observation-based estimates of cloud-free aerosol radiative forcing across China. J. Geophys. Res. Atmos. 115 (D00K18). https://doi.org/10.1029/2009JD013306.

Liu, Z., Vaughan, M.A., Winker, D.M., Kittaka, C., Getzewich, B.J., Kuehn, R.E., Omar, A., Powell, K., Trepte, C.R., Hostetler, C.A., 2009. The CALIPSO Lidar cloud and aerosol discrimination: version 2 algorithm and initial assessment of performance. J. Atmos. Ocean. Technol. 26, 1198–1213. https://doi.org/10.1175/2009JTECHA1229.1.

Mattis, I., Ansmann, A., Wandinger, U., Müller, D., 2003. Unexpectedly high aerosol load in the free troposphere over central Europe in spring/summer 2003. Geophys. Res. Lett. 30, 2178. https://doi.org/10.1029/2003GL018442.

Myhre, G., Shindell, D., Bréon, F.-M., Collins, W., Fuglestvedt, J., Huang, J., Koch, D., Lamarque, J.-F., Lee, D., Mendoza, B., Nakajima, T., Robock, A., Stephens, G., Takemura, T., Zhang, H., 2013. Anthropogenic and natural radiative forcing. In: Stocker, T.F., Qin, D., Plattner, G.-K., Tignor, M., Allen, S.K., Boschung, J., Nauels, A., Xia, Y., Bex, V., Midgley, P.M. (Eds.), Climate Change 2013: The Physical Science Basis. Contribution of Working Group I to the Fifth Assessment Report of the Intergovernmental Panel on Climate Change. Cambridge University Press, Cambridge, New York.

Noh, Y.M., Müller, D., Lee, H., Lee, K.H., Kim, K., Shin, S.K., Kim, Y.J., 2012. Estimation of radiative forcing by the dust and non-dust content in mixed east Asian pollution plumes on the basis of depolarization ratios measured with Lidar. Atmos. Environ. 61, 221–231.

Omar, A., Winker, D., Kittaka, C., Vaughan, M., Liu, Z., Hu, Y., Trepte, C., Rogers, R., Ferrare, R., Kuehn, R., Hostetler, C., 2009. The CALIPSO automated aerosol classification and Lidar ratio selection algorithm. J. Atmos.

Oceanic Technol. 26, 1994–2014. https://doi.org/10.1175/2009JTECHA1231.1.

Qiu, J., 2003. Broadband extinction method to determine aerosol optical depth from accumulated solar direct radiation. J. Appl. Meteorol. 42, 1611–1625.

Shimizu, A., Sugimoto, N., Matsui, I., Arao, K., Uno, I., Murayama, T., Kagawa, N., Aoki, K., Uchiyama, A., Yamazaki, A., 2004. Continuous observations of Asian dust and other aerosols by polarization lidars in China and Japan during ACE-Asia. J. Geophys. Res. 109. https://doi.org/10.1029/2002JD003253.

Spinhirne, J.D., Palm, S.P., Hart, W.D., Hlavka, D.L., Welton, E.J., 2005. Cloud and aerosol measurements from GLAS: overview and initial results. Geophys. Res. Lett. 32, L22S03. https://doi.org/10.1029/2005GL023507.

Vaughan, M., Young, S., Winker, D., Powell, K., Omar, A., Liu, Z., Hu, Y., Hostetler, C., 2004. Fully automated analysis of space-based lidar data: an overview of the CALIPSO retrieval algorithms and data products. Proc. SPIE 5575, 16–30.

WHO, 2013. HRAPIE, Health Risks of Air Pollution in Europe-HRAPIE Project. World Health Organisation, WHO Regional Office for Europe, Copenhagen.

Winker, D.M., Couch, R.H., McCormick, M.P., 1996. An overview of LITE: NASA's Lidar in-space technology experiment. Proc. IEEE 84, 164–180.

Wong, M.S., Nichol, J.E., Lee, K.H., 2008. A study of aerosol vertical profiles: scaling height and extinction coefficient. In: Proceedings of 2nd Faculty Postgraduate Research Conference, Hong Kong, January 2008. ISBN: 978-988-17311-1-1.

Wong, M.S., Nichol, J., Lee, K.H., 2009. Modeling of arosol vertical profiles using GIS and remote sensing. Sensors 9 (6), 4380–4389. https://doi.org/10.3390/s90604380.

CHAPTER

3

Remote Sensing of Aerosols From Space: Retrieval of Properties and Applications

Alaa Mhawish, Manish Kumar*, Akhila K. Mishra†, Prashant K. Srivastava*, Tirthankar Banerjee**

***Institute of Environment and Sustainable Development, Banaras Hindu University, Varanasi, India**

†Indian Institute of Remote Sensing, Dehradun, India

1 INTRODUCTION

The use of earth-observing satellites in air-quality management is although contemporary, but first began nearly four decades ago. In the late 1970s, satellite imagery from the GOES and Landsat satellites was utilized to identify haze (Lyons and Husar, 1976) and study population exposure to air pollution (Todd et al., 1979). However, the advanced scientific tools for sharing and using satellite data have improved the potential of satellite-retrieved information to be used for assessment, forecasting, and management of air quality. The concurrence of satellite and ground-based information is now widely accepted for urban-scale air-quality management among both the scientific and the policy-making communities. The most remarkable feature in applying satellite-derived information enlists the downscaling of synoptic and geospatial information to assess localized phenomena. The foremost applications include the measurement of gaseous pollutants, like ozone (O_3), sulfur dioxide (SO_2), nitrogen dioxide (NO_2), carbon monoxide (CO), formaldehydes (HCHO), etc. Airborne particulates are also well explored by satellite remote sensing due to their relevance to climate science, heterogeneous nature, and spatial variability. Fraser et al. (1984) for the first time accomplished successful retrieval of aerosol optical depth (AOD) using GOES observations over land and applied it to inspect a haze episode over the eastern United States. A sharp rise in aerosol remote-sensing information over the last decade has been observed and has been utilized in multiple ways including the estimation of ground-level aerosols (Dey et al., 2012), mortality studies (Hu, 2009; Corbett et al., 2007), crop simulations (Fang et al., 2011), identification of extreme

Remote Sensing of Aerosols, Clouds, and Precipitation
https://doi.org/10.1016/B978-0-12-810437-8.00003-7

© 2018 Elsevier Inc. All rights reserved.

events (Kumar et al., 2014) and for air-quality forecasting (Kumar et al., 2007). These multifaceted applications of remote-sensing technology have opened up a wide scope for interdisciplinary interventions in air-quality studies.

The management of near-surface air quality is essential due to its possible implications for public health, agricultural output, visibility, and aesthetic and cultural values. However, the intricacy in the availability of ground-based data makes the entire process of air-quality management difficult and uneconomic. Satellite-based observations reduce uncertainties in spatial distribution of air pollutants and the associated phenomena affecting them over synoptic and geospatial context. The estimation of ground-level pollutant concentration using space-based observations is one of the foremost applications of remote sensing, which has recently been used for air-quality management. Near-surface aerosols with highly heterogeneous behavior are being analyzed on various scales and resolutions using satellite-based observations. Researchers have exploited the dependence of aerosol optical properties on their size distributions for estimating near-surface concentrations. Improved associations were further assessed by Wang and Christopher (2003) between AOD and $PM_{2.5}$ in Alabama. However, such a relationship is highly region-specific and depends on various meteorological conditions. These assessments made a firm basis for the development of empirical relationships between satellite-derived properties and their ground-based counterparts. Studies reported the improved relationship between MODIS-derived AOD and near-surface $PM_{2.5}$ with a specific set of circumstances, including cloud-free conditions, low boundary layer heights, low relative humidity and with high aerosol loadings (Gupta et al., 2007). Additionally, the inclusion of time varying meteorological variables along with the chemistry transport model are reported to further improve the relationship.

Most of the epidemiological studies rely on near-surface measurements of air pollutants as human beings are subject to the toxicity of pollutants, most of which are present at the breathing level. However, sparse monitoring networks limit the applicability of surface-measured pollutant concentrations for exposure studies (Kumar et al., 2015a). Spatially resolved estimates of particulate matter using satellite AOD provide a better scope for assessing pollution-induced diseases and mortality. Hu (2009), using a geographically weighted regression derived near surface $PM_{2.5}$ concentrations in the eastern United States, inferred a significant hike in coronary diseases with an increase in $PM_{2.5}$. Using geospatial emission inventories, Corbett et al. (2007) revealed the mortality due to shipping-related particulate emissions. Evans et al. (2013) found distinctly high mortality estimates from satellite observations in comparison to the ground-based databases. However, considerable collinearity was observed between satellite-derived mortality estimates and chemical transport model simulations. Thus the regions with high susceptibility to diseases induced by air pollution may be conveniently mapped using remote sensing and geographic information systems (GIS) techniques, leading to the formulation of suitable mitigation plans. These techniques have been successfully implemented in different parts of the world.

The appropriateness of any air-quality management plan requires precise information about emission inventories, considering information on particulate sources, emission strengths, and rates (Banerjee et al., 2011). Satellite-based measurements of various aerosol products are extensively used to examine possible sources and to develop a spatial emission map. Additionally, with the advancement of remote-sensing observation in the last decades, techniques have been developed for using these observations as constraints on pollutant sources.

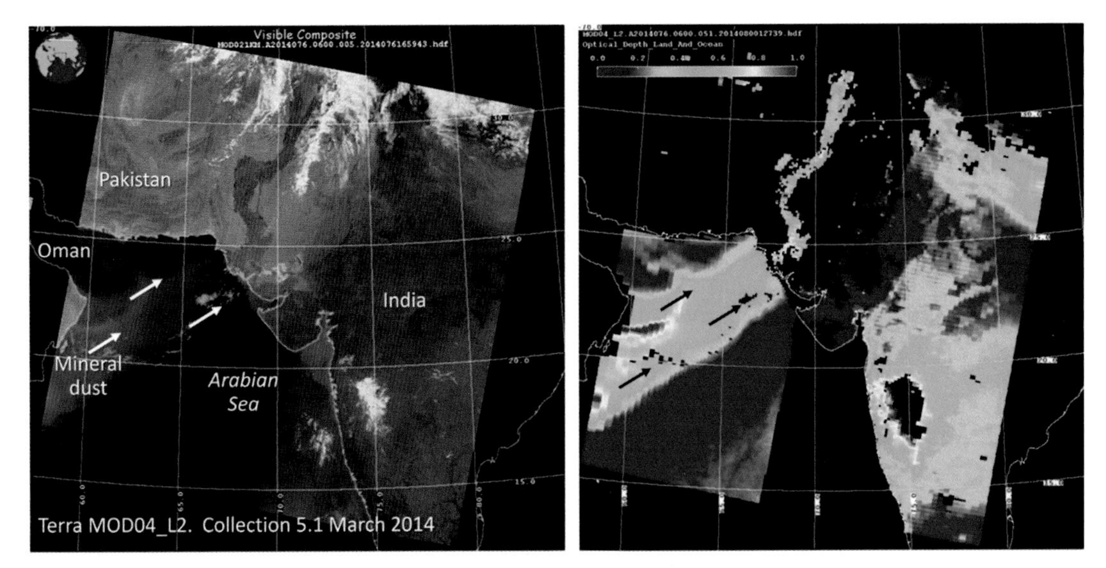

FIG. 1 Transboundary movement of mineral dust over South Asia. *Note*: Both true color and AOD images are courtesy of Terra MODIS. *Modified from Kumar, M., Tiwari, S., Murari, V., Singh, A.K., Banerjee, T., 2015b. Wintertime characteristics of aerosols at middle Indo-Gangetic plain: impacts of regional meteorology and long range transport. Atmos. Environ. 104, 162–175.*

Likewise, during summer and premonsoon months, the transboundary movement of airborne particulates originating in Middle East dry regions often causes dust storms over the Indian subcontinent (Fig. 1). Inverse modeling of MODIS AOD in conjunction with a few advanced models such as GOCART have been utilized by Dubovik et al. (2007) for the quantification of major aerosol sources. Wang et al. (2013) combined satellite-measured radiances and inverse modeling to spatially constrain sources of dust emissions in the Taklimakan and Gobi deserts. Wang et al. (2016) also reported a new approach to integrate OMI-retrieved SO_2 measurements and GEOS-Chem adjoint model simulations to constrain anthropogenic emissions of SO_2. Spatially resolved information on backscatter characteristics and classified aerosol types retrieved through the CALIPSO satellite constructively helps in apportioning its sources. Mazzoni et al (2007) determined the heights of smoke plumes from wildfires through MISR retrievals. Not limited to aerosols, MODIS and MISR aerosol products were also interpreted for annual growth rates of sulfur emissions over China (van Donkelaar et al., 2008).

The framework for managing air quality for a regional or national interest includes a forecasting mechanism that can take care of future emission scenarios and anticipated risks associated with it. Recent advancements in remote-sensing bring promising support to air-quality forecasting. Assessments of satellite imagery of smoke from MODIS, AIRS, and other fire detection maps are being made with the help of GIS to project the effects of wildfire and biomass burning. There are some specific web tools that collect all relevant information together, such as Infusing Satellite Data into Environmental Applications (IDEA) and the Hazard Mapping System Fire and Smoke Product (HMS). Also, some extreme events including dust storms, haze episodes, and smog become a serious nuisance to the regional air quality, affecting health, visibility, and climate. Several remote-sensing techniques have been documented in studies, which could help air-quality managers in minimizing risks associated with such extreme events.

Satellite remote sensing of aerosols does have some limitations. Technical and scientific issues such as cloud contamination, surface reflectance, sensitivity of instruments, and other atmospheric errors need to be considered while using remote-sensing techniques for atmospheric and climate science. Many satellites have passive sensors that are used to retrieve optical properties only during daytime and for the entire column, which may not be relevant to ground-level concentrations. Retrieving aerosol optical properties becomes additionally complicated due to real-time variations in aerosol microphysical properties influenced by existing meteorology and the source profile. Therefore, real-time measurement of aerosol properties always remains a challenge. Additionally, complexity in data retrieval methods, data access, and a low level of understanding from policymakers about the satellite data put limitations on its use. Resolution—be it spatial, spectral, temporal, or radiometric—regulates the applicability of a particular data product, but often there is a tradeoff between these. Likewise, polar-orbiting satellites offer higher resolution with less-frequent global coverage while geostationary satellites have lower resolution but almost daily global coverage. Therefore, often satellite data products are application-specific and the user should consider the limitations/quality of data products before the intended use.

2 AEROSOLS: HETEROGENEITY AND CLIMATIC IMPLICATIONS

The climate-governing properties of atmospheric aerosols have been well acknowledged (Ramana et al., 2010; Schwartz and Andreae, 1996; Hansen et al., 1997; Banerjee et al., 2017a). The complex aerosol chemistry mediated through its surface additionally supports multifaceted climatic implications. Among various atmospheric entities, aerosols constitute a major source of uncertainty in their role in climate perturbations mainly due to the associated heterogeneities (Murari et al., 2015).

The degree of heterogeneity in aerosol properties highly depends on geological and topographical features (Ramanathan and Ramana, 2005), source strengths, and meteorological factors (Banerjee et al., 2015). The complicacy in aerosol-climate interaction is governed by its multidimensional heterogeneity, mainly in space, time, chemical composition, size distribution, sources, vertical distribution, and mixing states, leading to different climatic impacts (Fig. 2).

A wide range of spatial heterogeneity has been reported in the global distribution of aerosols, which is responsible for the differential climatic impacts (van Donkelaar et al., 2015). Owing to their lower atmospheric residence time (typically hours to a few weeks), the distribution of aerosols often remains in proximity to their sources. Additionally, local geography and meteorology are the most influential factors in regulating aerosol properties. Therefore, often the global distribution of aerosol loading is specifically linked with isolated geographical regions. Likewise, some of the major recognized regions—typically called aerosol hotspots—are tropical Africa, Eastern China, the Indo-Gangetic Plain (IGP) of South Asia, the tropical region of Central America, Southeast Asia, and Mexico. Fig. 3 depicts the global distribution of columnar AOD (Fig. 3A) and fine-mode fractions (Fig. 3B) retrieved from the Terra-MODIS satellite for August 2015. The heterogeneity in aerosol optical and microphysical properties over a wide range of spatial and temporal scales obstructs the true estimates of aerosol-induced climate change. Likewise, the Sahara Desert in Northern Africa is fairly characterized by coarser mineral dusts (Jiménez et al., 2010) while the Southeast Asian region is typically dominated by secondary aerosols (Trivitayanurak et al., 2012). Singh et al. (2017a) reviewed the predominate sources of airborne

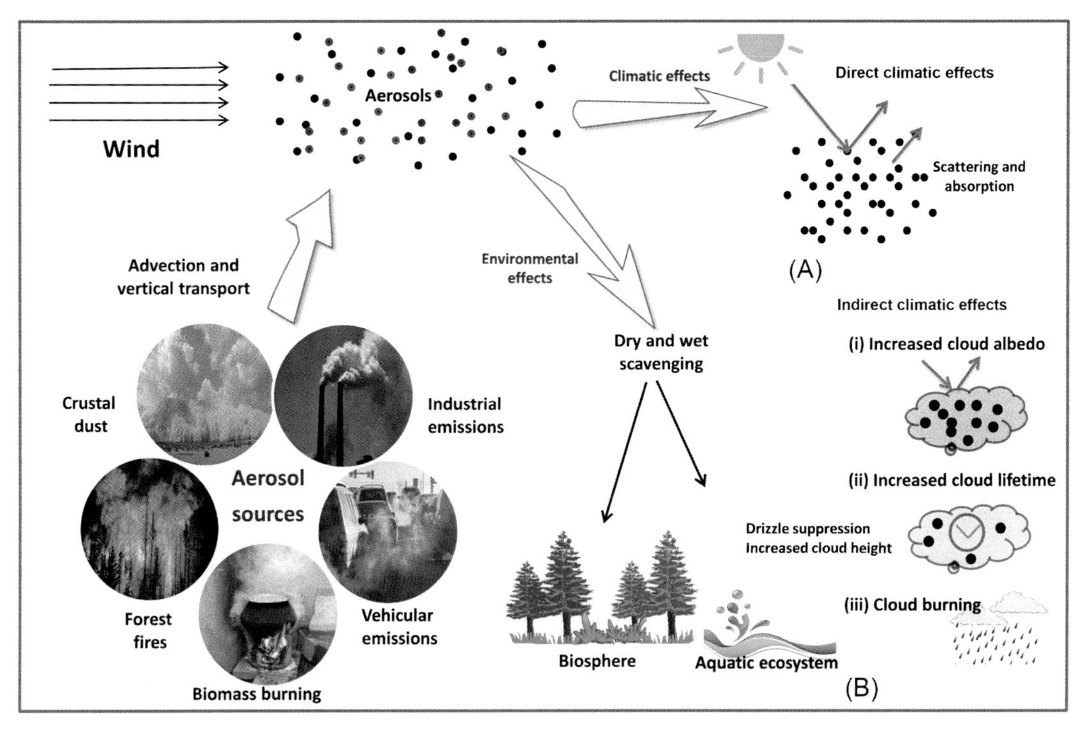

FIG. 2 Aerosol heterogeneity and climate-governing properties.

fine particulates over South Asia and reported significant contributions from vehicular and industrial sources with significant variabilities, both in terms of season and geography. The entire Indo-Gangetic basin in South Asia is characterized with mixed aerosols with dominance of coarser dusts at the upper basin while elevated proportions of finer aerosols and soot over central and lower parts (Fig. 9). Sen et al. (2017) concluded the presence of a very high particulate load, especially over the IGP, which reportedly undergoes long-range transport of aerosols from its source to the Indo-Himalayan Range and Bay of Bengal. Extremely high particulate loadings have also been reported over the middle IGP during winter (Kumar et al., 2015b, 2017a; Sen et al., 2014) while minimal concentrations were observed during monsoon season (Murari et al., 2015, 2017). Das et al. (2014) also observed a seasonality in dust transport in the Indian region with an enhanced dust loading (AOD > 0.5) over the oceans during monsoon season. Efforts were also made to characterize the particulate composition and transport over the IGP. The coarse transported dust from the great Indian Thar and other Middle East countries is rich in iron, calcium, silicon, aluminum, magnesium, lithium, and phosphorus (Moreno et al., 2012). Aerosols emitted from combustion sources such as biomass/refuse burning, forest fires, and vehicular emissions are characterized by soot, black carbon (BC), sulfates, nitrates, and ammonium ions (Murari et al., 2015).

The existence of aerosols in a wide-size spectrum (nm to μm) is another source of heterogeneity that influences the climate and is reported to play a crucial role in cloud microphysical properties. Aerosols of natural origin (mineral dust, sea salt) are usually coarser in size except for sulfate and carbonaceous aerosols, which are emitted from volcanic eruptions and

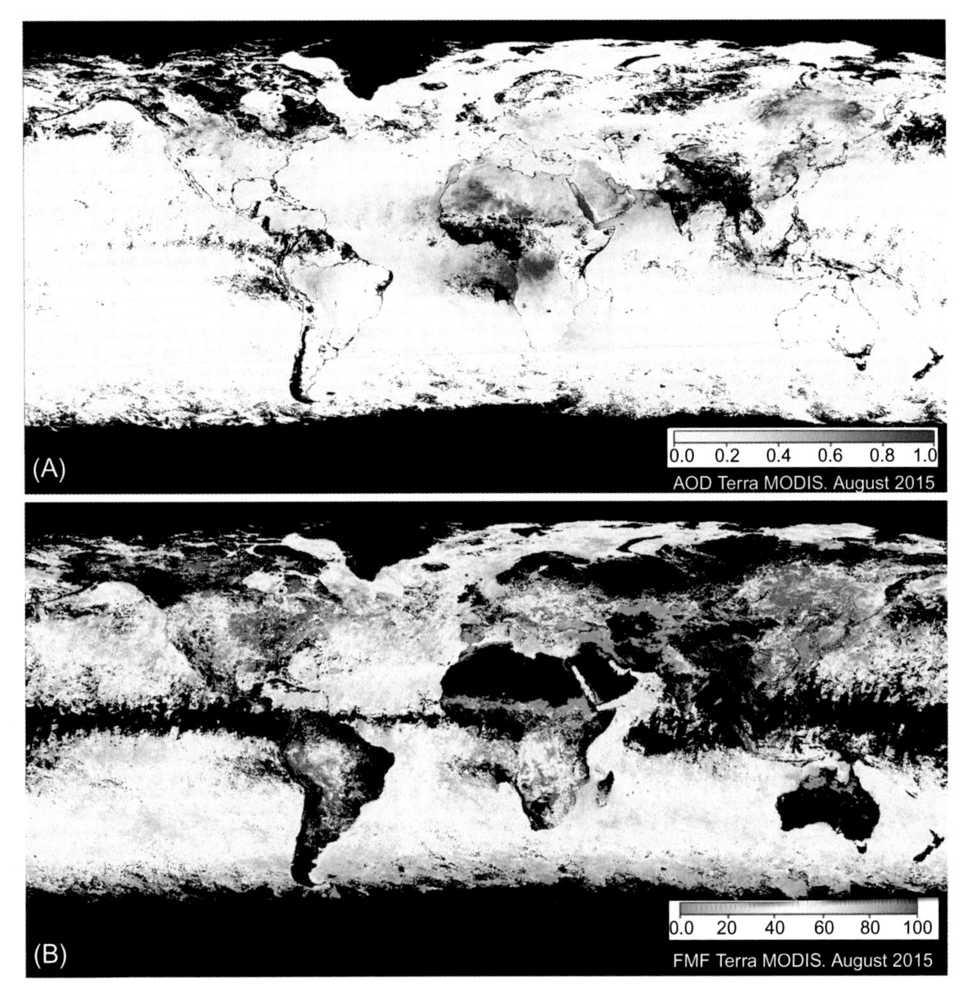

FIG. 3 (A and B) Mean global aerosol loading for August 2015 retrieved from Terra MODIS (A) AOD at 550 nm (B) fine-mode fractions (FMF).

forest fires. In contrast, most of the anthropogenic sources such as vehicular emissions, biomass burning, and fossil-fuel combustion release particulates with dominance of finer particulates. The information on mixing states of aerosols is one of the unresolved issues in atmospheric sciences due to their highly heterogeneous behavior on spatial and temporal scales. One of the peculiar impacts of varying mixing states was reported on the modification of radiative effects of sulfate and nitrate aerosols in the presence of BC. Global chemistry models have estimated that the net radiative impacts over a region are regulated by relative amounts of BC and sulfate (Ramana et al., 2010). Sulfate aerosols strongly reflect solar radiation while BC strongly absorbs it. However, the net radiative impact of BC is considered to be well amplified in the case of its internal mixing with sulfate (Moffet and Kimberly, 2009). These heterogeneities in aerosol types, mass loadings, composition, and transport make it extremely complicated for determining future projections of climatic impacts.

The direct effects of aerosols on climate refer to their interference with incoming solar radiation, either through scattering (cooling) or absorbance (warming). The response of aerosols to incoming solar insolation greatly depends on their composition and size. The brighter aerosols (SO_4^{2-}) scatter the solar radiation while the darker aerosols (BC, Brown Carbon) tend to absorb it. The direct effects of aerosols play significant roles in modifying Earth's radiation balance. The properties of aerosols in modifying Earth's radiative budget through changes in cloud microphysical properties come under indirect climatic effects. The vital consequences of aerosol indirect effects are based on the fact that even 5% enhancements in shortwave cloud forcing can recompense the increase in greenhouse gases from 1750 (Ramaswamy et al., 2001). The foremost climatic implications of indirect effects of aerosols result in suppressed precipitation with increased cloud lifetime and the amount of clouds (Albrecht, 1989). Further, an increase in cloud coverage leads to an enhanced albedo (Twomey, 1959), which in turn amends the Earth's radiation budget. With recent advancements in both in situ and satellite-based measurements, significant progress in recognizing the cloud-aerosol interaction has been made. However, quantification of the indirect effect still pose uncertainties in comparison to the direct effects (Schwartz and Andreae, 1996). With enhanced concentrations of anthropogenic aerosols, the evidence of reduced planetary albedo over clouds has been inferred from satellite observations, mostly in regions with high absorbing aerosols (Krüger and Graßl, 2004). A reduction in the albedo of light-absorbing species is recognized as a semidirect effect (Hansen et al., 1997). These collective impacts of aerosols, including direct, indirect, and semidirect effects, on the climate could have major socioeconomic concerns and pose serious threats to the sustainability of the planet (Banerjee et al., 2017a).

3 SATELLITE OBSERVATIONS FOR AEROSOL MONITORING: DEVELOPMENTS

Remote sensing has long been utilized for monitoring airborne particulates using various observational platforms. A complete historical background on the development of aerosol observations through remote sensing is available from Lee et al. (2009) and Kokhanovsky and de Leeuw (2009). However, recent advancements in satellite-retrieved aerosols have been briefly discussed here, especially in terms of polar-orbiting satellites. The polar-orbiting satellites are sun-synchronous satellites that regularly cover the entire globe, providing repetitive coverage on a periodic basis. Due to the large spatial and temporal variability of atmospheric aerosols, remote-sensing products deliver precise information about global distributions of aerosols. The very first of its kind AOD was generated by space-borne spectral measurement using the Multi Spectral Scanner (MSS) onboard the Earth Resources Technology Satellite (ERTS-1) (Griggs, 1975), while the first operational aerosol product was generated in 1978 using the radiometer (Advanced Very High Resolution Radiometer, AVHRR) onboard the TIROS-N satellite. The series of historical satellite sensors still in operation for monitoring Earth's atmosphere and climate are TOMS (Total Ozone Mapping Spectrometer, for ozone) and AVHRR (for weather system). AVHRR, initially launched in 1978, had multiple bands (0.63-11.5 μm) and was used to retrieve AOD and AE. It was subsequently improved and relaunched with NOAA-6∼16. The satellite Nimbus-7 was launched in 1978 having a Stratospheric Aerosol Measurement instrument (SAM-2) which functionally initiated retrieval of aerosol properties on a regular basis. TOMS was also launched in 1978 (with Nimbus-7) and in 1996 (with Earth Probe), it was associated with the measurements of the Absorbing Aerosol Index.

Initially, due to lack of scientific understanding on climatic implications of aerosols, limited efforts were made to retrieve aerosol optical properties from space. All satellite exploration during the initial stages of development in the 1990s was especially focused on retrieving trace gases (like GOME, Global Ozone Monitoring Experiment; SCIAMACHY, Scanning Imaging Absorption SpectroMeter for Atmospheric CHartographY) or land/sea surface temperature and reflectance (SeaWiFS, Sea-viewing Wide Field-of-view Sensor; MERIS, MEdium Resolution Imaging Spectrometer). GOME launched with the ERS-2 satellite in 1995, associated with ATSR-2 (Along-track Scanning Radiometer), which helped to retrieve the first operational and reliable aerosol properties from space. The sensor POLDER (with ADEOS) launched in 1996 was a promising instrument due to its capability for multiple-view measurements for a varied band with spectral polarization for reflected radiation. The finest version of POLDER (POLDER-3) was launched with Parasol in 2002 to measure the polarization of reflected light for identifying airborne fine particulates. It remained functional until 2013. This was subsequently followed by the launch of the Moderate Resolution Imaging Spectro-Radiometer (MODIS) onboard Terra (1999, with Multi-angle Imaging Spectro-Radiometer) and Aqua (2002), AATSR (with ENVISAT, 2002), and OMI (with Aura, 2004), providing multidimensional information on aerosol optical properties. The aerosol retrieval algorithms were further developed for application-based retrieval and to collocate both with ground-based observations. One of the significant inclusions for satellite-based retrieval of aerosol optical properties is CALIOP. It is a typical space-borne LIDAR onboard collocate CALIPSO satellite, which was launched in 2006. The CALIOP has two-wavelength (visible and IR band) polarization-sensitive LIDAR and passive imagers for retrieving the vertical distribution of both aerosols and clouds from the surface to 40 km. Another LIDAR-based sensor, the Geosciences laser altimeter system (GLAS), onboard ICESat was launched in 2003 to determine the elevation and position of each point measurement on the Earth and to identify cloud and aerosols. Most recently, the Visible Infrared Imaging Radiometer Suite (VIIRS) onboard the NPP-Suomi satellite was launched in 2011. It is a successor to MODIS and MISR in terms of retrieving aerosol optical properties from space-borne platforms.

4 SATELLITE RETRIEVAL OF AEROSOL PROPERTIES

4.1 Aerosol Retrieval Algorithms

With consequent development in sensor technology, satellites are frequently being used to obtain information on various products, both on global and regional scales. The information retrieved by satellites depends on the radiation retrieved by the sensor at the top of the atmosphere (TOA), mounted on the satellite on a specific orbit. Satellite-retrieved information may have varying attributes. In terms of information, the optical data may be useful for identifying spatiotemporal variations as well as classifying columnar properties and resolute vertical distributions. Retrieval of aerosol properties from remote-sensing satellites depends on the interaction of tiny airborne particulates with radiation. Upon interaction with particles, radiation is distributed in multiple directions, typically based on particle morphology and chemical composition. Aerosol retrieval algorithms use angular distributions of radiation scattering after proper isolation of various atmospheric and surface interferences.

The retrieval process fundamentally poses three constraints: cloud effects, surface reflectance, and molecular scattering. In principle, the retrieval of aerosol properties depends on the reflected radiation. However, in the presence of clouds, such reflection is dominated by the clouds themselves over aerosols. Any cloud effects may thereby reduce the possibility of retrieving minute variations in aerosols while excessive cloud screening may remove areas with higher aerosol loading. Thus, often cloud screening is the first and most important step in any aerosol-retrieval algorithm (Lee et al., 2009). Cloud screening may be achieved by various means such as identifying wavelength dependence in the visible to the thermal infrared (IR), spatiotemporal analysis, using the O_2 A-band ,or by using multiple concurrent observations (Ackerman et al., 1998). Contribution of land surface reflectance radiation to the TOA is the second important constraint in remote sensing that needs to be isolated based on recognizing land surface properties varying from a dark surface (ocean, forest) to brighter ones (bare soil, urban habitat, industrial area, permafrost). Considering variations in surface reflectance, it is essential to set a boundary upon which atmospheric contributions are measured (Kokhanovsky and de Leeuw, 2009). On this account, retrieval of aerosol properties over oceans is relatively easy and straightforward; upon land, however, multiple corrections need to be made to remove interference. The third significant step in satellite retrieval is to account for path radiance, which considers both molecular scattering and absorption by various atmospheric molecules and gases. A Radiative Transfer Model (RTM) is generally applied to account for path radiance for a specific sun-satellite geometry.

An aerosol-retrieval algorithm uses the radiance measured at the TOA and compares it to that calculated by the RTM over a range of aerosol models for recognizing the best model to retrieve aerosol properties. Radiative transfer calculations are made for a variety of discrete situations, including a set of wavelengths that are used in retrieval, viewing geometries (viewing and solar zenith angle, relative azimuth angle), surface pressure, and AOD ranges (Kokhanovsky and de Leeuw, 2009). The calculations are time-consuming; therefore they are made outside the actual algorithm and then stored in look-up tables (LUTs). Each aerosol model has its own LUTs and contains multiple parameters such as path reflectance (R_a); total transmittance from the TOA to surface (T_a) and surface to satellite (T_b); the spherical albedo (r); AOD (τ); aerosol phase functions ($P(\theta)$); scattering angle (θ); and single-scattering albedo (ω). The TOA reflectance (R), ratio of radiance received by sensor over that reaching to the TOA (E_0) may be represented as:

$$R = \frac{\pi I}{E_0 \mu_0} \tag{1}$$

where, I is the intensity of reflected light and μ_0 is the cosine of the observation angle. The TOA reflectance with albedo A and wavelength λ is presented as:

$$R = R_a + \frac{A T_a T_b}{1 - A_r} \tag{2}$$

Now, except surface albedo, all the other parameters depend on AOD (τ) which is, in fact, the integral of the extinction of light ($K^a_{ext} z, \lambda$; aerosol extinction coefficient at height z for wavelength λ) caused by airborne particulates over a vertical column (height of TOA, H) through the atmosphere. Therefore, AOD may be computed as:

$$\tau(\lambda) = \int_0^H K^a_{ext}(z, \lambda) dz \tag{3}$$

The spectral dependence of AOD from satellite measurements of spectral reflectance (R) requires precise isolation. The surface albedo needs to be retrieved or rejected using multiview observations (Kokhanovsky and de Leeuw, 2009). For most atmospheric conditions, the value of A_r remain <0.1, and therefore, for the dark land surfaces ($A \rightarrow 0$) the product A_r may be ignored. The choice of appropriate LUTs for path reflectance, spherical albedo, and for T_a and T_b either must be assumed or considered based on the region of interest. The LUTs are typically made by the vector radiation transfer equation for collection of aerosol models representing a specific area (Levy et al., 2007). The algorithm selects the LUT for the appropriate aerosol model that best fits the observed spectral reflectance by the satellite sensor. However, such consideration also depends on the climatology (Levy et al., 2007) derived from observations (Levy et al., 2007) or the transport model for a particular area (Curier et al., 2008).

4.2 Aerosol Optical Properties

The optical properties of aerosols are important for atmospheric visibility and climate-related studies. Advancements in satellite remote sensing have made it possible to retrieve several aerosol optical properties from space with global coverage. These remote-sensing techniques include both passive and active sensors. Some of the important optical properties of aerosols include AOD, the Angstrom exponent (α), the turbidity coefficient (β), the single-scattering albedo (SSA), the refractive index, the absorption and extinction coefficients, the phase function $P(\theta)$, the asymmetry parameter (g), the phase matrix, and the backscatter properties.

(i) Aerosol optical depth
AOD is defined as the integral of the extinction coefficient over a vertical column of atmosphere of the unit cross-section. The extinction coefficient refers to the depletion of radiance per unit path length (also termed attenuation). AOD depicts the extinction of solar radiation by aerosols in open atmosphere. Typically, aerosols—based on their morphology and chemical composition—attenuate the solar radiation either by absorbing or scattering. It is a dimensionless entity and it relates the amount of aerosols in the vertical column of atmosphere. A typical value of AOD as 0.01 represents an extremely clean environment while AOD values above 0.5 correspond to a hazy atmosphere, which may be extended to 5.0 at certain cases. The measurement of total optical depth is based on the Beer-Lambert-Bouguer law considering the following equation:

$$V(\lambda) = V_o(\lambda)d^2 \exp\left[-\tau(\lambda)_{TOT}{}^{*}m\right] \quad (4)$$

where V is the digital voltage measured at wavelength λ; V_o is the extraterrestrial voltage; d is the ratio of the average to the actual Earth-Sun distance; τ_{TOT} is the total optical depth; and m is the optical air mass (Holben et al., 1998). The optical depth (τ) for wavelength λ with α as the extinction coefficient of the atmosphere having the path z as an inconvenient unit, can be defined as:

$$\tau_\lambda = \int_z^\infty \alpha(\lambda, z')dz' \quad (5)$$

At surface ($z=0$), the τ will be τ^* by definition while at the TOA, $\tau=0$. Since radiation is subject to attenuation due to other wavelength-dependent trace gases, water vapor, and Rayleigh scattering, AOD can be obtained after subtracting the optical depths from all other components from the total optical depth.

So, the equation for computing AOD turns to:

$$\begin{aligned}\tau(\lambda)_{\text{Aerosol}} = \tau(\lambda)_{TOT} - \tau(\lambda)_{\text{water vapor}} - \tau(\lambda)_{\text{Rayleigh}} \\ - \tau(\lambda)O_3 - \tau(\lambda)NO_2 - \tau(\lambda)CO_2 \\ - \tau(\lambda)CH_4 \end{aligned} \quad (6)$$

where $\tau(\lambda)$ is the respective optical depths from different atmospheric components.

The primary applications of AOD in atmospheric and environmental sciences include atmospheric correction of remotely sensed surface features, monitoring of sources and sinks of aerosols, monitoring volcanic eruptions and forest fires, monitoring air quality, health and epidemiological studies, climate change studies, and radiative transfer studies.

(ii) Angstrom exponent (α)
The Angstrom exponent (α) is an optical property of aerosols that provides the information on aerosol size distribution. The Angstrom exponent is estimated from AOD, typically from 440 to 870 nm. It is the negative slope of AOD for a particular wavelength in logarithmic scale. It can be calculated from two or more wavelengths (like λ and λ_0) using a least squares fit.

$$\tau = \tau_{\lambda_0}\left(\frac{\lambda}{\lambda_0}\right)^{-\alpha} \tag{7}$$

$$\alpha = \frac{\ln^{\tau_1}/_{\tau_2}}{\ln^{\lambda_2}/_{\lambda_1}} \tag{8}$$

where α is the Angstrom exponent; τ is the AOD; and λ is the wavelength. The values of a above 2.0 designate finer aerosols such as smoke and sulfates whereas values close to zero specify the presence of coarser particles such as desert dust.

(iii) Turbidity coefficient (β)
Turbidity coefficient (β) is a dimensionless quantity that measures the opacity of a vertical column of atmosphere (Utrillas et al., 2000). The Angstrom turbidity coefficient indicates the total aerosol content of the atmosphere in the zenith direction. It can be defined as the aerosol optical thickness corresponding to a wavelength of 1 μm. Since the Angstrom formula is just an approximation, it may not be valid over an extensive spectral range. However, the Angstrom formula seems to provide a good parameterization of spectral AOD between 400 and 670 nm bands (Martinez-Lozano et al., 1998). The Angstrom turbidity coefficient can be calculated as:

$$\beta = \tau \cdot \lambda^{\alpha} \tag{9}$$

where β is the turbidity coefficient; τ is the AOD; λ is the wavelength in micrometers; and α is the wavelength exponent representing the aerosol size distribution (Angstrom, 1964). Turbidity coefficients <0.1 are associated with a relatively clear atmosphere while values >0.2 are associated with a relatively polluted atmosphere.

(iv) Asymmetry parameter and phase function
The asymmetry parameter (g) represents the degree of asymmetry of the angular scattering and is a measure of the preferred scattering direction (forward or backward) for the light coming across the aerosols. It is equal to the mean value of μ (the cosine of the scattering angle), weighted by the angular scattering phase function $P(\mu)$. The asymmetry parameter corresponding to +1 represents the scattering strongly peaked in the forward direction (scattering by larger particles) whereas −1 denotes entirely backscattered light. Asymmetry parameters equal to 0 represent an evenly distributed scattering mostly from smaller particles (Rayleigh scattering). Asymmetry parameter may be defined as:

$$g = {}^1/_2 \int_{-1}^{1} d(\cos\theta) p(\theta) \cos\theta \tag{10}$$

where θ is the angle between incident light and scattering direction and $P(\theta)$ is the angular distribution of scattered light (the phase function). The asymmetry parameter strongly depends on the particulate size. Phase function is defined as the energy scattered per unit solid angle in a given direction to the average energy in all directions. Phase functions may be expressed as:

$$p(\theta) = \frac{4\pi}{C_{sca}} \frac{dC_{sca}}{d\Omega} \tag{11}$$

$$C_{sca} = \int_{4\pi} d\Omega \frac{dC_{sca}}{d\Omega} \tag{12}$$

where $dC_{sca}/d\Omega$ is the differential cross-section and C_{sca} is total scattering cross-section.

(v) Single-scattering albedo (ω)
A single-scattering albedo (SSA) is the measure of effectiveness of light scattering due to atmospheric aerosols relative to its total extinction. It can be defined as the ratio of scattering and extinction coefficients (scattering + absorption) at a given wavelength. SSA is a dimensionless entity with values ranging from 0 (dominance of absorbing aerosols) to 1 (light extinction mostly due to scattering aerosols). It is an important optical parameter for aerosol radiative transfer studies for its use in the calculation of scattering phase function and as an input to the OPAC model for the calculation of aerosol radiative forcing. Not limited to this, SSA can be used to characterize the type of aerosols. Eq. (13) describes the mathematical expression for SSA:

$$\omega = \frac{\sigma_s}{\sigma_s + \sigma_\alpha} \tag{13}$$

where σ_s and σ_a are the aerosol scattering and absorption coefficients, respectively.

(vi) Mass extinction coefficient (B_{ext})
The mass extinction coefficient (B_{ext}) of aerosol represents how strongly particulates absorb/scatter radiation at a given wavelength per unit mass in a constant volume. The mass extinction coefficient depicts the area of extinction for a unit mass of the aerosol (usually in units of (m^2/g)). Mathematically, it can be expressed as:

$$B_{ext} = \frac{3QM}{4\rho rMd} \tag{14}$$

where B_{ext} is the aerosol mass extinction efficiency; Q is the extinction coefficient; M is the total aerosol mass loading (dry mass + mass of water taken by the aerosols); ρ is the particle density; r is the particle effective radius; and Md is the dry aerosol mass (Chin et al., 2002).

5 SATELLITE AEROSOL DATABASE

Recent advancements in space sciences and remote-sensing technologies have endowed enormous prospects for studying aerosol and other atmospheric properties using satellite-based measurements. Though these techniques deliver a diverse range of information, retrieving them for applications is complicated and requires specific skills, tools, and computation facilities. Additionally, due to extremely large datasets, variation in data types, metadata, and the availability of various platforms for data access make the selection criteria even more critical. Several web tools and Internet-based resources are now available that offer ready-to-use aerosol products to end users with different resolutions, data types, and applications. These tools provide information on global aerosol distribution and properties in the form of imagery, maps, and easily executable files that can be processed further for different air-quality applications.

The web tools for aerosol-related information offer direct access to available satellite data products of various resolutions (raw or fully processed), time series, data analysis, software, and programming tools. The satellite products available from NASA are managed through NASA's Earth Observing System Data and Information System (EOSDIS). This not only provides data to the end users but also allows spatial subsetting to large global data files. Various other platforms including the Goddard Earth Sciences Data and Information Services Center (GES DISC) and the Langley Research Center Atmospheric Science Data Center (LaRC ASDC) provide information about different gaseous and aerosol products. Some specific web tools for exceptional events like fire information and hazard mapping are also available. These are being used in collocation with the information available through other tools. Also, clubbing spatially resolved data with GIS has made data analysis relatively easy. Other options for graphical representation of data are also available through selected searches and selection of required parameters on spatial and temporal scales. NASA's Fire Information for Resource Management System (FIRMS) provides fire-related information in the form of shaped files of predefined grids, which helps in accessing, analyzing, and forecasting fire events on a long-term basis. Some other popular tools available are LAADS Web, Worldview and Giovanni. Table 1 presents different web-based tools and resources for retrieving, analyzing and applying satellite-retrieved aerosol data products that are most commonly used by the scientific communities.

In recent years the diverse applications of multipurpose satellite data by several agencies have led to the development of common platforms for data access and use. NASA, NOAA, and EPA have initiated a partnership for the assessment, management, and prediction of air quality by infusing various satellite measurements for public benefit. All relevant information has been provided on a common platform known as IDEA (Infusing satellite Data into Environmental air quality Applications). Along with resources for data availability, there are some specific web tools (like ARSET) that provide online training for the application of satellite retrieval for aerosols and other air quality parameters.

5.1 Retrieval of Aerosol Properties From MODIS

The Moderate Resolution Spectroradiometer (MODIS) sensor flew onboard the sun-synchronous, near-polar satellites Terra and Aqua from 1999 and 2002, respectively. The MODIS instrument observed the earth from a 700-km altitude (±55 degrees view scan) with a swath of about 2330 km. This means it had nearly global coverage every 1–2 days on a daily basis with 16 days of repeat cycle (http://modis.gsfc.nasa.gov). The different orbital paths of MODIS-Terra and MODIS-Aqua gives different overpass times at morning (Terra descending node 10.30 local time) and afternoon (Aqua ascending node 13.30 local time) over the equator, which allows MODIS to observe the earth two times during the day. MODIS instruments measured in the 36-wavelength band (0.4–14.4 μm), with ground spatial resolution at nadir 0.25×0.25 km (bands 1–2), 0.5×0.5 km (bands 3–7), and 1×1 km (bands 8–36). The data retrieved by MODIS is organized into 5-minute sections called granules. It is designed to be spectrally stable (better than 2 nm) and adequately sensitive related to aerosol properties with absolute radiance calibrated to within 2% (Guenther et al., 2002). Originally intended for climatic applications, it has wide applicability in air quality, especially in terms of aerosol columnar optical depth and size distributions (Fig. 3A and B).

MODIS uses three separate algorithms for retrieving aerosol optical properties over land and oceans: the Dark Target (DT) algorithm over land (Kaufman et al., 1997; Levy et al., 2007), the DT algorithm over ocean (Tanré et al., 1997), and the Deep Blue (DB) algorithm over land (Hsu et al., 2004). The DT algorithm is used to retrieve aerosol optical properties over oceans and dark land (vegetated area). It is based on the concept that in the visible band within an atmospheric *window*, aerosol tends to brighten while vegetation appears dark (Kaufman et al., 1997). Such spectral contrast is purposefully used for retrieving aerosol properties. It was originally developed by Kaufman et al. (1997) over land and by Tanré et al. (1997) over ocean. The principal products for DT algorithms include the total AOD (τ) at 0.55 μm with an estimate of fine-mode fractions (FMF, η) to the total optical depth. However, there are some basic constraints in using the DT algorithm by single-view sensors (SeaWiFS, VIIRS, AVHRR, and

TABLE 1 Satellite Data Access, Visualization, and Analysis Web Resources for the End User

Database Type	Agency	Database/Resource	Website
Online training and webinars	NASA	Applied remote sensing training	http://arset.gsfc.nasa.gov/
Data, tools, and modeling	NASA	Air Quality Applied Sciences Team	http://acmg.seas.harvard.edu/aqast/
Search, access, and downloading files, with options for subsetting	NASA	Reverb	http://reverb.echo.nasa.gov/reverb
Visualization and downloading data	NASA	Land Atmosphere Near-real-time Capability for EOS	https://earthdata.nasa.gov/data/near-real-time-data/
Data and imagery	NASA	Goddard Earth Sciences Data and Information Services Center	http://disc.sci.gsfc.nasa.gov
Web-based interactive visualization and analysis	NASA	Giovanni	http://disc.sci.gsfc.nasa.gov/giovanni/
Worldview		An interactive visualization and analysis web tool	https://earthdata.nasa.gov/labs/worldview/
Data and imagery	NASA	Mirador	http://mirador.gsfc.nasa.gov
Data and imagery	NASA	Langley Research Center Atmospheric Science Data Center	http://eosweb.larc.nasa.gov
MODIS and VIIRS data and imagery	NASA	Level 1 and Atmosphere Archive and Distribution System	http://ladsweb.nascom.nasa.gov
True color imagery and smoke	NOAA	Hazard Mapping System Fire and Smoke Product	http://www.ospo.noaa.gov/Products/land/hms.html
data and imagery for fire locations	NASA	Fire Information for Resource Management System	http://earthdata.nasa.gov/data/near-real-time-data/firms
Data visualization	NOAA-NASA-EPA	Infusing Satellite Data into Environmental Applications	http://www.star.nesdis.noaa.gov/smcd/spb/aq/
Software package		International MODIS/AIRS Processing Package	http://cimss.ssec.wisc.edu/imapp/ideai_v1.0.shtml
Data and modeling	EPA	Remote Sensing Information Gateway	http://ofmpub.epa.gov/rsig/rsigserver?index.html
Data Analysis	EPA	Exceptional Event Decision Support System	http://www.datafed.net
Data Acquisition	ESA	Earth Online	https://earth.esa.int/web/guest/data-access
Browse and download imagery of EOS satellites	NASA	NASA Earth Observations	http://neo.sci.gsfc.nasa.gov/

Modified from Duncan, B.N., Prados, A.I., Lamsal, L.N., 2014. Satellite data of atmospheric pollution for U.S. air quality applications: examples of applications, summary of data end-user resources, answers to FAQs, and common mistakes to avoid. Atmos. Environ. 94, 647–662.

MODIS) over desert and semidesert regions. In particular, TOA reflectances received by satellite sensors at 620–850 nm (red to far red) are usually overshadowed by surface reflectance, specifically over a bright land mass (desert area). Further, AOD retrieved on the mid-visible wavelength is mostly low around the world (Hsu et al., 2004). This leads to aerosol retrievals by single-view sensors (SeaWiFS, VIIRS, AVHRR, and MODIS) using the DT algorithm being practically unsuitable for scientific exploration.

The DB algorithm was originally developed for retrieving aerosol optical properties over bright reflecting surfaces to fill the gaps in the DT algorithm, which was especially designed to retrieve aerosols over dark surfaces (Hsu et al., 2004; Levy et al., 2007). The DB algorithm used the blue channels (412 and 470 nm) whereas the surface reflectance at shorter wavelength channels is much lower than the longer wavelength channels (Hsu et al., 2004). The DB algorithm retrieves aerosol for cloud-free and snow/ice-free pixels at nominal spatial resolution 1×1 km and before aggregate to 10 km. In contrast, the DT algorithm aggregates pixels at nominal 500 m–10 km (20×20 pixels) and 3 km resolution (6×6 pixels), and retrieves aerosol after screening for cloud and snow/ice (Levy et al., 2013). The latest update of the aerosol MODIS retrieval algorithms over land and oceans known as collection 6 (C6) for all three algorithms over land and ocean (Levy et al., 2013; Sayer et al., 2013; Hsu et al., 2013). DT algorithms over land and oceans were updated several times, with the major update known as C5. This is considered the second generation of DT algorithms (Levy et al., 2007; Remer et al., 2008). However, the DT algorithms were updated without change in the principle of the C5. The major changes include quality assurance (QA) assignment, the assumption of Rayleigh optical depth and gases, and updating the cloud mask that allows the detection of thin cirrus cloud over oceans (DT-ocean) and retrieving heavy smoke over land (DT-land). In addition to refinement and improvement of aerosol retrieval over land and ocean, DT C6 algorithms also include aerosol data products at 3 km resolution in addition to the standard aerosol data products at 10 km. The 3 km products have the advantage in retrieving aerosols over a small scale, which allows local variability in aerosols (Kumar et al., 2015b, 2017). The first generation of the DB algorithm known as C5 was developed to retrieve aerosols over only bright surfaces. The second generation of the DB algorithm known as C6 was extended to retrieve aerosols over the entire land mass (except snow/ice surfaces), not only over bright surfaces. The improvement in surface reflectance determination, cloud screening, and updated aerosol model schemes allows the algorithm to increase the spatial retrieval coverage to vegetated surfaces (Hsu et al., 2013). In order to fill the gaps and improve visualizations of DT and DB algorithms over land, C6 includes a new dataset (Dark_Target_Deep_Blue_Optical_ Depth_550_-Combined) based on merged DT and BD products at a standard resolution of 10 km. The data merged based on the NDVI pixel value at spatial resolution 0.25 degrees for each month. For the pixel that has NDVI $\leq$ 0.2, DB data are used. For the pixel having NDVI $\geq$ 0.3, DT data are used. For the transition region ($0.2 \leq \text{NDVI} \leq 0.3$), the data that has the higher QA flag are used, and if both have QA = 3 the mean value of DT and DB data are considered (Levy et al., 2013).

The MODIS algorithm products have been evaluated and compared with ground-measured AERONET aerosol data on a global scale. Remer et al. (2013) evaluated DT 3 km AOD and reported it to be 0.01–0.02 higher than 10-km AOD over land while retrieving a proportionally higher number of very low AODs over ocean. The expected error (EE) of the DT algorithm over land at 3 km resolution is $\pm(0.05 + 0.20$ AOD) and $\pm(0.05 + 0.15$ AOD) at 10 km resolution (Levy et al., 2013; Remer et al., 2013). The EE of DT algorithm at 10 and 3 km resolution over ocean is ±(0.03+0.05 AOD) (Levy et al., 2013; Remer et al., 2013). Sayer et al. (2013) evaluated the highest quality (QA=3) DB AOD against AERONET at a global scale and reported the EE of DB AOD is ±(0.03 +0.15 AOD). Further, Sayer et al. (2014) compared Aqua's second generation DB, DT, and merged dataset with respect to AERONET and concluded that neither algorithm consistently outperformed the other and, in many cases, all the algorithms are equally suitable for quantitative applications.

Multispectral sensing over a wide range of wavelengths has made MODIS a popular sensor among researchers with its extensive applications, such as in air quality studies (Kumar et al., 2017a), long-term climatology (Gupta et al., 2013), meteorological applications (Yu et al., 2013), and studying extreme events (Youn et al., 2011; Kumar et al., 2016, 2017b). Based on the interaction of the majority of aerosols in the visible range of solar spectra, MODIS aerosol products have been widely used at 550 nm. Initial applications of MODIS data collections for aerosol-related studies include distribution of aerosol load, size distribution, and radiative impacts on various spatial and temporal scales. Gupta et al. (2013) using MODIS collection 5 Level 2 observations reported around 80% of days have unhealthy air quality in Lahore. The alarming levels of pollution leading to unhealthy conditions greatly depend on the seasonality of aerosol loadings over any region. Based on 5 years of observations of MODIS AOD and the Angstrom exponent, Che et al. (2013) noted a clear enhancement of aerosol load during spring and summer over the Taklimakan desert in China due to dust events.

Over the Indo-Gangetic Plain, the highest aerosol loadings have been reported during winter due to improved source strength and stable meteorological conditions (Kumar et al., 2015b; Murari et al., 2015; Sen et al., 2014, 2017). Fig. 4 represents the 5-years variations (2011–15) of MODIS AOD and FMF over the entire Indian subcontinent, which clearly identifies the spatial pattern of aerosol loading. Sen et al. (2017) compared AOD both over the IGP and the Indo-Himalayan Plain (IHP) during the winter and summer monsoons of 2014–2015 at multiple locations and concluded a continental outflow from the IGP to the Bay of Bengal. Recent advancements in space technology have opened up its utility in multidisciplinary research, including the estimation of

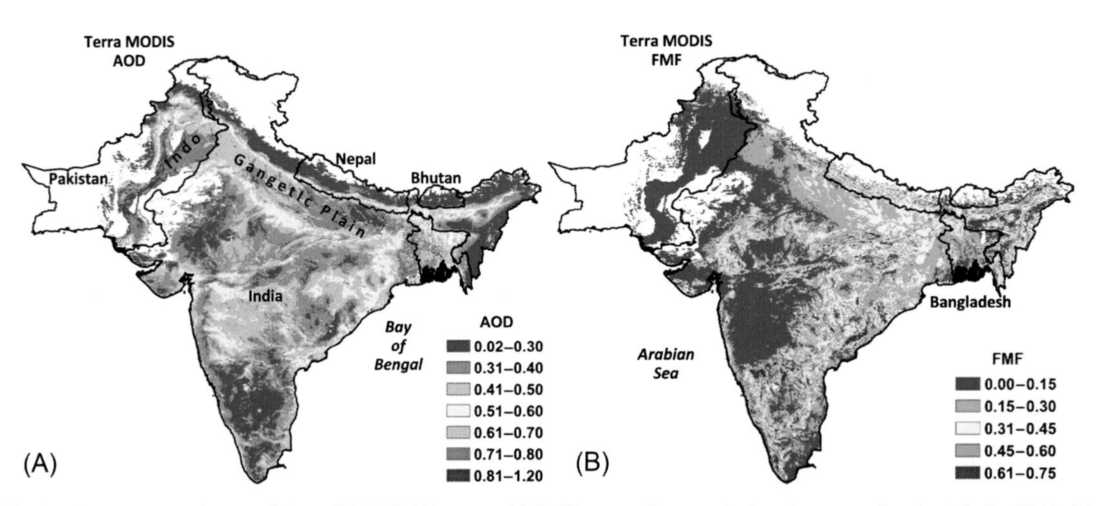

FIG. 4 Comparative figure of Terra MODIS (A) mean AOD (B) mean fine-mode fractions over South Asia for 2011–15.

population exposure to air pollution and mortality assessments based on satellite retrievals. Hu (2009), using collection-5 MODIS AOD data and its collation with near surface $PM_{2.5}$, reported an increase in cancer-induced mortality with an increase in $PM_{2.5}$ concentrations over the United States. The global estimates of mortality attributed to $PM_{2.5}$ concentrations based on MODIS were found comparable to the chemical transport model simulations for anthropogenic $PM_{2.5}$ (Evans et al., 2013). MODIS-based retrievals of aerosol properties have also been used for crop modeling, vegetation, and other climatological studies.

5.2 Aerosol Properties From MISR

The basic complication for a single-view satellite sensor (like MODIS) is the requirement for absolute reflectance or its spectral deviation at multiple wavelengths (Hsu et al., 2006). However, flexibility increases for the multiview satellite sensors (like MISR), which can scan a particular scene with multiple viewing angles and thereby reduce assumptions about surface reflectance. MISR is one of five sensors onboard the Terra satellite, which observes the earth from a polar orbit approximately 700 km above the surface. The sensor has nine cameras, each one fixed at a particular viewing zenith angle in a long-track direction with four spectral bands (446, 558, 672, and 866 nm). The cameras are arranged in different view angles, one at nadir and the other symmetrical views at 26.1 degrees, 45.6 degrees, 60.6 degrees, and 70.5 degrees aft. Each camera in the MISR sensor measures the reflected radiance in four different spectral bands (446, 558, 672, and 866 nm) in a cross-track with spatial resolution varying from 250 m (nadir) to 275 m (off-nadir). MISR has a global coverage of nine days with repeat coverage between 2 and 9 days based on the latitude. The sensor takes measurements in two modes: local and global. In the local mode, which is more of a specific type, each of the nine cameras obtains images at the four spectral bands (i.e., 36 channels) at a spatial resolution of 250–275 m, having a swath width varying from 378 (nadir) to 413 km (off-nadir). In the global mode, the nadir camera retrieves an image in four spectral bands at a spatial resolution of 250 m while the other eight cameras obtain images only by the red band with 250–275 m resolution. MISR products are further radiometrically calibrated, geolocated, and averaged to uniform resolution of 1.1 km (Level 1B2) and are available over both land and ocean at 17.6×17.6 km resolution with 16×16 homogenous arrays.

The aerosol retrieval algorithm of MISR takes advantages of the aerosol composition and relates its physicochemical and optical properties more fundamentally with the TOA radiance received by the sensor. It provides a generic representation in terms of SSA, aerosol size, and phase functions. This allows the user to directly interpret information about aerosol sources and enables comparison with various aerosol types over a region (Fig. 5). The MISR aerosol algorithm retrieves AOD and aerosol type by analyzing MISR TOA radiances from 16×16 pixel patches of 1.1-km resolution. To improve the retrieval algorithm, a series of theoretical experiments were carried out before its launch to determine the acceptability of the retrieval process under varying AODs and microphysical properties (Kahn et al., 2001). The aerosol retrieval process has three basic ancillary datasets: (1) the Terrestrial Atmosphere and Surface Climatology (TASC) dataset for climatological, meteorological, and ozone; (2) Aerosol Climatology Product (ACP) consisting of the aerosol physical and optical properties (APOP) file and a tropospheric aerosol mixture file having information on aerosol microphysical and scattering characteristics of 21 different aerosol compositions; and (3) a Simulated MISR Ancillary Radiative Transfer (SMART) dataset for large dark water bodies (for oceans) and spectrally black surfaces (for land). The fundamental process in aerosol retrieval involves comparing MISR-observed TOA radiances with modeled

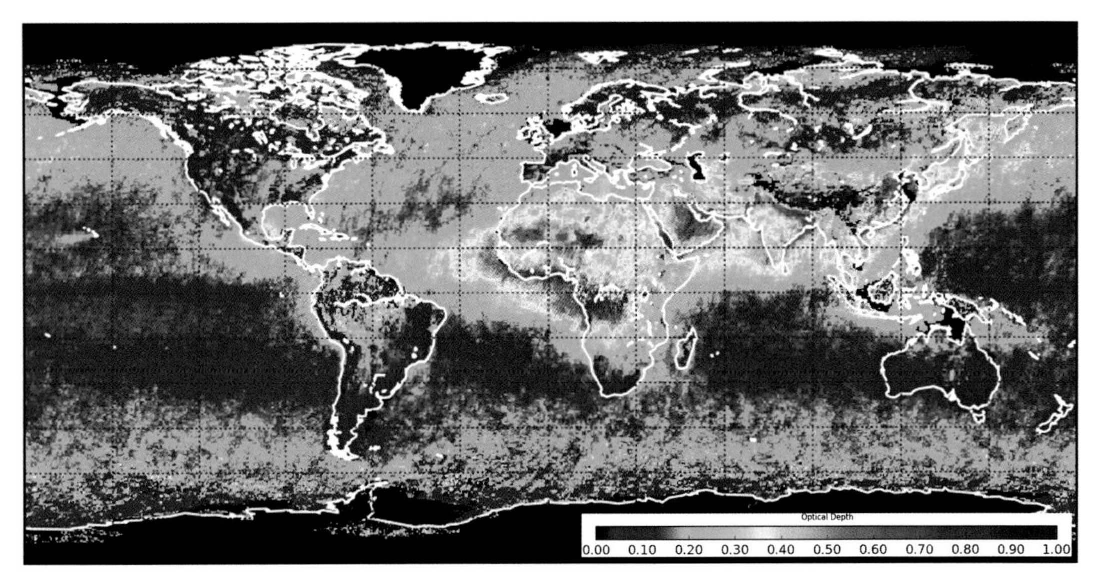

FIG. 5 Global mean (2011–2015) aerosol optical depth retrieved from MISR.

reflectance derived from a coupled atmosphere/surface RTM. The MISR RTM is defined by its base, top and scale heights, and optical depth. The RT calculations are performed using a scalar code based on a matrix operator technique including a correction for Rayleigh polarization effects. Further, within the SMART database containing multiple RT parameters (function of AOD, aerosol component, and view angles), RT calculations are carried out for individual aerosol components required to retrieve aerosols. The MISR under generic retrievals has the capability of distinguishing aerosols in 3–5 size groups and 2–4 compositional groups (by means of SSA; Kahn et al., 1998). It can also distinguish between spherical and nonspherical particulates (Kahn et al., 1997). Under climatological retrievals, MISR has the capability of identifying sea salt, BC, and dust within 20% or more of each component's AOD fraction. Kahn and Gaitley (2015) evaluated the MISR Version 22 aerosol-type self consistency in comparison to the Aerosol Robotic Network aerosol-type climatology and reported aerosol-type differentiation decreases below 0.15 or 0.20 AOD.

Kahn et al. (2010) assessed the quality of MISR version 22 (V22) aerosol products over land and oceans, comparing the MISR data with coincident observations from 81 globally distributed AERONET sites for 8 years. The study reported that most of the MISR AOD retrievals fall within 20% of AOD of the paired validation data, and about 50%–55% are within 10% × AERONET AOD, except sites where dust and smoke dominate. Most of the validation studies reported that MISR has greater accuracy in retrieving aerosol optical properties. Other studies also compared MISR-MODIS AOD against AERONET AOD and revealed that the MISR-AERONET comparison shows strong correlation compared to the MODIS-AERONET. Bibi et al. (2015) observed a higher degree of association with AERONET over oceans in comparison to land whereas Qi et al. (2013) reported better performance of MISR data over selected sites compared to MODIS. Additionally, MISR aerosol products have been utilized for the assessment of long-term global trends of aerosol distribution. Mehta et al. (2016) reported increasing trends in AOD over Asian and

African land masses and surrounding regions while declining aerosol trends were reported over Europe, South America, and North America. A better predictability of near-surface $PM_{2.5}$ was observed by You et al. (2015) using MISR AOD (72%) over MODIS AOD (67%), opening up an improved scope for estimating ground-level particulates using MISR aerosol retrievals. In a unique approach to link MISR-derived AOD with BC aerosols, Zeeshan and Oanh (2015) reported a high capacity of MISR AOD in estimating daily BC in Bangkok, Thailand.

5.3 Aerosol Properties From AATSR

The Advanced Along-Track Scanning Radiometer (AATSR) is a remote-sensing instrument onboard the satellite ENVISAT (ENVIronmental SATellite) of the European Space Agency (ESA). The mission is the continuation of ATSR-1 and ATSR-2 on the European Remote Sensing (ERS) satellite. It has been enlisted as part of a series of advanced instruments designed primarily for the measurement of sea surface temperature (SST), while products like AOD (at 555, 659, 865, and 1610 nm) and AE are also explored. With a data resolution of 1 km at the nadir, AATSR provides measurements of reflected and emitted radiation for a wide range of wavelengths. The most peculiar feature of AATSR includes the use of a conical scan for a dual view of Earth's surface, onboard calibrations, and mechanical coolers for the maintenance of the infrared detectors. AATSR was launched in March 2002, following the ATSR-2, and was in operation through May 2012. It was primarily launched as a geophysical ocean sensor for the measurement of SST. However, it was also utilized for other land applications. AATSR provided information on SST, clouds, aerosols, vegetation, and snow cover in addition to calibrated reference radiances and imagery. The swath width of AATSR is approximately 512 km wide, assuring average global coverage every five days over the equator with more frequent at higher latitude. Although, AATSR does not provide daily global coverage, it has an extended data product of 20 years of aerosol properties over land. At the nadir, retrieval is made for single pixels (1×1 km) and subsequently groups of 10×10 pixels are averaged to reduce noise and errors.

Two different aerosol retrieval algorithms have been developed for AATSR, the first using the single view over oceans (Veefkind and de Leeuw, 1998) while the other uses the double view over land (Veefkind et al., 1998). The special dual-view algorithm is used for the compensation of land reflectance over land. The dual-view capacity of AATSR over land provides improved cloud screening (de Leeuw et al., 2007). This algorithm has been validated with AERONET data (Robles Gonzales et al., 2000) and also from other aerosol-retrieval instruments such as MISR and MERIS. The retrieval algorithm is based on the construction of an aerosol model with two different aerosol types. The reflectance at the TOA is computed using RTM and further compared with the reflectance measured by the AATSR device. The model that results in the minimum discrepancy between the modeled and measured TOA reflectance was used for the determination of aerosol optical properties. The algorithm used for AATSR aerosol retrieval separates the surface bidirectional reflectance from the atmospheric aerosols without recourse to a priori information of the land surface properties (North et al., 1999). The ratio of surface reflectance at the nadir and forward viewing angles is well correlated between bands. North et al. (1999) further considered the variation of the diffuse fraction of light with wavelength. However, the algorithm is subject to discrepancies in retrieving AODs under cloudy conditions. Over the ocean, the aerosol properties are directly retrieved using single view. The accuracy of the AATSR algorithms was estimated at 0.03 over the ocean and 0.05 over land in comparison to AERONET (de Leeuw et al., 2007).

AATSR observations have varying applications, especially over the regions with varying aerosol compositions like the Indian Ocean (Robles-Gonzalez et al., 2006). The main aerosol products of AATSR are AOD and aerosol types, while the aerosol mixing ratio is also available in principle. Bevan et al. (2012) presented a long-term (2003–2009) global dataset of aerosol properties from the AATSR for a wide variety of surfaces including ocean, vegetated land, and desert. The AATSR datasets were compared with AERONET and the Maritime Aerosol Network (MAN), and with MODIS and MISR aerosol products. Agreement with AERONET ($r=0.80$; $\Delta\tau=\pm0.025\pm0.4\tau$) and MAN ($r=0.97$; $\Delta\tau=\pm0.04$) was reported to be high while monthly AODs from MODIS and MISR represented close agreement over most of the regions. The global mean values of AOD at 550 nm over land and ocean were reported to be 0.195 and 0.137, respectively. Bevan et al. (2009) used 13 years (1995–2004) of AATSR AOD over the Amazon forest to recognize the role of aerosols in biosphere-climate interactions. A decreasing trend in dry-season AOD (1995–2000) and subsequent increase (2000–2004) was explained in terms of deforestation practices over the region.

5.4 Aerosol Properties From CALIOP

The Cloud-Aerosol LIDAR with Orthogonal Polarization (CALIOP) is a space-borne LIDAR onboard the Cloud-Aerosol LIDAR and Infrared Pathfinder Satellite Observation (CALIPSO) satellite. It retrieves the vertical distribution of aerosols and clouds on a global scale (Winker et al., 2009; Kittaka et al., 2011; Kumar et al., 2016). The active laser provides a high vertical resolution aerosol profile in cloud-free conditions as well as above lower-lying clouds and below optically thin clouds. It was developed within NASA's Earth System Science Pathfinder (ESSP) program in collaboration with the French space agency Centre National d'Études Spatiales (CNES) for observing global distribution and properties of aerosols and clouds. CALIPSO was launched in April 2006 with the CloudSat satellite as part of the NASA A-train. It has a 705 km sun-synchronous polar orbit with an equator crossing at 13:30 local time. CALIPSO carries two sensors: (1) an active LIDAR CALIOP; and (2) two passive sensors, the Imaging Infrared Radiometer (IIR) and Wide Field Camera (WFC), which obtain data in the infrared and visible spectral regions, respectively. All the sensors work independently and continuously except the WFC sensor, which only works in daylight. The LIDAR provides high-resolution measurements of aerosol and cloud vertical distribution, which improves understanding about the effect of aerosols and clouds over climate (Winker et al., 2009). It has a swath with practically zero width with a vertical resolution of 30–60 m, having successive footprints spaced by 333 m along the orbit track. It measures the aerosol vertical profile with an elastic laser backscatter at 1064 nm with a parallel and cross-polarized return signal at 532 nm. However, the daytime profile records a slightly lower signal-to-noise ratio over nighttime due to solar background illumination. Within −0.5 to 8.2 km altitude, both the 532 and 1064 nm signals have a horizontal resolution of 330 m and vertical resolution varying from 30 to 60 m, respectively. Above 8.2–20.2 km, both profiles are averaged to 60 m vertical and 1 km horizontal resolution. The *level 0* data are reconstructed, unprocessed with full resolution while reconstructed, geolocated, and altitude-registered Level 1 data are calibrated before being processed for Level 2.

The CALIPSO aerosol-retrieval process is derived on the basis of cluster analysis of a multiyear (1993–2002) AERONET dataset. Six AERONET aerosol clusters were considered based on observed physical and optical properties (Omar et al., 2005). The AERONET-derived parameters were further adjusted to develop a more accurate LIDAR ratio related to actual observations. Out of six, only three (biomass burning, polluted continental, and polluted dust) were further adopted as CALIPSO aerosol models. Further, marine and background/clean continental aerosol models were developed both by direct measurements of particle size distributions and refractive indices or through adjusted model parameters. The CALIPSO aerosol-retrieval algorithm specifically uses attenuated backscatter and volume depolarization ratio coupled with surface type, layer, and altitude to determine the aerosol type (Fig. 6). The aerosol layers are identified by a set of algorithms known as selective iterative boundary locator (SIBYL) applied to 532-nm attenuated backscatter profiles. A scene classification algorithm

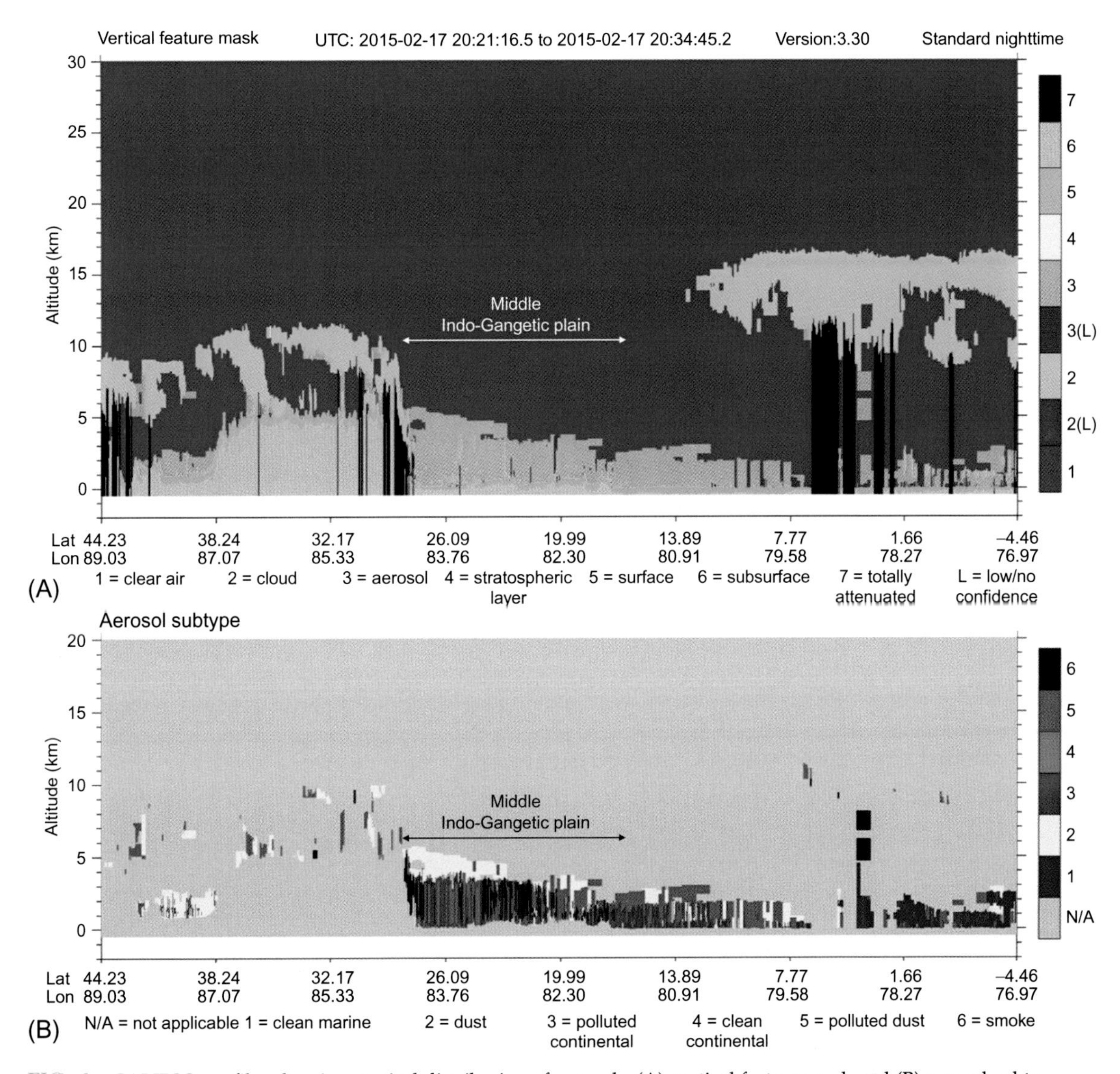

FIG. 6 CALIPSO profiles showing vertical distribution of aerosols, (A) vertical feature mask and (B) aerosol subtype.

(SCA) classifies the different layers of aerosols by type. Further, particle backscatter and extinction coefficients are retrieved using a hybrid extinction retrieval algorithm (HERA).

CALIPSO vertical profiles have been extensively used in atmospheric sciences, especially for studying various aerosol properties, their transport, sources, and radiative impacts. Yu et al. (2015) quantified trans-Atlantic dust transport from long-term CALIPSO LIDAR measurements. The study has estimated an outflow of nearly 182 Tga^{-1} dust per annum from the coast of North Africa. Observations taken by Ganguly et al. (2009) from multiple satellite instruments, including CALIPSO, confirmed the presence of huge amounts of aerosol haze over the Gangetic basin during winters. Over Bangkok, Bridhikitti (2013) recognized a uniform layer of smoke originating from biomass burning, resulting in high AOD over the entire region. Kumar et al. (2015a) used CALIPSO altitude-orbit cross-section profiles (level 2 version 3.30) to identify winter-specific nonspherical coarse particulates at relatively higher altitude and dominance of spherical fine particulates at lower altitude over the middle IGP (Fig. 6). In another study, Kumar et al. (2016) recognized the firework-specific emissions of aerosols and their vertical distribution over the middle IGP, which is reportedly dominated by smoke and polluted continental aerosols, especially at high altitudes (3.0–3.8 km).

5.5 Aerosol Properties From POLDER

Polarization and Directionality of the Earth's Reflectances (POLDER) is a passive optical imaging radiometer developed by CNES in collaboration with the LOA atmospheric optics laboratory in Lille, France. POLDER was designed to study cloud and aerosol properties in respect to climate systems. The first POLDER instrument was launched in August 1996 onboard an ADEOS-I satellite, which failed due to a technical error in June 1997. The next mission (POLDER 2) was launched in December 2002 aboard an ADEOS-II satellite; this, too, ended prematurely. The third-generation instrument was launched onboard PARASOL satellite in December 2004; this continued to operate until 2013. It measures the intensity, direction, and polarization of light reflected by the Earth and its atmosphere. POLDER scans between 443 and 1020 nm. Shorter wavelengths (i.e. 443–565 nm) are used to measure ocean color while longer wavelengths (670–1020 nm) are applied for studying vegetation and water vapor. POLDER is comprised of a two-dimensional CCD detector with a wide field view in cross-track (±51 degrees) and along-track (±43 degrees) directions (Deschamps et al., 1994). Polarization measurements are performed at 490, 670, and 0.865 nm. With an altitude of 705 km, it provides images of 2100 km × 1600 km resulting in global coverage within two days (Tanré et al., 2011).

Aerosol parameters are retrieved at 18.5 km × 18.5 km resolution using different algorithms over oceans and land. The inversion scheme developed by Deuzé et al. (1999) and Herman et al. (2005) was used to derive a number of parameters over oceans with dark surfaces in red and near IR spectral regions. The aerosol-retrieving algorithm utilizes the total and polarized radiances at 670 and 865 nm, assuming the size distribution follows two log-normal aerosol size distributions (reff < 0.5 μm and reff > 1.0 μm) with consideration of nonabsorbing particles in both modes. The total radiance (L) was given by:

$$L(\mu s, \mu v, \phi_v) = \eta L^f(\mu s, \mu v, \phi_v) + (1-\eta) L^c(\mu s, \mu v, \phi_v) \tag{15}$$

where η is the radiance weighting factor; $L^f(\mu s, \mu v, \phi_v)$ and $L^c(\mu s, \mu v, \phi_v)$ are the radiances of the fine (f) and coarse (c) modes, respectively; $\mu s = \cos(\theta_s)$ with θ_s the solar zenith angle; $\mu_v = \cos(\theta_s)$ for the viewing zenith angle; and ϕ_v the relative azimuth angle. The scattering by finer spherical particles generates highly polarized light, which is better at estimating the presence of aerosols in comparison to total radiances. That is why only accumulation-mode aerosols are considered for the algorithm and coarser mode aerosols are ignored. Although there is almost no polarization from coarser aerosols, some misinterpretations could be made during extreme events that must be taken into account. The refractive index used is considered between 1.47 and 0.01 corresponding to aerosols from biomass burning or pollution events (Dubovik et al., 2002).

The aerosol retrievals from POLDER over oceans include total AOD, aerosol effective radius, the Angstrom exponent, finer AOD, and nonspherical AOD. Huang et al. (2015) studied the aspect ratios of Saharan and Asian dust using collocated MODIS Deep Blue product and the PARASOL level 2 Earth radiation. The mean aspect ratio for Saharan dust and two branches of Asian dust were reported as 2.5, 1.25, and 2.5, respectively. Apart from aerosols, POLDER data has also been widely used for multilayered cloud identifications.

5.6 Aerosol Properties From OMI

The Ozone Monitoring Instrument (OMI) is one of four sensors onboard the Aura satellite launched in July 2004. The OMI project is a joint effort by the Netherland Agency for Aerospace Program (NIVR), the Finnish Meteorological Institute (FMI), and NASA. It continues in the footprint of NASA's TOMS and ESA's GOME (Global Ozone Monitoring Experiment) instrumental records of atmospheric parameters and generates hyperspectral images by retrieving solar backscatter radiation in the visible and ultraviolet (UV) bands. The sun-synchronous orbit of OMI with a wide swath of 2600 km allows global coverage each day with a spatial resolution of 13×24 km at nadir and 28×150 km at the extreme of the view angle. The OMI measures the backscattering solar radiation from the earth's surface and atmosphere in multiple bands (270–500 nm). The key feature of OMI is the use of UV [UV-1 (270–314 nm), UV-2 (306–380 nm)], and the visible channel (350–500 nm) to measure backscatter radiation. Different algorithms have been used to retrieve data for various tropospheric air pollutants including ozone (O_3; both tropospheric and columnar), nitrogen dioxide (NO_2), sulfur dioxide (SO_2) and aerosols with the capability of distinguishing dust, smoke, and sulfates. It helps to detect volcanic ash, BrO, formaldehyde, and OClO and, thereby, plays an important role in understanding atmospheric chemistry.

For retrieving aerosols through OMI, two types of aerosol-retrieval algorithms are in use: the Near-UV aerosol algorithm (OMAERUV) and the Multi-Wavelengths Algorithm (OMAERO) having a pixel resolution of 13×24 km at nadir. The OMAERUV uses two UV wavelengths (354 and 388 nm) for retrieving aerosol extinction and absorption optical depth (AAOD), whereas the OMAERO is a multiwavelength algorithm having 19 channels (330–500 nm). The benefits of using the near-UV band include the ability to retrieve aerosol properties over a heterogeneous land surface even for highly bright surface areas. This primarily occurs because of relatively less surface reflectance over a bright land surface within the UV band. Using the near-UV band also enables OMI to retrieve absorbing aerosols (BC and desert dust) by utilizing the large sensitivity of these aerosols over the near-UV spectral band (Torres et al., 2007; Kumar et al., 2016). OMAERUV provides AOD, AAOD, and the aerosol index (AI) at two wavelengths (354 and 388 nm). However, due to cloud contamination, OMI AOD is often more reliable near the aerosol source under clear sky conditions while the AAOD is less affected by cloud contamination and, therefore, more reliable. The OMAERUV algorithm measures the Lambert Equivalent Reflectivity (LER) (Rx388) at 388 nm under the assumption that only

Rayleigh scattering occurs in the atmosphere and is surrounded by an opaque Lambertian reflector of reflectance Rx388 (Torres et al., 2007). The primary step of the OMAERUV algorithm is the calculation of the UV Aerosol Index (*UVAI*) (Torres et al., 2007):

$$UVAI = -100\log_{10}\left[\frac{I_{354}^{obs}}{I_{354}^{calc}\left(R_{354}^{*}\right)}\right] \tag{16}$$

where I_{354}^{obs} is the TOA at 354 nm as observed by the sensor and I_{354}^{calc} is calculated by assuming an LER of (R_{354}^{*}). A multiwavelength algorithm (OMAERO) uses information over 19 spectral channels over near-UV and the visible wavelength to derive aerosol properties under cloud-free conditions (Fig. 7). At these spectral channels, Raman scattering and gas absorption occur only for the O_2-O_2 absorption band at

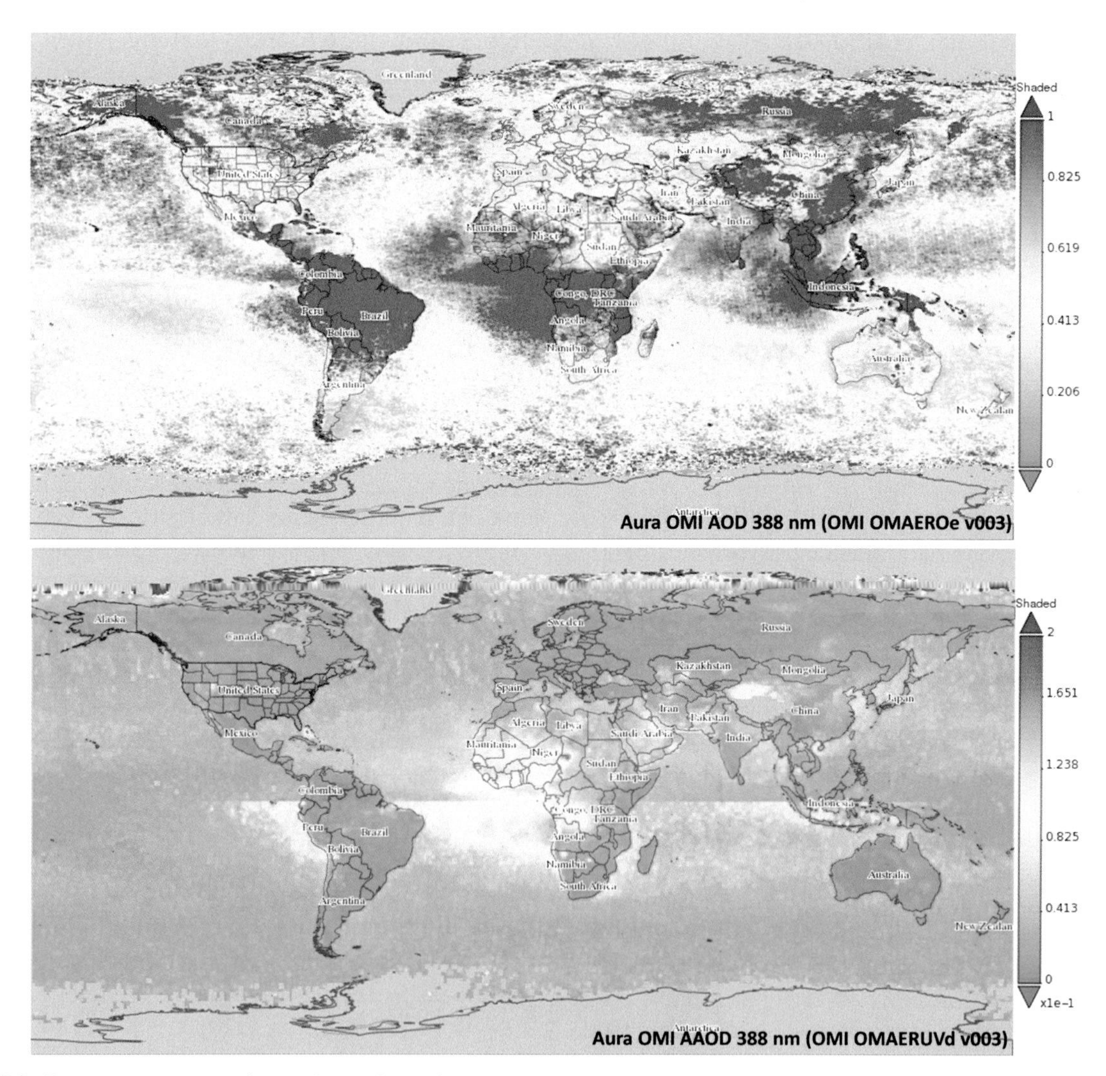

FIG. 7 Global variation of Aura OMI AOD and AAOD (at 388 nm) for 2015.

477 nm. The absorption band at 477 nm is used to enhance the sensitivity of the OMI reflectance measurement to aerosol layer and cloud height (Torres et al., 2007). Additionally, the inclusion of the near-UV band helps to separate weak and strong absorbing aerosols and therefore, functions better over the aerosol product from MODIS and MISR. The OMAERO algorithm provides AOD, aerosol type, SSA, aerosol layer height, and aerosol size distribution.

The OMI AOD and SSA retrievals have been previously compared with AERONET observations (Torres et al., 2007; Ahn et al., 2014). Both OMAERUV and OMAERO algorithms were also compared against MODIS (Curier et al., 2008) and MISR observations (Ahn et al., 2008), which reveal good agreement in terms of identifying major emission sources. Ahn et al. (2014) compared OMAERUV AOD with AERONET AOD at 44 sites globally and revealed that 65% of OMAERUV AOD agrees with AERONET AOD observations. The high sensitivity of OMI near-UV measurements to the absorption aerosols was used as complementary information on absorption aerosols when compared to MODIS and MISR. Curier et al. (2008) compared OMI-OMAERO and MODIS-retrieved AOD over Europe and surrounding oceans and concluded agreement between scenes both over ocean and land. Satheesh et al. (2009) used OMI-MODIS joint retrievals to improve the accuracy of retrieved aerosol products over the tropical North Atlantic. The results showed a good agreement between OMI and MODIS-predicted AODs in the UV range. Additionally, OMI has also been explored for retrieving trace gases around the world, like Shukla et al. (2017) used the Aura OMI-Differential Optical Absorption Spectroscopy (OMI-DOAS) algorithm to identify consistent and clear seasonal trends in the columnar ozone (2005–2015) with summertime maxima and wintertime minima over the middle IGP.

5.7 Aerosol Properties From VIIRS

The Visible Infrared Imaging Radiometer Suite (VIIRS) is a remote-sensing cross-track scanning radiometer onboard the Suomi National Polar-Orbiting Partnership (Suomi NPP) satellite. VIIRS was launched in October 2011 as a successor to AVHRR and MODIS, enabling a new generation of moderate resolution-imaging capabilities with high radiometric accuracy and spatial resolution. Using its 22 imaging and radiometric bands covering wavelengths from 0.41 to 12.5 μm, VIIRS accomplishes operational environmental monitoring and numerical weather forecasting. A dual-gain band allow VIIRS to retain high signal-to-noise ratio (SNR) at low radiance, thus making it suitable for concurrent application over land, ocean, and atmosphere. It has a scan width of about 3000 km (±56 degrees) allowing it to provide daily global coverage. VIIRS has an equator crossing time identical to MODIS (13:30 LT), with 86 seconds granule size having pixel resolution of 0.75 km at nadir and 1.5 km at edge (Jackson et al., 2013). VIIRS uses three different bands—imagery, moderate resolution (M-bands), and the day-night band—and its main products include clouds, sea surface temperature, ocean color, polar wind, vegetation, aerosol, fire, snow, and ice. The primary aerosol data include AOD and aerosol type retrieved mainly between 412 and 2250 nm. The VIIRS retrieved AOD at 388 nm is shown in Fig. 8, which closely resembles the MODIS-derived AOD over the same region (Fig. 4).

VIIRS aerosol retrievals are made over M-bands (0.412–12.016 μm) while AOD is specifically retrieved at 550 nm. It has the capability of retrieving the Angstrom exponent and aerosol type over land and can distinguish aerosol fine and coarse mode fractions over ocean (Jackson et al., 2013). The level 2 product of VIIRS available for end users is designated as the environmental data record (EDR), which includes aerosol optical thickness (AOT), the

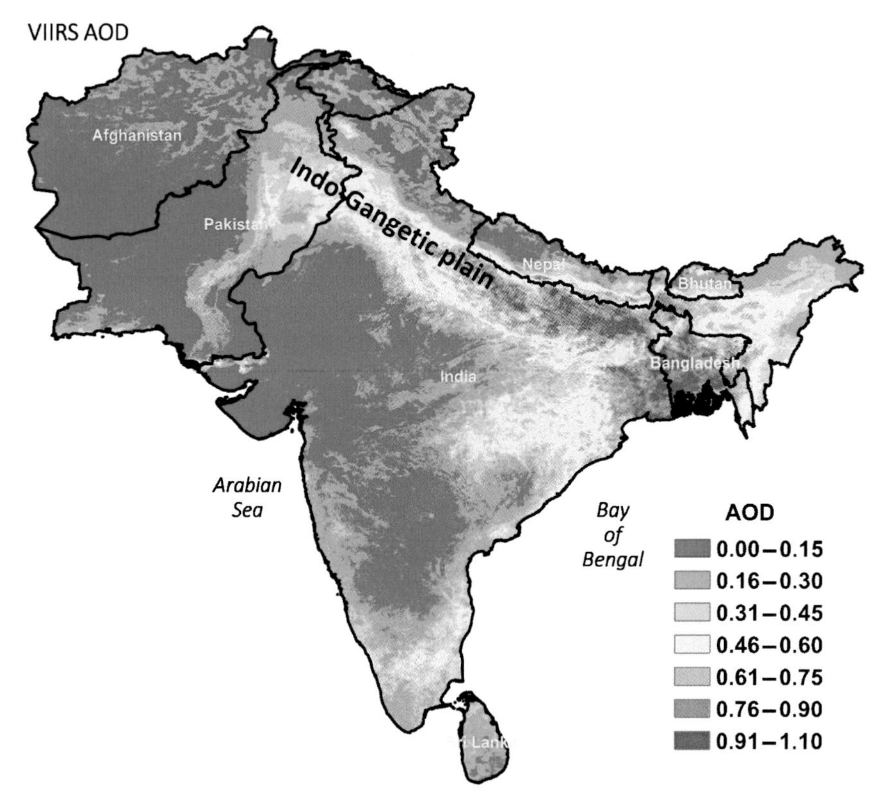

FIG. 8 Winter-specific columnar AOD (at 388 nm) over South Asia retrieved from VIIRS.

aerosol particle size parameter (APSP), and suspended matter (SM). The SM EDR includes classification of aerosol types with categories including crustal dust, smoke, volcanic ash, and sea salt. However, aerosol retrieval is performed only during daytime for cloud-free pixels and over dark land surfaces. The aerosol algorithm of VIIRS mostly resembles the MODIS collection 3 algorithm (Kaufman et al., 1997; Tanré et al., 1997) with specific changes over land. The algorithms over ocean also have some modifications in comparison to their MODIS counterpart. The VIIRS algorithm pursues an approximation of vector radiation transfer model (RTM) version 1.1 (6S-V1.1) (Kotchenova and Vermote, 2007). The spectral reflectance measured at satellite level $\rho_{toa}(\tau_A)$ is calculated using the equation:

$$\rho_{toa}(\tau_A) = \mathrm{Tg_{og}Tg O_3}\Big[\rho_A(\tau_A)\mathrm{TgH_2O(UH_2O/2)} + \rho_R(P) + \mathrm{TgH_2O(UH_2O)}\cdot\Big\{T_{R+A}(\tau_A,\theta_s)(T_{R+A}(\tau_A,\theta_v)\frac{\rho_{\mathrm{surf}}}{1 - S_{R+A},\rho_{\mathrm{surf}}} + (\text{nonlambertian terms})\Big\}\Big] \quad (17)$$

where τ_A is AOT; θ_s and θ_v are the solar and view zenith angles, respectively; P is surface pressure; ρ_R and ρ_A denote the atmospheric path reflectance due to molecular and aerosol scattering; T_{r+a} is the total transmittance attributed to molecules and aerosols; TgO_3 and TgH_2O are transmittances from ozone and water vapor, respectively; UH_2O represents total columnar water vapor; Tg_{og} is the two-way transmittance from O_2, CO_2, CH_4, and N_2O); ρ_{surf} is the Lambertian surface reflectance; and S_{r+a} is the spherical albedo of the atmosphere.

A number of assumptions have been made for the calculation of spectral reflectance, which can be found elsewhere (Jackson et al., 2013). The reflectance in 412, 445, 488, 672, and 2250 nm bands have been used for aerosol retrievals over land. The land inversion approach used for retrieving AOD utilized an empirically derived relationship between surface reflectance in 488 and 672 nm bands. Reflectance in other bands is used for choosing an appropriate model among a set of candidates. The change in ratio of the TOA reflectance in the blue and red bands from the surface value in the presence of aerosols is the primary basis of AOD retrieval over land.

The accuracy of global VIIRS AOD has been evaluated in comparison to MODIS, AERONET, and the Maritime Aerosol Network (MAN) by Liu et al. (2014). A slight difference between VIIRS global mean AOD at 550 nm from MODIS AOD was recorded at 0.01 and 0.03 over ocean and land, respectively. Validation with AERONET and MAN measurements experienced biases of 0.01 over ocean and 0.01 over land. Similar efforts of validating global VIIRS AOD with AERONET AOD by Huang et al. (2016) exhibited an overall bias of 0.02 over oceans and −0.0008 over land. Meng et al. (2015) on the other hand reported the spatiotemporal variability of VIIRS AOD over China and observed a difference of −0.0032 from MODIS-derived AOD. Oliva and Schroeder (2015) applied the VIIRS 375 fire-detection product to assess its performance for direct burned area estimation for different ecologically sensitive areas. The results were compared with Landsat-8 supervised burned area classification and a higher accuracy was observed in forests with fires of a lengthy duration. However, errors were reported over grasslands, savannas, and agricultural areas. The limitations of MODIS AOD in terms of nonavailability of nighttime AOD for air-quality studies can be compensated by using the VIIRS day/night band. Wang et al. (2016) showed quantitative changes in the intensity of light during night that can reflect the variations in surface $PM_{2.5}$. Other prominent applications of VIIRS-based retrievals include estimation of chlorophyll-a based on ocean color index (Wang and Son, 2016) as well as validation of land surface temperature (Guillevic et al., 2014) and sea surface temperature (Gladkova et al., 2015).

5.8 Aerosol Properties From OCM

The Oceansat satellite was launched by the Indian Space Research Organisation (ISRO). The Oceansat-2 OCM (Ocean Colour Monitor) sensor, launched in 2009, is the successor to the Oceansat-1 OCM. The sensor details are provided in Table 2. The OCM sensor is designed for ocean color studies. The spectral bands of the Oceansat-1 OCM sensor were similar to those of the SeaWiFS sensor, while the Oceansat-2 OCM sensor is identical to the Oceansat-1 OCM sensor except in bands 6 and 7. Band 6 of Oceansat-1 had an effective wavelength at 670 nm, and for Oceansat-2, it shifted to 620 nm. This change was made to improve the quantification of suspended sediment concentration. Band 7 of Oceansat-1 had an effective wavelength at 765 nm, which was shifted to 745 nm for Oceansat-2 in order to avoid oxygen absorption. The overview of sensor details is given in Nagamani et al. (2008). Oceansat-2 acquires the data every two days at a spatial resolution of 360 m. The OCM sensor has been used for studying ocean phytoplankton, suspended sediments, and AOD over the ocean by many researchers (Nagamani et al., 2008; Mishra et al., 2008). In addition to its capability of studying the ocean surface, the OCM sensor also has the potential to study land surface features. Studies on coastal mangroves and NDVI have been carried out using the OCM Sensor (Nayak et al., 2001). The enhanced vegetation

TABLE 2 Oceansat-2 Bands, Effective Wavelength and Extraterrestrial Solar Radiation

Oceansat 2 (OCM) Bands	Wavelength Range (μm)	Effective Wavelength (μm)	Extraterrestrial Solar Radiation ($Wm^{-2}\mu m^{-1}$) ($\bar{E}_0$)
Band-1	0.402–0.422	0.412	1728.15
Band-2	0.433–0.453	0.443	1852.11
Band-3	0.480–0.500	0.490	1972.10
Band-4	0.500–0.520	0.510	1866.97
Band-5	0.545–0.565	0.555	1827.81
Band 6	0.610–0.630	0.620	1657.65
Band-7	0.725–0.755	0.740	1289.70
Band-8	0.845–0.885	0.865	952.073

index (EVI) has been retrieved from the OCM Sensor (Mishra and Dadhwal, 2008; Mishra, 2014).

The first step for retrieval of AOD from Oceansat-2 is to identify the high-reflecting pixels like clouds and snow in the image. In the absence of a thermal band, a threshold approach can be adopted for identifying the cloud-and snow-contaminated pixels. The second step is to differentiate the pixel of the image as vegetated and sparsely vegetated or a sand-dominated pixel based on the NDVI value. The surface reflectance of the pixel with NDVI value >0.1 can be determined following the approach by Hsu et al. (2013). The surface reflectance values for the pixel having NDVI values <0.1 can be determined following the procedure provided by von Hoyningen-Huene et al. (2003). However, surfaces having sparse or no vegetation need further approximation using statistical interpolation and the spectral relationship of white sand (Ouillon et al., 2002). The AOD derived from the Oceansat-2 can be used along with Terra- and Aqua-derived AOD to study the diurnal variation. The AOD over densely populated areas can be correlated with the concentrations of fine particulate matter.

6 AEROSOL REMOTE SENSING OVER THE INDO-GANGETIC PLAIN, SOUTH ASIA

The Indo-Gangetic Plain (IGP) has always been a center of discussion when it comes to air quality. Being a highly populated and fertile region starting from the plains of the Indus River over Pakistan to the Bay of Bengal over India, it covers an area of 255 million hectares of land. The presence of a huge number of polluting sources—industries (Banerjee et al., 2011), domestic biomass burning (Murari et al., 2016; Sen et al., 2016, 2017), high vehicular density with limited emission control (Murari et al., 2015), open refuse and waste burning practices (Banerjee and Srivastava, 2012; Banerjee et al., 2017b) and resuspension of crustal dusts (Banerjee et al., 2015)—are some common sources of air pollution. Additionally, the post-harvest burning of agricultural residue (Rastogi et al., 2016) and transboundary movement of aerosols (Kumar et al., 2015b, 2017; Sen et al., 2017) substantiate regional pollution loading. Due to the limited ground-based aerosol monitoring network, satellite remote sensing has long been used to measure the spatiotemporal variation of aerosols and the climatic impact

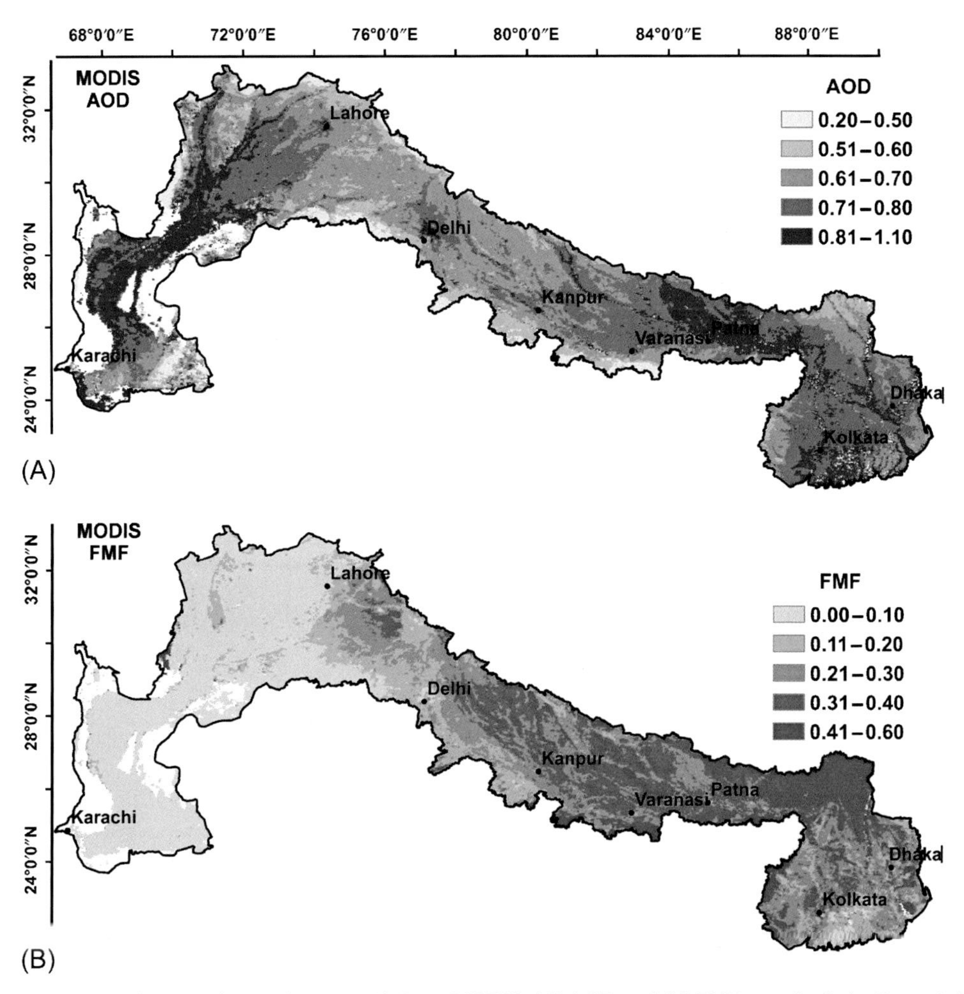

FIG. 9 Comparative figures of mean (2011–2015) Terra MODIS, (A) AOD and (B) FMF over the Indo-Gangetic Plain.

over the region. Fig. 9 represents the characteristic variation of AOD for 2 years (2014–2015; Fig. 9A) with corresponding FMFs over the entire IGP (Fig. 9B). Both the upper and middle IGP exhibit a high columnar aerosol loading while the prevalence of FMF is quite diverse. Relatively high coarser particulate loading was recognized over the upper IGP while dominance of fine particulates is only evident in the lower region. Several case studies (Table 3) on aerosol remote sensing over the IGP have documented the contribution of crustal sources (Banerjee et al., 2015) and biomass burning for aerosol emission (Rastogi et al., 2016; Kaskaoutis et al., 2014; Vadrevu et al., 2011), variation in aerosol composition (Kumar et al., 2015b, 2017; Singh et al., 2017a), estimation of surface $PM_{2.5}$ using satellite observation (Dey et al., 2012), and estimating aerosol radiative forcing (Dey and Tripathi, 2007; Kumar et al., 2017) over the region. For the following subsection, emphases were made to discuss some major conclusions on aerosol properties retrieved through satellite remote sensing over the entire IGP.

TABLE 3 Aerosol Remote Sensing Over the Indo-Gangetic Plain, South Asia

Location	Type of Study	Satellite Aerosol Parameters Used	References
Lahore	Trend analysis	MODIS-AOD	Gupta et al. (2013)
Patiala	Biomass burning event	MODIS-AOD, MODIS-Active Fire Count, OMI-Aerosol Index	Sharma et al. (2014)
Hisar	Source-specific radiative forcing	MODIS-AOD	Raman et al. (2011)
New Delhi	AOD-PM relation	MODIS-AOD	Kumar et al. (2007)
New Delhi	Aerosol characteristics and radiative forcing	MODIS-AOD, MISR-AOD, CALIPSO-Extinction Coefficient	Srivastava et al. (2014)
Agra	AOD-PM relation	MODIS-AOD	Chitranshi (2015)
Kanpur	Aerosol climatology	MODIS-AOD, MISR-AOD	Prasad and Singh (2007)
Allahabad	Fog event	MODIS-False color composite	Badarinath et al. (2011)
Varanasi	AOD-PM relation	MODIS-AOD	Kumar et al. (2015b)
Varanasi	Temporal variation	MODIS-AOD	Tiwari and Singh (2013)
Varanasi	Fireworks event	MODIS-AOD, Fine-Mode Fraction, CALIPSO-Extinction Coefficient, Backscatter profiles	Kumar et al. (2016)
Patna	Aerosol climatology	MODIS-AOD	Ramachandran et al. (2012)
Kolkata	Temporal variation	MODIS-AOD, MISR-AOD	Mehta (2015)
Dhaka	Temporal variation	MODIS-AOD	Mamun et al. (2014)

A long-term decadal assessment of MODIS-based aerosol distribution was made by Ramachandran et al. (2012) throughout India, including selected stations over the IGP. MODIS Terra level-2 AOD was used to delineate their season-specific trends. An increasing trend of AOD at a few stations such as New Delhi depicted the upsurge of anthropogenic sources of pollution, including fossil fuel- and biomass-burning activities. Prasad and Singh (2007) also reported similar evidence of increasing satellite-derived (MODIS and MISR) aerosol loading during 2000–2005 over major cities of the IGP. Studies clearly indicate the impact of natural sources and anthropogenic activities, especially biomass burning, resulting in different climatic implications. The long-term indirect effects of the aerosol burden on water and the ice-cloud effective radius were assessed by Tripathi et al. (2007) using MODIS-derived AOD. The intensity of indirect effects for clouds were reported highest during winter while its spatial extent was reported maximum during monsoon.

Post-harvest specific agro-residue burning is an important event that critically affects air quality over the IGP. Remote-sensing techniques have been extensively used to study the biomass burning evidence and its impact on regional air quality. Vadrevu et al. (2011) highlighted the remotely sensed fire products and their potential to characterize the agricultural residue burning events and their impacts on air quality.

The assessment of fire counts from MODIS data revealed the bimodal activities in fire intensity, fire radiative energy, and optical depths pertaining to wheat and rice residue burning. Kaskaoutis et al. (2014), using a synergy of both ground-based and satellite observations (MODIS and OMI), evaluated the impacts of rice residue burning over northern India postmonsoon and found evidence that it generated a thick aerosol layer (below 2–2.5 km) over the IGP. Strong AOD gradients during postharvest were also observed before subsequently being transported from the upper to lower Gangetic basin. Sharma et al. (2010) established significant impacts of crop residue burning on aerosol properties using ground as well as multisatellite data with high AOD and Angstrom exponent values. In a recent review on organic aerosols over the IGP, Singh et al. (2017b) concluded the contribution of biomass burning, especially crop residue burning over the upper IGP, which contribute massive aerosol loading to both the middle and lower IGP regions through long-range transport. Similar findings have also been reported by Sen et al. (2017) recognizing the long-range transport of aerosols from the upper IGP and IHP region to the Bay of Bengal, thereby potentially intruding upon the cloud formation process.

Estimation of near-surface particulates using satellite-retrieved data is one of the foremost applications of aerosol remote sensing. Based on the highly heterogeneous nature of aerosols over the IGP and uneven meteorological conditions throughout the year, it is quite a challenging task for researchers. Dey et al. (2012), using MISR-retrieved AOD, estimated surface $PM_{2.5}$ for the Indian subcontinent accounting for both composition and aerosol vertical distribution. The study revealed the exposure of nearly 51% of the subcontinent's population to high airborne particulate concentration. Interestingly, some rural areas were reported to have higher aerosol loading compared to urban areas. Relation between MODIS AOD and surface $PM_{2.5}$ in New Delhi was reported by Kumar et al. (2007). The empirical estimation of near-surface aerosol masses using satellite-based AOD appears to be helpful. However, it is quite complicated over the IGP with varying sources and meteorological conditions. Exemplifying range of association ($r = 0.46$–0.54) between AOD and near-surface particulates was reported by Kumar et al. (2015b) during winter.

Satellite-retrieved information has opened up a new way of characterizing aerosol subtypes with distinct spatial and vertical resolution. Several studies have been reported over the IGP using remote-sensing techniques to identify various aerosol types. Using CALIPSO-based measurements, Kumar et al. (2015a,b) identified a variety of aerosol layers at different altitudes over the middle IGP. A sharp deviation and episodic nature were reported in the aerosol subtypes that clearly indicate the dominance of different sources. The application of prevailing air masses for the classification of aerosol types was made by Tiwari et al. (2015) at Varanasi on the basis of cluster analysis of air mass. The contribution of polluted dust, polluted continental, and black and organic carbon-enriched aerosols were mainly observed at the central part of the IGP, which highlights the extent of human-induced pollution over the region. Aerosol extinction properties were used by Verma et al. (2016) to categorize the possible aerosol subtypes at the lower Gangetic basin. They reported the influence of dust and polluted dust during daytime whereas polluted continental and smoke during nighttime. Srivastava et al. (2014) reported the presence of dust over the national capital of Delhi throughout the year, with an aerosol layer mostly confined within 2 km during winter and postmonsoon before elevating to 6 km during premonsoon and monsoon seasons.

Some extreme events like dust storms, haze, and fog over the IGP are tracked using remote-sensing techniques. Northern India usually experiences premonsoon dust storms that significantly modify the regional aerosol characteristics for a shorter period of time. Kumar et al. (2014), using WRF-Chem simulations based on MODIS and AERONET retrievals, observed an increase (>50%) in local-to-regional AOD and a decrease (>70%) in the AE during a premonsoon dust storm. The study on aerosol radiative impacts during the dust storm showed a cooling effect at the surface and top of the atmosphere, while warming the atmosphere. Kumar et al. (2016) used MODIS-AOD and CALIPSO aerosol vertical profiles to identify fireworks-induced particulate pollution over the entire Indian region. These case studies referring to the application of satellite remote sensing for studying aerosol characteristics, distribution, transport, and radiative impacts over the IGP provide extremely useful information on aerosol-induced climate and air quality changes over the region.

7 CONCLUSIONS AND FUTURE PROSPECTS

Retrieval of aerosol optical properties from satellite-based remote sensing has been used most efficiently to complement ground-based measurements and to link aerosol properties with climate change. Although retrieval of aerosol microphysical properties from satellite-based sensors is relatively new, it poses the potential to provide a generic description of existing atmospheric profiles. Application of remote-sensing techniques has emerged with several advantages like large spatial coverage, real-time data, reliability, the use of multiple platforms, and the identification of aerosol properties with varying vertical resolutions. With multiple validation and an intercomparison network, remote-sensing information has emerged with much more reliability and, thereby, provides a scope of extensive applications in multidisciplinary fields. There are numerous examples of the successful application of satellite-retrieved aerosol properties in identifying transboundary movement, aerosol source and receptors, variation in particle morphology, physiochemical properties, and computing radiative impacts both in the TOA and at surface.

Further, numerous satellites have been launched with a diverse field of applications under a common platform with near-simultaneous observations to provide a holistic view of the entire atmosphere. Likewise, NASA's international Afternoon Constellation (A-Train), including OCO-2, GCOM-W1, Aqua, CALIPSO, CloudSat, and Aura, provide near-simultaneous observation of multiple parameters related to cloud, aerosol, trace gases, and other climate variables. Additionally, some application-based satellites are also in line to be launched. For example, ESA's EarthCARE satellite is scheduled to be launched in 2018 and is expected to advance our understanding on aerosol-cloud interaction using high-performance LIDAR and radar technology. The Earth Explorer Atmospheric Dynamics Mission (ADM-Aeolus) will provide global observations of wind to improve weather forecasts. The Sentinel series is focusing on retrieving trace gas and aerosol properties with improved near-real time monitoring. There are few overambitious projects scheduled to be launched under Earth observing system. Most notably, the Stratospheric Aerosol and Gas Experiment (SAGE III) on the International Space Station; the Geostationary Coastal and Air Pollution Events (GEO-CAPE) for identifying natural and anthropogenic sources of aerosols; the Aerosol-Cloud-Ecosystems (ACE) for measuring aerosol and cloud types and properties; the Global Atmosphere Composition Mission (GACM) for ozone and related gases for intercontinental air quality; the Pre-Aerosol, Clouds, and Ocean Ecosystem (PACE) for extended data records on clouds and aerosols (http://eospso.nasa.gov/future-missions): and the Tropospheric Emissions: Monitoring of Pollution (TEMPO) for ozone precursors, aerosols, and clouds. The TEMPO (for North America) will further be a part of the global air quality monitoring constellation, including GEMS (Geostationary Environment Monitoring Spectrometer) for Asia-Pacific and Sentinel-4 for Europe. The NEMO (Next-Generation Earth Monitoring and Observation) bus will be the next evolution to the Generic Nanosatellite Bus (GNB) technology jointly developed by the Indian Space Research organization (ISRO) and the Space Flight laboratory at the University of Toronto to provide information on aerosol and earth systems. Another flagship project called the Multi-Angle Imager for Aerosols (MAIA) is scheduled to be launched by NASA in January 2020. MAIA is a multiview satellite sensor, currently under development by NASA's Earth Science Division. It contains two push-broom spectropolarimetric cameras capable of measuring radiance in 12 spectral bands. This will help to provide information on aerosol loading, size, and compositions at much finer scale and thereby has potential for broader applications like assessing the impact of aerosols on human health, agricultural productivity, and hydrological cycles.

Therefore, it may well be projected that the use of satellite-retrieved aerosol properties for applications in climate and ambient air quality will be further developed in near future. This will further improve our knowledge on earth, the atmosphere and climate. Observations of atmospheric aerosols and trace gases from satellite sensors will be combined with health information to determine the toxicity of airborne particulates and the subsequent development of a sustainable habitat. Consequently, the availability of near real-time remote sensing data will precisely help in air quality forecasting and developing an effective health-advisory service.

Acknowledgments

The authors are grateful to the agencies that provided the satellite data on the earth system's science. Authors wish to acknowledge Terra/Aqua-MODIS and VIIRS data from NASA's LAADS Web, Terra-MISR from ASDC/NASA, CALIPSO-CALIOP from Atmospheric Science Data Center at NASA Langley Research Center, Aura-OMI data from GES DISC, and OCM-Oceansat-2 from ISRO. Additionally, the authors also wish to thank all the scientists who have devoted their life to developing remote-sensing science.

References

Ackerman, S.A., Strabala, K.I., Menzel, W.P., Frey, R.A., Moeller, C.C., Gumley, L.E., 1998. Discriminating clear sky from clouds with MODIS taken from the MAS during the SUCCES experiment. J. Geophys. Res 103 (D24), 32141–32157.

Ahn, C., Torres, O., Bhartia, P.K., 2008. Comparison of ozone monitoring instrument UV aerosol products with Aqua/Moderate Resolution Imaging Spectroradiometer and Multiangle Imaging Spectroradiometer observations in 2006. J. Geophys. Res. 113 (D16), 1–13.

Ahn, C., Torres, O., Jethva, H., 2014. Assessment of OMI near-UV aerosol optical depth over land. J. Geophys. Res. Atmos. 119 (5), 2457–2473.

Albrecht, B., 1989. Aerosols, cloud microphysics, and fractional cloudiness. Science 245, 1227–1230.

Angstrom, A., 1964. The parameters of atmospheric turbidity. Tellus 16, 64–75.

Badarinath, K.V.S., Kharol, S.K., Chand, T.K., Latha, K.M., 2011. Characterization of aerosol optical depth, aerosol mass concentration, UV irradiance and black carbon aerosols over Indo-Gangetic plains, India, during fog period. Meteorol. Atmos. Phys. 111 (1–2), 65–73.

Banerjee, T., Srivastava, R.K., 2012. Plastics waste management and resource recovery in India. Int. J. Environ. Waste Manag. 10 (1), 90–111.

Banerjee, T., Barman, S.C., Srivastava, R.K., 2011. Application of air pollution dispersion modeling for source-contribution assessment and model performance evaluation at integrated industrial estate-Pantnagar. Environ. Pollut. 159, 865–875.

Banerjee, T., Murari, V., Kumar, M., Raju, M.P., 2015. Source apportionment of airborne particulates through receptor modeling: Indian scenario. Atmos. Res. 164–165, 167–187.

Banerjee, T., Kumar, M., Singh, N., 2017a. Aerosol, climate and sustainability. In: Reference Module in Earth Systems and Environmental Sciences. Encyclopaedia of Anthropocene, Elsevier. https://doi.org/10.1016/B978-0-12-409548-9.09914-0.

Banerjee, T., Kumar, M., Mall, R.K., Singh, R.S., 2017b. Airing 'clean air' in Clean India Mission. Environ. Sci. Pollut. Res. 24 (7), 6399–6413.

Bevan, S.L., North, P.R.J., Grey, W.M.F., Los, S.O., Plummer, S.E., 2009. Impact of atmospheric aerosol from biomass burning on Amazon dry-season drought. J. Geophys. Res. Atmos. 114 (D9).

Bevan, S.L., North, P.R.J., Los, S.O., Grey, W.M.F., 2012. A global dataset of atmospheric aerosol optical depth and surface reflectance from AATSR. Remote Sens. Environ. 116, 199–210.

Bibi, H., Alam, K., Chishtie, F., Bibi, S., Shahid, I., Blaschke, T., 2015. Intercomparison of MODIS, MISR, OMI, and CALIPSO aerosol optical depth retrievals for four locations on the Indo-Gangetic plains and validation against AERONET data. Atmos. Environ. 111, 113–126.

Bridhikitti, A., 2013. Atmospheric aerosol layers over Bangkok Metropolitan Region from CALIPSO observations. Atmos. Res. 127, 1–7.

Che, H.Z., Wang, Y.Q., Sun, J.Y., Zhang, X.C., Zhang, X.Y., Guo, J.P., 2013. Variation of aerosol optical properties over the taklimakan desert in China. Aerosol Air Qual. Res. 13, 777–785.

Chin, M., Ginoux, P., Kinne, S., Torres, O., Holben, B.N., Duncan, B.N., Martin, R.V., Logan, J.A., Higurashi, A., Nakajima, T., 2002. Tropospheric aerosol optical thickness from the GOCART model and comparisons with satellite and sun photometer measurements. J. Atmos. Sci. 59, 461–483.

Chitranshi, S., Sharma, S.P., Dey, S., 2015. Satellite-based estimates of outdoor particulate pollution (PM10) for Agra City in northern India. Air Qual. Atmos. Health 8 (1), 55–65.

Corbett, J.J., Winebrake, J.J., Green, E.H., Kasibhatla, P., Eyring, V., Lauer, A., 2007. Mortality from ship emissions: a global assessment. Environ. Sci. Technol. 41 (24), 8512–8518.

Curier, R.L., Veefkind, J.P., Braak, R., Veihelmann, B., Torres, O., 2008. Retrieval of aerosol optical properties from OMI radiances using a multiwavelength algorithm: application to Western Europe. J. Geophys. Res. Atmos. 113, 1–16.

Das, S., Dey, S., Dash, S., 2014. Impacts of aerosols on dynamics of Indian summer monsoon using a regional climate model. Clim. Dynam. 44, 1685–1697.

De Leeuw, G., Schoemaker, R., Curier, L., Bennouna, Y., Timmermans, R., Schaap, M., Koelemeijer, R., 2007. AATSR derived aerosol properties over land. In: ENVISAT Symposium, ESA SP-636, July.

Deschamps, P.Y., Buriez, J.C., Bréon, F.M., Leroy, M., Podaire, A., Bricaud, A., Sèze, G., 1994. The POLDER mission: instrument characteristics and scientific objectives. IEEE Trans. Geosci. Remote Sens. 32 (3), 598–615.

Deuzé, J.L., Herman, M., Goloub, P., Tanré, D., Marchand, A., 1999. Characterization of aerosols over ocean from POLDER/ADEOS-1. Geophys. Res. Lett. 26 (10), 1421–1424.

Dey, S., Tripathi, S.N., 2007. Estimation of aerosol optical properties and radiative effects in the Ganga basin, northern India, during the wintertime. J. Geophys. Res. Atmos 112 (D3), 1–16.

Dey, S., Di Girolamo, L., van Donkelaar, A., Tripathi, S.N., Gupta, T., Mohan, M., 2012. Variability of outdoor fine particulate ($PM_{2.5}$) concentration in the Indian subcontinent: a remote sensing approach. Remote Sens. Environ. 127, 153–161.

Dubovik, O., Holben, B., Eck, T.F., Smirnov, A., Kaufman, Y.J., King, M.D., Tanré, D., Slutsker, I., 2002. Variability of absorption and optical properties of key aerosol types observed in worldwide locations. J. Atmos. Sci. 59 (3), 590–608.

Dubovik, O., Lapyonok, T., Kaufman, Y., Chin, M., Ginoux, P., et al., 2007. Retrieving global sources of aerosols from MODIS observations by inverting GOCART model. Atmos. Chem. Phys. Discuss. 7 (2), 3629–3718.

Evans, J., van Donkelaar, A., Martin, R.V., Burnett, R., Rainham, D.G., Birkett, N.J., Krewski, D., 2013. Estimates of global mortality attributable to particulate air pollution using satellite imagery. Environ. Res. 120, 33–42.

Fang, H., Liang, S., Hoogenboom, G., 2011. Integration of MODIS LAI and vegetation index products with the CSM–CERES–Maize model for corn yield estimation. Int. J. Remote Sens. 32 (4), 1039–1065.

Fraser, R.S., Kaufman, Y.J., Mahoney, R.L., 1984. Satellite measurements of aerosol mass and transport. Atmos. Environ. 18, 2577–2584.

Ganguly, D., Ginoux, P., Ramaswamy, V., Winker, D.M., Holben, B.N., Tripathi, S.N., 2009. Retrieving the composition and concentration of aerosols over the Indo-Gangetic basin using CALIOP and AERONET data. Geophys. Res. Lett. 36 (13), 1–5.

Gladkova, I., Kihai, Y., Ignatov, A., Shahriar, F., Petrenko, B., 2015. SST Pattern Test in ACSPO clear-sky mask for VIIRS. Remote Sens. Environ. 160, 87–98.

Griggs, M., 1975. Measurements of atmospheric aerosol optical thickness over water using ERTS-1 data. J. Air Pollut. Control. Assoc. 25, 622–626.

Guenther, B., Xiong, X., Salomonson, V.V., Barnes, W.L., Young, J., 2002. On-orbit performance of the earth observing system moderate resolution imaging spectroradiometer; first year of data. Remote Sens. Environ. 83 (1–2), 16–30.

Guillevic, P.C., Biard, J.C., Hulley, G.C., Privette, J.L., Hook, S.J., Olioso, A., Csiszar, I., 2014. Validation of land surface temperature products derived from the visible infrared imaging radiometer suite (VIIRS) using ground-based and heritage satellite measurements. Remote Sens. Environ. 154, 19–37.

Gupta, P., Christopher, S.A., Box, M.A., Box, G.P., 2007. Multiyear satellite remote sensing of particulate matter air quality over Sydney, Australia. Int. J. Remote Sens. 28 (20), 4483–4498.

Gupta, P., Khan, M.N., da Silva, A., Patadia, F., 2013. MODIS aerosol optical depth observations over urban areas in Pakistan: quantity and quality of the data for air quality monitoring. Atmos. Pollut. Res. 4 (1), 43–52.

Hansen, J., Sato, M., Ruedy, R., 1997. Radiative forcing and climate response. J. Geophys. Res. 102, 6831–6864.

Herman, M., Deuzé, J.L., Marchand, A., Roger, B., Lallart, P., 2005. Aerosol remote sensing from POLDER/ADEOS over the ocean: Improved retrieval using a nonspherical particle model. J. Geophys. Res. D: Atmos. 110 (10), 1–11.

Holben, B.N., Eck, T.F., Slutsker, I., Tanré, D., Buis, J.P., Setzer, A., Vermote, E., 1998. AERONET—a federated instrument network and data archive for aerosol characterization. Remote Sens. Environ. 66 (1), 1–16.

Hsu, N.C., Tsay, S.-C., King, M.D., Herman, J.R., 2004. Aerosol properties over bright-reflecting source regions. IEEE Trans. Geosci. Remote Sens. 42 (3), 557–569.

Hsu, N.C., Tsay, S., King, M.D., Member, S., Herman, J.R., During, A., 2006. Deep blue retrievals of Asian aerosol properties during ACE-Asia. IEEE Trans. Geosci. Remote Sens. 44 (11), 3180–3195.

Hsu, N.C., Jeong, M.J., Bettenhausen, C., Sayer, A.M., Hansell, R., Seftor, C.S., Tsay, S.C., 2013. Enhanced deep blue aerosol retrieval algorithm: the second generation. J. Geophys. Res. Atmos. 118 (16), 9296–9315.

Hu, Z., 2009. Spatial analysis of MODIS aerosol optical depth, $PM_{2.5}$ and chronic coronary heart disease. Int. J. Health Geogr. 12 (8), 27.

Huang, X., Yang, P., Kattawar, G., Liou, K.-N., 2015. Effect of mineral dust aerosol aspect ratio on polarized reflectance. J. Quant. Spectrosc. Radiat. Transf. 151, 97–109.

Huang, J., Kondragunta, S., Laszlo, I., Liu, H., Remer, L.A., Zhang, H., Petrenko, M., 2016. Validation and expected error estimation of Suomi-NPP VIIRS aerosol optical thickness and Ångström exponent with AERONET. J. Geophys. Res. Atmos. 121 (12), 7139–7160.

Jackson, J.M., Liu, H., Laszlo, I., Kondragunta, S., Remer, L.A., Huang, J., Huang, H.C., 2013. Suomi-NPP VIIRS aerosol algorithms and data products. J. Geophys. Res. Atmos. 118 (22), 12673–12689.

Jiménez, E., Linares, C., Martínez, D., Díaz, J., 2010. Role of Saharan dust in the relationshipbetween particulate matter and short-term daily mortality among the elderly in Madrid (Spain). Sci. Total Environ. 408 (23), 5729–5736.

Kahn, R.A., Gaitley, B.J., 2015. An analysis of global aerosol type as retrieved by MISR. J. Geophys. Res. Atmos. 120 (9), 4248–4281.

Kahn, R., West, R., McDonald, D., Rheingans, B., Mishchenko, M., 1997. Sensitivity of multi-angle remote sensing observations to aerosol sphericity. J. Geophys. Res. 102, 16861–16870.

Kahn, R., Banerjee, P., McDonald, D., Diner, D.J., 1998. Sensitivity of multiangle imaging to aerosol optical depth and to pure-particle size distribution and composition over ocean. J. Geophys. Res. Atmos. 103 (D24), 32195–32213.

Kahn, R.A., Banerjee, P., McDonald, D., 2001. Sensitivity of multi-angle imaging to natural mixtures of aerosols over ocean. J. Geophys. Res. 106, 18219–18238.

Kahn, R.A., Gaitley, B.J., Garay, M.J., Diner, D.J., Eck, T.F., Smirnov, A., Holben, B.N., 2010. Multiangle imaging spectroradiometer global aerosol product assessment by comparison with the Aerosol Robotic Network. J. Geophys. Res. Atmos. 115 (D23), 1–28.

Kaskaoutis, D.G., Kumar, S., Sharma, D., Singh, R.P., Kharol, S.K., Sharma, M., Singh, A.K., Singh, S., Singh, A., Singh, D., 2014. Effects of crop residue burning on aerosol properties, plume characteristics, and long-range transport over northern India. J. Geophys. Res. Atmos. 119, 5424–5444.

Kaufman, Y.J., Tanré, D., Remer, L.A., Vermote, E.F., Chu, A., Holben, B.N., 1997. Operational remote sensing of tropospheric aerosol over land from EOS moderate resolution imaging spectroradiometer. J. Geophys. Res. 102 (D14), 17051–17067.

Kittaka, C., Winker, D.M., Vaughan, M.A., Omar, A., Remer, L.A., 2011. Intercomparison of column aerosol optical depths from CALIPSO and MODIS-Aqua. Atmos. Meas. Tech. 4 (2), 131–141.

Kokhanovsky, A., de Leeuw, G. (Eds.), 2009. Satellite Aerosol Remote Sensing Over Land. Springer, p. 388. ISBN 978-3-540-69396-3.

Kotchenova, S.Y., Vermote, E.F., 2007. Validation of a vector version of the 6S radiative transfer code for atmospheric correction of satellite data. Part II. Homogeneous Lambertian and anisotropic surfaces. Appl. Optics 46 (20), 4455–4464.

Krüger, O., Graßl, H., 2004. Albedo reduction by absorbing aerosols over China. Geophys. Res. Lett. 31 (2). L02108.

Kumar, N., Chu, A., Foster, A., 2007. An empirical relationship between $PM_{2.5}$ and aerosol optical depth in Delhi Metropolitan. Atmos. Environ. 41 (21), 4492–4503.

Kumar, R., Barth, M.C., Pfister, G.G., Naja, M., Brasseur, G.P., 2014. WRF-Chem simulations of a typical pre-monsoon dust storm in northern India: influences on aerosol optical properties and radiation budget. Atmos. Chem. Phys. 14, 2431–2446.

Kumar, M., Singh, R.S., Banerjee, T., 2015a. Associating airborne particulates and human health: exploring possibilities. Environ. Int. 84, 201–202.

Kumar, M., Tiwari, S., Murari, V., Singh, A.K., Banerjee, T., 2015b. Wintertime characteristics of aerosols at middle Indo-Gangetic plain: impacts of regional meteorology and long range transport. Atmos. Environ. 104, 162–175.

Kumar, M., Singh, R.K., Murari, V., Singh, A.K., Singh, R.S., Banerjee, T., 2016. Fireworks induced particle pollution: a spatio-temporal analysis. Atmos. Res. 180, 78–91.

Kumar, M., Raju, M.P., Singh, R.K., Singh, A.K., Singh, R.S., Banerjee, T., 2017a. Wintertime characteristics of aerosols over middle Indo-Gangetic Plain: Vertical profile, transport and radiative forcing. Atmos. Res. 183, 268–282.

Kumar, M., Raju, M.P., Singh, R.S., Banerjee, T., 2017b. Impact of drought and normal monsoon scenarios on aerosol induced radiative forcing and atmospheric heating rate in Varanasi over middle Indo-Gangetic Plain. J. Aerosol Sci. 113, 95–107.

Lee, K.H., Li, Z., Kim, Y.J., Kokhanovsky, A., 2009. Atmospheric aerosol monitoring from satellite observations : a history of three decades. Atmos. Biol. Environ. Monit. 13–38.

Levy, R.C., Remer, L.A., Dubovik, O., 2007. Global aerosol optical properties and application to moderate resolution imaging spectroradiometer aerosol retrieval over land. J. Geophys. Res. 112 (D13), 13210.

Levy, R.C., Mattoo, S., Munchak, L.A., Remer, L.A., Sayer, A.M., Patadia, F., Hsu, N.C., 2013. The Collection 6 MODIS aerosol products over land and ocean. Atmos. Meas. Tech. 6 (11), 2989–3034.

Liu, H., Remer, L.A., Huang, J., Huang, H.-C., Kondragunta, S., Laszlo, I., Jackson, J.M., 2014. Preliminary evaluation of S-NPP VIIRS aerosol optical thickness. J. Geophys. Res. Atmos. 119 (7), 3942–3962.

Lyons, W.A., Husar, R.B., 1976. SMS/GOES visible images detect a synoptic-scale air pollution episode. Mon. Weather Rev. 104, 1623–1626.

Mamun, M.I., Islam, M., Mondol, P.K., 2014. The seasonal variability of aerosol optical depth over Bangladesh based on satellite data and HYSPLIT model. Am. J. Remote Sens. 2 (4), 20–29.

Martinez Lozano, J.A., Utrillas, M.P., Tena, F., Cachorro, V.E., 1998. The parameterization of the atmospheric aerosol optical depth using the Angstrom power law. Sol. Energy 63, 303–311.

Mazzoni, D., Logan, J.A., Diner, D., Kahn, R., Tong, L., Li, Q., 2007. A data-mining approach to associating MISR smoke plume heights with MODIS fire measurements. Remote Sens. Environ. 107, 138–148.

Mehta, M., 2015. A study of aerosol optical depth variations over the Indian region using thirteen years (2001–2013) of MODIS and MISR Level 3 data. Atmos. Environ. 109, 161–170.

Mehta, M., Singh, R., Singh, A., Singh, N., Anshumali, 2016. Recent global aerosol optical depth variations and trends—a comparative study using MODIS and MISR level 3 datasets. Remote Sens. Environ. 181, 137–150.

Meng, F., Cao, C., Shao, X., 2015. Spatio-temporal variability of Suomi-NPP VIIRS-derived aerosol optical thickness over China in 2013. Remote Sens. Environ. 163, 61–69.

Mishra, A.K., 2014. Retrieval of EVI from Oceansat 2 data and comparison with MODIS derived EVI. J. Indian Soc. Remote Sens. 42 (4), 877–883.

Mishra, A.K., Dadhwal, V.K., 2008. Comparison of enhanced vegetation index from IRS-P4 (OCM) and MODIS on Aqua. Int. J. Geoinform. 4 (4), 57–66.

Mishra, A.K., Dadhwal, V.K., Dutt, C.B.S., 2008. Analysis of marine aerosol optical depth retrieved from IRS-P4 OCM sensor and comparison with the aerosol derived from SeaWiFS and MODIS sensor. J. Earth Syst. Sci. 117 (1), 361–373.

Moffet, R.C., Kimberly, P., 2009. In-Situ Measurements of the mixing state an optical properties of soot with implications for radiative forcing estimates. Proc. Natl. Acad. Sci. 106 (29), 11872–11877.

Moreno, T., Kojima, T., Querol, X., Alastuey, A., Amato, F., Gibbons, W., 2012. Natural versus anthropogenic inhalable aerosol chemistry of trans-boundary East Asian atmospheric outflows into western Japan. Sci. Total Environ. 424, 182–192.

Murari, V., Kumar, M., Barman, S.C., Banerjee, T., 2015. Temporal variability of MODIS aerosol optical depth and chemical characterization of airborne particulates in Varanasi, India. Environ. Sci. Pollut. Res. 22, 1329–1343.

Murari, V., Kumar, M., Singh, N., Singh, R.S., Banerjee, T., 2016. Particulate morphology and elemental characteristics: variability at middle Indo-Gangetic plain. J. Atmos. Chem. 73 (2), 165–179.

Murari, V., Kumar, M., Mhawish, A., Barman, S.C., Banerjee, T., 2017. Airborne particulate in Varanasi over middle Indo-Gangetic Plain: variation in particulate types and meteorological influences. Environ. Monit. Assess. https://doi.org/10.1007/s10661-017-5859-9.

Nagamani, P.V., Chauhan, P., Dwivedi, R.M., 2008. Development of chlorophyll-a algorithm for ocean colour monitor onboard OCEANSAT-2 satellite. IEEE Geosci. Remote Sens. Lett. 5 (3), 527–531.

Nayak, S.R., Sarangi, R.K., Rajawat, A.S., 2001. Application of IRS-P4 OCM data to study the impact of cyclone on coastal environment of Orissa. Curr. Sci. 80 (9), 1208–1213.

North, P.R.J., Briggs, S.A., Plummer, S.E., Settle, J.J., 1999. Retrieval of land surface bidirectional reflectance and aerosol opacity from ATSR-2 multiangle imagery. IEEE Trans. Geosci. Remote Sens. 37 (1), 526–537.

Oliva, P., Schroeder, W., 2015. Assessment of VIIRS 375m active fire detection product for direct burned area mapping. Remote Sens. Environ. 160, 144–155.

Omar, A.H., Won, J.G., Winker, D.M., Yoon, S.C., Dubovik, O., McCormick, M.P., 2005. Development of global aerosol models using cluster analysis of Aerosol Robotic Network (AERONET) measurements. J. Geophys. Res. D: Atmos. 110 (10), 1–14.

Ouillon, S., Lucas, Y., Gaggelli, J., 2002. Hyperspectral detection of sand. In: Proc. 7th Int. conf. Remote Sensing for Marine and Coastal Environments, pp. 681–682.

Prasad, A.K., Singh, R.P., 2007. Changes in aerosol parameters during major dust storm events (2001–2005) over the Indo-Gangetic Plains using AERONET and MODIS data. J. Geophys. Res 112 (D9), 1–18.

Qi, Y., Ge, J., Huang, J., 2013. Spatial and temporal distribution of MODIS and MISR aerosol optical depth over northern China and comparison with AERONET. Chin. Sci. Bull. 58 (20), 2497–2506.

Ramachandran, S., Kedia, S., Srivastava, R., 2012. Aerosol optical depth trends over different regions of India. Atmos. Environ. 49, 338–347.

Raman, R.S., Ramachandran, S., Kedia, S., 2011. A methodology to estimate source-specific aerosol radiative forcing. J. Aerosol Sci. 42 (5), 305–320.

Ramana, M.V., Ramanathan, V., Feng, Y., Yoon, S.-C., Kim, S.-W., Carmichael, G.R., Schauer, J.J., 2010. Warming influenced by the ratio of black carbon to sulphate and the black-carbon source. Nat. Geosci. 3, 542–545.

Ramanathan, V., Ramana, M.V., 2005. Persistent, widespread and strongly absorbing haze over the Himalayan foot hills and the Indo-Gangetic plains,. Pure Appl. Geophys. 162, 1609–1626.

Ramaswamy, V., et al., 2001. Radiative forcing of climate change. Climate Change 2001: The Scientific Basis, Contribution of Working Group I to The Third Assessment Report of The Intergovernmental Panel on Climate Change. Cambridge University Press, New York, NY. pp. 349–416.

Rastogi, N., Singh, A., Sarin, M.M., Singh, D., 2016. Temporal variability of primary and secondary aerosols over northern India: impact of biomass burning emissions. Atmos. Environ. 125, 396–403.

Remer, L.A., Kleidman, R.G., Levy, R.C., Kaufman, Y.J., Tanré, D., Mattoo, S., Martins, J.V., Ichoku, C., Koren, I., Yu, H., Holben, B.N., 2008. Global aerosol climatology from the MODIS satellite sensors. J. Geophys. Res. Atmos. 113 (D14), 1–18.

Remer, L.A., Mattoo, S., Levy, R.C., Munchak, L.A., 2013. MODIS 3 km aerosol product: algorithm and global perspective. Atmos. Meas. Tech. 6 (7), 1829–1844.

Robles Gonzales, C., Veefkind, J.P., De Leeuw, G., 2000. Aerosol optical depth over Europe in August 1997 derived from ATSR-2 data. Geophys. Res. Lett. 27 (7), 955–958.

Robles-Gonzalez, C., De Leeuw, G., Decae, R., Kusmierczyk-Michulec, J., Stammes, P., 2006. Aerosol properties over the Indian Ocean Experiment (INDOEX) campaign area retrieved from ATSR-2. J. Geophys. Res. Atmos. 111 (D15), 1–10.

Satheesh, S.K., Torres, O., Remer, L.A., Babu, S.S., Vinoj, V., Eck, T.F., Holben, B.N., 2009. Improved assessment of aerosol absorption using OMI-MODIS joint retrieval. J. Geophys. Res. Atmos. 114, 1–10.

Sayer, A.M., Hsu, N.C., Bettenhausen, C., Jeong, M.J., 2013. Validation and uncertainty estimates for MODIS Collection 6 "Deep Blue" aerosol data. J. Geophys. Res. Atmos. 118 (14), 7864–7872.

Sayer, A.M., Munchak, L.A., Hsu, N.C., Levy, R.C., Bettenhausen, C., Jeong, M.-J., 2014. MODIS Collection 6 aerosol products: comparison between Aqua's e-Deep Blue, Dark Target, and "merged" data sets, and usage recommendations. J. Geophys. Res. Atmos. 119, 13965–13989.

Schwartz, S.E., Andreae, M.O., 1996. Uncertainty in climate change caused by aerosols. Science 272, 1121–1122.

Sen, A., Ahammed, Y.N., Arya, B.C., Banerjee, T., et al., 2014. Atmospheric fine and coarse mode aerosols at different environments of India and the Bay of Bengal during winter-2014: implications of a coordinated campaign. MAPAN J. Metrol. Soc. India 29 (4), 273–284.

Sen, A., Ahammed, Y.N., Banerjee, T., Chatterjee, A., Choudhuri, A.K., Das, T., Deb, N.C., Dhir, A., Goel, S., Khan, A.H., Mandal, T.K., 2016. Spatial variability in ambient atmospheric fine and coarse mode 20 aerosols over Indo-Gangetic plains, India and adjoining oceans during the onset of summer monsoons, 2014. Atmos. Pollut. Res. 7 (3), 521–532.

Sen, A., Abdelmaksoud, A.S., Ahammed, Y.N., Banerjee, T., Bhat, M.A., Chatterjee, A., Choudhuri, A.K., Das, T., Dhir, A., Dhyani, P.P., Gadi, R., 2017. Variations in particulate matter over Indo-Gangetic Plains and Indo-Himalayan Range during four field campaigns in winter monsoon and summer monsoon: role of pollution pathways. Atmos. Environ. 154, 200–224.

Sharma, A.R., Kharol, S.K., Badarinath, K.V.S., Singh, D., 2010. Impact of agriculture crop residue burning on atmospheric aerosol loading—a study over Punjab State, India. Ann. Geophys. Atmos. Hydrospheres Space Sci. 28 (2), 367–379.

Sharma, M., Kaskaoutis, D.G., Singh, R.P., Singh, S., 2014. Seasonal variability of atmospheric aerosol parameters over Greater Noida using ground sunphotometer observations. Aerosol Air Qual. Res. 14 (3), 608–622.

Shukla, K., Srivastava, P.K., Banerjee, T., Aneja, V.P., 2017. Variation of ground-level and columnar ozone at middle Indo-Gangetic Plain: impacts of seasonality and precursor gases. Environ. Sci. Pollut. Res. 24 (1), 164–179.

Singh, N., Murari, V., Kumar, M., Barman, S.C., Banerjee, T., 2017a. Fine particulates over South Asia: review and meta-analysis of PM 2.5 source apportionment through receptor model. Environ. Pollut. 223, 121–136.

Singh, N., Mhawish, A., Deboudt, K., Singh, R.S., Banerjee, T., 2017b. Organic aerosols over Indo-Gangetic

Plain: sources, distributions and climatic implications. Atmos. Environ. 157, 59–74.

Srivastava, P., Dey, S., Agarwal, P., Basil, G., 2014. Aerosol characteristics over Delhi national capital region: a satellite view. Int. J. Remote Sens. 35 (13), 5036–5052.

Tanré, D., Kaufman, Y.J., Herman, M., Mattoo, S., 1997. Remote sensing of aerosol properties over oceans using the MODIS/EOS spectral radiances. J. Geophys. Res. Atmos. 102 (D14), 16971–16988.

Tanré, D., Bréon, F.M., Deuzé, J.L., Dubovik, O., Ducos, F., François, P., Waquet, F., 2011. Remote sensing of aerosols by using polarized, directional and spectral measurements within the A-Train: the PARASOL mission. Atmos. Meas. Tech. 4 (7), 1383–1395.

Tiwari, S., Singh, A.K., 2013. Variability of aerosol parameters derived from ground and satellite measurements over Varanasi located in the Indo-Gangetic Basin. Aerosol Air Qual. Res. 13, 627–638.

Tiwari, S., Srivastava, A.K., Singh, A.K., Singh, S., 2015. Identification of aerosol types over Indo-Gangetic Basin: implications to optical properties and associated radiative forcing. Environ. Sci. Pollut. Res. 22 (16), 12246–12260.

Todd, W.J., George, A.J., Bryant, N.A., 1979. Satellite-aided evaluation of population exposure to air pollution. Environ. Sci. Technol. 13, 970–974.

Torres, O., Tanskanen, A., Veihelmann, B., Ahn, C., Braak, R., Bhartia, P.K., Levelt, P., 2007. Aerosols and surface UV products from ozone monitoring instrument observations: an overview. J. Geophys. Res. Atmos. 112 (D24), 1–14.

Tripathi, S.N., Pattnaik, A., Dey, S., 2007. Aerosol indirect effect over Indo-Gangetic plain. Atmos. Environ. 41 (33), 7037–7047.

Trivitayanurak, W., Palmer, P.I., Barkley, M.P., Robinson, N.H., Coe, H., Oram, D.E., 2012. The composition and variability of atmospheric aerosol over Southeast Asia during 2008. Atmos. Chem. Phys. 12, 1083–1100.

Twomey, S.A., 1959. The nuclei of natural cloud formation. Part II: the supersaturation in natural clouds and the variation of cloud droplet concentrations. Geofisica Pura e Applicata 43, 227–242.

Utrillas, M.P., Pedro's, R., Martinez-Lozano, J.A., Tena, F., 2000. A new method for determining the Angstrom turbidity coefficient from broadband filter measurement. Am. Meteorol. Soc. 39 (6), 863–874.

Vadrevu, K.P., Ellicott, E., Badarinath, K.V.S., Vermote, E., 2011. MODIS derived fire characteristics and aerosol optical depth variations during the agricultural residue burning season, north India. Environ. Pollut. 159 (6), 1560–1569.

van Donkelaar, A., Martin, R.V., Leaitch, W.R., Macdonald, A.M., Walker, T.W., Streets, D.G., Zhang, Q., Dunlea, E.J., Jimenez, J.L., Dibb, J.E., Huey, L.G., Weber, R., Andreae, M.O., 2008. Analysis of aircraft and satellite measurements from the Intercontinental Chemical Transport Experiment (INTEX-B) to quantify long-range transport of East Asian sulfur to Canada. Atmos. Chem. Phys. 8, 2999–3014.

van Donkelaar, A., Martin, R.V., Brauer, M., Boys, B.L., 2015. Use of satellite observations for long term exposure assessment of global concentrations of fine particulate matter. Environ. Health Perspect. 123, 135–143.

Veefkind, P., De Leeuw, J., 1998. A new algorithm to determine the spectral aerosol optical depth from satellite radiometer measurements. J. Aerosol Sci. 29 (10), 1237–1248.

Veefkind, J.P., De Leeuw, G., Durkee, P.A., 1998. Retrieval of Aerosol Optical Depth over Land using two-angle view Satellite Radiometry during TARFOX. Geophys. Res. Lett. 25 (16), 3135–3138.

Verma, S., Priyadharshini, B., Pani, S., Kumar, D.B., Faruki, A.R., Bhanja, S.N., Mandal, M., 2016. Aerosol extinction properties over coastal West Bengal Gangetic plain under inter-seasonal and sea breeze influenced transport processes. Atmos. Res. 167, 224–267.

von Hoyningen-Huene, W., Freitag, M., Burrows, J.B., 2003. Retrieval of aerosol optical thickness over land surfaces from top-of-atmosphere radiance. J. Geophys. Res. Atmos. 108 (D9), 4260.

Wang, J., Christopher, S.A., 2003. Intercomparison between satellites derived aerosol optical thickness and PM 2.5 mass: implications for air quality studies. Geophys. Res. Lett. 30, 2095.

Wang, M., Son, S., 2016. VIIRS-derived chlorophyll-a using the ocean color index method. Remote Sens. Environ. 182, 141–149.

Wang, J., Kessner, A., Aegerter, C., Sharma, A., Judd, L., Wardlow, B., You, J., Shulski, M., Irmak, S., Kilic, A., Zeng, J., 2016. A multi-sensor view of the 2012 central plains drought from space. Front. Environ. Sci. 4, 45. https://doi.org/10.3389/fenvs.2016.00045.

Wang, H., Zhang, L., Cao, X., Zhang, Z., Liang, J., 2013. A-train satellite measurements of dust aerosol distributions over northern China. J. Quant. Spectrosc. Radiat. Transf. 122, 170–179.

Winker, D.M., Vaughan, M.A., Omar, A., Hu, Y., Powell, K.A., Liu, Z., Young, S.A., 2009. Overview of the CALIPSO mission and CALIOP data processing algorithms. J. Atmos. Ocean. Technol. 26 (11), 2310–2323.

You, W., Zang, Z., Pan, X., Zhang, L., Chen, D., 2015. Estimating $PM_{2.5}$ in Xi'an, China using aerosol optical depth: a comparison between the MODIS and MISR retrieval models. Sci. Total Environ. 505, 1156–1165.

Youn, D., Park, R.J., Jeong, J.I., Moon, B.K., Yeh, S.W., Kim, Y.H., Woo, J.H., Im, E.G., Jeong, J.H., Lee, S.J.,

Song, C.K., 2011. Impacts of aerosols on regional meteorology due to Siberian forest fires in May 2003. Atmos. Environ. 45, 1407–1412.

Yu, H., Chin, M., Bian, H., Yuan, T., Prospero, J.M., Omar, A.H., Zhang, Z., 2015. Quantification of trans-Atlantic dust transport from seven-year (2007–2013) record of CALIPSO lidar measurements. Remote Sens. Environ. 159, 232–249.

Yu, H., Remer, L.A., Kahn, R.A., Chin, M., Zhang, Y., 2013. Satellite perspective of aerosol intercontinental transport: from qualitative tracking to quantitative characterization. Atmos. Res. 124, 73–100.

Zeeshan, M., Oanh, N.T.K., 2015. Relationship of MISR component AODs with black carbon and other ground monitored particulate matter composition. Atmos. Pollut. Res. 6 (1), 62–69.

Further Reading

Bellouin, N., Boucher, O., Tanré, D., Dubovik, O., 2003. Aerosol absorption over the clear-sky oceans deduced from POLDER-1 and AERONET observations. Geophys. Res. Lett. 30 (14), 1748.

Deuzé, J.L., Bréon, F.M., Devaux, C., Goloub, P., Herman, M., Lafrance, B., Tanré, D., 2001. Remote sensing of aerosols over land surfaces from POLDER-ADEOS-1 polarized measurements. J. Geophys. Res. Atmos. 106 (D5), 4913–4926.

Duncan, B.N., Prados, A.I., Lamsal, L.N., 2014. Satellite data of atmospheric pollution for U.S. air quality applications: examples of applications, summary of data end-user resources, answers to FAQs, and common mistakes to avoid. Atmos. Environ. 94, 647–662.

Kaufman, Y.J., Koren, I., Remer, L.A., Tanre, D., Ginoux, P., Fan, S., 2005. Dust transport and deposition observed from the Terra-Moderate Resolution Imaging Spectroradiometer (MODIS) spacecraft over the Atlantic Ocean. J. Geophys. Res. 110, D10S12.

Mekler, Y., Quenzel, H., Ohring, G., Marcus, I., 1977. Relative atmospheric aerosol content from ERS observations. J. Geophys. Res. 82, 967–972.

Muhammad, Z., Oanh, K.N., 2015. Relationship of MISR component AODs with black carbon and other ground monitored particulate matter composition. Atmos. Pollut. Res. 6 (1), 62–69.

CHAPTER

4

Remote Sensing of Heavy Aerosol Pollution Episodes: Smoke and Dust

Sonoyo Mukai

The Kyoto College of Graduate Studies for Informatics (KCGI), Kyoto, Japan

1 INTRODUCTION

In this section, we focus on a remote sensing application for extreme concentrations of atmospheric aerosols—specifically heavy air pollutions—which is called a heavy aerosol episode or simply an episode, since it is an infrequently occurring phenomenon. Such aerosol episodes deteriorate air quality, environment, and ecosystems. Further, dense concentrations of aerosols in the atmosphere can prevent aerosol monitoring from surface-level sun/sky photometers; however, satellites can still be used to observe the Earth's atmosphere from space. As such, aerosol remote sensing via satellites is known to be useful and effective even during aerosol episodes; however, before attempting to retrieve aerosol properties from satellite measurements, the efficient algorithms for aerosol retrieval need to be considered. The characteristics and distributions of atmospheric aerosols are known to be complicated, due to both natural factors and human activities. In urban areas, small anthropogenic aerosols dominate because of emissions from diesel vehicles and other similar industrial and urban activities. Increased emissions of anthropogenic particles cause increased concentrations of harmful air pollutants, i.e., increases in suspended particulate matter (SPM) in the atmosphere. In particular, large cities suffer from heavy air pollution episodes (Kahn et al., 2004; Nakata et al., 2015).

Turning to naturally originating aerosols, a primary example is desert dust or dust storms, which are considered to be the most dynamic natural phenomena, injecting high volumes of mineral dust into the atmosphere, which are then transported over long distances; such phenomena cause an enhanced atmospheric turbidity (Sokolik et al., 2001). Further, biomass burning plumes due to large-scale forest fires and agriculture burns have caused severe air pollution (Crutzen and Meinrat, 1990). Overall, it is highly likely that large-scale aerosol episodes due to a mixture of both natural factors and human activities occur in select locations.

In addition to the above, atmospheric aerosols have various impacts on global environmental change and climate by directly absorbing and/or scattering solar radiation, and by indirectly modifying the optical properties and lifetimes of clouds, as well as meteorology (Pérez et al., 2006). Given

Remote Sensing of Aerosols, Clouds, and Precipitation
https://doi.org/10.1016/B978-0-12-810437-8.00004-9

© 2018 Elsevier Inc. All rights reserved.

these impacts, determining accurate estimates of aerosol properties and emissions is an increasingly urgent subject in relation to global climate problems. As an example, forest fires increase due to global warming and climate change, and vice versa. This negative cycle decreases the quality of the global environment and human health.

The 5th Intergovernmental Panel on Climate Change' (IPCC) report on global warming (https://www.ipcc.ch/report/ar5/wg1/) emphasizes the importance of observing aerosol characteristics and their temporal and spatial variations, indicating the warming effect of black carbon aerosols versus the cooling effect of other aerosol types; however, aerosol properties of hazardous air pollutants are still insufficiently understood. Aerosol distribution varies seasonally due to factors such as emissions, photochemical reactions, and wind direction (Littmann, 1991; Kinne et al., 2003).

2 DETECTION OF AEROSOL EPISODES

2.1 Dust Storms

Fig. 1 shows the global distribution of aerosol optical thickness (AOT) at a wavelength of 0.55 μm, as measured by the Moderate Resolution Imaging Spectroradiometer (MODIS) sensor on board Aqua, an Earth-observing satellite (King et al., 1992). In the figure, the AOT values are averaged over the four seasons, i.e., winter (DJF), spring (MAM), summer (JJA), and fall (SON), from 2003 to 2015. Numerical AOT values are shown using a rainbow color scale ranging from violet to red, where red indicates extreme mass concentrations of aerosols. At a glance, we observe that red is dominant in spring and summer in the Northern Hemisphere, and in fall in the central part of South America.

In general, AOT values are high over desert areas. As noted in the previous section, desert dust is the first candidate for naturally originating aerosols. These dust storms are the most dynamic natural phenomena, introducing high volumes of mineral dust into the atmosphere, with these dust particles widely transported from the source desert region to other regions (Tegen and Miller, 1998; Sokolik et al., 2001). During transportation, the original mineral dust particles become contaminated with anthropogenic aerosols due to human activities and/or biomass burning plumes and the like. Huge amounts of soil particles are swept up in the strong dry winds across the large desert regions. For Asia, dense soil dust originating in the Gobi and Taklamakan deserts is transported to

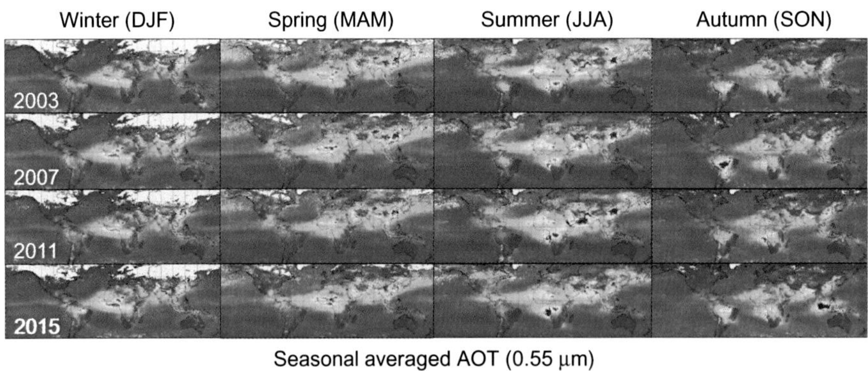

FIG. 1 Global distribution of seasonally averaged AOT (0.55 μm) from 2003 to 2015 compiled from MODIS products of MYD_M3 Collection 6.

eastern and central Asia on the westerly winds, especially in the spring (Mukai et al., 2004). In Japan, this so-called "yellow sand," "tuchifuru," or "kasumi" are used as a spring season word in poems consisting of seventeen syllables, called haikus. Despite the beauty of these poems, Asian dust has caused severe negative impacts on social life and human health.

Fig. 1 suggests various issues regarding aerosols from the global and seasonal change perspectives, but we postpone such perspectives at present. We first introduce an algorithm to detect dust aerosols from space. More specifically, we propose the aerosol vapor index (AVI) for detecting mineral-dust aerosols based on the difference between brightness temperatures at infrared wavelengths provided by the National Aeronautics and Space Administration (NOAA)/ Advanced Very High Resolution Radiometer (AVHRR) (Ackerman, 1997). Here, AVI is defined with space-based data given by the Terra or Aqua/MODIS sensor (King et al., 1992) as:

$$\mathrm{AVI} = T(\mathrm{B32}) - T(\mathrm{B31}) \quad (1)$$

where T represents a brightness temperature at Band 31 (i.e., 10.78–11.28 μm) or Band 32 (i.e., 11.77–12.27 μm) of the MODIS sensor; more specifically,

$$T = C2/\lambda \times \ln\left[\left(C1/\left(\lambda^5 \times R \times 10^6\right)\right) + 1.0\right] \quad (2)$$

where R represents the radiance (in $\mathrm{W/m^2/sr/m}$) at wavelength λ (in μm) and constants $C1$ and $C2$ are set to $1.1910439 \times 10^{-16}$ (K/m) and 1.4387686×10^{-2} ($\mathrm{sr/m^3/W}$), respectively. Water vapor transmittance is usually larger in Band 31 than in Band 32; hence, AVI takes on a negative value. However, if mineral dust exerts an inverse influence, the AVI values become positive for dense dust episodes. Given this behavior, AVI can be used to detect the presence of dust aerosols.

However, AVI has the following two problems: (1) desert areas exhibit similar characteristics to those of dust aerosols and (2) dust aerosols mixed into the clouds cannot be detected using AVI alone. Given these problems, another index should be incorporated into attempts to detect dust aerosols in order to complement AVI. From Mukai et al. (2010), this additional index is the yellow dust index (YDI) and is defined as:

$$\mathrm{YDI} = \frac{R(\mathrm{B4}) - R(\mathrm{B3})}{R(\mathrm{B4}) + R(\mathrm{B3})} \quad (3)$$

where $R(\mathrm{B3})$ and $R(\mathrm{B4})$ represent the MODIS data at Band 3 (i.e., 0.459–0.479 μm) and Band 4 (i.e., 0.545–0.565 μm), respectively. YDI has positive values for light reflected by mineral dust, which absorbs blue-wavelength radiation. Therefore, combining YDI with AVI looks promising for solving the second problem described above. Note that wavelength channels at 1.6 and 2.1 μm can be used for dust identification because water is absorbed at these wavelengths but dust is not there.

Next, we can use temperature to overcome the first problem of desert areas exhibiting similar features to those of dust aerosols. Here, brightness temperature T (12 μm) is much higher for deserts than for dust aerosols or clouds; hence, dust aerosols can be detected from the satellite using the following two steps: (1) detect dust aerosols using $\mathrm{AVI} \geq 0.4$ and T (12 μm) < 294 K and (2) detect dust mixed with clouds using $\mathrm{YDI} \geq -0.01$ and T (12 μm) < 268 K. The associated algorithms for detecting dust aerosols have been described in detail in Mukai et al. (2010).

Using the Aqua/MODIS data collected on April 18, 2006 over Japan, Fig. 2B shows a dust aerosol map generated using the aforementioned algorithms. In the figure, yellow and white denote dust aerosols and cloud, respectively, while the red square indicates the Osaka site of the National Aeronautics and Space Administration (NASA)/ Aerosol Robotic Network (AERONET). From the figure, we observe that Osaka was covered with dust aerosols (refer to Fig. 2B, which shows a true-color image from Aqua/MODIS). Further, ground-based measurements from the Osaka site are shown in Fig. 2C, where the AOT values (0.675 μm) from NASA/AERONET data are

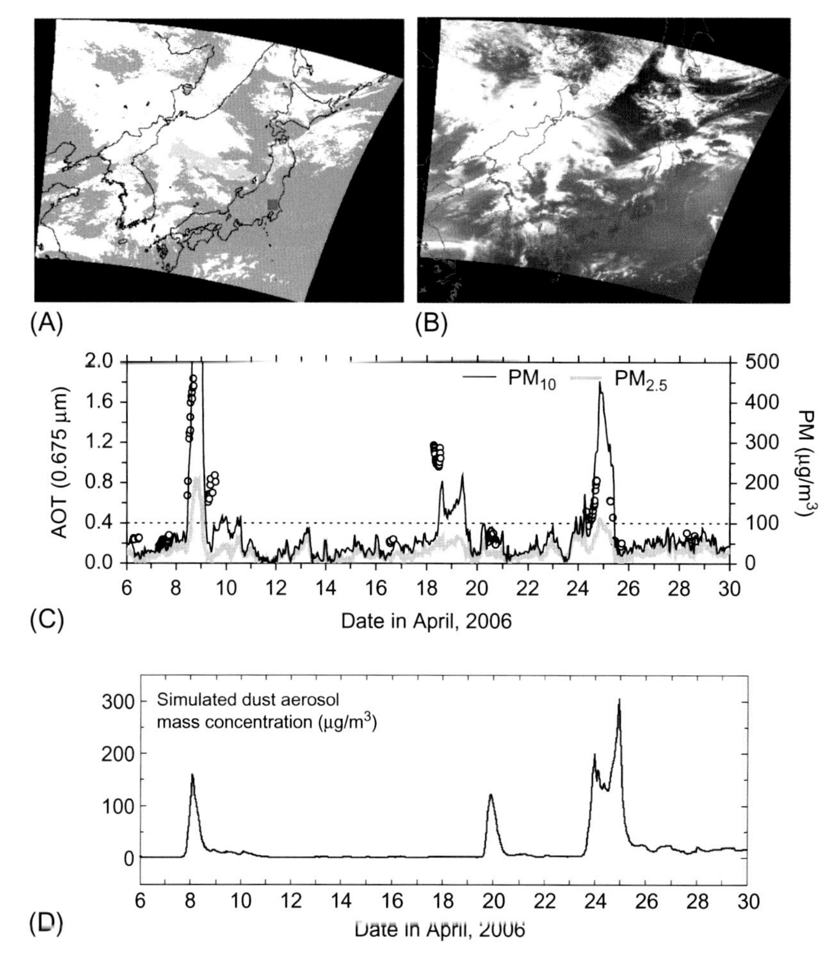

FIG. 2 Dust episode over Japan in April 2006. The *red square* indicates the position of the Osaka NASA/AERONET site. (A) Map of dust aerosols detected using our proposed combined algorithms with Aqua/MODIS data collected on April 18, 2006 over Japan. *Yellow* and *white* represent dust aerosols and cloud, respectively. (B) True-color image from Aqua/MODIS obtained on April 18, 2006. (C) Ground-based measurements obtained at Osaka, where *open circles* and *black* and *gray curves* denote AOT (0.675) for AERONET data and PM_{10} and $PM_{2.5}$ levels from the SPM-613D sampler, respectively. (D) Dust aerosol mass concentration as simulated using the SPRINTARS mode. *From Mukai, M., Sano, I., Iizuka, T., Yokomae, T., Mukai, S., 2010. Detection and analysis of dust aerosol particles over the East Asia, J. Remote Sens. Soc. Jpn. 30 (1), 1–10.*

denoted by open circles and particulate matter (PM) values for PM_{10} and $PM_{2.5}$ from the Kimoto/SPM-613D sampler are represented by black and gray solid curves, respectively. Finally, Fig. 2D presents the dust aerosol mass concentration simulated using a global climate model called SPRINTARS (Takemura et al., 2002, 2005). Results obtained using the above algorithm for detecting dust aerosols from space coincide with other measurements and model simulations. Osaka in Japan is far away from the source of soil dust, i.e., desert areas in China, and hence, dust aerosol episodes are detected here as shown in Fig. 2, but are not as severe as dust storms. A dust storm, i.e., the core of dust episodes, defined with AOT (0.55 μm) ≥ 4, should be detected.

Fig. 3 presents images obtained by Aqua/MODIS over the Badain Jaran Desert in northern

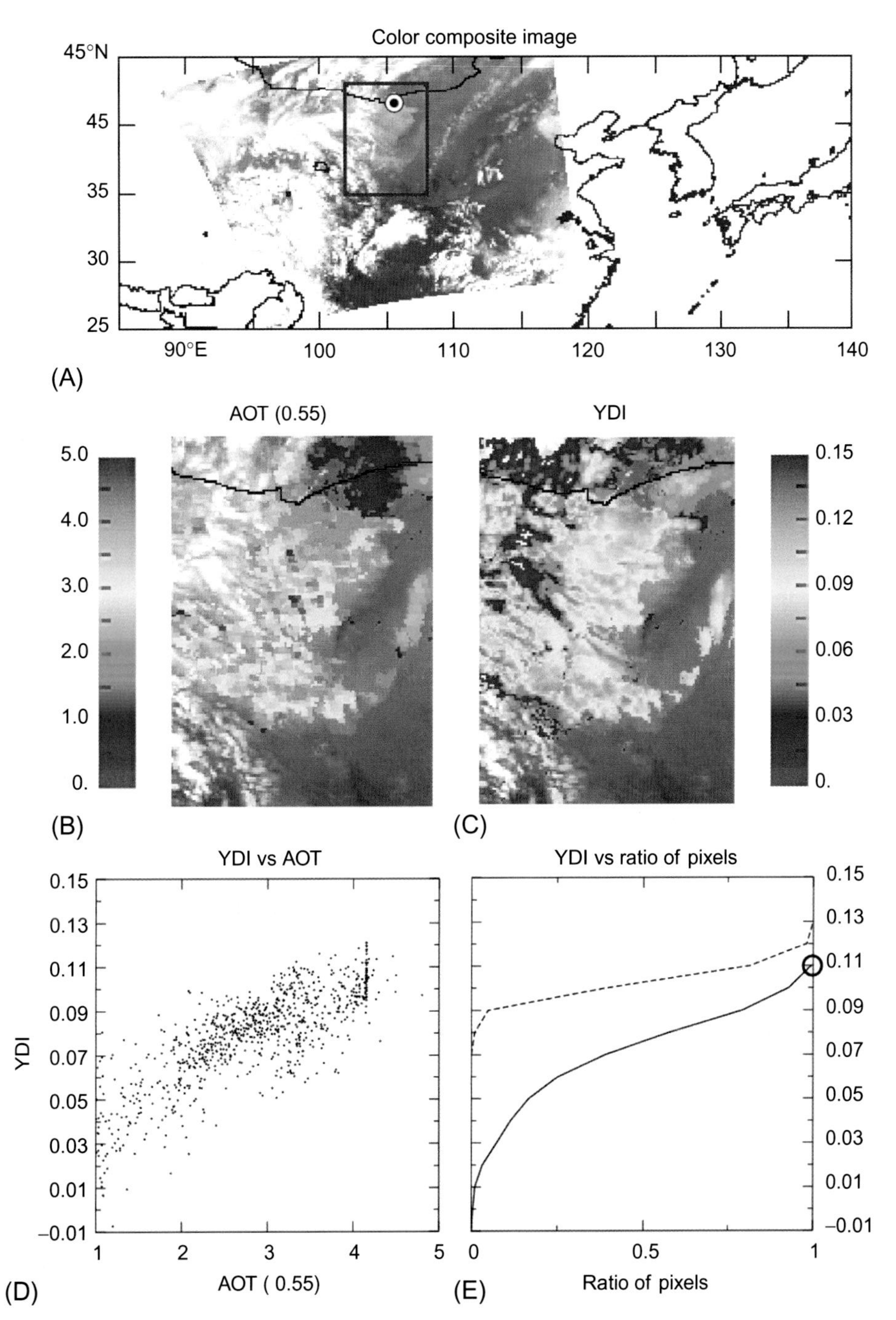

FIG. 3 Detection of a core of dust episodes from Aqua/MODIS data over the Badain Jaran Desert in northeast China on April 10, 2006. (A) Color composite image. (B) Distribution of AOT (0.55 μm) derived using the Deep Blue algorithm within the red square in (A). (C) Same as (B), but for YDI from Eq. (3). (D) Scatter plot of YDI versus AOT (0.55 μm). (E) Values of r-above (4.0, y^*) *(dashed curve)* and r-below (4.0, y^*) *(solid curve)* defined by Eqs. (4), (5), respectively. *From Mukai, M., Sano, I., Iizuka, T., Yokomae, T., Mukai, S., 2010. Detection and analysis of dust aerosol particles over the East Asia, J. Remote Sens. Soc. Jpn. 30 (1), 1–10.*

China on April 10, 2006. Fig. 3A is a color composite image in which the blue curve indicates the coast and the black curve represents the boundary. Fig. 3B and C shows the distributions of AOT (0.55) and YDI (i.e., Eq. 3) derived from Aqua/MODIS data over the area contained within the red square in Fig. 3A. Further, Fig. 3D presents a scatter plot of YDI of Fig. 3C versus AOT (0.55) of Fig. 3B, indicating that YDI increases with AOT. We thereby calculate:

$$r-\text{above}(\tau^*, y^*) = \text{pxl}(\text{AOT}(0.55) \geq \tau^* \text{and YDI} \leq y^*)/\text{pxl}(\text{AOT}(0.55) \geq \tau^*) \quad (4)$$

$$r-\text{below}(\tau^*, y^*) = \text{pxl}(\text{AOT}(0.55) < \tau^* \text{and YDI} \leq y^*)/\text{pxl}(\text{AOT}(0.55) < \tau^*) \quad (5)$$

where pxl represents the number of pixels satisfying the conditions within the parentheses. Next, we define the τ^* value as 4 for the core of the dust episode. The dashed and solid curves shown in Fig. 3E indicate the values of r-above (4, y^*) and r-below(4, y^*) obtained from Fig. 3D, respectively. The solid curve in Fig. 3E crosses the vertical axis for a pixel ratio of 1.0 at YDI = 0.11, as indicated by the circle, suggesting that all pixels satisfying AOT (0.55 μm) < 4.0 have YDI values lower than 0.11. As a result, we note that a core of the dust episode satisfies YDI ≥ 0.11.

2.2 Biomass Burning Plumes

Biomass burning aerosol plumes generated by large-scale forest fires or agriculture burns have caused severe air pollution (Mukai et al., 2014); further, forest fires increase due to global warming and climate change, and vice versa. This negative cycle decreases the quality of global environment and human health. Fig. 4 shows the global distribution of Aqua/MODIS products in odd months from 2003 to 2015. In the figure, red dots on the images in the left-hand column represent monthly accumulated hotspots provided by the Rapid Fire Response System of Aqua/MODIS (MYD14 Collection 5) (Justice et al., 2002; Giglio et al., 2003), whereas the monthly averaged values of AOT (0.55 μm) (MYD_M3 Collection 6) are shown in the right-hand images. Note that the AOT values in the figure were compiled based on the values from Land and Ocean, Dark Target, and Deep Blue aerosol retrieval algorithm modes. Also note that the AOT values (0.55 μm) averaged seasonally were already shown in Fig. 1.

From Fig. 4, we observe that AOT (0.55 μm) values over desert areas (ref. the left-hand images) are high in a similar fashion as that of the seasonal change highs shown in Fig. 1. Comparing Fig. 4 with Fig. 1, we observe that AOT values depend on the month. It is quite natural that AOT values vary with time and place. The distribution of hotspots in Fig. 4 implies that biomass burning plumes have occurred in tropical areas such as central Africa, America, and South Asia. It is evident from both sets of hotspots and AOT distributions in the figure that AOT values are not always determined by biomass burning plumes or desert dust because of the variance in emission sources of aerosols, as described above.

Moreover, biomass burning plume is a seasonal phenomenon peculiar to a particular region. Fig. 5 presents monthly averaged AOT (0.55 μm) values from 2003 to 2015 compiled from MODIS products at five selected areas denoted by A, B, C, D, and E in the upper geographic map of the figure. From Figs. 4 and 5, we at least observe that high AOT (0.55 μm) values are caused by biomass burning plumes in July in region A of China, in September in region B of Indonesia, in March in region C of Africa, in September in region D of Africa, and again in September in region E of Brazil. To validate the satellite measurements presented in Figs. 4 and 5, we present in Fig. 6 the monthly averaged AOT (0.44 μm) values collected by ground-based NASA/AERONET radiometers in 2015 at six AERONET sites (i.e., the upper map of Fig. 6) corresponding to areas A, B, C, D, and E in Fig. 5; here, monthly variations are similar to those shown in Fig. 5. Note that the

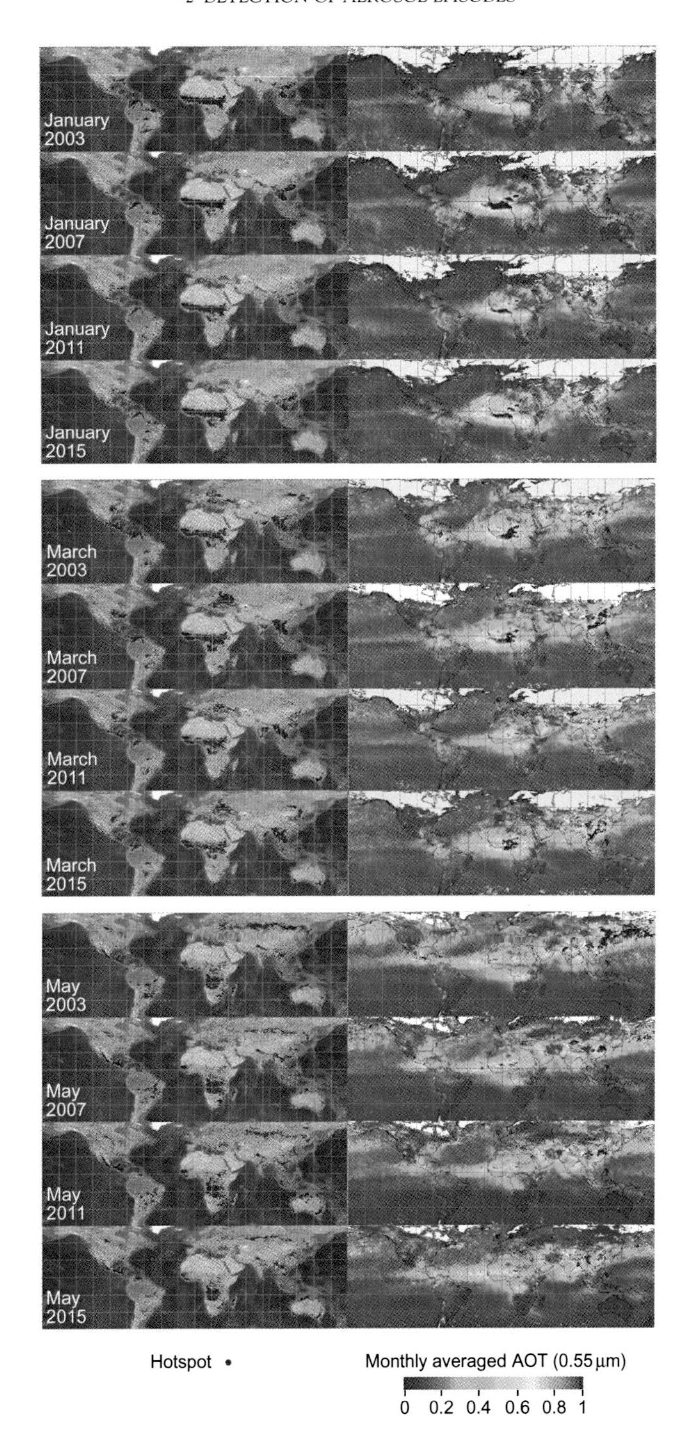

FIG. 4 **See legend on next page.**

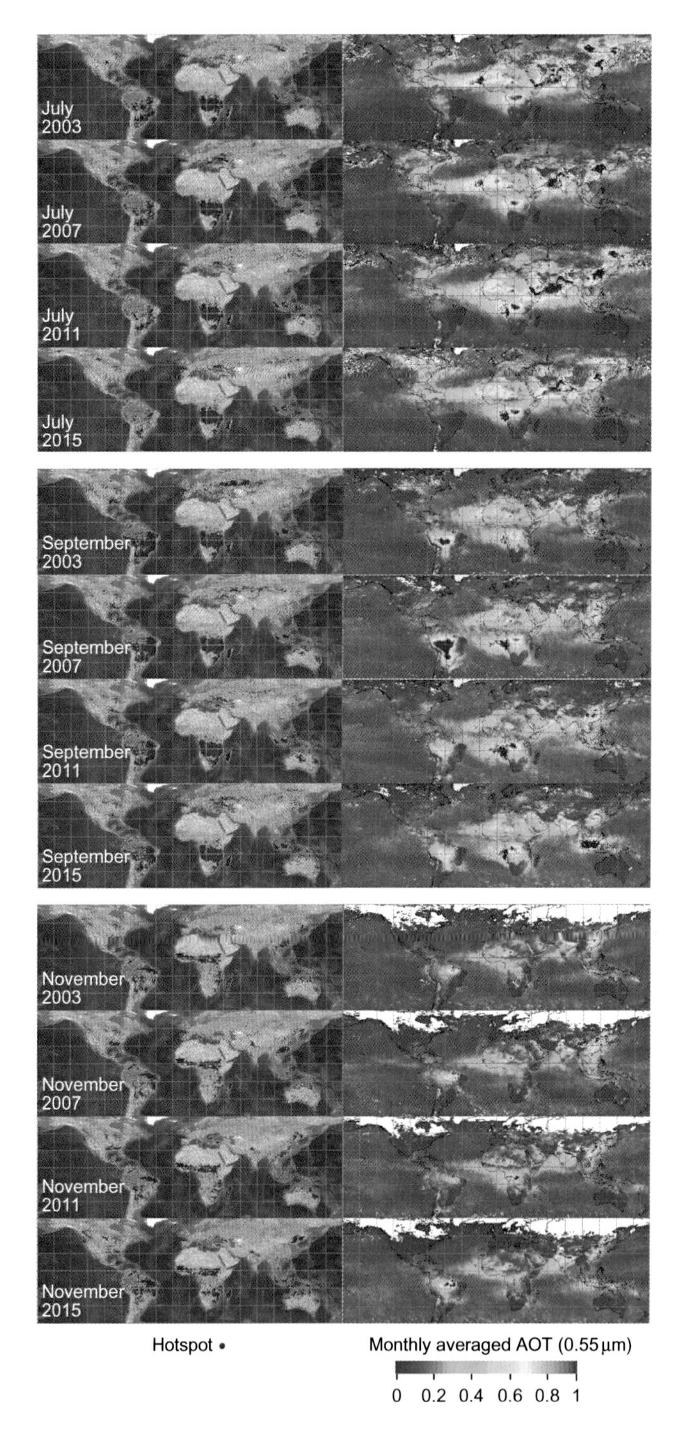

FIG. 4 Global distribution of monthly accumulated hotspots (left) and monthly averaged AOT (0.55 μm) (right) derived from Aqua/MODIS measurements in odd months from 2003 to 2015, compiled from MYD14 Collection 5 and MYD08_M3 Collection 6, respectively.

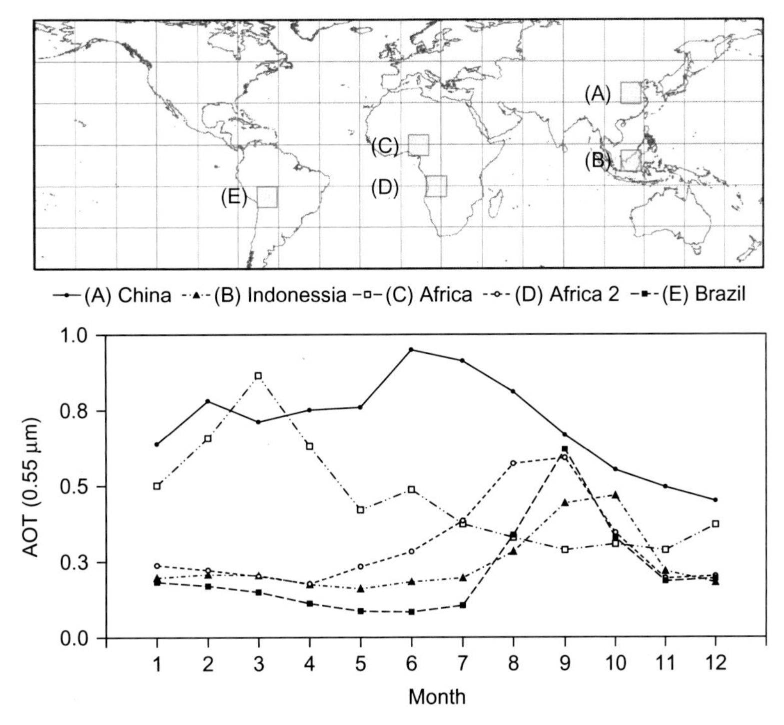

FIG. 5 Monthly averaged AOT (0.55 μm) from 2003 to 2015, compiled from MODIS products of MYD_M3 Collection 6 at five selected areas denoted by A, B, C, D, and E in the upper geographic map.

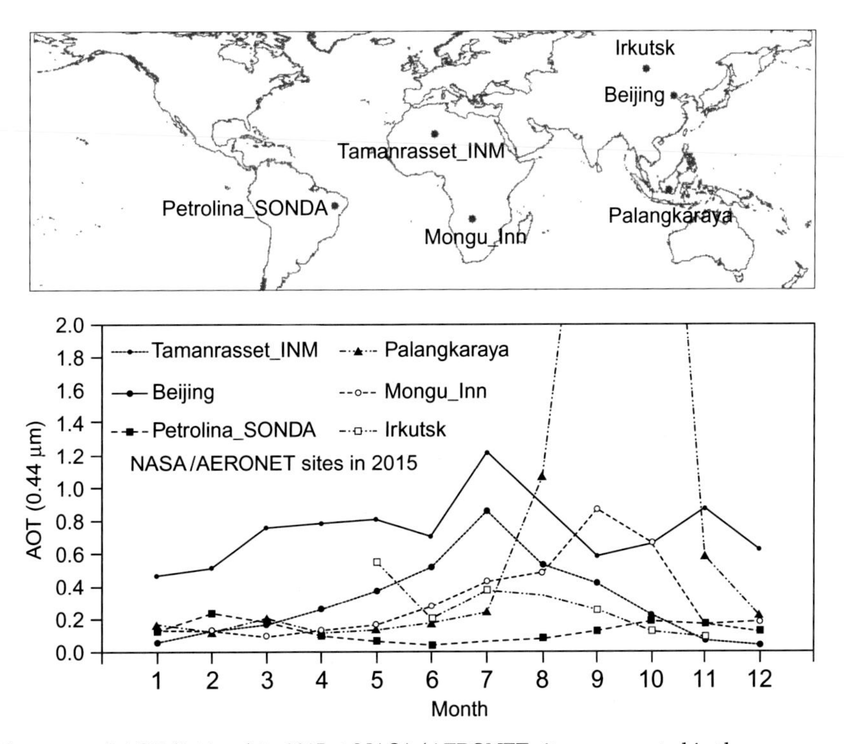

FIG. 6 Monthly averaged AOT (0.44 μm) in 2015 at NASA/AERONET sites represented in the upper map.

wavelength is different between Figs. 5 and 6; thus, the short-wavelength radiometry must be more sensitive to biomass burning aerosols. Regardless, the satellite products are well-validated via the ground-based measurements. Hence, we further note that mass concentrations of aerosols are frequently governed by spatial and/or temporal variations in biomass burning plumes.

Referring back to the biomass burning in the Advanced Earth-Observing Satellite (ADEOS) era, ADEOS-2 represents a short-term Japanese mission in the framework of the International Earth Observation (IEOS) system from December 2002 to October 2003. ADEOS-2/GLI has observing channels in the near-ultraviolet (UV) (i.e., Band 1: 0.38 μm) and violet (i.e., Band 2: 0.40 μm) wavelengths, which are sensitive to detecting biomass burning aerosols. Major forest fires in Siberia usually occur in May (see Fig. 4), and the forest fires in May of 2003 were the strongest. This large forest fire in Siberia had a strong influence on aerosol concentrations in the atmosphere. Fig. 7A and B shows color composite images from May 20, 2003 over Siberia, as observed by ADEOS-2/GLI and Polarization and Directionality of the Earth's Reflectances (POLDER), respectively (Sano et al., 2009). Both images definitely indicate a heavy haze atmosphere, which must have originated from the large forest fires around Lake Baikal.

In general, aerosol retrieval over land, based on satellite data, is accomplished via three- or two-channel algorithms in the visible to near-infrared wavelengths (Remer et al., 2005); however, it is difficult to distinguish between absorbing (e.g., biomass burning) and nonabsorbing (e.g., sulfate) particles. Thus, usual retrieval algorithms have adopted only sulfate-type nonabsorbing particles like standard aerosols. This limitation can be overcome using near-UV and violet data (Höller et al., 2004; Cyranoski and Fuyuno, 2005; Flores et al., 2014). Using shorter wavelength data is an approach that has been introduced to detect absorbing aerosols, such as plumes from biomass burning (Hsu et al., 1996; Herman et al., 1997; Torres et al., 1998).

We briefly describe our algorithm based on the combined use of near-UV and violet data here, as it has already been presented in detail in our previous chapter based on ADEOS-2/GLI measurements (Sano et al., 2009). Our algorithm is a method for detecting absorbing

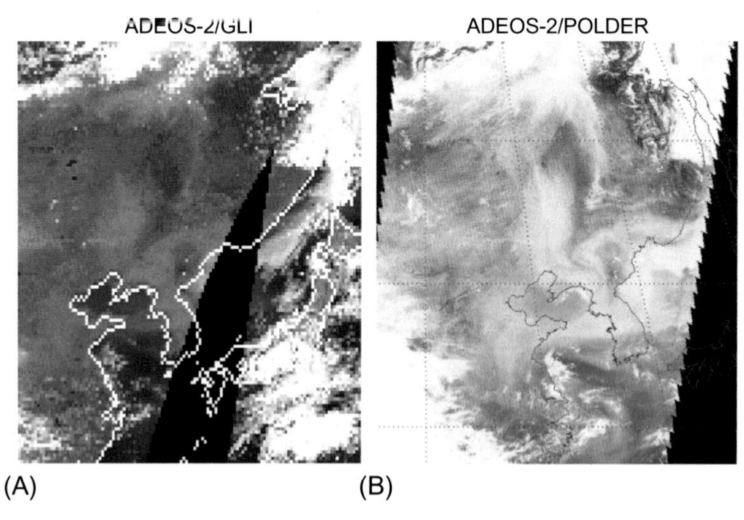

FIG. 7 Color composite images from May 20, 2003, captured by (A) ADEOS-2/GLI and (B) ADEOS-2/POLDER.

particles based on the ratio of reflectance at 0.40 (Band 2/GLI) and 0.38 μm (Band 1/GLI), hereafter called the absorbing aerosol index (AAI). Fig. 8 shows numerical AAI values cited from our previous study. The solid curve in Fig. 8 shows a histogram of AAI from the GLI data, which are the same data as those shown in Fig. 7A, indicating that the mean and standard deviation values of AAI are 0.907 and 0.095, respectively. The dashed curve represents the same type of data as the solid curve, but for the minimum values of AAI during May 2003, which represent values on ordinary days; the mean and standard deviation values of 0.870 and 0.019, respectively, can therefore be considered the standard values for ordinary clear days. Accordingly, AAI is an indicator for absorbing particles, such as biomass burning aerosols. From Fig. 8, we observe that AAI for biomass burning aerosols has a value higher than 0.9, signifying that AAI ≈ 0.9 must be the threshold of a biomass burning plume.

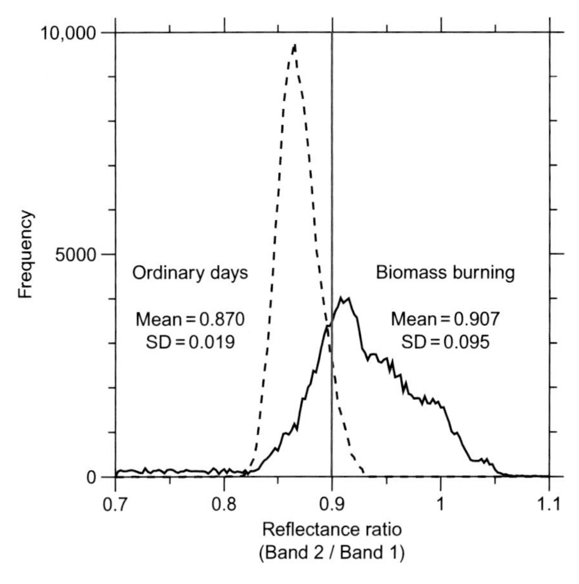

FIG. 8 The AAI histogram showing the ratio of reflectance at 0.40 μm (Band 2/GLI) and 0.38 μm (Band 1/GLI) over northeast Asia, derived from ADEOS-2/GLI data. The *solid* and *dashed curves* represent the histogram of the image shown in Fig. 7A on May 20, 2003, and the ordinary days during May 2003, respectively.

Fig. 9 shows the global distribution of AAI derived from ADEOS-2/GLI on May 20, 2003. We observe from the figure, that the AAI values are high across almost all of northeast Asia. Images in Fig. 10 show the same data as in Fig. 9, but for the enlarged region denoted by the yellow solid line in Fig. 9 from May 17 to May 20, 2003. From Fig. 10, we observe that an air mass with an AAI greater than 0.9 moves farther along with the wind, indicating that biomass burning aerosols were transported far away from their origin in Siberia.

FIG. 9 Global distribution of AAI calculated from ADEOS-2/GLI measurements on May 20, 2003.

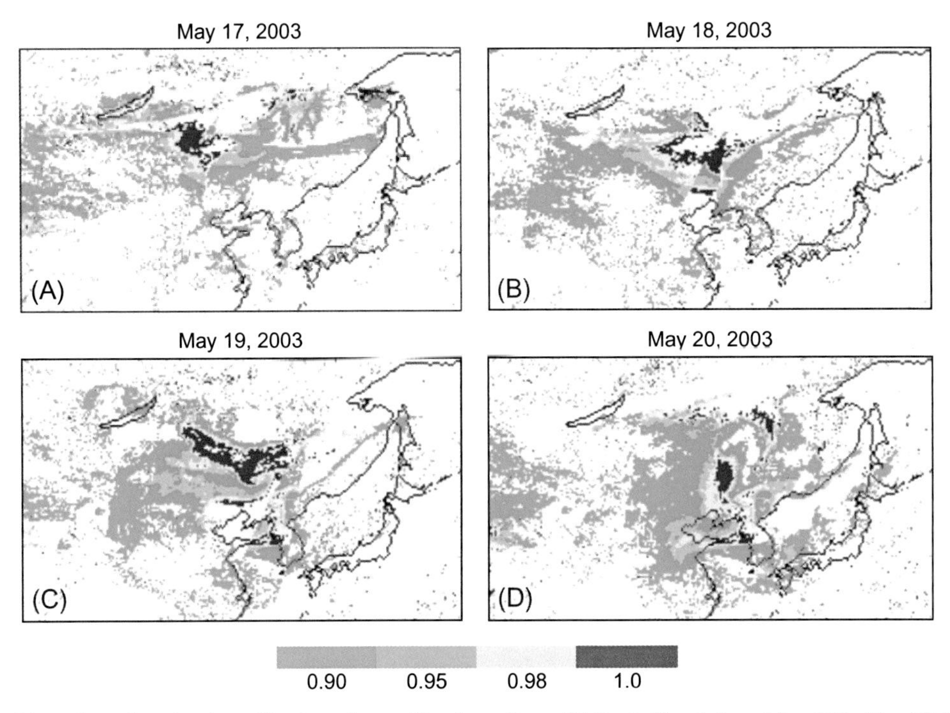

FIG. 10 The enlarged region from Fig. 9, as denoted by the *yellow solid line* in Fig. 9, from May 17 to May 20, 2003.

3 AEROSOL RETRIEVAL FRAMEWORK

3.1 Aerosol Remote Sensing Algorithm

We designed our aerosol retrieval algorithm to be consistent with satellite measurements. Radiation observed by the space-borne sensor at the top of the atmosphere (TOA) is simulated via computer, with simulation results compared with the satellite measurements to estimate the optimized aerosol model. Fig. 11 shows a basic framework for aerosol retrieval from satellite data. The retrieval process is divided into three parts, i.e., satellite data processing (*S*), numerical model simulations (*M*), and radiation simulation calculations (*R*). Aerosol properties, such as AOT, refractive index (m), and Angstrom exponential (α) determined by the size distribution of aerosols, are retrieved based on the optimized comparison of satellite measurements (denoted by DB for satellite reflectance) with the numerical values of radiation simulations in the Earth-atmosphere–surface model (recorded as a look-up table (LUT) for simulated reflectance). The model of the Earth's atmosphere is based on the Air Force Geophysics Laboratory (AFGL) code, which provides molecular scattering/absorption distributions with height (Kneizys et al., 1988). The multiple scattering calculations represented by the "R.T. simulation" box in Fig. 11 yield Rayleigh scatterings by molecules and Mie scatterings by spherical aerosols in the atmosphere. Note that theories other than Mie should be considered for irregularly shaped aerosols (Mischenko et al., 2000; Okada, 2008; Kokhanovsky, 2010; Ishimoto et al., 2012). The aerosol models are essentially estimated from the accumulated data by first using NASA/AERONET (Holben et al., 1998; Dubovik et al., 2002; Omar et al., 2005).

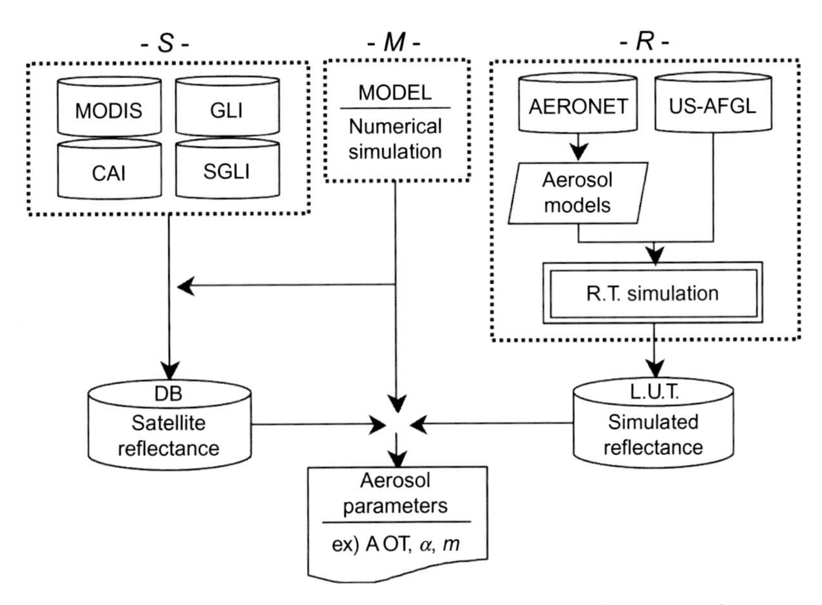

FIG. 11 Our basic framework for aerosol retrieval from space. The variables of AOT, α, and m represent aerosol optical thickness, angstrom index, and refractive index, respectively.

The satellite database system should be constructed so that it can respond to various space-based measurements. The satellite sensor measures the spectral radiance at TOA from visible to far-infrared wavelengths. In some cases, UV wavelength bands or polarization observation abilities are equipped with space-borne sensors that measure the upwelling radiance at TOA, and hence, the satellite data involves the radiance due to both atmospheric scattering and ground surface reflections. Accordingly, in satellite data analysis, separating atmospheric information from ground-surface information tends to be difficult. The reflectance behavior of the Earth's surface, including directional distribution and polarization, is both an indispensable and troublesome issue for radiation simulations at the global scale. As noted above, results of RT simulation are stored in an LUT, and then, the simulation is repeated after changing the aerosol and/or ground-surface models. This process is repeated until the LUT is complete. This LUT-based algorithm has proven to be limited for aerosol retrieval on a global scale when various shapes, compound aerosols, and ground surfaces are involved.

Given these limitations, more versatile algorithms for aerosol retrieval have been proposed; Generalized Retrieval Aerosol and Surface Properties (GRASP) is one such proposal (Dubovik and King, 2000; Dubovik et al., 2011). While most conventional procedures, including the LUT-based algorithms, assume knowledge of the characteristics of either aerosols or ground surfaces, GRASP can obtain both characteristics by limiting the range of unknown parameters of aerosols and ground surfaces as a constraint in optimization. Statistical optimization provides aerosol properties simultaneously with ground-surface reflection characteristics.

Further, multipixel analysis is implemented in GRASP by grouping a large number of pixels from one or several scenes. GRASP overcomes spatial and/or temporal inconsistencies that occur in conventional approaches by applying several generalization principles with the idea of developing a scientifically rigorous, versatile, practically efficient, transparent, and accessible algorithm (Dubovik et al., 2014). The features and advantages of GRASP are summarized as follows. GRASP is composed of two main independent modules, i.e., numerical inversion and the forward model. The numerical module is implemented as a statistically optimized fitting method based on the multiterm least-squares method. Thus, simultaneous retrieval of both aerosol and ground-surface characteristics is realized. The forward model is similar to the aforementioned conventional method in that it simulates atmospheric remote sensing measurements. Note that GRASP is still evolving according to its surrounding environment. For instance, it provides columnar and vertical aerosol properties via the statistically optimized multipixel inversion of GRASP and Generalized Aerosol Retrieval from Radiometer and Lidar Combined (GARRIC) data (Lopatin et al., 2013). Thus, GRASP is an excellent algorithm for aerosol retrieval; however, the time required for operation can be very high.

3.2 Aerosol Model

Aerosol properties can be represented using several parameters. The most basic parameter is spectral aerosol optical thickness AOT(λ) at wavelength λ. Further, the Ångström exponent α is defined from the spectral AOT(λ) (O'Neill et al., 2001), i.e.,

$$\alpha = -\ln(\mathrm{AOT}(\lambda_2)/\mathrm{AOT}(\lambda_1))/\ln(\lambda_2/\lambda_1) \quad (6)$$

Values of α are closely related to aerosol size distribution, i.e., small values of α indicate large particles, whereas large values indicate small particles. In general, values of α from approximately 0 to 1.0 indicate large particles, such as sea salt aerosols and soil dusts, whereas values in the range $1.0 < \alpha < 2.5$ indicate particles, such as sulfate and those associated with biomass burning (O'Neill et al., 2003). Detection of high α values almost always indicates contamination by small anthropogenic particles. Several other aerosol parameters, such as size distribution and refractive index, are also derived from AOT(λ).

According to the automatic classification of accumulated NASA/AERONET data, atmospheric aerosols are classified into the following six categories: (1) biomass burning (BB) is an aged smoke aerosol consisting primarily of soot and organic carbon; (2) rural (RU) is referred to as a clean continental aerosol; (3) continental pollution (CP) represents anthropogenic aerosols, including various species of sulfate- (SO_4^{2-}), nitrate- (NO_3^-), OC, ammonium (NH_4^+), and soot; (4) dirty pollution (DP) consisting of the same aerosol types as CP, but at significantly higher levels; (5) desert dust (DD) is assumed to be mostly mineral soil; and (6) polluted marine (PM) consists primarily of sea salt with traces of CP.

We first focus on the size of aerosol particles. Size distributions of these six aerosol types based on AERONET measurements have two modes for small (f) and large (c) particles in a bimodal log-normal distribution of particle volume, i.e.,

$$\frac{dV}{d\ln r} = \frac{V_f}{\sqrt{2\pi}\ln\sigma_f}\exp\left[-\frac{(\ln r - \ln r_f)^2}{2\ln^2\sigma_f}\right] + \frac{V_c}{\sqrt{2\pi}\ln\sigma_c}\exp\left[-\frac{(\ln r - \ln r_c)^2}{2\ln^2\sigma_c}\right] \quad (7)$$

Here, parameters V_f, r_f, and σ_f represent the volume concentration, mode radius, and standard deviation of small mode particles, respectively. The corresponding parameters for large mode particles are V_c, r_c, and σ_c, respectively. These six parameters are necessary to define aerosol size and are excessive for retrieving the optimized size of aerosols on a global scale. Therefore, proper simplification is required for the aerosol size distribution function of Eq. (7).

We therefore consider continental aerosols here, and hence, the maritime aerosol type (PM) is discarded for the present discussion.

An assemblage of five aerosol types {BB, RU, CP, DP, DD} is treated with the *k*-means method by using four characteristic parameters $\{r_f, \sigma_f, r_c, \sigma_c\}$. With respect to the size distribution function, these four parameters are iteratively assembled into k subclusters to minimize the following equation:

$$d_j = \sum_{i=1}^{4} \frac{\left(x_j(i) - c_j(i)\right)^2}{v_j(i)^2}, \quad j = 1, k \tag{8}$$

where variables $x_j(i)$, $c_j(i)$, and $v_j(i)$ denote data records of the characteristic parameter, mean value, and standard deviation for character (i) in category (j), respectively. Naturally, the category number (k) should be less than the original cluster number (i.e., $k < 5$). The central value of the cluster is iteratively calculated by averaging the same character for every class involved within category j. The value of d_j represents the Euclidean distance of data $x_j(i)$ from the central value of the category. In other words, the steps of assembling and updating the central value are simultaneously repeated until the termination condition is reached. In practice, we assume 0.1% as the difference between the updated central value and the previous central value. Finally, we obtain k subclusters ($k < 5$), with each such subcluster comprising similar characters.

As a result, we determine the optimum number of final categories to be two (i.e., $k = 2$), with the two obtained subclusters being {BB, RU, CP, DP} and {DD} (refer to Table 1). This statistical result is reasonable for understanding the optical and physical properties of aerosols. The former and latter clusters represent the small anthropogenic aerosols (AA) and mineral dusts, respectively. This result was somewhat obvious before cluster analysis, because a cursory glance at Table 1 reveals that the four characteristic parameters $\{r_f, \sigma_f, r_c, \sigma_c\}$, except volume concentration $\{V_f, V_c\}$, are similar to each other for {BB, RU, CP, DP} types. We can then use fixed values {0.144, 1.533, 3.607, 2.104} for $\{r_f, \sigma_f, r_c, \sigma_c\}$ for the AA-type aerosols.

Next, we should pay attention to volume concentrations $\{V_f, V_c\}$. Here, we introduce a unique parameter (f) of the fine particle fraction, defined as $f = V_f/(V_f + V_c)$. Accordingly, the size distribution function in Eq. (7) can be approximately expressed for continental anthropogenic aerosols in a simpler form defined by a unique variable of fine particle fraction (f) with the above fixed parameters.

To validate the proposed approximate size distribution function, Fig. 12 presents as *the dashed* curves each form for each aerosol type described by the parameters in Table 1; further, the approximate size distribution function is denoted by *the solid* curves with each optimized value of (f).

TABLE 1 Subclustering of the Continental Aerosol Types With Respect to Size Distribution Parameters Partly Cited From AERONET Works (Dubovik et al., 2002; Omar et al., 2005).

Size Distribution		**DD**	**BB**	**RU**	**CP**	**DP**	**AA**
Parameters	r_f	0.117	0.144	0.133	0.140	0.140	0.144
	σ_f	1.482	1.562	1.502	1.540	1.540	1.533
	V_f	0.077	0.040	0.013	0.032	0.032	
	r_c	2.834	3.733	3.590	3.556	3.556	3.607
	σ_c	1.908	2.144	2.104	2.134	2.134	2.104
	V_c	0.268	0.081	0.020	0.034	0.034	

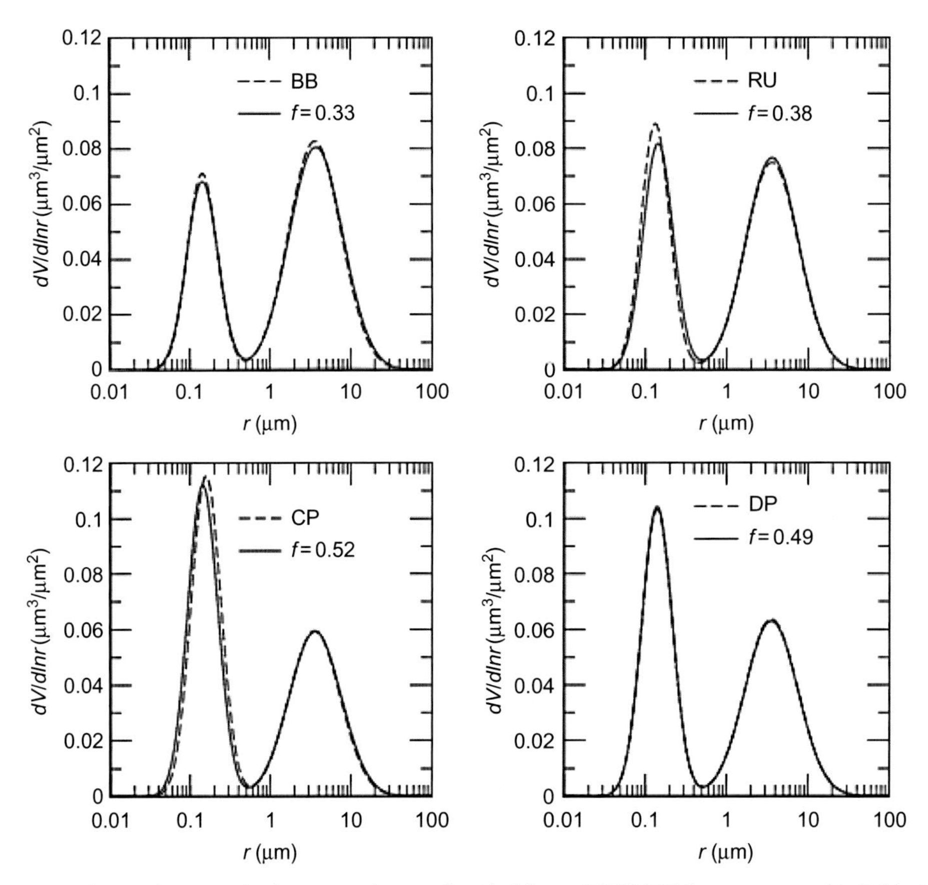

FIG. 12 Size distribution functions for four aerosol types classified from AERONET data represented in Table 1, and for an approximate form, denoted by the *dashed* and *solid curves*, respectively.

Another parameter of aerosol characteristics is the refractive index. It is reasonable to consider that a mixture of various aerosol types exists in nature. We adopt a simple internal homogeneous mixing of two components using the Maxwell Garnett mixing (MGM) rule. The MGM rule provides a complex refractive index calculated from the given equation (Chýlek and Srivastava, 1977; Bohren and Wickramasinghe, 1983), i.e.,

$$\varepsilon = \varepsilon_m \frac{(\varepsilon_j + 2\varepsilon_m) + 2g(\varepsilon_j - \varepsilon_m)}{(\varepsilon_j + 2\varepsilon_m) - g(\varepsilon_j - \varepsilon_m)} \tag{9}$$

where ε denotes electricity, subscripts m and j represent matrix and inclusion, respectively, and g is the volume fraction of the inclusions, which must be less than or equal to 0.4. Referring to the accumulated AERONET measurements, the values of the complex refractive index of AA-type aerosols seem to be weakly dependent on the wavelength, at least for the visible wavelength bands. We should also consider spectral absorption, i.e., the imaginary part of the refractive index, at near-UV wavelengths (Flores et al., 2014). We approximate $1.404 - 0.004i$ and $1.520 - 0.035i$ for matrix and inclusions, respectively, for applications in the visible bands data. In this approximation, the matrix and inclusions are assumed to be weak absorbing particles, such as CP-type, and strong absorbers, such as soot (Fuller et al., 1999; Syer et al., 2014).

The subclustering work of the AERONET products suggests that two types of aerosol properties, i.e., refractive index and size, are induced from the mixing of aerosol types with the volume fraction of inclusions (g) and an approximate bimodal log-normal distribution, defined by the fine particle fraction (f). In short, aerosol models are simply represented by two parameters, i.e., f and g from Eq. (9), for the size and the component of AA-type particles, such as biomass burning aerosols, respectively.

4 RETRIEVAL OF BIOMASS BURNING EPISODES

In radiative transfer calculations, AOT is a key parameter. It is reasonable to assume that AOT values are very high during aerosol episodes where incident solar radiation experiences multiple interactions with atmospheric aerosols due to the dense radiation field. Unfortunately, the precise simulation of multiple light-scattering processes requires very large computation times especially for aerosol episodes. Therefore, efficient algorithms are required for calculating the multiple scattering processes in an optically thick atmosphere model. Hence, the semiinfinite atmosphere model (i.e., AOT $\approx \infty$) is used for radiation simulations and applies to remote sensing of severe air pollution. Such cases have been treated with a radiation simulation method called the method of successive order of scattering (MSOS), which effectively calculates the upward intensity of radiation at TOA, i.e., reflectance of the optically thick atmosphere with AOT $>$ 5 approximately regarded as a semiinfinite atmosphere (Mukai et al., 2012, 2014). In other words, it is not necessary for the MSOS to take into account AOT. Furthermore, in the MSOS, the properties of ground-surface reflection can be ignored because the optical thickness of the atmosphere is too thick to affect the radiation interaction of the atmosphere with the ground surface.

In practice, application examples for satellite measurements in cases of biomass burning episodes have been shown. In Fig. 13, we show

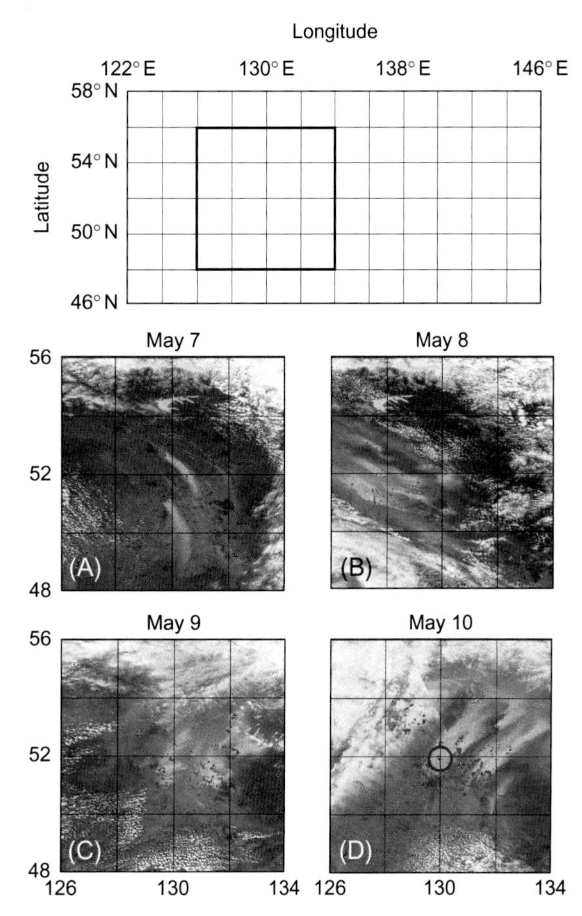

FIG. 13 Distribution of hotspots by Aqua/MODIS/MYD14 Collection 5 superimposed on the true color composite (RGB: Band1-4-3) images over Siberia (refer to the geographical map on top) from May 7 to May 10, 2016.

the true color composite images (RGB: Band1-4-3) from May 7 to May 10, 2016 of Aqua/MODIS, with the distribution of hotspots indicated by red dots (MYD14 Collection 5) over the enlarged area of northeastern Siberia (refer to the black square in the geographical map at the top of the figure). The hotspot distribution in the figure clearly suggests that the emission of a large volume of biomass of burning aerosols and gaseous particles is due to the forest fires at that time. The hazy smoke flowing out of the region of the fire's origin, through the wind, can thus be recognized.

We can assume very dense aerosol concentrations in the atmosphere for the present aerosol retrieval from Aqua/MODIS data on May 10, 2016 denoted by the blue open circle in Fig. 13D, where the value of AOT is very high (AOT (0.55 μm) > 5). Hence, the method of MSOS for the semiinfinite atmosphere model is used for radiation simulations. As noted above, the biomass burning aerosols are described solely by two parameters, i.e., f and g. We compare MODIS data (represented by the open circle in Fig. 13D) with simulated values of the reflected intensity, calculated based on radiation simulations, in a two-channel diagram with wavelengths of 0.46 and 0.55 μm. In radiation simulations in the Earth's atmosphere, we assume the atmosphere to be a semiinfinite, optically thick system consisting of AA-type aerosols. In the case of the semiinfinite atmosphere, we use the MSOS method (Mukai et al., 2012); in radiation simulations using MSOS, the two parameters f and g are desired. In other words, the size distribution takes the approximate form with a unique parameter f, while the complex refractive index is calculated using the Maxwell Garnett internal mixing rule, represented by Eq. (9), where g represents the volume fraction of the inclusions into the matrix. Parameter f represents the fine particle fraction of the approximate bimodal log-normal size distribution function. Thus, the two parameters, f and g, of aerosol properties for size distribution and refractive index, respectively, are sufficient to carry out radiation simulations for retrieving AA-type aerosols based on the MSOS. Accordingly, optimized values (f^*, g^*) for two parameters can be retrieved as estimated from the two wavelength planes. A detailed description of this procedure is presented by Mukai et al. (2015).

The dots in Fig. 14 denote the Aqua/MODIS data indicated by the open circles in Fig. 13D for May 10, 2016. Further, the desired aerosol properties are derived from the two-dimensional interpolation of the (f, g) coordinates, with step sizes of 0.01, to fit the MODIS data denoted by the dots. For clarity, the bottom shaded square

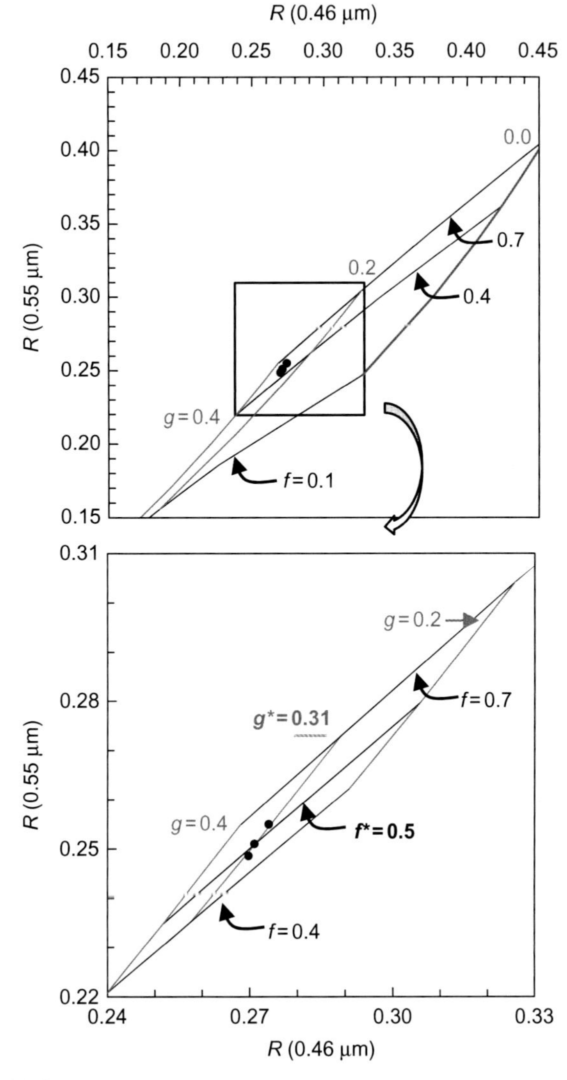

FIG. 14 Simulated values of the reflected intensity for our proposed aerosol models, described by two parameters (f, g) in a two-channel diagram of 0.46 μm and 0.55 μm, where the *red* and *black solid curves* denote the results for various values of parameters g and f, respectively. The *dots* denote Aqua/MODIS data acquired on June 10, 2016, at the target points in Fig. 13D. The *shaded lower figure* is the enlarged graph of a shaded portion of the *upper figure*.

demonstrates the enlarged portion of the top original figure in Fig. 14. From this enlarged figure, the optimized aerosol parameters are derived as (f^*, g^*) = (0.5, 0.31). The value of

$f=0.5$ suggests the dominance of fine particles, whereas $g=0.31$ corresponds to the absorbing carbonaceous aerosols.

5 MISCELLANEOUS-POLARIZATION REMOTE SENSING

The satellite polarization sensors POLDER-1, 2, and 3 onboard satellites ADEOS-1, -2, and Polarization & Anisotropy of Reflectances for Atmospheric Sciences coupled with Observations from a Lidar (PARASOL), respectively, have shown that the spectral photopolarimetry of the terrestrial atmosphere is very useful for observing Earth, and especially for aerosols (Kokhanovsky et al., 2015). It is impossible for atmospheric remote sensing to avoid the mixture case of aerosols and clouds, in particular, heavy aerosol pollution episodes. The POLDER sensors also collect both reflectance and polarization information at 16 different viewing angles (Deschamps et al., 1994). This multiangle polarization observation is useful and effective for cloud detection (Parol et al., 1999). Thus, polarization information measured by POLDER sensors has proven their usefulness in cloud and aerosol analysis, including difficult issues such as detecting cloud-top heights (Xianquiang et al., 2006) and retrieval of aerosols above cloud or snow (Waquet et al., 2013; Peers et al., 2015; Chang and Christopher, 2016).

Given these outstanding results, the POLDER heritage multiview polarization spectral imager, also known as the Multiviewing Multichannel Multipolarization (3MI), is planned for 2021 under the EPS follow-on system (EPS-SG) providing continuity of observations during the time frame from 2020 to 2040 (http://www.eumetsat.int). Note that EPS-SG denotes the European Organization for the Exploitation of Meteorological Satellites (EUMETSAT) Polar System-Second Generation. In Japan, the JAXA Global Change Observation Mission-Climate (GCOM-C) satellite boards the polarization sensor SGLI in 2017. Therefore, we intend to retrieve aerosol and cloud information from various perspectives, with a focus on polarization.

The polarization radiation field is described by Stokes vector $\mathbf{I}=(I, Q, U, V)$, where I is the intensity of light, Q and U define the magnitude and orientation of the linearly polarized fraction of light, respectively, and V is the magnitude and helicity of the circular polarization. First, we keep in mind the possibility or impossibility of polarization information in the light-scattering problems. Assuming that the incident light to the particle is natural, polarization is produced by the light-scattering process. The polarization signal decreases further with each scattering. Accordingly, the optical properties of aerosols or clouds are preserved in polarized radiance. On the contrary, total radiance is strongly influenced by multiple scattering.

Fig. 15 presents a numerical example for simulated reflectance vector $\mathbf{R}=(I, Q)$ in the meridian plane at a wavelength of 0.67 μm for a finite optical thickness atmosphere model represented by AOT (0.67 μm) consisting of BB-type aerosols (Omar et al., 2005, Section 3.2) with an incident solar angle of 0.5 degrees. The simulations are conducted on the assumptions that the aerosol shape is spherical and the incident solar light is natural; then, I and Q represent total radiance and linear polarized component of reflectance, respectively, from TOA. Fig. 15 shows that the polarized component ($=Q$) converges at approximately AOT (0.67 μm) $=3$, i.e., the polarization signal saturates here and the total radiance ($=I$) converges for AOT (0.67 μm) >5. Even in this simple calculation, it is shown that the aerosol characteristics are preserved in the polarized component. It is of interest to mention that the degree of polarization ($\approx Q/I$) decreases due to increasing of intensity (I) with AOD. But the polarized component ($\approx Q$) is preserved even in the thick atmosphere derived from the single and/or a few times of scattering process. Therefore, the polarized component (Q, or exactly $(Q^2+U^2)^{1/2}$ in general) is available for aerosol retrieval. We know that the information of the aerosol characteristics is involved just in the

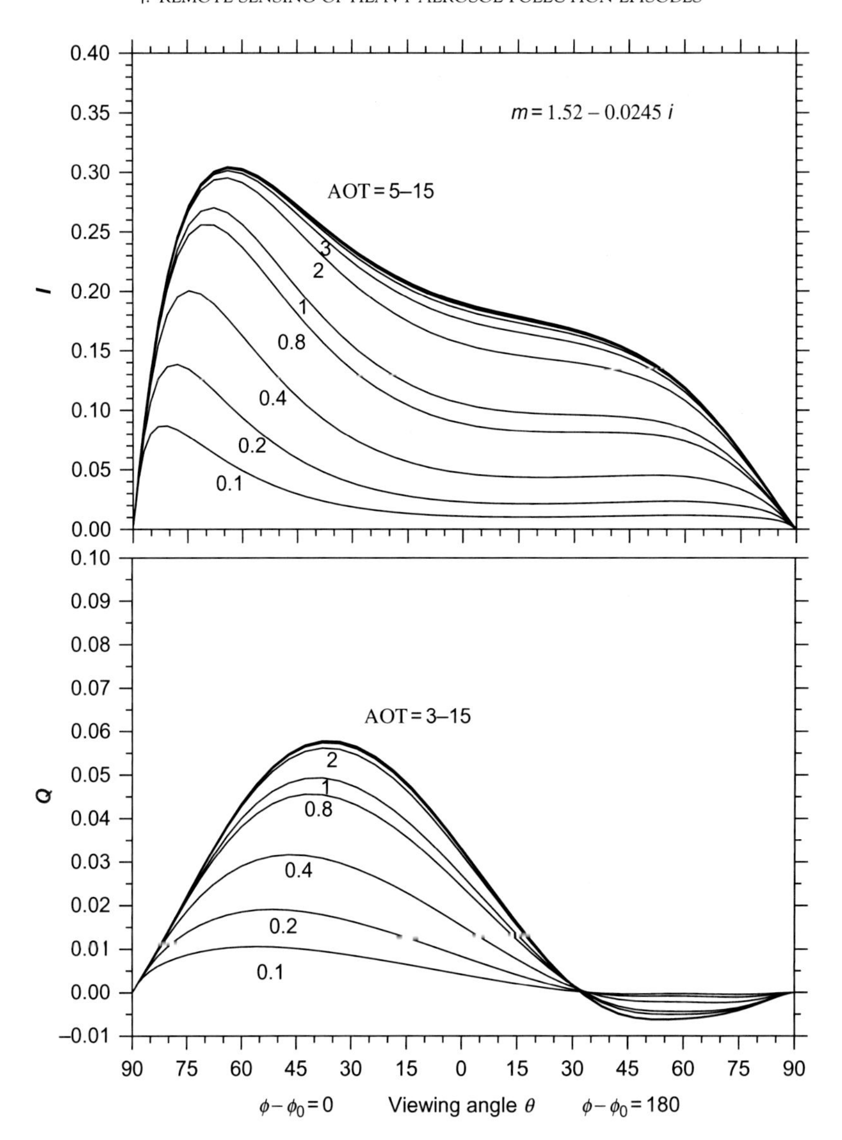

FIG. 15 Example of simulated reflectance vector $\mathbf{R} = (I, Q)$ in the meridian plane at a wavelength of 0.67 μm for a finite optical thickness atmosphere model represented by AOT (0.67 μm) consisting of BB-type aerosols (Omar et al., 2005, Section 3.2) with an incident solar angle of 0.5 degrees. Here, I and Q denote the total radiance and linear polarized component of reflectance at TOA, respectively.

single scattering behavior. On the contrary, atmospheric information, such as an aerosol optical thickness (AOT) of the Earth's atmosphere, is deduced from the multiple scattering effects.

As an example of polarization remote sensing, we describe here that the polarized component is helpful in determining cloud-top (or aerosol) heights. In wavelengths shorter than the blue visible frequency, LPR is primarily related to Rayleigh scattering of atmospheric molecules. The values of the polarized radiance are less sensitive to multiple scattering effects than those of total radiance, as indicated above. The polarized radiance must then be less affected from reflection by the bottom surface (i.e., cloud, land, or ocean). In other words, polarized radiance observed in the short-wavelength channels around a scattering angle ranging from 80 to 120 degrees is approximately decided by the atmospheric molecular optical thickness above the middle- or high-level clouds. This directional condition occurs because maximum polarization occurs at an angle $\cong$ 90 degrees in Rayleigh scattering. Further, specific directions, such as the sun glint region or water cloud rainbow ($\cong$ 140 degrees), are not appropriate for determining the optical thickness above the cloud (Buriez et al., 1997). Once the optical thickness above the cloud has been retrieved, the cloud-top height can be retrieved via the vertical profile of Rayleigh scattering optical thickness provided from the standard Earth's atmosphere model.

In retrieving aerosols above the cloud, the cloud top is often assumed as a Lambert reflecting surface due to the multiple scattering that occurs within a cloud. Note that the RT simulation method MSOS is useful not only for the heavy dense atmosphere shown in the previous section but also for the cloud. In particular for the retrieval of absorbing aerosols above the cloud, short-wavelength satellite data provided by GCOM-C/SGLI or 3MI/EPS-SG have proven to be useful. Further, the vector form of MSOS plays an important role in polarization measurements (Waquet et al., 2013; Peers et al., 2015; Kokhanovsky et al., 2015). In addition, MSOS is available for aerosol retrieval in the mixture case involving clouds and haze. We conclude that both aerosol and cloud information should be retrieved considering various perspectives by using polarization and intensity in the UV-IR spectral region.

Our attention should now be to polarization remote sensing. Several polarimetric missions to the planet Earth are scheduled for launch in the coming year. It is known that the satellite polarimetry is one of the most promising fields in aerosol/cloud remote sensing as mentioned above. Nevertheless, much more improvement of observation quality and advancement of retrieval algorithms for aerosols, clouds, and so on are earnestly required in this field. After the achievements of these issues, effective utilization of polarization remote sensing will be realized. It is possible to say that this field has grown even now.

6 CONCLUDING REMARKS

This study has investigated a series of flows on heavy aerosol pollutants due to dust storms and biomass burning plumes observed by satellites from the detection to the retrieval. The results are briefly summarized below.

1. Accumulated distribution of aerosol optical thickness (AOT) from space provides us with regional and/or temporal variations in the ecosystem as well as atmospheric aerosols.
2. The products obtained from the satellite apart from AOT, such as temperature, were also available for this study.
3. The spectral information obtained by satellites is useful for detection and retrieval of aerosol pollutants.
4. Algorithm design is an essential basis for satellite data analysis. Therefore, we have focused on retrieval algorithms herein.
5. The space-based data and/or the retrieved products should be validated with the ground-based data and/or numerical model simulations and vice versa. Data assimilation between various products upgrades the accuracy of each product.
6. Polarization is a strong tool for the present subject, and hence more improvements from both sides of the monitoring technology and the analytical theory are being developed now.

Satellite remote sensing provides us with systematic monitoring of the Earth with different spatial, spectral, and temporal resolutions. In addition, space observations are particularly useful for aerosol pollution episodes. Although, in the broad sense, the term "aerosol pollution" may denote various meanings, we restrict it to the atmospheric aerosol causing air pollution here. In other words, air pollution deteriorates the air quality. Good air quality refers to clean, clear, unpolluted air. Although clean air is a basic requirement for a better ecosystem, air pollution has become a serious problem worldwide. It is highly likely that large-scale air pollution will continue to occur because air pollution becomes severe due to both the increasing emissions of the anthropogenic aerosols and the complicated behavior of natural aerosols. Therefore, a synthetic study of aerosols from space, ground, and computers is desired.

References

Ackerman, S.A., 1997. Remote sensing aerosols using satellite infrared observations. J. Geophys. Res. 102, 17069–17080.

Bohren, C.F., Wickramasinghe, N.C., 1983. On the computation of optical properties of heterogeneous grains. Astrophys. Space Sci. 50, 461–472.

Buriez, J.-C., Vanbauce, F., Parol, F., Goloub, P., Herman, M., Bonnel, B., Fouquart, Y., Couvert, P., Seze, G., 1997. Cloud detection and derivation of cloud properties from POLDER. Int. J. Remote Sens. 18 (13), 2785–2813.

Chang, I., Christopher, A., 2016. Identifying absorbing aerosols above clouds from the spinning enhanced visible and infrared imager coupled with NASA A-train multiple sensors. IEEE Trans. Geosci. Remote Sens. 54 (6), 3163–3173.

Chýlek, P., Srivastava, V., 1977. Dielectric constant of a composite inhomogeneous medium. Phys. Rev. B 27, 5098–5106.

Crutzen, P.J., Meinrat, A.O., 1990. Biomass burning in the tropics: impact on atmospheric chemistry and biogeochemical cycles. Science 250 (4988), 1669.

Cyranoski, D., Fuyuno, I., 2005. Climatologists seek clear view of Asia's smog. Nature 434, 128. https://doi.org/10.1038/434128a.

Deschamps, P.Y., Breon, F.M., Leroy, M., Podaire, A., Bricaud, A., Buriez, J.C., Seze, G., 1994. The POLDER mission: instrument characteristics and scientific objectives. IEEE Trans. Geosci. Remote Sens. 32, 598–615.

Dubovik, O., King, M.D., 2000. A flexible inversion algorithm for retrieval of aerosol optical properties from Sun and sky radiance measurements. J. Geophys. Res. 505, 20673–20696.

Dubovik, O., Holben, B.N., Eck, T.F., Smirnov, A., Kaufman, Y.J., King, M.D., Tanré, D., Slutsker, I., 2002. Variability of absorption and optical properties of key aerosol types observed in worldwide locations. J. Atmos. Sci. 59, 590–608.

Dubovik, O., Herman, M., Holdak, A., Lapyonok, T., Tanré, D., Deuzé, J.L., Ducos, F., Sinyuk, A., Loptin, A., 2011. Stastistically optimized inversion algorithm for enhanced retrieval of aerosol properties from spectral multi-angle polaimetric satellite observations. Atmos. Meas. Tech. 4, 975–1018.

Dubovik, O., Lapyonok, T., Litvinov, P., Herman, M., Fuertes, D., Ducos, F., Lopatin, A., Chaikovsky, A., Torres, B., Derimian, Y., Huang, X., Aspetsberger, M., Federspiel, C., 2014. GRASP: a versatile algorithm for characterizing the atmosphere. SPIE Newsroom, 1–4. https://doi.org/10.1117/2.1201408.005558.

Flores, J.M., Zhao, D.F., Segev, L., Schlag, P., Kiendler-Scharr, A., Fuchs, H., Watne, A.K., Blivshtein, N., Mentel, T.F., Hallquist, M., Rudich, Y., 2014. Evolution of the complex refractive index in the UV spectral region in ageing secondary organic aerosol. Atmos. Chem. Phys. 14, 5793–5806.

Fuller, K.A., Malm, W.C., Kreidenweis, S.M., 1999. Effects of mixing on extinction by carbonaceous particles. J. Geophys. Res. 104 (D13), 15941–15954.

Giglio, L., Descloitres, J., Justice, C.O., Kaufman, Y.J., 2003. An enhanced contextual fire detection algorithm for MODIS. Remote Sens. Environ. 87, 273–282. https://doi.org/10.1016/S003404257(03)00184-6.

Herman, J.R., Bhartia, P.K., Torres, O., Hsu, N.C., Seftor, C.J., Celarier, E., 1997. Global distribution of UV-absorbing aerosols from Nimbus-7 TOMS data. J. Geophys. Res. 102, 16911–16922.

Holben, B.N., Eck, T.F., Slutsker, I., Tanré, D., Buis, J.P., Setzer, A., Vermote, E., Reagan, J.A., Kaufman, Y., Nakajima, T., Lavenu, F., Jankowiak, I., Smirnov, A., 1998. AERONET—a federated instrumentnetwork and data archive for aerosol characterization. Remote Sens. Environ. 66, 1–16.

Höller, R., Higurashi, A., Nakajima, T., 2004. The GLI 380-nm channel – application for satellite remote sensing of tropospheric aerosol. In: Proc. EUMETSAT Meteorological Satellite Conference.

Hsu, N.C., Herman, J.R., Bhartia, P.K., Seftor, C.J., Torres, O., Thompson, A.M., Gleason, J.F., Eck, T.F., Holben, B.N., 1996. Detection of biomass burning smoke from TOMS instruments. Geophys. Res. Lett. 23, 745–748.

Ishimoto, H., Masuda, K., Mano, Y., Orikasa, N., Uchiyama, A., 2012. Irregularly shaped ice aggregates in optical modeling of convectively generated ice clouds.

JQSRT 113, 632–643. https://doi.org/10.1016/j.jqsrt.2012.01.017.

Justice, C.O., Giglio, L., Korontzi, S., Owens, J., Morisette, J.T., Roy, D., Descloitres, J., Allesume, S., Petitcolin, F., Kaufman, Y.J., 2002. The MODIS fire products. Remote Sens. Environ. 83, 244–262. pii: S0034-4257(02)00076-7.

Kahn, R., Anderson, J., Anderson, T.L., Bates, T., Brechtel, F., Carrico, C.M., Clarke, A., Doherty, S., Dutton, E., Flagan, R., Frouin, R., Fukushima, H., Holben, B., Howell, S., Huebert, B., Jefferson, A., Jonsson, H., Kakashnikova, O., Kim, J., Kim, S.-W., Kus, P., Li, W.-H., Livingston, J.M., Mcnaughton, C., Merrill, J., Mukai, S., Murayama, T., Nakajima, T., Quinn, P., Redemann, J., Rood, M., Russell, P., Sano, I., Schmid, B., Seinfeld, J., Sugimoto, N., Wang, J., Welton, E.J., Won, J.-G., Yoon, S.-C., 2004. Environmental snapshots from ACE-Asia. J. Geophys. Res. 109, D19S14. https://doi.org/10.1029/ 2003JD004339.

King, M.D., Kaufman, Y.J., Menzel, W.P., Tanré, D., 1992. Remote sensing of cloud, aerosol, and water vapor properties from the Moderate Resolution Imaging Spectrometer (MODIS). IEEE Trans. Geosci. Remote Sens. 30, 2–27.

Kinne, S., Lohmann, U., Feichter, J., Schulz, M., Timmreck, C., Ghan, S., Easter, R., Chin, M., Ginoux, P., Takemura, T., Koch, D., Herzog, M., Penner, J., Pitari, J., Holben, B., Eck, T., Smirnov, A., Dubovik, O., Slutsker, I., Tanre, D., Torres, O., Mishchenko, M., Geogdzhayev, I., Chu, A., Kaufman, Y., 2003. Monthly averages of aerosol properties: A global comparison among models, satellite data and AERONET ground data. J. Geophys. Res. 108 (D20), 4634. https://doi.org/10.1029/ 2001JD 001253.

Kneizys, F.X., Shettle, E.P., Abreu, L.W., Chetwynd, J.H., Anderson, G.P., Gallery, W.O., Selby, J.E.A., Clough, S.A., 1988. Users Guide to LOWTRAN 7, AFGL-TR-88-0177. Air Force Geophysics Laboratory, Hanscom AFB, MA.

Kokhanovsky, A.A., 2010. Light Scattering Reviews 5. Springer-Verlag, Berlin, Heidelberg. ISBN 978-3-642-10335-3.

Kokhanovsky, A.A., Davis, A.B., Cairns, B., Dubovik, O., Hasekamp, O.P., Sano, I., Mukai, S., Rozanov, V.V., Litvinov, P., Lapyonok, T., Kolomiets, I.S., Oberemok, Y.A., Savenkov, S., Martin, W., Wasilewski, A., Di Noia, A., Stap, F.A., Rietjens, J., Xu, F., Natraj, V., Duan, M., et al., 2015. Space-based remote sensing of atmospheric aerosols: the multi-angle spectro-polarimetric frontier. Earth Sci. Rev. 145, 85–116. https://doi.org/10.1016/j.earscirev.2015.01.012.

Littmann, T., 1991. Dust storm frequency in Asia: climatic control and variability. Int. J. Climatol. 11, 393–412.

Lopatin, P., Dubovik, O., Chaikovsky, A., Goloub, P., Lapyonok, T., Tanré, D., Litvinov, P., 2013. Enhancement of aerosols characterization issuing synergy of lidar and sunphotometer coincident observations: the GARRLiC algorithm. Atmos. Meas. Tech. 6, 2065–2088. https://doi.org/10.5194/amt-6-2065-2013.

Mischenko, M.I., Hovenier, J.W., Travis, L.D., 2000. Light Scattering by Nonspherical Particles. Academic Press, San Diego, London.

Mukai, M., Nakajima, T., Takemura, T., 2004. A study of long-term trends in mineral dust aerosol distributions in Asia using a general circulation model. J. Geophys. Res. 109, D19204. https://doi.org/10.1029/2003JD004270.

Mukai, M., Sano, I., Iizuka, T., Yokomae, T., Mukai, S., 2010. Detection and analysis of dust aerosol particles over the East Asia. J. Remote Sens. Soc. Jpn. 30 (1), 1–10.

Mukai, S., Yokomae, T., Sano, I., Nakata, M., Kohkanovsky, A., 2012. Multiple scattering in a dense aerosol atmosphere. Atmos. Meas. Tech. Discuss. 5, 881–907.

Mukai, S., Yasumoto, M., Nakata, M., 2014. Estimation of biomass burning influence on air pollution around Beijing from an aerosol retrieval model. Sci. World J. Article ID 649648.

Mukai, S., Nakata, M., Yasumoto, M., Sano, I., Kokhanovsky, A., 2015. A study of aerosol pollution episode due to agriculture biomass burning in the east-central China using satellite data. Front. Environ. Sci. 3, 57. https://doi.org/10.3389/fenvs.2015.00057.

Nakata, M., Sano, I., Mukai, S., 2015. Air pollutants in Osaka (Japan). Front. Environ. Sci. Environ. Inf. 3, https://doi.org/10.3389/fenvs.2015.00018. Article 18.

Okada, Y., 2008. Efficient numerical averaging of light scattering properties with quasi-Monte-Carlo method. JQSRT 109, 1719–1742. https://doi.org/10.1016/j.jqsrt.2008.01.002.

Omar, A.H., Won, J.-G., Winker, D.M., Yoon, S.-C., Dubovik, O., McCormick, M.P., 2005. Development of global aerosol models using cluster analysis of Aerosol Robotic Network (AERONET) measurements. J. Geophys. Res. 110 (D10S14), 1–14.

O'Neill, N.T., Dubovik, O., Eck, T.F., 2001. Modified Angstrom exponent for the characterization of submicrometer aerosols. Appl. Opt. 40, 2368–2375.

O'Neill, N.T., Eck, T.F., Smirnov, A., Holben, B.N., Thulasiraman, S., 2003. Spectral discrimination of coarse and fine mode optical depth. J. Geophys. Res. 108, 4559. https://doi.org/10.1029/2002JD0029753.

Parol, F., Buriez, J.-C., Vanbauce, C., Couvert, P., Seze, G., Goloub, P., Cheinet, S., 1999. First results of the POLDER "Earth Radiation Budget and Clouds" operational algorithm. IEEE Trans. Geosci. Remote Sens. 37 (3), 1579–1612.

Peers, F., Waquet, F., Cornet, C., Dubuisson, P., Ducos, F., Goloub, P., Szczap, F., Tanré, D., Thieuleux, F., 2015. Absorption of aerosols above clouds from POLDER/PARASOL measurements and estimation of their radiative effect. Atmos. Chem. Phys. 15, 4179–4196.

Pérez, C., Nickovic, S., Pejanovic, G., Baldasano, J.M., Özsoy, E., 2006. Interactive dust-radiation modeling: a step to improve weather forecast. J. Geophys. Res

111 (D16206), 1984–2012. https://doi.org/10.1029/2005JD006717.

Remer, L.A., Kaufman, Y.J., Tanré, D., Mattoo, S., Chu, D.A., Martins, J.V., Li, R.-R., Ichoku, C., Levy, R.C., Kleidman, R.G., Eck, T.F., Vermote, E., Holben, B.N., 2005. The MODIS aerosol algorithm, products, and validation. J. Atmos. Sci. 62, 947–973. https://doi.org/10.5194/acp-15-417902015.

Sano, I., Okada, Y., Mukai, M., Mukai, S., 2009. Retrieval algorithm based on combined use of POLDER and GLI data for biomass aerosols. JRSSJ 29 (1), 54–59.

Sokolik, I.N., Winker, D.M., Bergametti, G., Gillette, D.A., Carmichael, G., Kaufman, Y.J., Gomes, L., Schuetz, L., Penner, J.E., 2001. Introduction to special section: outstanding problems in quantifying the radiative impacts of mineral dust. J. Geophys. Res. 106 (D16), 18015–18027. https://doi.org/10.1029/2000JD900498.

Syer, A.M., Hsu, N.C., Eck, T.F., Smirnov, A., Holben, B.N., 2014. AERONET-based models of smoke-dominated aerosol near source regions and transported over ocean, and implications for satellite retrievals of aerosol optical depth. Atmos. Chem. Phys. 14, 11493–11523.

Takemura, T., Nakajima, T., Dubovik, O., Holben, B.N., Kinne, S., 2002. Single-scattering albedo and radiative forcing of various aerosol species with a global three-dimensional model. J. Clim. 15, 333–352.

Takemura, T., Nozawa, T., Emori, S., Nakajima, T., Nakajima, T., 2005. Simulation of climate response to aerosol direct and indirect effects with aerosol transport-radiation model. J. Geophys. Res. 110. https://doi.org/10.1029/2004JD005029.

Tegen, I., Miller, R., 1998. A general circulation model study on the interannual variability of soil dust aerosol. J. Geophys. Res. 103 (D20), 25975–25995.

Torres, O., Bhartia, P.K., Herman, J.R., Ahmad, Z., Gleason, J., 1998. Derivation of aerosol properties from satellite measurements of backscattered ultraviolet radiation: theoretical basis. J. Geophys. Res. 103, 17099–17110.

Waquet, F., Cornet, C., Deuzé, J.-L., Dubovik, O., Ducos, F., Goloub, P., Herman, M., Lapyonok, T., Labonnote, L.C., Riedi, J., Tanré, D., Thieuleux, F., Vanbauce, C., 2013. Retrieval of aerosol microphysical and optical properties above liquid clouds from POLDER/PARASOL polarization measurements. Atmos. Meas. Tech. 6, 991–1016. https://doi.org/10.5194/amt-6-991-2013.

Xianquiang, H., Delu, P., Yan, B., Zhihua, M., 2006. Cloud-top height retrieval from polarizing sensor-POLDER. Proc. SPIE 6419, 641922. https://doi.org/10.1117/12.713397.

CHAPTER

5

Aerosol and Cloud Bottom Altitude Covariations From Multisensor Spaceborne Measurements

*Luca Lelli**, *Marco Vountas**

***University of Bremen, Bremen, Germany**

1 INTRODUCTION

Earth's climate is a complex perturbed system in which a wealth of chemical and physical processes takes place on a wide range of spatial and temporal scales. Global and regional climate regimes are increasingly changing, driven by changes of the constituents of the Earth-atmosphere system. It is understood that the human well-being may be subject to climate settings. This is particularly true for populations heavily dependent on favorable climate conditions. The demographic vulnerability to climate change may be enhanced by geographic and economic disparities (Samson et al., 2011).

Additionally, the revolving economic cycle and the energy consumption by emerging economies are giving rise to specific forcings and feedbacks to the climate system. Contemporary concerns include change of land and water use, change in mean temperatures and greenhouse gases emissions and injection of solid-phase particulate matter in the atmosphere (van Donkelaar et al., 2010) or low-volatility compounds in the gas phase (Andreae and Rosenfeld, 2008). Both classes can be categorized as precursors of a broader type of dry or moist aerosols. The aerosols may, in turn, change cloud properties (Twomey, 1977, 1991) and perturb cloud lifetime (Albrecht, 1989), as well the cloud precipitation efficiency (Stevens and Feingold, 2009). Therefore, aerosol particles are also important players within the hydrological cycle (Allen and Ingram, 2002), mediated by the clouds, which act as water reservoirs in the atmosphere.

Is it well-known that aerosol particles can perturb the precipitation efficiency of clouds in various ways (Rosenfeld et al., 2014, and references therein), from drying the atmospheric column via direct absorption of sunlight (Feingold, 2005) or by changing the size spectrum of cloud droplets (Twomey, 1977), thus modulating diffusion and coalescence processes. As a consequence, the vertical distribution of latent heat can also change and impact on the cloud and atmospheric dynamics.

Remote Sensing of Aerosols, Clouds, and Precipitation
https://doi.org/10.1016/B978-0-12-810437-8.00005-0

© 2018 Elsevier Inc. All rights reserved.

Another atmospheric constituent interacting with clouds is black carbon. Its direct effect on clouds, both in liquid and ice phase, takes place through perturbations of the temperature of the atmospheric column around the clouds themselves and their spatial distribution may change. No consensus has yet been reached on the role played by these black carbon-contaminated clouds in the climate system (Bond et al., 2013). This brief, yet incomplete, overview of the possible interactions between aerosols and clouds in the atmosphere portrays their structural trait of being proxies of many chemical and physical processes in the climate system. Therefore, monitoring cloud and aerosol properties together over time and space unveils the underlying processes of a changing climate.

Aerosol particles are broadly categorized for their size and chemical composition. Typical sizes span several orders of magnitude, ranging from the nanometer scale of the nucleation mode of nuclei throughout to giant particles of several tens of microns, serving as condensation or freezing nuclei for clouds (e.g., CCN or IN) in a water-saturated local environment. Satellite-based estimates of aerosol properties focus primarily on their ability to attenuate impinging sunlight throughout the tropospheric column (i.e., aerosol optical depth, AOD). Patterns of AOD are therefore pivotal in setting the spatio-temporal constraints of possible interactions with water vapor and clouds. Another quantity needed for spaceborne aerosol-cloud interaction studies is the Ångström exponent, which is an indicator of the particulate size spectrum, as the ability to nucleate a cloud droplet is a function of the aerosol particle size. Eventually, with the aid of the Absorbing Aerosol Index (AAI), which is defined as the ratio of two channels in the UV range of the spectrum, it is possible to isolate absorbing from nonabsorbing atmospheric aerosols and assess their relative abundances.

Aerosols that can perturbate clouds may be natural, such as mineral dust or sea-spray, or anthropogenic, like soot or particles emitted as a result of biomass burning and fossil fuel consumption and are, therefore, often well correlated with spatial and temporal patterns of human activities (van Donkelaar et al., 2010; Rosenfeld, 2006). Their impact onto clouds is manifold and the conceptual framework provided by Rosenfeld et al. (2008) illustrates a small, yet representative, subset of the possible effects that aerosols may exert on the macro- and micro-structure of clouds and their ability to precipitate.

Modern satellite observations offer the possibility to monitor such covariations between atmospheric particulate and clouds, as the recent body of literature increasingly shows (Kaufman and Nakajima, 1993; Bréon et al., 2002; Costantino and Bréon, 2010; L'Ecuyer et al., 2009; Bulgin et al., 2008; Chen et al., 2014, among others). They have advantages and drawbacks. On the one hand, the relatively long temporal coverage increases the robustness of the statistical analysis and the nearly global view allows the selective inspection of local different meteorological conditions, which can be considered natural laboratories of specific interaction regimes. On the other hand, remote sensing techniques rely on prior assumptions of radiative transfer and local optical properties that often hamper the full description of the complexity of real cloud habits or do not capture the actual changes in aerosol and cloud microphysics (Koren et al., 2010a).

Nonetheless, one can expect control of cloud development by aerosols as a result of the transition between pristine ($0 \leq \text{AOD} \lesssim 0.1$, CCN $\sim 100\ \text{cm}^{-3}$), slightly polluted ($0.1 \lesssim \text{AOD} \lesssim 0.3$, CCN $\sim 300\ \text{cm}^{-3}$), and heavily polluted environments ($\text{AOD} \gtrsim 0.3$) (Koren et al., 2010a). Convective (warm base, cold top, above the freezing level) and nonconvective (warm base, warm top, below the freezing level) clouds are seen to extend, albeit with distinct magnitudes, in the vertical and horizontal dimensions, since aerosols delay in-cloud collision-coalescence processes and rain formation, providing the extra time for the clouds to develop further. The occurrence of this covariation has been observed with a 10-year record of ground based data at the Atmospheric Radiation Measurement (ARM) site of Southern Great Plain (SGP) (Li et al., 2011), as well as with satellite data of the MODIS instrument onboard the Terra (Koren et al., 2010a,b) and Aqua (Koren et al., 2014) platforms. The mediating role in this perturbation of the primary indirect aerosol effect (Twomey, 1977) has already been corroborated (Li et al., 2011; Koren et al., 2014).

When studying invigoration of clouds by aerosols with satellite-based data, it has been a straightforward option to first look at changes of the upper boundary of clouds. Not only are vertical shifts of the higher portions of cloud bodies relevant for the microphysics in the ice phase (Fridlind et al., 2004), for the lifetime of stratiform anvils of deep convective clouds (Fan et al., 2013), for rain efficiency (Goren and Rosenfeld, 2014), for in-cloud evaporative processes (Christensen and Stephens, 2011), and for radiative forcing (Feingold et al., 2016), but they are also the first accessible cloud parameter by means of modern spaceborne retrieval algorithms. Recent studies focused on the relationship of AOD to vertical displacement of clouds by means of changes in cloud fraction (Gryspeerdt et al., 2014, 2016; Engström and Ekman, 2010). Hence, the bottom altitude of clouds has, so far, received less attention. Only recently, new methods have been developed (Zheng et al., 2015; Zheng and Rosenfeld, 2015) to infer updraft velocities at the cloud base, filling the gaps in the understanding of CCN activation in supersaturated environments.

For the above reasons, beside a brief description of the aerosol and cloud records generated from broadband measurements by the Advanced Along-Track Scanning Radiometer (AATSR) sensor (Section 2.1) within the European Climate Change Initiative (CCI) framework, we provide the description and the deployment, upon validation against ground-based ARM profiles, of a satellite-based dataset of cloud bottom height (CBH). The CBH dataset is inferred from measurements of reflected sunlight in the oxygen A-band in the near-infrared (NIR) by the polar orbiting sensor SCanning Imaging Absorption spectroMeter for Atmospheric CHartographY (SCIAMACHY) (Section 2.2), companion payload of AATSR on Envisat. The results of covariations of aerosol and cloud properties are given in Section 3 and potential implications as well as future avenues of advances in the spaceborne assessment of interactions are given in Section 4.

2 DATA AND METHODS

2.1 Aerosol and Cloud Record: CCI Datasets

Global datasets of aerosol and cloud properties are available as a result of the CCI initiated by the European Space Agency (Hollmann et al., 2013). Every CCI subproject consists of several phases and, namely, a round robin exercise between existing algorithms with the purpose of refinement and further development and a consolidation phase in which long-term datasets are generated for climate applications. This procedure is common to CCI working groups dealing with the retrieval of cloud and aerosol properties. We refer the interested reader to Holzer-Popp et al. (2013) and de Leeuw et al. (2015) for the aerosol round robin phase and to Popp et al. (2016) for the consolidation of the aerosol records, while information on the activity of the cloud working group can be found in Stengel et al. (2015).

The choice of the datasets used for this work is based on the aerosol and cloud records generated by the Oxford-RAL Retrieval of Aerosol and Cloud (ORAC) algorithm applied to measurements of the AATSR sensor. Details on the ORAC aerosol module can be found in

Thomas et al. (2009), while the ORAC cloud setup, also renamed Community Code for Climate (CC4CL), is described in Poulsen et al. (2012).

ORAC uses an optimal estimation retrieval scheme (Rodgers, 2000) designed for the retrieval of aerosol and/or cloud properties from AATSR (in general the whole ATSR family of dual-viewing instruments). It provides uncertainty propagation and includes the use of a priori knowledge by calculating all retrieved parameters as a function of all measurements simultaneously. The algorithm underwent several steps of validation of which the recent two assessments Popp et al. (2016) and Che et al. (2016) are briefly discussed below, as previous attempts were related to earlier versions of the ORAC AOD retrieval not actively used any more.

In Popp et al. (2016) ORAC's AOD quality (v3.02) was systematically assessed among two other (competing) AOD retrieval algorithms using AATSR. A score comparison indicated that ORAC over land performs slightly worse than one out of three algorithms while over ocean/coastal sites, differences between the three algorithms are small. The regional total scores show that no single AATSR retrieval is better than any other ATSR retrieval simultaneously in all regions.

In the recent technical note by Che et al. (2016), ORAC derived AOD (v3.04) has been specifically validated over China. It performs well, even at great AOD values, however, the stability decreases with increasing AOD, especially when AOD values are larger than one. Che et al. (2016) corroborate that the ORAC AOD product is not only consistent with AERONET but also with the CARSNET (China Aerosol Remote Sensing Network) product, another measurement network specifically designed for aerosol observations over China. In comparison, the ORAC product has in general better coverage than the other competing aerosol algorithms.

Similar to the aerosol module, the CC4CL for clouds (Poulsen et al., 2017a) uses an inverse model also based on optimal estimation. To obtain cloud parameters, which give the best fit between a prediction by a forward model and observed radiances, pixel-based measurement uncertainty estimates and relevant prior knowledge are taken into account. While these estimates are based on optimal estimation theory for most variables, cloud mask uncertainty is based on hit rate scores against measurements from the Cloud-Aerosol Lidar with Orthogonal Polarization (CALIOP). All pixel level uncertainties are propagated in a mathematically consistent way into daily products.

The CC4CL forward model is essentially represented by a radiative transfer model to simulate satellite radiances based on a parameterized cloud-atmosphere-surface model and prescribed observing conditions. The basic principle is the probability maximization of the retrieved state vector, conditional on the value of the measurements and any prior knowledge. Inherently, it is assumed that errors in the measurements, forward model, and a priori parameters are normally distributed with zero mean and covariances, respectively. Starting from some initial guess of the state and linearizing the forward model, the gradient of a cost function is computed. Using that, a state is selected which is predicted to have lower cost. The procedure is iterated until the change in cost between iterations is less than a certain threshold or the retrieval is given up after 40 iterations.

The essential part of the forward model can consist of three components: (i) a scattering cloud layer is located within (ii) a clear-sky atmosphere over (iii) a surface of known reflectance or emissivity (depending on the spectral region the AATSR channel is measuring, visible-to-shortwave or thermal infrared wavelengths). The surface is characterized by a bidirectional reflectance distribution function (BRDF) which is computed differently for ocean and land surfaces. The BRDF over ocean is computed using the methodology outlined by Sayer et al. (2011). The BRDF over land is a weighted sum of three different BRDF kernels. The weights for these kernels are provided by the 0.05° MODIS MCD43C1 BRDF auxiliary input.

Each measurement pixel is considered to be either fully cloudy or clear. A cloud is assumed to be a single, plane-parallel layer of either liquid or ice particles. The layer is assumed to be (geometrically) infinitely thin and is placed within the clear-sky atmosphere model. The cloud layer is parametrized in terms of cloud phase (i.e., ice or liquid); effective radius of the cloud particle size distribution; total (vertically integrated) optical depth of the cloud at a fixed wavelength of 0.55 μm; and cloud top pressure.

To the best of our knowledge (see Poulsen et al., 2017a for additional information), there are no recent peer-reviewed articles on the validation of CC4CL. Resorting to the official Product Validation and Intercomparison Report (Poulsen et al., 2017b), in which recommendations for the usage of the cloud product suite are given, cloud fraction and top pressure/height derived from AATSR possess moderate-to-high accuracy and stability and the found pixel-based bias against CALIPSO retrievals amounts to ~400 m for low-level clouds. However, the predecessor version Global Retrieval of ATSR cloud Parameters and Evaluation (GRAPE) has been validated before (Sayer et al., 2011; Poulsen et al., 2012). Here the authors reported that cloud-top heights from GRAPE compare well to ground-based data at four sites, particularly for shallow clouds.

Most of the differences between GRAPE and global cloud fields derived from other satellite instrumentation (Stubenrauch et al., 2013) are linked to differing sensitivities as a result of different photon penetration depths at distinct wavelengths in the clouds and treatment of multilayer cloud systems. The correlation coefficient between GRAPE and two MODIS cloud products is greater than 0.7 for most cloud properties, except for liquid and ice cloud effective radius, which also show biases between the datasets. GRAPE underestimates liquid cloud water path relative to microwave radiometers by up to 100 g m^{-2} near the Equator and overestimates by around 50 g m^{-2} in the storm tracks. Lelli et al. (2016) have compared GRAPE cloud products with cloud data derived from SCIAMACHY and found the consistent tendency of underestimation of geometrical cloud top by AATSR measurements.

In summary, the main advantage of using the ORAC aerosol and cloud datasets, instead of other datasets generated by competitive algorithms, is the inherent consistency provided by the optimal estimation scheme shared by both modules, such that the aerosol and the cloud records can be regarded as two nearly orthogonal sets of the same atmospheric scenario.

2.2 Cloud Bottom Height Record

While crucial for aviation safety and atmospheric science in general, the determination of the CBHs by ground-based measurements is well established and is mainly based on ceilometers (e.g., Eberhard, 1986). Retrieval of CBH using passive satellite-borne instrumentation is, however, not common. Meerkötter and Zinner (2007) and Meerkötter and Bugliaro (2009) used an algorithm based on adiabatic lifting to find CBH from data of a polar orbiter (AVHRR) and geostationary satellite (SEVIRI) for convective clouds. Both sensors are broadband radiometers, which can still profit from a relatively fine spatial footprint, but with the drawback of a coarse spectral resolution. Recently, Seaman et al. (2017) presented a global assessment of an operational CBH product calculated from measurements of the Visible Infrared Imaging Radiometer Suite (VIIRS) employing an algorithm originally developed for MODIS (Hutchison, 2002), that derives CBH relating the optical depth of a cloud with its microphysical properties, such as liquid water content and effective droplet radius. The mean accuracy for water liquid clouds, evaluated against colocated cloud geometric profiles of Cloudsat (Stephens et al., 2002, 2008), is reported to range between −1.2 km and +0.3 km, whether a filter on cloud top altitude is applied.

Conversely, we derive the altitude of the bottom of cloud layers from measurements of the absorption strength of molecular oxygen (the A-band, 758–772 nm) by the SCIAMACHY instrument (Bovensmann et al., 1999) at a nominal spectral resolution of 0.4 nm. SCIAMACHY is on board the same satellite (Envisat) as AATSR, making the synergistic use of both instruments advantageous, because the scene sensed by them is already colocated and artifacts arising from lags in overpass time between sensors can be avoided. This is beneficial for the assessment of aerosol-cloud covariations, as it has been already shown that the significance of found relationships depends on the different meteorological conditions of the air masses carrying either polluted or cloud fields. For this reason, back-trajectory models have been employed to make such masses coincide with and single out the influence of meteorology (Avey et al., 2007; Bréon et al., 2002).

In particular, the A-band has historically shown sensitivity to the vertical displacement of the upper boundary of a cloud as a result of the screening efficiency of the oxygen column below it (Yamamoto and Wark, 1961; Saiedy et al., 1965; Fischer and Grassl, 1991). Only later on, methods to infer also the bottom layer of cloud layers have been developed (Rozanov and Kokhanovsky, 2004, 2006), exploiting the fact that clouds, in reality, are not purely reflective Lambertian surfaces but scattering objects. In this way, when photon penetration throughout and below the cloud is modeled, both scattering by water droplets and absorption by gas molecules can be accounted for. Thus, the fit of the A-band delivers both cloud top and geometrical thickness, because in-cloud and above-cloud absorptances together are able to reproduce the actual top-of-atmosphere-measurement. This is the rationale of the Semi-Analytical CloUd Retrieval Algorithm (SACURA) (Rozanov and Kokhanovsky, 2004), which has been deployed to generate the CBH record used in this work. More details on the quality flagging scheme required to populate realistic cloud records from pixel-based retrievals can be found in Lelli et al. (2012, 2016).

In general, cloud top height is the most accurate parameter. Rozanov and Kokhanovsky (2004) and Lelli et al. (2012, 2014) have reported a model error of ±250 m for low and mid-level clouds irrespective of changes in optical density of the cloud and radiance calibration drifts, while the error introduced by the forward model can increase up to 1 km for high-level and thin clouds (i.e., CTH $\geq$ 12 km and COT $\lesssim$ 5). The accuracy of the retrieved CBH for a dark and moderately bright underlying ground is also barely dependent on COT and ranges between −200 m and 350 m for CBH values varying between 0 and 10 km, as can be seen in Fig. 1 (taken from Lelli et al., 2011). Similar to CTH, the error in CBH grows beyond 1 km for high-level and thin clouds. This error assessment was performed assuming a single-layer water cloud, 1 km geometrically thick, represented by a modified Gamma droplet distribution and effective radius 6 μm (Kokhanovsky, 2006).

Due to the coarse nominal footprint size of SCIAMACHY (60 × 40 km^2), the heterogeneous cloud system is likely to intervene in the field-of-view (FOV) of the sensor, such that the assumption in the forward model of a single-layered cloud is disputable. Lelli et al. (2012, Fig. 4, p. 1557) have provided an error assessment, based on model data, in which a low-level water cloud (COT 10), placed between 3 and 4 km, is shadowed by a second cloud, placed between 13 and 14 km, of increasing optical depth (from 0 to 10), in both warm (liquid droplet) and cold (ice crystals) thermodynamic phase. As multi-layered scenes are often characterized by a lower

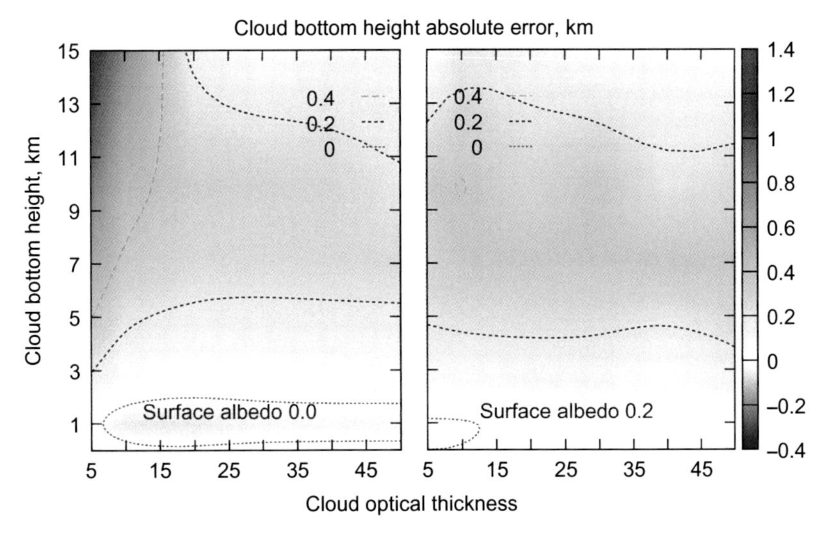

FIG. 1 Absolute error (km) in cloud bottom altitude for a single-layer water cloud, 1 km geometrically thick, for a dark (left) and moderately bright (right) surface. The sun elevation is 60° zenith for a satellite nadir view. *From Lelli, L., Kokhanovsky, A.A., Rozanov, V.V., Burrows, J.P., 2011. Radiative transfer in the oxygen A-band and its application to cloud remote sensing. AAPP Phys. Math. Nat. Sci. 89 (S1). https://doi.org/10.1478/C1V89S1P056.*

warm liquid cloud with overlying cirrus clouds, referring to the bottom panel of Fig. 4 in Lelli et al. (2012), one can see that the influence of a flatter phase function of ice crystals (as compared to the more oscillating phase function of water droplets) induces a steeper increase in CTH. As a result, in the water case, errors in CBH of the lowermost cloud layer do not exceed 1 km, given an upper layer of COT = 6, whereas in the ice, case COT shall not exceed 3. Thus, one can conclude that CBH can also be retrieved in the case of multilayered cloud systems, given that the upper cloud does not become optically thick enough to fully screen the lowermost one.

Validation of the satellite-based CBH retrievals has been carried out against ground-based data collected at the ARM research facilities of SGP, Barrow (North Slope Alaska, NSA), and Nauru Island (Tropical Western Pacific, TWP) and is provided in Fig. 2. This is the same dataset analyzed with the methodology used for the comparison presented by Lelli et al. (2016). Briefly, ground-based CBH is inferred from the lowermost altitude of the cloud layer identified by the vertical profiles of a micropulsed lidar device during the full time window of SCIAMACHY operation (2002–2012). From the website http://www.archive.arm.gov/, among all data streams available, the one named "30smplcmask" (ARM Climate Research Facility, 1996) has been selected for the comparison. This datastream is not only the longest available among all ARM facilities, therefore being beneficial to the maximization of the statistics of usable colocations, but it is also generated with the same algorithm (Wang and Sassen, 2001), warranting the same spectral sensitivity to different parts of the cloud bodies.

The lidar profiles are subsequently colocated with the SCIAMACHY overpass time and the median CBH is calculated over one hour of collected profiles of single and multilayered scenes, in order to take into account the full footprint size of the instrument. Then, only those satellite pixels have been retained, whose distance of their center from the lidar does not exceed 20 km at either side of the footprint

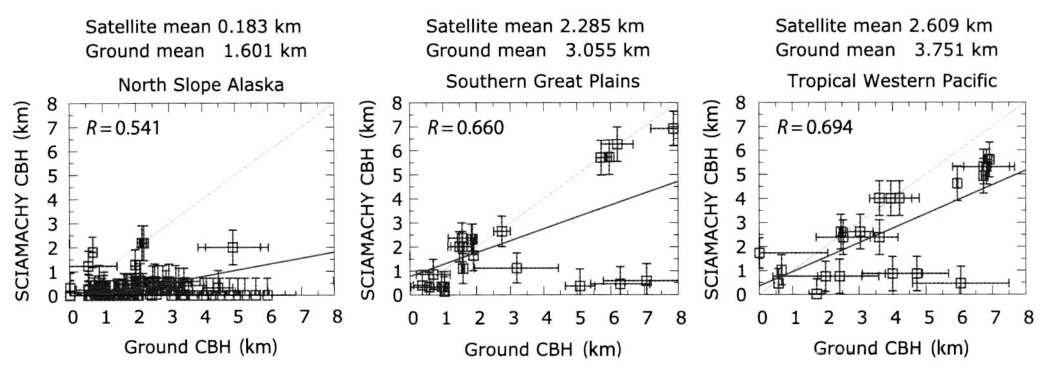

FIG. 2 Scatterplots of time-averaged ground-based lidar-derived (x-axis) and SCIAMACHY-derived cloud bottom altitudes (y-axis) for three ground-base facilities: (left) North Slope Alaska (elevation 8 m; 71.32° N, 156.62° W), Southern Great Plain (elevation 320 m; 36.6° N, 97.48° W), and Tropical Western Pacific (elevation 7.1 m; 0.52° S 166.62° E). For the meaning of the error bars see explanation in the text. The NSA case points to a suboptimal performance of the spaceborne algorithm above bright surfaces and for extreme observational geometries and is discarded in the analysis.

(i.e., maximal allowed spatial footprint extent equals 40 km along the SCIAMACHY flight direction). The selection criteria of the satellite CBH retrievals include a bias threshold in concurrent retrieved CTH of 2 km and a COT threshold of 5 (Lelli et al., 2012). In this way, together with information from the lidar ground mask, we exclude overlying cirrus clouds, which might degrade satellite-derived CBH accuracy. The error bars along the x-axis of Fig. 2 represent the variability of the lidar-derived CBH within a SCIAMACHY footprint, whereas the y-axis error bars represent the sum of systematic errors introduced by the forward model with respect to surface albedo, height, optical thickness, and heterogeneity of the sensed cloud. Eventually, the total number of usable colocations amounts to 133 at NSA, 25 at SGP, and 26 at TWP.

The inspection of Fig. 2 reveals the general tendency of the algorithm to underestimate CBH. Especially in the presence of a bright underlying surface, as can be expected at the NSA site, CBH is placed close to the ground (−1.4 km) as a result of the similar spectral signatures shared by snow- and ice-capped-surfaces with the clouds in the visible and NIR bands. Therefore, such retrievals must be excluded from any analysis. Looking at the scatterplots for SGP and TWP, the correlation with ground CBH increases up to 0.66 and 0.69, respectively, with a mean bias of −0.77 km and −1.14 km. However, it can be noticed that the occurrence of outliers is correlated with larger error whiskers, meaning higher variability in the ground CBH and pointing to a residual heterogeneity of clouds which could not be removed with the procedure described above. Therefore it becomes clear that the spaceborne CBH record is more reliable when homogeneous and stratified cloud systems are sensed and less reliable for systems characterized by open cellular and broken conditions typical of a turbulent and unsteady atmosphere.

Keeping this in mind, the left plot in Fig. 3 shows seasonal maps of SCIAMACHY CBH for the year 2008 and zonal averages in the right plot, as selected with the criteria described above, with mean values given in Table 1. Since the reported values are defined above mean sea level, topographic features appear over elevated mountainous areas such the Tibetan Plateau, and the Andean and Rocky Mountains. Generally, CBH synoptic features resemble the characteristics of CTH fields (Lelli et al., 2012),

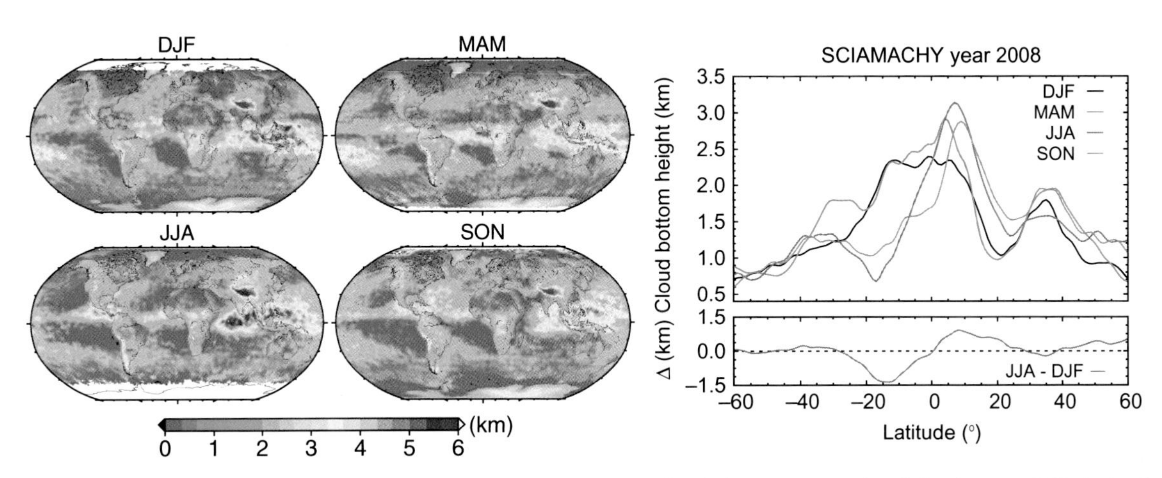

FIG. 3 (Left) Boreal seasonal maps of cloud bottom altitude (km, defined above mean sea level) for the year 2008, derived from measurements of the oxygen A-band by the SCIAMACHY sensor. (Right) Zonal means of CBH (top) and the difference between summer and winter months (bottom). Mean values provided in Table 1.

TABLE 1 Seasonal SCIAMACHY Mean CBH (km) for the Year 2008

Season	Mean CBH (km)	Counts (10^6)
Winter (DJF)	1.52 ± 0.74	2.37
Spring (MAM)	1.66 ± 0.87	2.61
Summer (JJA)	1.51 ± 0.78	2.74
Autumn (SON)	1.52 ± 0.79	2.63

Reported values are between ±60° latitude to exclude bright surfaces.

spatially and temporally correlated with deep convective clouds along the tropics, with stratiform low-level cloud decks over the oceans, with subsidence belts in the mid latitudes and with the storm cloud alleys streaming across the North Atlantic in the late summer and autumn months. It also follows that CBH connects with the oscillation of the InterTropical Convergence Zone (ICTZ), slightly northward of the equatorial parallel about 10° N. The sinusoidal curve plotted in the bottom panel of Fig. 3 (right plot) shows the difference in CBH between the summer and winter months, as evidence of this meridional atmospheric circulation. Conversely to CTH, which exhibits wider zonal oscillations in both hemispheres, stretching to 40° S and 40° N (Lelli et al., 2012, Fig. 18), likely as a result of stronger air divergence in the upper layers of the atmosphere, CBH is mainly confined within the tropical belt of 30° N–20° S. Quantitatively, CTH is also seen to temporally vary between ±4 km, whereas CBH is bounded to a maximum of ±1.5 km, thus revealing less variability across the year.

3 RESULTS

Even if total AOD can be regarded as a proxy for CCN (Andreae, 2009), its significance can be questioned (Bréon et al., 2002), because in this case, total light extinction by aerosols is dominated by the coarse mode. One possible way to overcome this limitation is to use information on the aerosol size spectrum by means of the aerosol index (AI), which is the product of AOD and the Ångstrom exponent (Deuzé et al., 2001). Another possibility is to use aerosol optical thickness due only to the fine mode (AOD_{fm}), which is provided in the aerosol CCI dataset.

For this analysis, and to improve data quality, the daily L3 data of all variables are filtered upfront for those locations where their standard deviation exceeds the mean value (Andersen and Cermak, 2015; Andersen et al., 2016) and are aggregated to monthly means. Then, we subset the multiyear (2003–2009) records between the winter (DJF) and summer (JJA) months. Fig. 4 shows seasonal absolute values of AOD_{fm} (top row) and CBH (mid), together with their correlation in the bottom row. AOD_{fm} patterns show typical features of human and natural activity such as outflows from populated industrialized regions (e.g., the Indian subcontinent, Eastern China, Northern East America) or from regions of biomass burning or lightning-induced fire activity off the central African continent. Due to the contrast between land and sea masses in some regions, we limit our investigation to only areas above water.

The correlation shows distinct patterns of synoptic features. The steady season-independent negative wide areas of the southern oceans (e.g., Benguela and Chilean gulfs) as well as the Northeast Pacific display a strong negative correlation. Defining the sensitivity of CBH to AOD_{fm}, upon linearization of the latter with the natural logarithm (i.e., $d\,CBH/d\ln(AOD_{fm})$), as the slope of the linear fit between the two standardized quantities and ensuring the applicability of linear

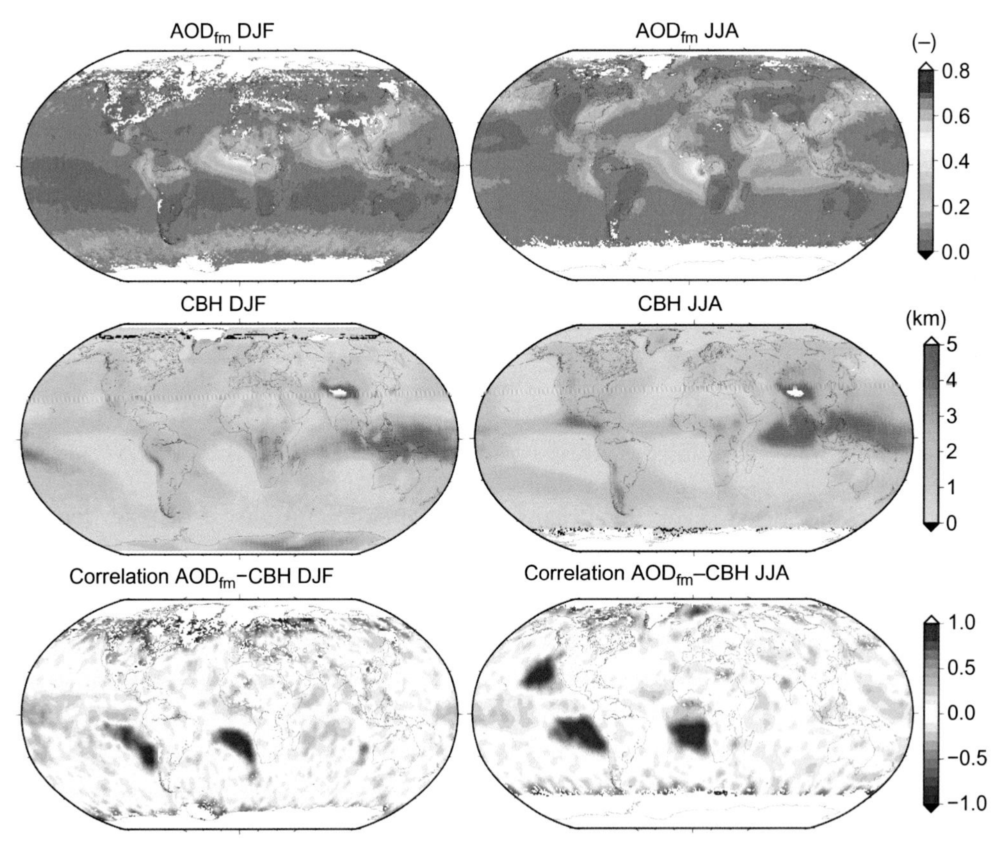

FIG. 4 AATSR fine mode aerosol optical thickness (AOD_{fm}) at 550 nm (top), SCIAMACHY cloud bottom height (middle), and their correlation (bottom) for winter (left) and summer (right) months. Data spans years 2003–2009.

regression methods (e.g., Gryspeerdt et al., 2014), we find barely any sensitivity in both seasons. This points to a modulating effect of meteorology that overshadows any possible aerosol influence on CBH, as confirmed with MODIS (on Aqua) measurements by Engström and Ekman (2010). The authors find a controlling effect of a 10-meter wind speed onto correlations of AOD and cloud fraction, especially strong in such regions. In this case, cloud fraction occurrence can be regarded as a reliable approximation for a height-integrated quantity as the altitude of the cloud bottom, as clouds are mostly stratocumuli confined in the lowest layers of the atmosphere.

It becomes also clear that the study region considered by Andersen and Cermak (2015), the Southeast Atlantic (10° S–20° S, 0° E–11° E), poses challenges with respect to confounding meteorological factors. Among these, it can also be supposed that sea surface temperature (SST) exerts a modulating effect on the bottom of the clouds. Therefore, in prior analysis, the correlation of ocean temperatures with CBH was calculated and, indeed, most of the CBH variations have been found to be strongly seasonally linked to SST. We have made use of the AATSR-derived SST dataset generated within the ESA CCI SST initiative in its version 1.1 (Merchant et al., 2014), as this is the natural choice for reasons of spatio-temporal colocation and sensor spectral characteristics. For this reason, CBH anomalies, instead of absolute values, have been computed, subtracting the mean climatological seasonal cycle and other regions of interests have been selected that satisfy the following criteria: first, AOD_{fm} and CBH of Fig. 4 shows a clear annual or seasonal signal; and second, we do not filter data to only low-level stratocumuli in order to be able to capture variations which may be induced by internal cloud updrafts and dissipation processes in a pristine environment (Rosenfeld et al., 2008; Koren et al., 2010b). However, relaxing the constraint on cloud heights has the likely consequence of retaining scenes of convective and broken cloud systems. It has been suggested that such situations occur during the transition between a clean CCN-free and a slightly polluted regime ($AOD_{fm} \leq 0.1$) as a result of a nonlinear aerosol-precipitation mechanism (Rosenfeld et al., 2006; Koren et al., 2014). This would imply that the quality of the SCIAMACHY CBH can decrease, if scene heterogeneity becomes prominent.

Fig. 5 shows the selected regions of interest, while Table 2 lists the respective geographical coordinate. In Fig. 6, the extracted time series are shown. Table 3 shows correlations of CBH, SST, and AOD_{fm} as functions of regions of interest and seasons. The comparison of the first column (CBH-SST) with the second column (CBH anomaly-SST) shows that in the winter as well

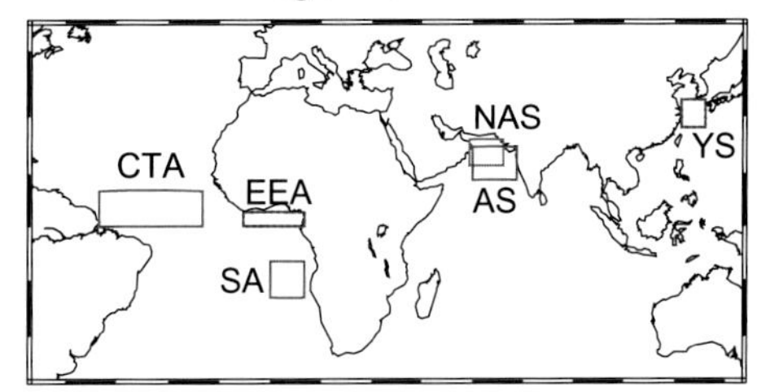

FIG. 5 Regions of interest (see Table 2 for the coordinates) selected for the investigation of CBH and AOD_{fm} covariations.

TABLE 2 Regions of Interest for the Investigation of CBH and AOD_{fm} Covariations

Region	Coordinates
Southeast Atlantic (SA)	10° S–20° S, 0° E–11° E
Central Tropical Atlantic (CTA)	10° N–0° S, 20° W–50° W
Equatorial East Atlantic (EEA)	4° N–6° S, 8° W–10° E
Arabian Sea (AS)	22° N–13° N, 60° E–73° E
North Arabian Sea (NAS)	24° N–17° N, 59° E–69° E
Yellow Sea (YS)	34° N–27° N, 122° E–129° E

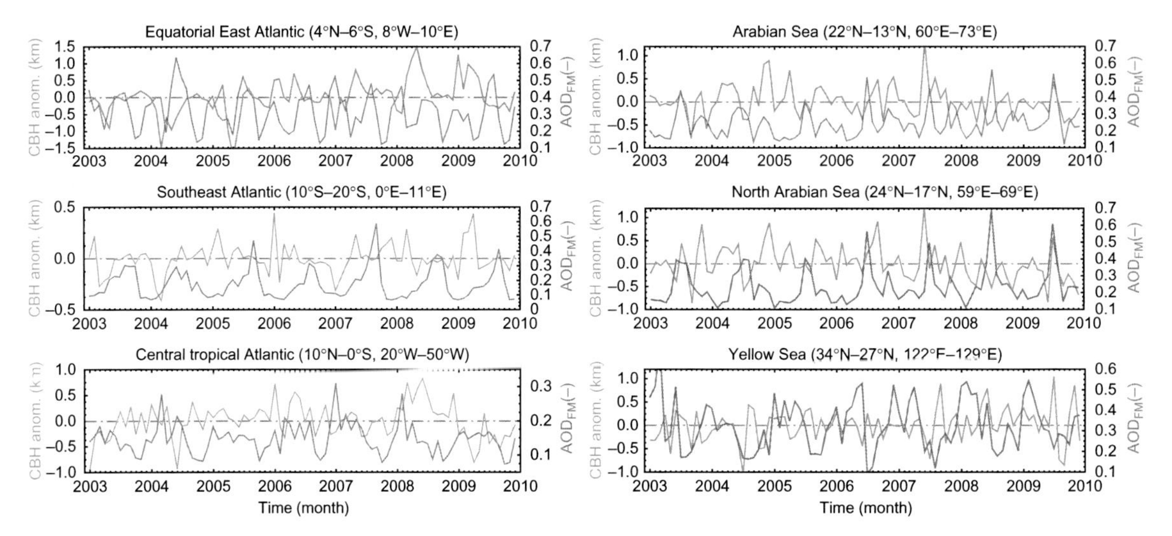

FIG. 6 Time series of CBH anomaly and AOD_{fm} for the six regions of interest from Fig. 5.

TABLE 3 Correlations of Variables Under Study as Functions of Seasons (DJF Left/JJA Right) and Regions of Interest

Region	CBH-SST	CBH an.-SST	CBH an.-AOD_{fm}	AOD_{fm}-SST
Southeast Atlantic (SA)	0.62/0.15	0.13/−0.26	−0.04/−0.02	0.07/−0.69
Central Tropical Atlantic (CTA)	0.76/0.58	0.37/0.42	−0.00/−0.33	−0.14/0.05
Equatorial East Atlantic (EEA)	0.59/0.79	0.15/0.43	−0.63/−0.48	−0.29/−0.65
Arabian Sea (AS)	0.45/0.65	0.12/0.01	−0.10/−0.02	0.70/−0.11
North Arabian Sea (NAS)	0.28/0.25	0.88/0.04	−0.07/−0.19	0.69/−0.43
Yellow Sea (YS)	−0.20/−0.31	−0.16/−0.18	−0.19/0.04	−0.19/−0.71

CBH anomaly corresponds to the actual CBH after subtraction of the climatological mean.

as in the summer months, the deseasonalized CBH correlates less with SST, giving us reasonable confidence that, in this way, the influence of SST is lessened. We note that the SA region has an unphysical low correlation between CBH and SST, which might point to poor statistics due to the persistent (at satellite overpass time) marine stratocumulus clouds, which can considerably hamper the cloud clearance for SST retrievals.

Looking at the covariations of deseasonalized CBH and AOD_{fm}, we note that all values are, notwithstanding different magnitudes, negative. This would imply a lower CBH for an increase in fine-mode aerosol loading. In particular, EEA stands out as a unique region among all others under consideration because in both seasons it displays a stable negative relationship, which is also discernible in the uppermost time series of Fig. 6. The double-peaked annual aerosol signal has its maxima centered in December-January and July-August when, concurrently, the CBH anomaly is seen to be steady about the zero baseline. This effect is even more pronounced within the SA. The AOD_{fm} maxima in JJA are coinciding with zero-changes in CBH, although the seasonal correlation coefficients alone do not provide any hints to plausible covariations as they are negligibly small.

CTA (the region studied in Gryspeerdt et al., 2014) is typically characterized by relatively low AOD_{fm}, with a seasonal inclusion of wind-blown, long-range transported Saharan dust (Knippertz and Stuut, 2014), and by mature systems of convective clouds. Heterogeneity can be expected and, as explained by Lelli et al. (2012), multilayeredness leads to an overestimation of CBH up to 5 km for mid-level clouds and a total optical thickness of 20. It is unlikely that aerosols, especially with a relatively weak maxima AOD_{fm} of 0.3, would be the cause of the negative correlation (−0.33) in summer months.

The YS region shows irregular patterns and a noisy time series for CBH as well for AOD_{fm}. The latter can partially be traced back to lowered data quality, as might be the case for total AOD greater than one (Che et al., 2016). Additionally, the YS is a known atmosphere-surface system of enhanced biological, tidal, and turbulent activity (Ye et al., 2016). These very specific settings will likely have an additional impact on the quality of the retrieved pair of CBH and AOD_{fm}, such that any further investigation is, at the moment, likely unfeasible.

Focusing on the Arabian Sea as a whole (AS) or on its coastal part (NAS), one can see rather significant levels of AOD_{fm} up to 0.7 with relatively strong emissions in summer months. This loading gets higher closer to land (i.e., injection sources) as well as the temperature contrast between advected air warmed by the neighboring deserts and the relatively cold sea water. This meteorological configuration often leads to temperature inversions, which modulate the convective cloud activity (Bhat, 2006). Only within NAS, we observe a negative correlation of −0.2 in JJA that is not observed in the wider AS or in DJF. It must be stressed that, despite the strong surface winds blowing from land masses over the ocean and the resulting injection of hygroscopic sea-salt particles, AOD_{fm} should not be influenced by it. Additionally, the AS-NAS region is characterized, over the course of the year, by near-surface10-meter winds blowing along the southwest-northwest direction (see, for instance, Fig. 5 in (Jia et al., 2017)), therefore minimizing the presence of coarse-mode dust particles originating from the deserts.

Fig. 7 shows the extent of the modulating effect of AOD_{fm} over NAS onto the cloud droplet effective radius (CER) and cloud height (CH). The seasonal values of these quantities (given in Table 4) are plotted also for the SA region (top row), which we use as a calibrating region for the influence of aerosol particles on clouds. If the well-established Twomey indirect effect (i.e., the decrease of CER in a polluted environment) is seen over SA during summer (e.g., Bulgin et al., 2008), then the same relationship should also hold over NAS.

While CER over SA decreases from 13.00 to 11.23 μm from a pristine ($AOD_{fm} = 0.08$ in DJF) to a polluted ($AOD_{fm} = 0.25$ in JJA) case, thus confirming the Twomey effect at play, the same does not hold for NAS. In this region, CER is seen to increase (from 12.78 in DJF to 16.82 μm in JJA) with larger values of AOD_{fm} (from 0.17 in DJF to 0.37 in JJA). Concurrently, clouds over NAS are also seen to rise in summer months (CH changes from 3.41 to 5.84 km), suggesting that their geometrical thickness increases, because the CBH anomaly is stable about the zero (mid row, left plot in Fig. 6). It can be deduced that turbulent mixing due to the strong westerly Somali jets (Jia et al., 2017, Fig. 5, p. 4920) is the driving mechanism, because the wind fields in the pre- and post-monsoon months (March-to-May and September-to-November) rotate, respectively blowing AOD_{fm} away from the Indian subcontinent over water during DJF (low aerosol loading) and toward the Indian subcontinent during JJA (high aerosol loading). This would also explain the higher variance in both summer scatterplots for the NAS region.

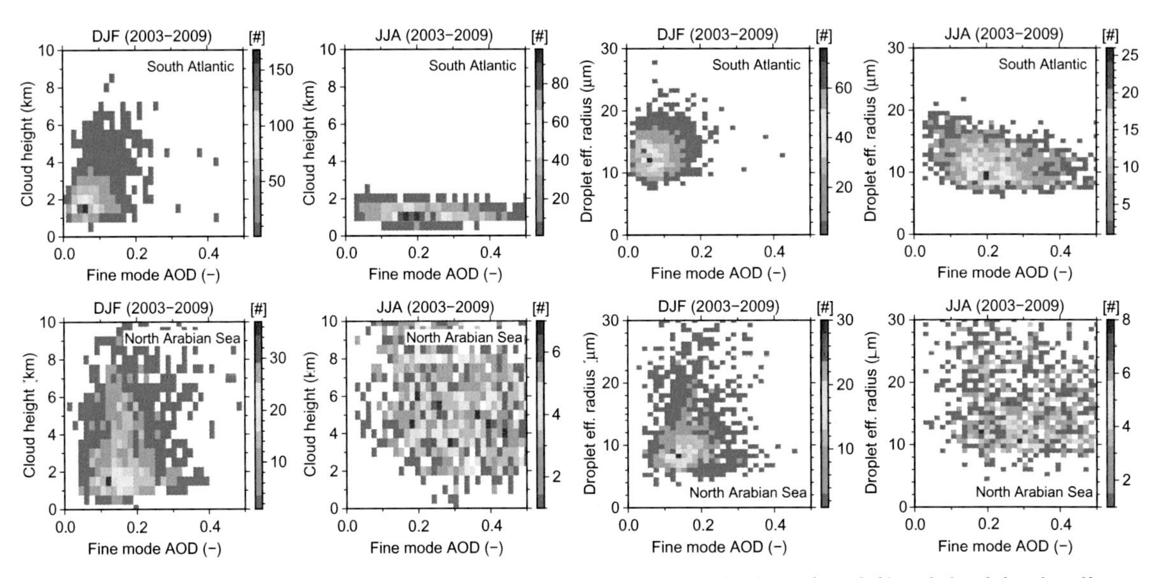

FIG. 7 Seasonal scatterplots of AOD_{fm} vs. cloud height (km, first and second column from left) and cloud droplet effective radius (μm, third and fourth column) for the South Atlantic (SA, top row) and North Arabian Sea (NAS, bottom row). Values are given in Table 4.

TABLE 4 Seasonal Values of Cloud Droplet Effective Radius (CER, μm), Cloud Height (CH, km), and AOD_{fm} for the South Atlantic (SA) and the North Arabian Sea Regions

Region	Season	AOD_{fm}	CER (μm)	CH (km)
Southeast Atlantic (SA)	DJF	0.08 ± 0.04	13.00 ± 2.34	2.24 ± 1.09
	JJA	0.25 ± 0.17	11.23 ± 2.58	1.15 ± 0.26
North Arabian Sea (NAS)	DJF	0.17 ± 0.07	12.78 ± 6.9	3.41 ± 2.21
	JJA	0.37 ± 0.18	16.82 ± 6.4	5.84 ± 2.38

Given the arguments above, we speculate that a clearer (yet probably not causal) linear relationship between CBH and aerosols is possible only in the presence of larger values of AOD_{fm}, closer to the sources and accumulated by winds blowing along the southwest-northwest direction. While not contributing to the onset of the Twomey effect, such fine-mode aerosol particles can serve as a stabilizing factor of the local thermodynamic lapse rate through the absorption of solar radiation and warming of the air column.

4 SUMMARY AND CONCLUSIONS

In this chapter we have made use of a unique synergistic combination of datasets inferred from seven years of measurements of two spaceborne sensors, namely the AATSR and the SCIAMACHY, both payloads of the Envisat platform. More precisely, the aerosol and cloud datasets have been created from AATSR within the framework of the ESA CCI, while the spaceborne record of CBH has been derived from measurements of the absorption strength of oxygen in the NIR by SCIAMACHY. CBH, being to date the longest available dataset with a passive remote sensing technique, has been validated against ground lidar profiles operated at three ARM facilities, showing good correlations as long as the surface is not highly reflective and the sensed cloud systems are not highly heterogeneous.

In order to investigate possible covariations of fine mode aerosol loading (AOD_{fm}) and CBH, six oceanic regions have been carefully selected to account for potential regimes of specific cloud and aerosol interactions. Moreover, seasonal subsetting between winter and summer months has been carried out in order to minimize the impact of temperature changes on the above-mentioned covariations. It has been found that not all the selected regions showed clear patterns, neither in the temporal nor in the spatial domains. In general, an increase in AOD_{fm} tends to decrease the CBH, with this effect especially true in those regions closer to emission sources and with less influence by humidification on aerosol particles. Yet, the regions of Equatorial East and the Southeast Atlantic exhibit the distinctive feature of a regular coincidence of elevated AOD_{fm} with zero-change CBH, being this likely attributable to a competing effect of aerosols with air updraft at the cloud base. This is seen especially above the Arabian Sea. This region is characterized by regular outflows of absorbing aerosols from the Indian subcontinent during boreal summer months (June-September), when low-level tropospheric winds also contribute to the accumulation of AOD_{fm} closer to the sources (see, for instance, Turner and Annamalai, 2012; Jia et al., 2017). The presence of absorbing aerosols, in turn, can stabilize the atmosphere column, so that convection over open waters is inhibited.

However, we caution the reader that this work also has caveats. In fact, it is well-known that correlation does not imply causation, especially for those relationships that are not only highly nonlinear in nature (such as the ones involving AOD_{fm} as condensation nuclei in a water vapor-saturated environment), but also because one of the most challenging tasks is to unambiguously single out the superimposed modulation of meteorological variables onto the sought-after relationships.

Second, the sensors used in this work are payloads of a Sun-synchronous polar orbiting satellite and the revisit time can vary from 1.5 days of SCIAMACHY up to 3 days for AATSR. This implies a considerable temporal undersampling of short time scale interactions that cannot be resolved with the monthly averaged datasets used in this study. Nonetheless, the advantage of a global coverage and long-term monitoring of covariations will be beneficial to further work from the perspective of climate applications. Additionally, since the accuracy of passive satellite products cannot compete with the accuracy of ground-based instrumentation, there is the need of additional validation, especially against active techniques (despite the comparatively poor spatial coverage), in order to analyze additional meteorological regimes and to confirm the correlations provided in this study.

Acknowledgments

L. Lelli has been financially supported by the European Space Agency (ESA) via the Living Planet Fellowship for the STAR-CLINT (STatistics of AeRosol and CLoud INTeractions) project and by the L2 UVN/Sentinel-4 project, managed by the German Aerospace Agency (DLR). Additional funding comes from the German Science Foundation (DFG) in the framework of the Transregional Collaborative Project TR 172 AC^3 (ArctiC Amplification: Climate relevant Atmospheric and surfaCe processes and feedback mechanisms).

References

Albrecht, B.A., 1989. Aerosols, cloud microphysics, and fractional cloudiness. Science 245 (4923), 1227–1230.

Allen, M.R., Ingram, W.J., 2002. Constraints on future changes in climate and the hydrologic cycle. Nature 419 (6903), 224–232. https://doi.org/10.1038/nature01092.

Andersen, H., Cermak, J., 2015. How thermodynamic environments control stratocumulus microphysics and interactions with aerosols. Environ. Res. Lett. 10 (2), 024004. http://stacks.iop.org/1748-9326/10/i=2/a=024004.

Andersen, H., Cermak, J., Fuchs, J., Schwarz, K., 2016. Global observations of cloud-sensitive aerosol loadings in low-level marine clouds. J. Geophys. Res. Atmos. 121 (21), 12,936–12,946. https://doi.org/10.1002/2016JD025614.

Andreae, M.O., 2009. Correlation between cloud condensation nuclei concentration and aerosol optical thickness in remote and polluted regions. Atmos. Chem. Phys. 9 (2), 543–556. https://doi.org/10.5194/acp-9-543-2009.

Andreae, M.O., Rosenfeld, D., 2008. Aerosol-cloud-precipitation interactions. Part 1. The nature and sources of cloud-active aerosols. Earth Sci. Rev. 89 (1–2), 13–41. https://doi.org/10.1016/j.earscirev.2008.03.001.

ARM Climate Research Facility, 1996. Cloud Mask from Micropulse Lidar (30SMPLCMASK1ZWANG). Southern Great Plains (SGP) Central Facility, Lamont, OK; North Slope Alaska (NSA) Central Facility, Barrow AK; Tropical Western Pacific (TWP) Central Facility, Nauru Island. https://doi.org/10.5439/1027736 Atmospheric Radiation Measurement. Updated monthly. Compiled by C. Sivaraman and L. Riihimaki. Data Archive: Oak Ridge, Tennessee, USA. Data set last accessed: January 2014.

Avey, L., Garrett, T.J., Stohl, A., 2007. Evaluation of the aerosol indirect effect using satellite, tracer transport model, and aircraft data from the international consortium for atmospheric research on transport and transformation. J. Geophys. Res. Atmos. 112 (D10), D10S33. https://doi.org/10.1029/2006JD007581.

Bhat, G.S., 2006. Near-surface temperature inversion over the Arabian Sea due to natural aerosols. Geophys. Res. Lett. 33 (2), L02802. https://doi.org/10.1029/2005GL024157.

Bond, T.C., Doherty, S.J., Fahey, D.W., Forster, P.M., Berntsen, T., DeAngelo, B.J., Flanner, M.G., Ghan, S., Kärcher, B., Koch, D., Kinne, S., Kondo, Y., Quinn, P.K., Sarofim, M.C., Schultz, M.G., Schulz, M., Venkataraman, C., Zhang, H., Zhang, S., Bellouin, N., Guttikunda, S.K., Hopke, P.K., Jacobson, M.Z., Kaiser, J.W., Klimont, Z., Lohmann, U., Schwarz, J.P., Shindell, D., Storelvmo, T., Warren, S.G., Zender, C.S., 2013. Bounding the role of black carbon in the climate system: a scientific assessment. J. Geophys. Res. Atmos. 118 (11), 5380–5552. https://doi.org/10.1002/jgrd.50171.

Bovensmann, H., Burrows, J.P., Buchwitz, M., Frerick, J., NoÃńl, S., Rozanov, V.V., Chance, K.V., Goede, A.P.H., 1999. SCIAMACHY: mission objectives and measurement modes. J. Atmos. Sci. 56 (2), 127–150. https://doi.org/10.1175/1520-0469(1999)056⟨0127:SMOAMM⟩2.0.CO;2.

Bréon, F.M., Tanré, D., Generoso, S., 2002. Aerosol effect on cloud droplet size monitored from satellite. Science 295 (5556), 834–838. https://doi.org/10.1126/science.1066434.

Bulgin, C.E., Palmer, P.I., Thomas, G.E., Arnold, C.P.G., Campmany, E., Carboni, E., Grainger, R.G., Poulsen, C., Siddans, R., Lawrence, B.N., 2008. Regional and seasonal variations of the Twomey indirect effect as observed by the ATSR-2 satellite instrument. Geophys. Res. Lett. 35 (2), L02811. https://doi.org/10.1029/2007GL031394.

Che, Y., Xue, Y., Mei, L., Guang, J., She, L., Guo, J., Hu, Y., Xu, H., He, X., Di, A., Fan, C., 2016. Technical note: Intercomparison of three AATSR Level 2 (L2) AOD products over China. Atmos. Chem. Phys. 16 (15), 9655–9674. https://doi.org/10.5194/acp-16-9655-2016.

Chen, Y.C., Christensen, M.W., Stephens, G.L., Seinfeld, J.H., 2014. Satellite-based estimate of global aerosol-cloud radiative forcing by marine warm clouds. Nat. Geosci. 7 (9), 643–646. https://doi.org/10.1038/ngeo2214.

Christensen, M.W., Stephens, G.L., 2011. Microphysical and macrophysical responses of marine stratocumulus polluted by underlying ships: evidence of cloud deepening. J. Geophys. Res. Atmos. 116 (D3), D03201. https://doi.org/10.1029/2010JD014638.

Costantino, L., Bréon, F.M., 2010. Analysis of aerosol-cloud interaction from multi-sensor satellite observations. Geophys. Res. Lett. 37 (11), L11801. https://doi.org/10.1029/2009GL041828.

Deuzé, J.L., Bréon, F.M., Devaux, C., Goloub, P., Herman, M., Lafrance, B., Maignan, F., Marchand, A., Nadal, F., Perry, G., Tanré, D., 2001. Remote sensing of aerosols over land surfaces from polder-adeos-1 polarized measurements. J. Geophys. Res. Atmos. 106 (D5), 4913–4926. https://doi.org/10.1029/2000JD900364.

van Donkelaar, A., Martin, R.V., Brauer, M., Kahn, R., Levy, R., Verduzco, C., Villeneuve, P.J., 2010. Global estimates of ambient fine particulate matter concentrations from satellite-based aerosol optical depth: development and application. Environ. Health Perspect. 118, 847–855. https://doi.org/10.1289/ehp.0901623.

Eberhard, W.L., 1986. Cloud signals from lidar and rotating beam ceilometer compared with pilot ceiling. J. Atmos. Oceanic Technol. 3 (3), 499–512. https://doi.org/10.1175/1520-0426(1986)003⟨0499:CSFLAR⟩2.0.CO;2.

Engström, A., Ekman, A.M.L., 2010. Impact of meteorological factors on the correlation between aerosol optical depth and cloud fraction. Geophys. Res. Lett. 37 (18), L18814. https://doi.org/10.1029/2010GL044361.

Fan, J., Leung, L.R., Rosenfeld, D., Chen, Q., Li, Z., Zhang, J., Yan, H., 2013. Microphysical effects determine macrophysical response for aerosol impacts on deep convective clouds. Proc. Natl. Acad. Sci. U. S. A. 110 (48), E4581–E4590. https://doi.org/10.1073/pnas.1316830110.

Feingold, G., 2005. On smoke suppression of clouds in Amazonia. Geophys. Res. Lett. 32 (2), L02804. https://doi.org/10.1029/2004GL021369.

Feingold, G., McComiskey, A., Yamaguchi, T., Johnson, J.S., Carslaw, K.S., Schmidt, K.S., 2016. New approaches to quantifying aerosol influence on the cloud radiative effect.

Proc. Natl. Acad. Sci. U. S. A. 113 (21), 5812–5819. https://doi.org/10.1073/pnas.1514035112.

Fischer, J., Grassl, H., 1991. Detection of cloud-top height from reflected radiances within the oxygen a band, part 1: Theoretical studies. J. Appl. Meteorol. 30, 1245–1259. https://doi.org/10.1175/1520-0450(1991)030⟨1245:DOCTHF⟩2.0.CO;2.

Fridlind, A.M., Ackerman, A.S., Jensen, E.J., Heymsfield, A.J., Poellot, M.R., Stevens, D.E., Wang, D., Miloshevich, L.M., Baumgardner, D., Lawson, R.P., Wilson, J.C., Flagan, R.C., Seinfeld, J.H., Jonsson, H.H., VanReken, T.M., Varutbangkul, V., Rissman, T.A., 2004. Evidence for the predominance of mid-tropospheric aerosols as subtropical anvil cloud nuclei. Science 304 (5671), 718–722. https://doi.org/10.1126/science.1094947.

Goren, T., Rosenfeld, D., 2014. Decomposing aerosol cloud radiative effects into cloud cover, liquid water path and Twomey components in marine stratocumulus. Atmos. Res. 138, 378–393. https://doi.org/10.1016/j.atmosres.2013.12.008.

Gryspeerdt, E., Stier, P., Grandey, B.S., 2014. Cloud fraction mediates the aerosol optical depth-cloud top height relationship. Geophys. Res. Lett. 41 (10), 3622–3627. https://doi.org/10.1002/2014GL059524.

Gryspeerdt, E., Quaas, J., Bellouin, N., 2016. Constraining the aerosol influence on cloud fraction. J. Geophys. Res. Atmos. 121 (7), 3566–3583. https://doi.org/10.1002/2015JD023744.

Hollmann, R., Merchant, C.J., Saunders, R., Downy, C., Buchwitz, M., Cazenave, A., Chuvieco, E., Defourny, P., De Leeuw, G., Forsberg, R., Holzer-Popp, T., Paul, F., Sandven, S., Sathyendranath, S., Van Roozendael, M., Wagner, W., 2013. The ESA climate change initiative: satellite data records for essential climate variables. Bull. Am. Meteorol. Soc. 94 (10), 1541–1552. https://doi.org/10.1175/BAMS-D-11-00254.1.

Holzer-Popp, T., de Leeuw, G., Griesfeller, J., Martynenko, D., Klüser, L., Bevan, S., Davies, W., Ducos, F., Deuzé, J.L., Graigner, R.G., Heckel, A., von Hoyningen-Hüne, W., Kolmonen, P., Litvinov, P., North, P., Poulsen, C.A., Ramon, D., Siddans, R., Sogacheva, L., Tanré, D., Thomas, G.E., Vountas, M., Descloitres, J., Griesfeller, J., Kinne, S., Schulz, M., Pinnock, S., 2013. Aerosol retrieval experiments in the ESA Aerosol_cci project. Atmos. Meas. Tech. 6 (8), 1919–1957. https://doi.org/10.5194/amt-6-1919-2013.

Hutchison, K.D., 2002. The retrieval of cloud base heights from MODIS and three-dimensional cloud fields from NASA's EOS Aqua mission. Int. J. Remote Sens. 23 (24), 5249–5265. https://doi.org/10.1080/01431160110117391.

Jia, J., Ladstätter-Weißenmayer, A., Hou, X., Rozanov, A., Burrows, J.P., 2017. Tropospheric ozone maxima observed over the Arabian Sea during the pre-monsoon. Atmos. Chem. Phys. 17 (8), 4915–4930. https://doi.org/10.5194/acp-17-4915-2017.

Kaufman, Y.J., Nakajima, T., 1993. Effect of Amazon smoke on cloud microphysics and albedo-analysis from satellite imagery. J. Appl. Meteorol. 32 (4), 729–744. https://doi.org/10.1175/1520-0450(1993)032⟨0729:EOASOC⟩2.0.CO;2.

Knippertz, P., Stuut, J.B.W., 2014. Mineral Dust (A Key Player in the Earth System). Springer, Netherlands. https://doi.org/10.1007/978-94-017-8978-3.

Kokhanovsky, A.A., 2006. Cloud optics. Atmospheric and Oceanographic Sciences Library, vol. 34 Springer, Dordrecht, Netherlands. https://doi.org/10.1007/1-4020-4020-2.

Koren, I., Feingold, G., Remer, L.A., 2010. The invigoration of deep convective clouds over the Atlantic: aerosol effect, meteorology or retrieval artifact? Atmos. Chem. Phys. 10 (18), 8855–8872. https://doi.org/10.5194/acp-10-8855-2010.

Koren, I., Remer, L.A., Altaratz, O., Martins, J.V., Davidi, A., 2010. Aerosol-induced changes of convective cloud anvils produce strong climate warming. Atmos. Chem. Phys. 10 (10), 5001–5010. https://doi.org/10.5194/acp-10-5001-2010.

Koren, I., Dagan, G., Altaratz, O., 2014. From aerosol-limited to invigoration of warm convective clouds. Science 344 (6188), 1143–1146. https://doi.org/10.1126/science.1252595.

L'Ecuyer, T.S., Berg, W., Haynes, J., Lebsock, M., Takemura, T., 2009. Global observations of aerosol impacts on precipitation occurrence in warm maritime clouds. J. Geophys. Res. Atmos. 114 (D9), D09211. https://doi.org/10.1029/2008JD011273.

de Leeuw, G., Holzer-Popp, T., Bevan, S., Davies, W.H., Descloitres, J., Grainger, R.G., Griesfeller, J., Heckel, A., Kinne, S., Klüser, L., Kolmonen, P., Litvinov, P., Martynenko, D., North, P., Ovigneur, B., Pascal, N., Poulsen, C., Ramon, D., Schulz, M., Siddans, R., Sogacheva, L., Tanré, D., Thomas, G.E., Virtanen, T.H., von Hoyningen Huene, W., Vountas, M., Pinnock, S., 2015. Evaluation of seven European aerosol optical depth retrieval algorithms for climate analysis. Remote Sens. Environ. 162, 295–315. https://doi.org/10.1016/j.rse.2013.04.023.

Lelli, L., Kokhanovsky, A.A., Rozanov, V.V., Burrows, J.P., 2011. Radiative transfer in the oxygen A-band and its application to cloud remote sensing. AAPP Phys. Math. Nat. Sci. 89 (S1). https://doi.org/10.1478/C1V89S1P056.

Lelli, L., Kokhanovsky, A.A., Rozanov, V.V., Vountas, M., Sayer, A.M., Burrows, J.P., 2012. Seven years of global retrieval of cloud properties using space-borne data of GOME. Atmos. Meas. Tech. 5 (7), 1551–1570. https://doi.org/10.5194/amt-5-1551-2012.

Lelli, L., Kokhanovsky, A.A., Rozanov, V.V., Vountas, M., Burrows, J.P., 2014. Linear trends in cloud top height from passive observations in the oxygen A-band. Atmos.

Chem. Phys. 14 (11), 5679–5692. https://doi.org/10.5194/acp-14-5679-2014.

Lelli, L., Weber, M., Burrows, J., 2016. Evaluation of SCIAMACHY ESA/DLR cloud parameters version 5.02 by comparisons to ground-based and other satellite data. Front. Environ. Sci. https://doi.org/10.3389/fenvs.2016.00043.

Li, Z., Niu, F., Fan, J., Liu, Y., Rosenfeld, D., Ding, Y., 2011. Long-term impacts of aerosols on the vertical development of clouds and precipitation. Nat. Geosci. 4 (12), 888–894. https://doi.org/10.1038/NGEO1313.

Meerkötter, R., Bugliaro, L., 2009. Diurnal evolution of cloud base heights in convective cloud fields from MSG/SEVIRI data. Atmos. Chem. Phys. 9 (5), 1767–1778. https://doi.org/10.5194/acp-9-1767-2009.

Meerkötter, R., Zinner, T., 2007. Satellite remote sensing of cloud base height for convective cloud fields: a case study. Geophys. Res. Lett. 34 (17), L17805. https://doi.org/10.1029/2007GL030347.

Merchant, C.J., Embury, O., Roberts-Jones, J., Fiedler, E., Bulgin, C.E., Corlett, G.K., Good, S., McLaren, A., Rayner, N., Morak-Bozzo, S., Donlon, C., 2014. Sea surface temperature datasets for climate applications from Phase 1 of the European Space Agency Climate Change Initiative (SST CCI). Geosci. Data J. 1 (2), 179–191. https://doi.org/10.1002/gdj3.20.

Popp, T., de Leeuw, G., Bingen, C., Brühl, C., Capelle, V., Chedin, A., Clarisse, L., Dubovik, O., Grainger, R., Griesfeller, J., Heckel, A., Kinne, S., Klüser, L., Kosmale, M., Kolmonen, P., Lelli, L., Litvinov, P., Mei, L., North, P., Pinnock, S., Povey, A., Robert, C., Schulz, M., Sogacheva, L., Stebel, K., Stein Zweers, D., Thomas, G., Tilstra, L.G., Vandenbussche, S., Veefkind, P., Vountas, M., Xue, Y., 2016. Development, production and evaluation of aerosol climate data records from European satellite observations (aerosol_cci). Remote Sens. 8 (5), 421. https://doi.org/10.3390/rs8050421.

Poulsen, C.A., Siddans, R., Thomas, G.E., Sayer, A.M., Grainger, R.G., Campmany, E., Dean, S.M., Arnold, C., Watts, P.D., 2012. Cloud retrievals from satellite data using optimal estimation: evaluation and application to ATSR. Atmos. Meas. Tech. 5 (8), 1889–1910. https://doi.org/10.5194/amt-5-1889-2012.

Poulsen, C., McGarragh, G., Thomas, G., Christensen, M., Povey, A., Grainger, D., Proud, S., Hollmann, R., 2017. ESA Cloud Climate Change Initiative (ESA Cloud cci) data: Cloud cci ATSR2-AATSRL3C-L3U-L2 CLD PRODUCTS v2.0. https://doi.org/10.5676/DWD/Esa:Cloud_cci/ATSR2-AATSR/.

Poulsen, C., Thomas, G.E., Siddans, R., Povey, A., McGarragh, G., Schlundt, C., Grainger, R.G., 2017. Product Validation and Intercomparison Report (PVIR). http://www.esa-cloud-cci.org/sites/default/files/documents/public/Cloud_cci_D4.1_PVIR_v4.1.pdf.

Rodgers, C.D., 2000. Inverse Methods for Atmospheric Sounding: Theory and Practice. World Scientific, London.

Rosenfeld, D., 2006. Aerosols, clouds, and climate. Science 312 (5778), 1323–1324. https://doi.org/10.1126/science.1128972.

Rosenfeld, D., Kaufman, Y.J., Koren, I., 2006. Switching cloud cover and dynamical regimes from open to closed Benard cells in response to the suppression of precipitation by aerosols. Atmos. Chem. Phys. 6 (9), 2503–2511. https://doi.org/10.5194/acp-6-2503-2006.

Rosenfeld, D., Lohmann, U., Raga, G.B., O'Dowd, C.D., Kulmala, M., Fuzzi, S., Reissell, A., Andreae, M.O., 2008. Flood or drought: how do aerosols affect precipitation? Science 321 (5894), 1309–1313. https://doi.org/10.1126/science.1160606.

Rosenfeld, D., Andreae, M.O., Asmi, A., Chin, M., de Leeuw, G., Donovan, D.P., Kahn, R., Kinne, S., KivekŠs, N., Kulmala, M., Lau, W., Schmidt, K.S., Suni, T., Wagner, T., Wild, M., Quaas, J., 2014. Global observations of aerosol-cloud-precipitation-climate interactions. Rev. Geophys. 52 (4), 750–808. https://doi.org/10.1002/2013RG000441.

Rozanov, V.V., Kokhanovsky, A.A., 2004. Semianalytical cloud retrieval algorithm as applied to the cloud top altitude and the cloud geometrical thickness determination from top-of-atmosphere reflectance measurements in the oxygen a band. J. Geophys. Res. Atmos. 109 (D5), D05202. https://doi.org/10.1029/2003JD004104.

Rozanov, V.V., Kokhanovsky, A.A., 2006. Determination of cloud geometrical thickness using backscattered solar light in a gaseous absorption band. IEEE Geosci. Remote Sens. Lett. 3 (2), 250–253. https://doi.org/10.1109/LGRS.2005.863388.

Saiedy, F.H., Hilleary, D.T., Morgan, W.A., 1965. Cloud-top altitude measurements from satellites. Appl. Opt. 4 (4), 495–500. https://doi.org/10.1364/AO.4.000495.

Samson, J., Berteaux, D., McGill, B.J., Humphries, M.M., 2011. Geographic disparities and moral hazards in the predicted impacts of climate change on human populations. Glob. Ecol. Biogeogr. 20 (4), 532–544. https://doi.org/10.1111/j.1466-8238.2010.00632.x.

Sayer, A.M., Poulsen, C.A., Arnold, C., Campmany, E., Dean, S., Ewen, G.B.L., Grainger, R.G., Lawrence, B.N., Siddans, R., Thomas, G.E., Watts, P.D., 2011. Global retrieval of ATSR cloud parameters and evaluation (grape): dataset assessment. Atmos. Chem. Phys. 11 (8), 3913–3936. https://doi.org/10.5194/acp-11-3913-2011.

Seaman, C.J., Noh, Y.J., Miller, S.D., Heidinger, A.K., Lindsey, D.T., 2017. Cloud-base height estimation from VIIRS. Part I: Operational algorithm validation against

CloudSat. J. Atmos. Oceanic Technol. 34 (3), 567–583. https://doi.org/10.1175/JTECH-D-16-0109.1.

Stengel, M., Mieruch, S., Jerg, M., Karlsson, K.G., Scheirer, R., Maddux, B., Meirink, J.F., Poulsen, C., Siddans, R., Walther, A., Hollmann, R., 2015. The clouds climate change initiative: assessment of state-of-the-art cloud property retrieval schemes applied to {AVHRR} heritage measurements. Remote Sens. Environ. 162, 363–379. https://doi.org/10.1016/j.rse.2013.10.035.

Stephens, G.L., Vane, D.G., Boain, R.J., Mace, G.G., Sassen, K., Wang, Z., Illingworth, A.J., O'Connor, E.J., Rossow, W.B., Durden, S.L., Miller, S.D., Austin, R.T., Benedetti, A., Mitrescu, C., Team, T.C.S., 2002. The Cloudsat mission and the A-Train. Bull. Am. Meteorol. Soc. 83 (12), 1771–1790. https://doi.org/10.1175/BAMS-83-12-1771.

Stephens, G.L., Vane, D.G., Tanelli, S., Im, E., Durden, S., Rokey, M., Reinke, D., Partain, P., Mace, G.G., Austin, R., L'Ecuyer, T., Haynes, J., Lebsock, M., Suzuki, K., Waliser, D., Wu, D., Kay, J., Gettelman, A., Wang, Z., Marchand, R., 2008. Cloudsat mission: performance and early science after the first year of operation. J. Geophys. Res. Atmos. 113 (D8), D00A18. https://doi.org/10.1029/2008JD009982.

Stevens, B., Feingold, G., 2009. Untangling aerosol effects on clouds and precipitation in a buffered system. Nature 461 (7264), 607–613.

Stubenrauch, C.J., Rossow, W.B., Kinne, S., Ackerman, S., Cesana, G., Chepfer, H., Getzewich, B., Di Girolamo, L., Guignard, A., Heidinger, A., Maddux, B., Menzel, P., Minnis, P., Pearl, C., Platnick, S., Riedi, J., Sun-Mack, S., Walther, A., Winker, D., Zeng, S., Zhao, G., 2013. Assessment of global cloud datasets from satellites: project and database initiated by the GEWEX radiation panel. Bull. Am. Meteorol. Soc. https://doi.org/10.1175/BAMS-D-12-00117.1.

Thomas, G.E., Poulsen, C.A., Sayer, A.M., Marsh, S.H., Dean, S.M., Carboni, E., Siddans, R., Grainger, R.G., Lawrence, B.N., 2009. The GRAPE aerosol retrieval algorithm. Atmos. Meas. Tech. 2 (2), 679–701. https://doi.org/10.5194/amt-2-679-2009.

Turner, A.G., Annamalai, H., 2012. Climate change and the South Asian summer monsoon. Nat. Clim. Change 2 (8), 587–595. https://doi.org/10.1038/nclimate1495.

Twomey, S., 1977. The influence of pollution on the shortwave albedo of clouds. J. Atmos. Sci. 34 (7), 1149–1152. https://doi.org/10.1175/1520-0469(1977)034⟨1149:TIOPOT⟩2.0.CO;2.

Twomey, S., 1991. Aerosols, clouds and radiation. Atmos. Environ. A. Gen. Top. 25 (11), 2435–2442.

Wang, Z., Sassen, K., 2001. Cloud type and macrophysical property retrieval using multiple remote sensors. J. Appl. Meteorol. 40 (10), 1665–1682. https://doi.org/10.1175/1520-0450(2001)040⟨1665:CTAMPR⟩2.0.CO;2.

Yamamoto, G., Wark, D.Q., 1961. Discussion of letter by A. Hanel: determination of cloud altitude from a satellite. J. Geophys. Res. 66 (5), 3596.

Ye, H., Li, J., Li, T., Shen, Q., Zhu, J., Wang, X., Zhang, F., Zhang, J., Zhang, B., 2016. Spectral classification of the yellow sea and implications for coastal ocean color remote sensing. Remote Sens. 8 (4). https://doi.org/10.3390/rs8040321.

Zheng, Y., Rosenfeld, D., 2015. Linear relation between convective cloud base height and updrafts and application to satellite retrievals. Geophys. Res. Lett. 42 (15), 6485–6491. https://doi.org/10.1002/2015GL064809.

Zheng, Y., Rosenfeld, D., Li, Z., 2015. Satellite inference of thermals and cloud-base updraft speeds based on retrieved surface and cloud-base temperatures. J. Atmos. Sci. 72 (6), 2411–2428. https://doi.org/10.1175/JAS-D-14-0283.1.

CHAPTER

6

Cloud-Aerosol-Precipitation Interactions Based of Satellite Retrieved Vertical Profiles of Cloud Microstructure

Daniel Rosenfeld

The Hebrew University of Jerusalem, Jerusalem, Israel

1 INTRODUCTION

The vertical evolution of cloud drop size distribution in convective clouds is a key property that dominates the precipitation forming processes in both the water and mixed phases. It also determines the radiative properties of cloud tops, which are used by satellites for retrieving cloud properties and radiative effects. Therefore, satellite retrieval of clouds' vertical microphysical profiles is a major objective, which is the subject of this chapter.

Aircraft measurements of vertical profiles of cloud drop concentrations (N_d), cloud drop effective radius (r_e), and cloud liquid water content (*CLW*) have been conducted extensively for marine stratocumulus (e.g., Wood, 2005; Painemal and Zuidema, 2011). The nearly adiabatic vertical profile (Fig. 1) has been the basis for satellite retrieval of N_d, *CLW*, and cloud geometrical depth (e.g., Han et al., 1998; Szczodrak et al., 2001). Uncertainty with respect to the adiabadicity assumption is a limiting factor in the accuracy of the satellite retrievals (Merk et al., 2016). Direct satellite retrievals of vertical profiles of marine stratocumulus are not available. The closest to such a vertical profile is exploiting the fact that satellite retrieval is based on 1.6, 2.1, and 3.7 μm, which have different weighting functions within the cloud, where the shorter wavelengths are less absorbing and arrive from deeper below the cloud top (Chang and Li, 2002). However, it was shown that the application of this principle is highly sensitive to small changes in the cloud reflectance, rendering the methodology insufficiently accurate with the present satellites (King and Vaughan, 2012).

Aircraft measurements of vertical microphysical profiles of developing deep convective clouds were conducted in the Amazon (Andreae et al., 2004; Braga et al., 2017), the summer monsoon clouds in India (Konwar et al., 2012; Prabha et al., 2011), the summer thunderstorms in West Texas (Rosenfeld and Woodley, 2000), the severe hailstorms at the lee of the Andes in Argentina (Rosenfeld et al., 2006), and the winter convective clouds in California (Rosenfeld et al., 2014a). Although the CLW of these deep convective clouds is far below adiabatic, it was found that the vertical profiles of r_e were nearly adiabatic, as

Remote Sensing of Aerosols, Clouds, and Precipitation
https://doi.org/10.1016/B978-0-12-810437-8.00006-2

© 2018 Elsevier Inc. All rights reserved.

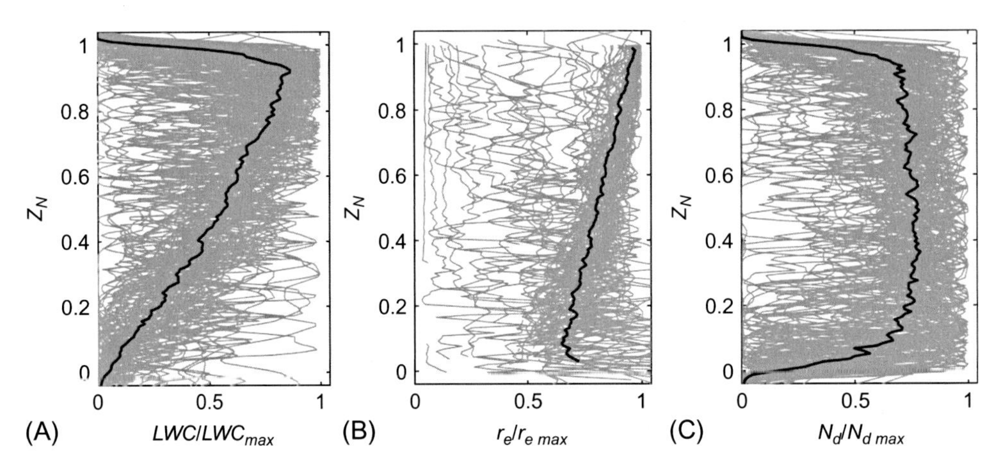

FIG. 1 Normalized vertical profiles of (A) *LWC*, (B) r_e, and (C) N_d. Values of $Z_N=0$ indicates the cloud base, whereas $Z_N=1$ indicates the cloud top. *Black lines* indicate the median profiles. *From Painemal, D., Zuidema, P., 2011. Assessment of MODIS cloud effective radius and optical thickness retrievals over the Southeast Pacific with VOCALS-REx in situ measurements. J. Geophys. Res. Atmos. 116 (D24).*

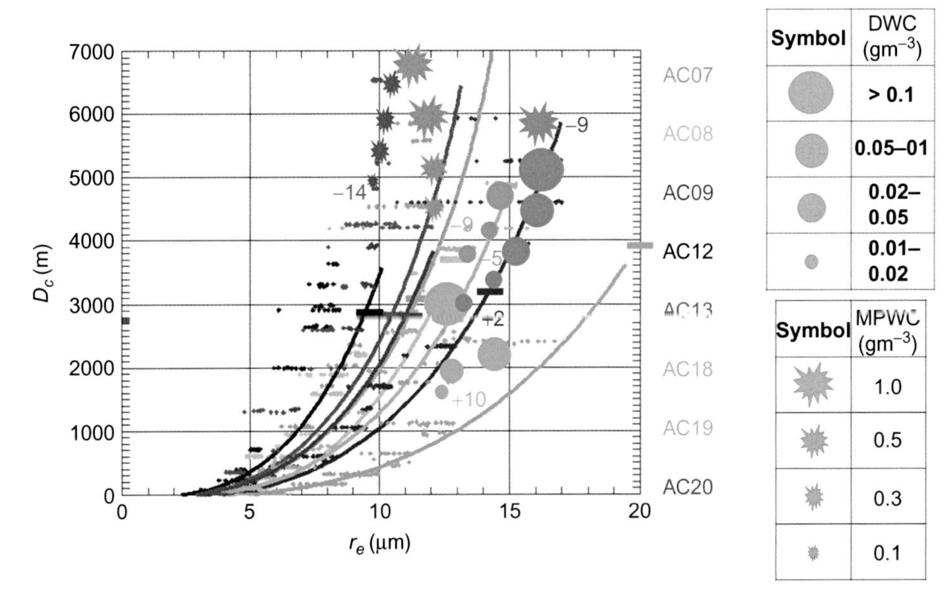

Symbol	DWC (gm^{-3})
	> 0.1
	0.05–01
	0.02–0.05
	0.01–0.02

Symbol	MPWC (gm^{-3})
	1.0
	0.5
	0.3
	0.1

FIG. 2 Aircraft-measured cloud droplet effective radius (r_e) *(colored dots)* and estimated cloud droplet adiabatic effective radius (r_{ea}) *(colored lines)* as a function of cloud depth (D_c) for all flights in the campaign (indicated by colors), as described in Wendisch et al. (2016). The height of 0°C is indicated by a horizontal bar across the r_{ea} line. The *circles* indicate the approximate values of drizzle water content (DWC) calculated from the CCP-CIP data, the range of DWC values is indicated in the table at the upper-right side of the figure. The *star symbols* indicate approximate mixed phase drizzle water content (MPWC) values calculated from the CCP-CIP data (indicated in the table at the bottom-right side of the figure). The temperature in degree celsius of rain or ice initiation (D_r and D_i, respectively) is indicated by the colored numbers close to circle or star symbols. *From Braga, R.C., Rosenfeld, D., Weigel, R., Jurkat, T., Andreae, M.O., Wendisch, M., Pöschl, U., Voigt, C., Mahnke, C., Borrmann, S., Albrecht, R.I., Molleker, S., Vila, D.A., Machado, L.A.T., 2017. What determines the height for warm rain initiation and cloud glaciation in convective clouds over the Amazon basin? Submitted to ACP.*

shown in Fig. 2 for the clouds in the Amazon (Freud et al., 2008, 2011; Braga et al., 2017). The fundamental reasons for that are discussed in the next sections. This rendered the vertical evolution of r_e as a fundamental property that characterizes the whole convective cloud cluster that develops under similar thermodynamic and aerosol conditions.

This motivates satellite retrievals of the same. Satellite retrieval of the vertical profile of r_e of a cloud cluster can be done by using the cloud top temperature (T) as a surrogate for height, assuming that all convective cloud bases have the same temperature. This is the case when the clouds develop from a well-mixed boundary layer. Satellite retrieved $r_e(T)$ was introduced by Rosenfeld and Lensky (1998), and subsequently applied extensively for documenting the impacts of aerosols on convective clouds and precipitation forming processes. These applications will be reviewed after laying the physical background for the methodology and observed relationships.

2 PHYSICAL CONSIDERATIONS IN RETRIEVING VERTICAL MICROPHYSICAL PROFILES

Remote sensing of aerosols, clouds, and precipitation require very different wavebands and methods for each of these three components of the highly interactive system of aerosols, clouds, and precipitation. Most of the cloud-active aerosols are submicron, thus requiring a respectively short wavelength in the visible spectrum to be measured. The cloud particles size (few to 50 μm) and phase (water or ice) are well suited to be measured by analyzing combinations of visible and IR wavebands. Because precipitation occurs within and below the clouds, the precipitation size particles, larger than a diameter of 0.1 mm, can be detected only by radiation that can penetrate the clouds while interacting with the precipitation particles. Such is the microwave (both passive and active, i.e., radar), which is used at wavelengths of several mm to 10 cm. Satellite radar measurements are very limited in spatial and temporal coverage. Passive microwave is limited in spatial resolution and does not resolve the vertical structure of the precipitation.

These widely varying measurement requirements make it is very difficult to observe simultaneously aerosols, clouds, and precipitation for understanding their interactions. Furthermore, the very presence of clouds obscure the clear air below them in which the aerosols reside. The clouds also obscure the precipitation within and below them. Therefore, it is very appealing to infer both aerosols and precipitation based on retrieving the vertical profiles of cloud microstructure and phase, which relies on the widely available coverage of the visible and IR wavebands.

Aerosols along with updrafts and temperatures determine cloud composition, which in turn determines precipitation forming processes. The response of the cloud to the precipitation affects the cloud dynamics and lifecycle as well as the scavenging of the aerosols from the air, eventually feeding back to the cloud-aerosol-precipitation interactions. Therefore, measuring the cloud composition and thermodynamic properties with visible and IR radiances can be used for inferring information on cloud-active aerosols on one side and on precipitation forming processes on the other side, and to combine them to understand the system as a whole.

To understand how visible and IR satellite measurements are applied to documenting aerosols-cloud-precipitation interactions, we first must elucidate some of the cloud physical processes, and then show how they are exploited in the satellite retrievals.

3 THE MICROSTRUCTURE OF VERTICAL PROFILES OF ADIABATIC CONVECTIVE CLOUDS

Cloud drops nucleate on cloud condensation nuclei (CCN) aerosols when vapor supersaturation S exceeds a critical value, which corresponds to the solubility and radius of the aerosol particles (Köhler, 1936). A maximum in S (S_{max}) is reached a small distance (few to tens of m) above the cloud base, where the number of cloud drop concentration (N_{db}) is determined. Additional growth of the cloud droplets above the height of S_{max} increases further the integrated drop surface area and leads to the decrease of S, thus preventing nucleation of additional cloud droplets above the cloud base. S_{max} is determined by N_{db} and cloud base updraft (W_b) according to Eq. (1) (Pinsky et al., 2012).

$$S_{max} = C(P,T)W_b^{3/4}N_{db}^{-1/2} \quad (1)$$

where C is a coefficient that depends on cloud base pressure (P) and temperature (T). Because of the existence of S_{max}, all the condensation of vapor within a rising cloud parcel leads to the growth of the cloud droplets that were nucleated at cloud base. In other words, the adiabatic cloud liquid water (CLW_a) content is divided to N_{db} droplets, which were all nucleated at cloud base. Therefore, the relationship between cloud drop mean volume radius (r_v) and LWC_a in an adiabatic cloud without drop coalescence is given by Eq. (2):

$$r_v = \left(\frac{3Q_l}{4\pi\rho_w N_{db}}\right)^{1/3} \quad (2)$$

where q_L is the CLW_a and ρ_w is water density. The N_{db} is determined by cloud base updraft (W_b) and the spectrum of CCN(S) (Twomey, 1959). The cloud drop concentration N_d remains constant (in units of mixing ratio) with height and equals N_{db} as long as there is no loss or gain of cloud droplets. In a rising adiabatic cloud, parcel cloud drops do not evaporate, but drop coalescence can occur and can reduce N_d and respectively increase r_v according to Eq. (2).

Cloud drop coalescence can decrease N_d. Its rate depends on r_v and on the spectrum width of the cloud drop size distribution. A common measure of the spectrum width is the ratio between cloud drop effective radius (r_e) and r_v, where r_e is given by Eq. (3).

$$r_e = \frac{\int_0^\infty r^3 n(r)dr}{\int_0^\infty r^2 n(r)dr} \quad (3)$$

where r is cloud drop radius and $n(r)$ is the concentration of cloud droplets having radius r. This statistical moment for defining r_e is used because this is the statistical moment of cloud drop size distribution that is measurable by satellites, using methods such as Nakajima and King (1990). The average ratio between r_v and r_e for aircraft full passes in nonprecipitating convective clouds was found to obey Eq. (4) with remarkably small variations of less than 1% between different flights at different times and locations. The variability cloud drop size distributions in cloud segments of 1 second flight (~100 m) was larger and varied between 1.04 and 1.17 for the 5th and 95th percentiles, respectively. (Freud and Rosenfeld, 2012)

$$r_e = 1.08 r_v \quad (4)$$

This leaves coalescence rate to depend, on average, almost exclusively on r_v. The coalescence kernel K [cm^3s^{-1}] is defined by Eq. (5):

$$K = \pi(R+r)^2[v_t(R) - v_t(r)]E(R,r) \quad (5)$$

where R and r are the radii of the collecting (larger) and collected (smaller) cloud drops, $v_t(R)$ and $v_t(r)$ are their respective terminal fall velocities, and E is the coalescence efficiency between these two drops. A scaling analysis shows that $K \alpha r_v^5$, is as follows:

$$\pi(R+r)^2 \propto r_v^2; \quad [v_v(R) - v_t(r)] \propto r_v^2; \quad E(R,r) \propto r_v^1$$

The combined sum of the powers of r_v amounts to 5. This was calculated for aircraft measured cloud drop spectra, which showed good agreement with the theoretical expectation, as evident in Fig. 3. The inset in Fig. 3 shows the same data

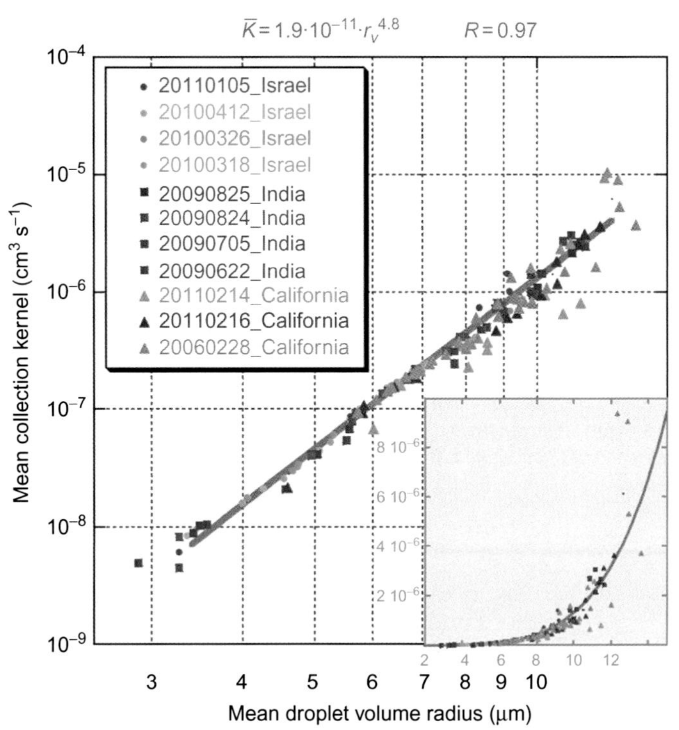

FIG. 3 The rate of production of rain water by cloud droplet coalescence as a function of cloud droplet mean volume radius. Each point is the calculation based on one aircraft measured droplet size distribution in a cloud pass at a location and date as shown in the legend. The inset is the same figure, but in linear scale. Note the sharp increase starting at r_v of 12 μm, which is r_e of 13 μm. *From Freud, E., Rosenfeld, D., 2012. Linear relation between convective cloud drop number concentration and depth for rain initiation, J. Geophys. Res. 117, D02207, doi:10.1029/2011JD016457.*

in linear scale. The fifth power is manifested as a negligible drop coalescence rate for $r_v < 11$ μm, but very fast acceleration when r_v increases beyond 12 μm. This means that warm rain processes become active when r_e exceeds 13–14 μm (Rosenfeld and Gutman, 1994; Gerber, 1996; Rangno and Hobbs, 2005; Suzuki et al., 2010; Freud and Rosenfeld, 2012; Rosenfeld et al., 2012a). This is demonstrated by aircraft observations in Fig. 4. Satellite observations are shown in Fig. 4, where CLOUDSAT radar echoes detect little rain in clouds with top $r_e < 10$ μm, while showing well developed rain when $r_e > 15$ μm, which intensifies further for even larger r_e (Suzuki et al., 2010). In this regard, the r_e threshold for rain initiation is valid equally to shallow and deep convection. Fig. 5 demonstrates it nicely by simulations of open cells of marine stratocumulus, showing that they start raining significantly only when r_e at their tops exceeds 12–14 μm (Rosenfeld et al., 2012a). This was also shown observationally to occur in marine stratocumulus (Goren and Rosenfeld, 2015).

These physical relationships create a situation where the knowledge of cloud base P, T, and N_{db} is sufficient for calculating the vertical evolution of the cloud's CLW_a, T, r_e, N_d and coalescence rate. In convective clouds, T is uniquely related to height (H) above cloud base, so that the knowledge of $T(H)$ can be used for expressing the vertical microphysical profile of the cloud, like $CLW_a(T)$ and $r_e(T)$. This is a very useful outcome for satellite retrievals, because r_e and T can be retrieved directly at various heights of the cloud top above its base. The $r_e(T)$ can be used for calculating N_{db} in adiabatic clouds, as long as the cloud drop coalescence did not cause an underestimation of N_{db} appreciably. Coalescence is not much of a concern for the lower parts of

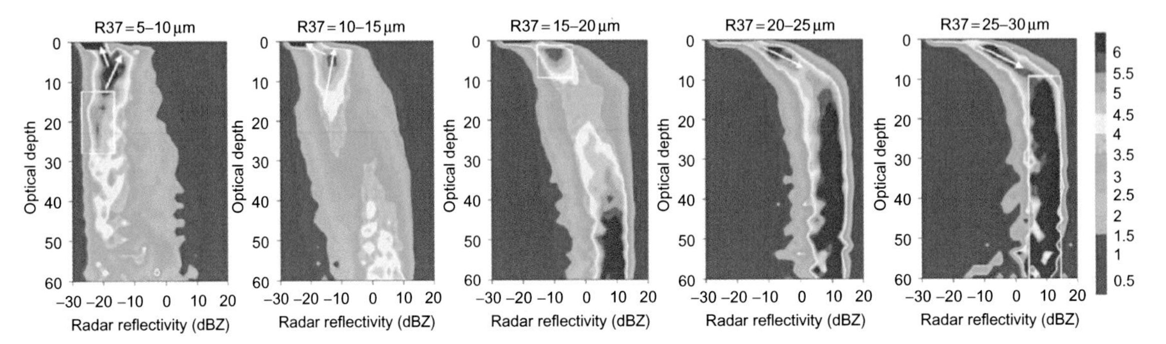

FIG. 4 Frequencies of rain radar reflectivities (dBZ) as a function of cloud optical depth for different classes of cloud top droplet effective radius. The reflectivity was measured by the CLOUDSAT radar, and the r_e was measured by MODIS using the 3.7 μm waveband. The figure shows practically no rain for $r_e < 10$ μm, and rain develops for clouds with progressively smaller optical depth when r_e increases beyond 15 μm. *From Suzuki, K., Nakajima, T.Y., Stephens, G.L., 2010. Particle growth and drop collection efficiency of warm clouds as inferred from joint CloudSat and MODIS observations. J. Atmos. Sci. 67(9), 3019–3032, © American Meteorological Society. Used with permission.*

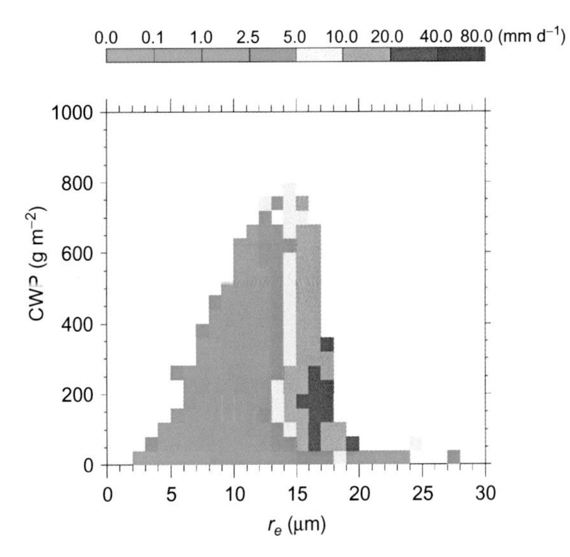

FIG. 5 Maximum rain intensity in the vertical as a function of cloud top droplet effective radius and cloud water path, for simulated open cells of marine stratocumulus. Note that significant rain develops only when cloud top r_e exceeds 13–14 μm, regardless of cloud water path. *From Rosenfeld, D., Wang, H., Rasch, P.J., 2012. The roles of cloud drop effective radius and LWP in determining rain properties in marine stratocumulus. Geophys. Res. Lett. 39, L13801, doi:10.1029/2012GL052028.*

the clouds, below where the height r_e reaches 14 μm. A cloud parcel model with 2000 size bins (Khain et al., 2000) was used for demonstrating these relationships (Rosenfeld et al., 2002). The results (Fig. 6) show the following key features:

- Rain water fraction is negligible for $r_e < 14$ μm, but increases sharply above it, marking $H(r_e = 14\ \mu m)$ as the height for rain initiation, H_r (panel D).
- r_e increases with H in a manner that agrees with nearly linear increase of cloud drop volume ($\alpha\ r_e^3$) with a height or CLW_a above cloud base (panel C).
- For $H > H_r$, dr_e/dH increases sharply (panel C).
- Below H_r, N_d remains nearly constant with H except for a slight decrease due to the dilution of air with height (panel B).
- Above H_r, N_d decreases sharply due to the strong drop coalescence.
- A strong reduction of N_d above H_r can be followed by a jump in N_d at certain heights due to secondary drop nucleation. This happens when S exceeds S_{max} due to the

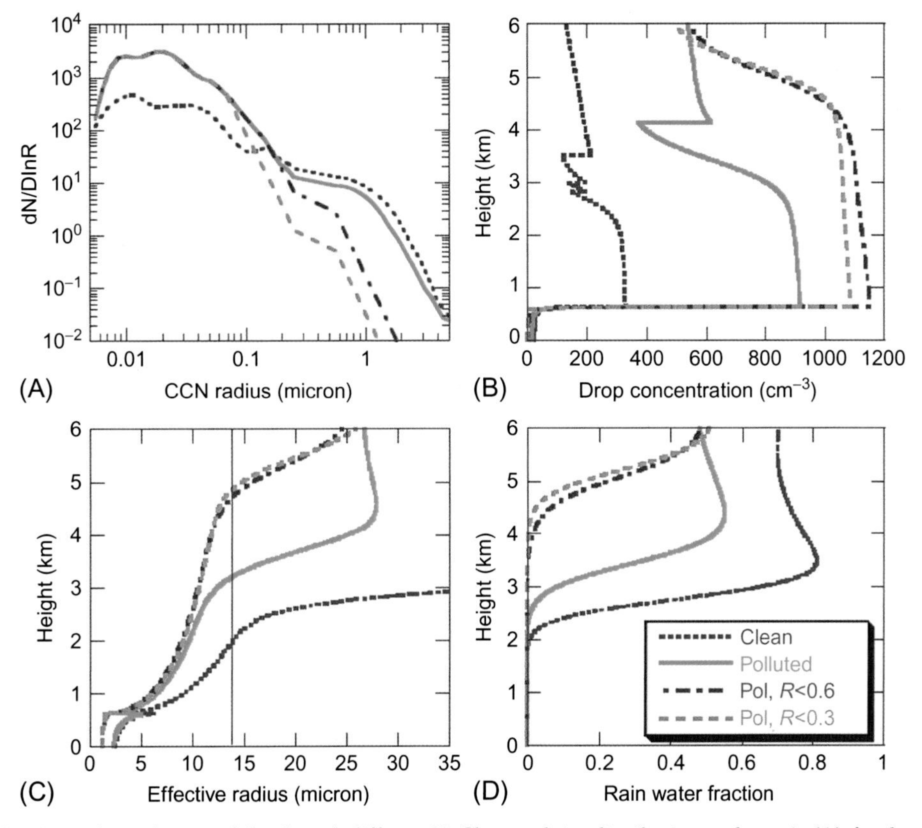

FIG. 6 Cloud parcel simulations of clouds with different NaCl aerosol size distribution as shown in (A), for clean *(blue)*, polluted with sea spray giant CCN *(green)*, polluted with truncated aerosols at radius of 0.6 and 0.3 μm *(red and purple*, respectively). The concentrations of drops with radius > 0.5 μm (B), effective drop radius (C), and fraction of cloud water converted to rain drops (D), as a function of height above cloud base. Note the sharp acceleration of growth of r_e with height at $r_e = 14$ μm (C), associated with initiation of rain water (D). This occurs at greater heights for clouds with more drop concentrations. Note also the secondary drop nucleation after initial drop concentration is depleted by coalescence (B). *From Rosenfeld, D., Lahav, R., Khain, A., Pinsky, M., 2002. The role of sea spray in cleansing air pollution over ocean via cloud processes. Science 297(5587), 1667–1670.*

decrease in the condensation rate associated with the reduced total drop surface area due to the drop coalescence. S can also increase with height because of vertical acceleration of the updraft speed (panel B).

- More polluted air has more N_{db} for the same W_b, and a respectively smaller r_e for a given $H < H_r$.
- Larger N_{db} results in greater H_R.
- Abundance of supermicron CCN leads to smaller N_{db}, larger r_e, and lower H_r.

The cloud depth for rain initiation, D_r, is defined as H_r-cloud base height. In rising adiabatic convective cloud parcels, D_r increases linearly with N_{db}, due to the following considerations:

$$CLW_a \propto D_r^b \tag{6}$$

where b is very near unity, and becomes appreciably lower than that only for clouds with a base colder than ~0°C and $D > \sim 3$ km.

$$CLW_a \propto r_v^3 N_{db} \tag{7}$$

Therefore, the right hand sides of Eqs. (6), (7) have nearly linear relationships:

$$D_r \propto \left(r_v^3 N_{db}\right)^{1/b} \tag{8}$$

Eq. (8) shows that D_r increases linearly with N_{db}. In fact, it can be generalized that the cloud depth for reaching any $r_e < 14\ \mu m$ is linear with N_{db} (Freud and Rosenfeld, 2012; Braga et al., 2017).

The parcel model simulations in Fig. 6 show that $r_e(H)$ is not affected by variability in updraft speeds above cloud base, but the rate of increase of r_e with H above H_r decreases with increasing updraft speed there, because there is less time for conversion of cloud droplets to rain drops when the parcel rises faster (Rosenfeld et al., 2007).

The relationships presented here for adiabatic clouds represent rather simple and unique relationships between $r_e(T)$, N_{db}, and H_r, which would have been very useful for satellite retrievals of N_{db} and H_r if water clouds were really adiabatic. It would have been useful even for retrieving updraft speeds at heights $> H_r$. To what extent do deviations from the adiabatic assumption in real world clouds complicate these relationships and what is the impact on their usefulness for satellite retrievals of cloud properties? This will be addressed next.

4 THE OBSERVED MICROSTRUCTURE OF CONVECTIVE CLOUDS VERTICAL PROFILES

Aircraft-measured vertical profiles of convective clouds show remarkably tight relationships between CCN, N_{db}, r_e(H), and H_r in agreement with the theoretical expectations from adiabatic clouds. Fig. 7 shows flights in contrasting

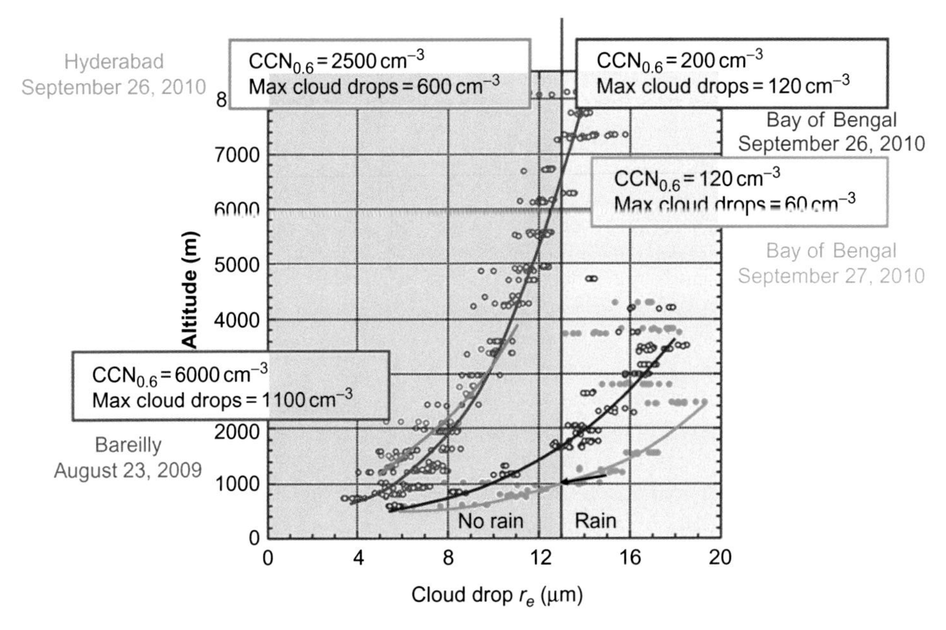

FIG. 7 Aircraft measurements of vertical profiles of cloud drop effective radius in environments with various CCN concentrations. The measurements were done in the summer monsoon convective clouds over India and the adjacent sea. Each point represents r_e during 1-s flight path, which is nearly 100 m. Rain was observed to initiate around r_e of 13 μm. Note the tight relationships between r_e and height. D. Rosenfeld served as the flight scientist, and collected and analyzed the flight data into this figure.

aerosol conditions that were executed in the Cloud Aerosol Interaction and Precipitation Enhancement Experiment (CAIPEEX) that was conducted by the Indian Institute of Tropical Meteorology during the summer monsoon seasons of 2009 and 2010, with the author of this chapter serving as the flight scientist. The measurements plotted in Fig. 7 show the following features:

- Clouds have smaller N_{db} when developing in air mass with lower CCN concentrations.
- The $r_e(H)$ produces well defined relationships of increasing r_e with H.
- The r_e grows faster with H for smaller N_{db}.
- H_r is marked here as the height where r_e reaches 13 μm. H_r is higher for clouds with larger CCN and N_{db}.

In fact, it was observed that the aircraft measured $r_e(H)$ in convective clouds in numerous field campaigns, and was close to the theoretically expected adiabatic r_e as calculated based on N_{db} and cloud base T and P. The r_e was near adiabatic despite the intense mixing of the cloud with the ambient air and CLW being far from adiabatic (Freud et al., 2011; Braga et al., 2017). This phenomenon was first observed in marine stratocumulus by Pawlowska et al. (2000). Burnet and Brenguier (2007) explained this in terms of the premoistening of ambient air parcels that are entrained into the clouds. When evaporation rate is much larger than mixing rate, the cloud droplets bordering the dry parcel evaporate completely while moistening the entrained air until near saturation. Then, the continued mixing of the moistened air with the cloud does not lead to any additional evaporation of the cloud droplets. The outcome is a decrease in CLW and N_d while r_e is conserved. This inhomogeneous mixing nature of the clouds was demonstrated vividly by Beals et al. (2015), who showed sharp borders between cloudy and clear air filaments down to the centimeter scale in mixing clouds, while cloud drop size was being preserved in the cloud filaments.

5 IMPACTS OF CCN AND UPDRAFTS ON VERTICAL MICROPHYSICAL PROFILES OF CONVECTIVE CLOUDS

The vertical microphysical profiles and precipitation forming processes in the growing phase of convective clouds can be deduced based on the satellite retrieved $r_e(T)$ profiles. Fig. 7 demonstrates the tight relationships between H and r_e in convective clouds. The unique relationship between H and T in convective clouds therefore result in similarly tight relationships between the satellite retrieved T and r_e. Since $r_e(T)$ for a given T and P at cloud base is determined by N_{db}, and N_{db}, which in turn is determined by CCN(S) and W_b, and the $r_e(T)$ for a given W_b is determined by CCN(S).

Satellite data can be used for retrieving r_e and T of cloud surfaces that are exposed to the satellite view and to direct solar illumination. The full $r_e(T)$ function is composed of many pairs of T and r_e measurements in a cloud cluster that contains cloud elements at various heights, as illustrated in Fig. 8. The procedure assumes that:

1) Cloud T and r_e near its surface is similar to these properties well within the cloud at the same height.
2) Cloud top T and r_e are the same as these properties at the same height within taller clouds, as long as precipitation is not falling through that cloud volume.
3) All clouds in the observed cluster have similar $r_e(T)$, so that tracking with time $r_e(T)$ of a growing single cloud is identical to compose $r_e(T)$ from a cloud ensemble at a single time.

The validity of assumptions 1 and 2 was demonstrated by the evidence for the dominance of inhomogeneous mixing in non precipitating growing convective clouds. The validity of assumption 3 was validated by using a rapid scan of geostationary data (Lensky and Rosenfeld, 2006). In practice, the T-r_e relationships are obtained by calculating the mean for each 1°C

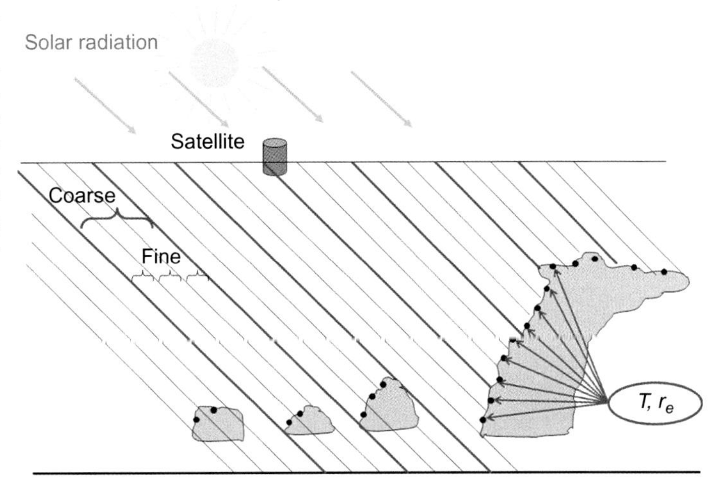

FIG. 8 Illustration of composing $r_e(T)$ from high resolution satellite retrievals from a field of convective clouds. Note that high resolution (*thin black lines*, representing NPP/VIIRS 375 m data) is required to resolve the vertical structure of the small boundary layer clouds. Smaller resolution (*red lines*, representing MODIS 1 km data) cannot resolve that. Backscatter solar illumination is required for avoiding cloud shadows and for minimizing the three-dimensional effects on the retrievals.

interval of T for all the cloudy pixels within a selected area. Because there are many pixels with a distribution of r_e for any given T, the median is associated with T to form $r_e(T)$.

The shape of $r_e(T)$ was used for identification of five microphysical zones in the clouds based on the dominance of different kinds of processes in the growth of cloud drops and hydrometeors. The five satellite-retrieved vertical microphysical zones are (Rosenfeld and Lensky, 1998; Rosenfeld and Woodley, 2003):

1. *Condensation-dominated growth of cloud droplets*: This is where $r_e < 14$ μm, and warm rain processes have not progressed significantly (Suzuki et al., 2010). The depth of this region is determined mainly by N_{db} and cloud base temperature. Because N_{db} is dominated by CCN concentrations (Nccn), except for updraft-limited conditions, the depth of the condensational growth zone is dominated by Nccn.
2. *Coalescence-dominated growth of cloud droplets*: Cloud droplets grow by coalescence into raindrops much faster than they would just by condensation, even in an adiabatic cloud parcel. Therefore, this zone is reached when the retrieved r_e exceeds its adiabatic value. The onset of coalescence zone occurs typically beyond $r_e = 14$ μm, where it is often evident in addition to the acceleration of $dr_e/-dT$.
3. *Rainout zone*: When the warm rain is fully developed at the top of growing convective clouds, the largest drops that fall from the cloud tops balance the additional growth by coalescence. This is manifested as a stabilization of r_e with further vertical growth at the size of about 25 μm.
4. *Mixed phase zone*: The development of the ice phase is indicated by an apparent additional increase of the indicated r_e, when the retrieval algorithm assumes a water cloud. The indicated increase in r_e is the result of ice absorbing more strongly the solar radiation at the 3.7 μm band. It is also caused by ice particles becoming much larger than the cloud droplets in the mixed phase clouds in which they are formed.
5. *Glaciated zone*: When all cloud drops are glaciated, or at least when the radiative signature is dominated by the ice phase, the r_e reaches its peak size, because there is no effective mechanism for additional growth of the ice particles. Above that height, the

retrieved r_e remains stable or decreases, due to secondary ice nucleation aloft (Yuan et al., 2010).

The shapes of the $r_e(T)$ relationships that define these microphysical zones are illustrated in Fig. 9. Actual examples of a wide range of conditions, as retrieved from the new high resolution NPP/VIIRS satellite Imager, are shown in Rosenfeld et al. (2014b). Two examples for deep tropical clouds are given in Fig. 10. Pristine clouds with developed rainout zone are shown in Fig. 10A, and its $r_e(T)$ is shown by the *blue line* in Fig. 11. Polluted clouds with strongly suppressed coalescence are shown in Fig. 10B and the *red line* in Fig. 11, respectively. The inferred microphysical zones can be validated by surface and space-borne radar measurements (L'Ecuyer et al., 2009; Suzuki et al., 2010, 2011). Not all microphysical zones are necessarily realized in any given convective vertical profile. They are determined by a combination of Nccn, ice nuclei number concentration (NIN), updraft speed, and cloud base and top temperatures.

In tropical convective clouds with warm bases, the depth of the condensation-dominated growth zone is determined by N_d. The coalescence zone and formation of warm rain can be delayed to above the freezing level in highly polluted atmospheres. For example, aircraft measurements have documented clouds with base

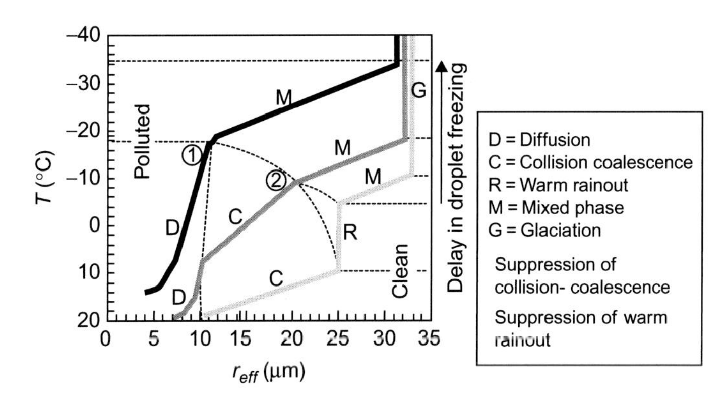

FIG. 9 Conceptual diagram describing five microphysical stages (droplet growth by diffusion, collision-coalescence, warm rainout, ice-water mixed phase, and glaciated phase) common to deep convective clouds and their response to the concentration of pollution aerosols. The dashed line divides the parameter space to the microphysical zones, as denoted in the legend. For example, areas 1 and 2 show the domains of cloud drop growth by diffusion and coalescence, respectively. The *bottom curve* shows the case of a maritime environment with low CCN concentration and ample warm rain processes; the *middle curve* corresponds to moderately polluted continental clouds, where the larger number of CCN aerosols helps to suppress the warm rain process and glaciation starts at slightly lower temperatures; the *top curve* represents extremely polluted cases where the very large number of CCNs produce smaller and more numerous droplets at the cloud base, suppressing the beginning of collision-coalescence processes and postponing the start of droplet freezing to lower temperatures. Intermediate situations to the above curves are also possible. Note that glaciogenic cloud seeding is mostly targeted to the region of mixed-phase clouds ("M" in the figure). Hygroscopic seeding may be applied to clouds with deep condensational growth region ("D" in the figure). *Figure from Martins, J.V., Marshak, A., Remer, L.A., Rosenfeld, D., Kaufman, Y.J., Fernandez-Borda, R., Koren, I., Correia, A.L., Zubko, V., Artaxo, P., 2011. Remote sensing the vertical profile of cloud droplet effective radius, thermodynamic phase, and temperature. Atmos. Chem. Phys. 11, 9485–9501, doi:10.5194/acp-11-9485-2011. http://www.atmos-chem-phys.net/11/9485/2011/acp-11-9485-2011.pdf who adapted it from Rosenfeld, D., Woodley, W.L., 2003. Closing the 50-year circle: from cloud seeding to space and back to climate change through precipitation physics. In Wei-Kuo Tao, Robert Adler (Eds.), Cloud Systems, Hurricanes, and the Tropical Rainfall Measuring Mission (TRMM). 234p., pp. 59–80, Meteorological Monographs 51, AMS (Chapter 6).*

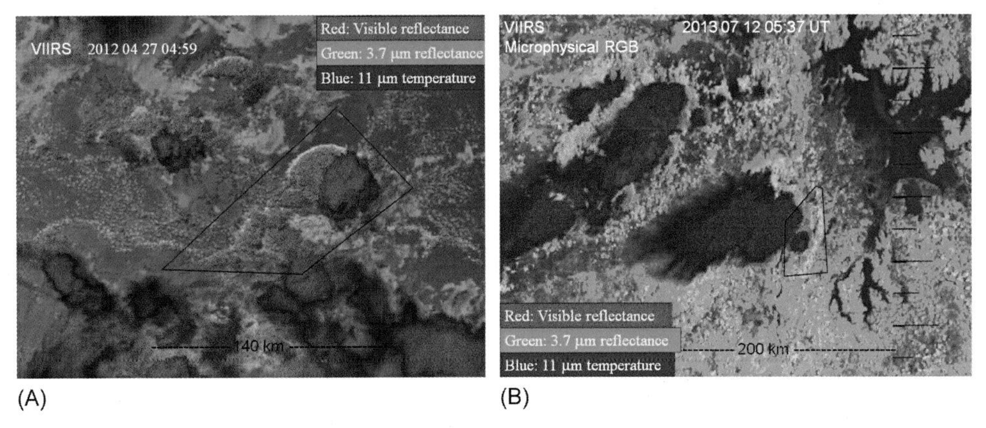

FIG. 10 Microphysical rendering of deep convective clouds in pristine and polluted clouds. The *red* is modulated by the visible reflectance; the *green* is modulated by 3.7 μm solar reflectance, where more *green* is smaller cloud particles; and *blue* is modulated by temperature. The top panel (A) shows tropical marine clouds over Indonesia. Its $r_e(T)$ is shown by the *blue line* in Fig. 9. The bottom panel (B) shows polluted clouds over southeast China. Its $r_e(T)$ is shown by the *red line* in Fig. 9. *From Rosenfeld, D., Liu, G., Yu, X., Zhu, Y., Dai, J., Xu, X., Yue, Z., 2014. High resolution (375 m) cloud microstructure as seen from the NPP/VIIRS Satellite imager. Atmos. Chem. Phys. 14, 2479–2496, doi:10.5194/acp-14-2479-2014.*

temperatures of 25°C over the Indo-Gangetic plains, which did not produce warm rain and whose r_e remained below 14 μm up to the −5°C isotherm level (Konwar et al., 2012; Khain et al., 2013). An example is shown in Fig. 7 by the aircraft measurements over Bareilly. Precipitation was initiated in these clouds as supercooled raindrops that froze at −12°C to −17°C. This means that such clouds with warmer tops would rain very little naturally unless seeded with IN or giant CCN.

In situ aircraft measurements have documented that cloud drops that were mostly formed at cloud base reached the homogeneous ice nucleation temperature with updrafts that did not exceed 15 ms^{-1} in clouds with high and cool bases and with large N_d, (Rosenfeld and Woodley, 2000). These cloud droplets did not freeze earlier because they remained too small to produce supercooled drizzle and rain drops, which freeze much faster than small cloud droplets. Apparently, ice nuclei concentrations could not have been very high in this situation, which would have otherwise allowed the cloud drops to freeze heterogeneously before reaching the homogeneous ice nucleation temperature. The cloud drops therefore froze into similarly small ice particles, which at these heights had no mechanism to coalesce and precipitate, but rather detrained from the anvil and evaporated.

6 APPLICATION OF $R_E(T)$ TO OBSERVE ANTHROPOGENIC AEROSOLS SUPPRESSING PRECIPITATION

The Tropical Rainfall Measuring Mission (TRMM) satellite was well equipped for measuring the relationships between $r_e(T)$ and its impact on precipitation (Simpson et al., 1988). TRMM had onboard the Visible Infrared Scanner (VIIRS) which can be used for constructing $r_e(T)$, Precipitation Radar (PR) for constructing vertical profile of precipitation radar reflectivity, TRMM Microwave Imager (TMI) which can be

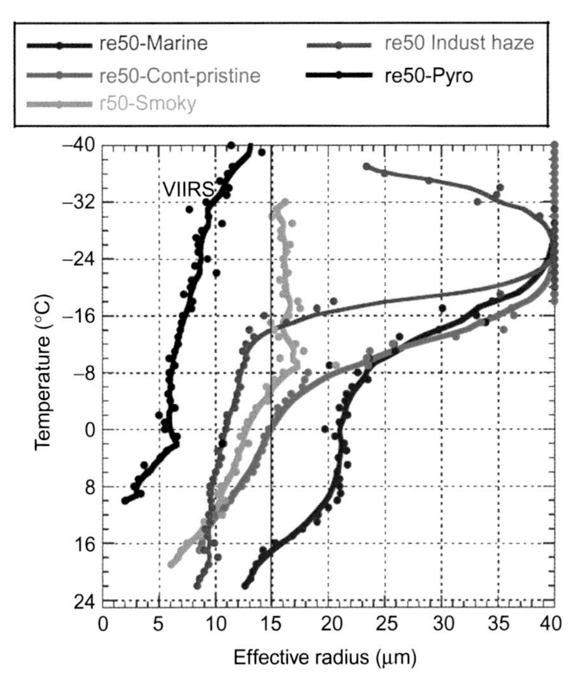

FIG. 11 Actual $r_e(T)$ profiles as measured in convective clouds by the NPP/VIIRS high resolution (375 m) Imager. Each point represents the median r_e for a given T in the analyzed area of the convective cloud cluster. The various lines are: *Blue*—marine convective clouds over sea near Indonesia; note the rainout zone between 9 and −6°C and the glaciation at −20°C. *Green*—pristine tropical continental cloud over eastern Australia. *Orange*—clouds growing in smoke from forest fires in Indonesia. Note the suppressed coalescence up to above the 0°C isotherm as indicated by $r_e < 14$ μm up to that height. This field of clouds contains a growing tower that did not glaciate up to −32°C. *Red*—clouds in heavy haze of air pollution in southeast China. Note the very small r_e strongly suppress coalescence up to −15°C, where the sudden increase of r_e indicates the development of mixed phase, and a full glaciation at −20°C. *Black*—pyro cumulonimbus over central Siberia, with extremely small r_e up to the homogeneous ice nucleation temperature of −38°C. *From Rosenfeld, D., Liu, G., Yu, X., Zhu, Y., Dai, J., Xu, X., Yue, Z., 2014. High resolution (375 m) cloud microstructure as seen from the NPP/VIIRS Satellite imager. Atmos. Chem. Phys. 14, 2479–2496, doi:10.5194/acp-14-2479-2014.*

used for assessing the amount of cloud water content, and Lightning Imaging Sensor (LIS). Convective clouds produced lightning that was proportional to the collisions of graupel with ice crystals in supercooled clouds. The application of TRMM analyses to the clouds in the Amazon under contrasting aerosol conditions revealed dramatic aerosol effects. Three cases are shown in Fig. 12:

1. Pristine conditions of the so called Green Ocean Amazon (panels A–C). Panel B shows very large r_e that exceeds the 14 μm rain threshold already near the cloud base. Respectively, all the clouds precipitate, including the smallest ones, as evident by the white stippling in panel A and the PR cross section in panel C. The rain rate reflectivities are modest and do not reach 45 dBZ. Very few lightning flashes were observed.
2. Smoke from deforestation affecting modestly developed clouds (panels D–G). The smoke is seen in the TOMS aerosol index in panel E. The r_e remains below the 14 μm rain threshold up to the −10°C isotherm (panel G). These clouds do not precipitate, as indicated by the lack of radar rain mask in panel D and lack of radar echoes in the vertical cross section in panel F for clouds with tops reaching 6 km. These clouds have very high cloud water path, as indicated by the very low TMI 85 GHz brightness temperature. This means that there is ample cloud water that does not precipitate because the cloud droplets are too small to coalesce efficiently. A similar situation was documented during forest fires in Indonesia (Rosenfeld, 1999).
3. Smoky deep convective clouds (panels H–J). The r_e reaches the 14 μm rain threshold at the 0°C isotherm, thus suppressing warm rain and making all the cloud water available for mixed phase precipitation forming processes. The cloud water is accreted on graupel and releases latent heating while freezing that

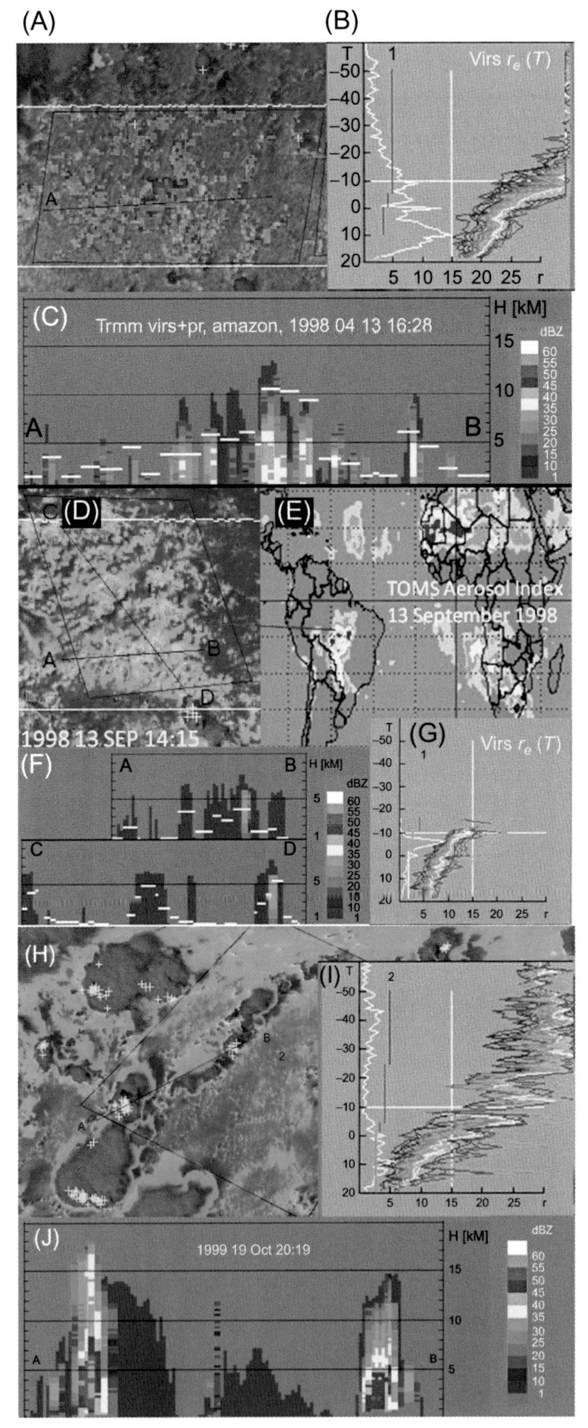

FIG. 12 TRMM observations of aerosol effects on cloud microphysical, precipitation vertical profiles, and electrification in the Amazon under contrasting conditions. The first case (panels A–C) shows a pristine situation with large r_e, where every small cloud precipitates. The second case (panels D–G) shows a smoky situation where rain is suppressed in clouds with tops up to 6 km. The third case (panels H–J) shows a smoky situation with deep clouds with very strong reflectivities reaching great height and having many lightning flashes. The VIIRS images rendered by the microphysical RGB are given in panels (A), (D), and (H). The $r_e(T)$ of the rectangles shown in these panels are given in panels (B), (G), and (I), respectively. The flashes detected by Lightning Imaging Sensor (LIS) are marked as yellow crosses. The swath width of 220 km of the TRMM Precipitation Radar (PR) is delimited between the two *white lines* in panels (A) and (D). The white stippling in these panels masks the rainy areas. Vertical cross sections of the PR reflectivity are given in panels (C), (F), and (J), for the *black lines* in panels (A), (D), and (H), respectively. The *dark gray area* delineates clouds without precipitation echoes, as measured by the VIIRS. The *white bold lines* on the PR cross sections in panels (C) and (F) represent the brightness temperatures of 85 GHz channel of the TRMM Microwave Imager (TMI), mapped to the height of that isotherm. Lower temperatures, which show greater heights, indicate greater rain intensity or larger amount of cloud water path. *Modified from Rosenfeld, D., Woodley, W.L., 2003. Closing the 50-year circle: from cloud seeding to space and back to climate change through precipitation physics. In Wei-Kuo Tao, Robert Adler (Eds.), Cloud Systems, Hurricanes, and the Tropical Rainfall Measuring Mission (TRMM). 234p., pp. 59–80, Meteorological Monographs 51, AMS (Chapter 6).* *© American Meteorological Society. Used with permission.*

invigorates the convection (Rosenfeld et al., 2008a). This causes very strong radar reflectivies in excess of 55 dBZ that reach large heights (panel J). The intense mixes phase precipitation electrifies the clouds, as evident by the numerous lightning flashes. This is in sharp contrast with the large r_e, dearth of lightning, and low reflectivities of the pristine case 1. Similar contrasts were documented in a tropical cyclone that one of its spiral cloud bands ingested polluted air from mainland China and showed a similar contrast between the polluted and pristine cloud bands (see Fig. 7 in Rosenfeld et al., 2012b).

A similar analysis of TRMM data for shallow extra-tropical clouds documented urban and industrial aerosols suppressing precipitation also there (Rosenfeld, 2000). Excessive amounts of CCN can suppress precipitation even in very deep convective clouds in the extra-tropics, such as in the case of pyro-Cb. Satellite retrievals documented greatly reduced r_e to nearly 10 μm at the homogeneous ice nucleation temperature of −38°C, showing that the cloud water was freezing into similarly small ice particles, as shown by the black line in Fig. 11 (Rosenfeld et al., 2007, 2014b). This leaves no effective mechanism for precipitation forming processes by drop coalescence, mixed phase, or ice aggregation. Indeed, radar measurements of the Chisholm fire storm show that precipitation echoes in the pyro-Cb were much weaker than in natural less vigorous natural Cb nearby (Rosenfeld et al., 2007).

7 APPLICATION OF $R_E(T)$ TO OBSERVE LARGE HYGROSCOPIC AEROSOLS RESTORING PRECIPITATION

In contrast to submicron aerosols, supermicron aerosols can act as giant CCN (GCCN), which serve as embryos of rain drops (e.g., Johnson, 1982; Bruintjes, 1999). The GCCN can also reduce cloud base S_{max} and respectively N_d, as illustrated in Fig. 6B. This causes a respective increase in r_e and a decrease in the height for rain initiation. TRMM observations of clouds that formed in a polluted air mass over the Bay of Bengal and the Indian Ocean during the winter monsoon shows that $r_e(T)$ becomes larger (the curve moving to the right) and the height for rain initiation becomes progressively lower with the distance the air travels over ocean. TRMM PR observations validated this inference (see Fig. 1 of Rosenfeld et al., 2002). Rosenfeld et al. (2002) interpreted these observations as the result of sea spray GCCN initiating rain that washes down some of the submicron CCN by precipitating the cloud droplets that they nucleated. This is a major process by which a polluted continental air is cleansed and becomes a pristine marine air mass.

One may claim that the change in thermodynamic conditions between land and ocean surfaces can contribute to the same process by lowering cloud base height and providing more moisture that accelerates the rain forming processes as well. However, observations of the response of clouds to a sea salt dust storm over the desiccated sea bed of the Aral Sea showed a similar effect on $r_e(T)$, making the rate of growth of r_e with decreasing T much steeper (Rudich et al., 2002). The effects of desert dust might be mixed, because the desert dust contains many GCCN, but often they are not very hygroscopic.

8 IMPACTS OF AEROSOLS ON CLOUDS' GLACIATION TEMPERATURE AND MIXED PHASE PRECIPITATION

The most conspicuous examples of impacts of ice forming aerosols on cloud glaciation are observed in clouds that are seeded for precipitation enhancement by dispersing AgI into them from airplanes. The glaciogenic seeded

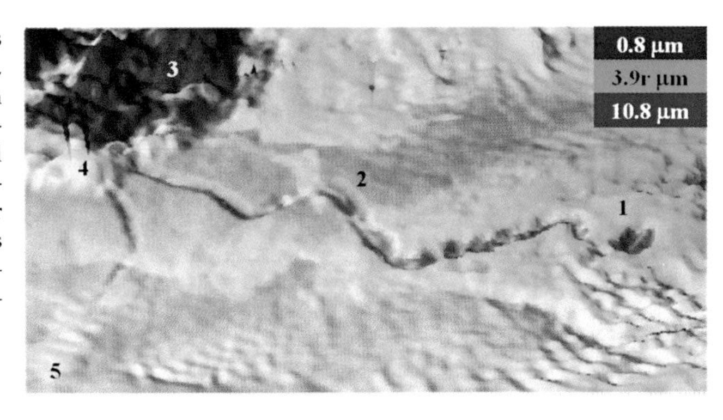

FIG. 13 Seeding track in supercooled layer clouds over central China with top temperature of ~ −15°C, as rendered based on NOAA-14 AVHRR data on 0735 UTC March 14, 2000. The three-dimensional rendering of cloud topography is done based on cloud top temperature. The color of this surface is microphysical RGB, where *yellow* is supercooled water cloud and *red* is ice clouds. The figure width is ~200 km. *From Rosenfeld D, Yu, X., Dai, J., 2005. Satellite retrieved microstructure of AgI seeding tracks in supercooled layer clouds. J. Appl. Meteorol. 44, 760–767.*

parts of the clouds are very evident in the form of a path of ice cloud within a deck of supercooled water cloud with top temperatures near −15°C, as shown in Fig. 13 (Rosenfeld et al., 2005). The seeding track appears there as the red channel. The background clouds glaciate naturally when they grow taller and colder, as evident in point 3. In several places (e.g., at point 4) the seeding track is hidden below a higher layer of supercooled water cloud. The aircraft started seeding just below cloud tops at point 1 of Fig. 13, 97 min before the satellite overpass time. It reached points 2, 4, and 5 by 53, 36, and 14 min before the satellite overpass time, respectively. The time is important because it shows that the glaciation becomes evident at cloud top only ~20 min after seeding, and that the glaciated track kept widening for the next hour to a width of 10–13 km. The channel was measured to be ~1.5 km deep at point 2. The deepening rate was calculated to be ~40 cm s^{-1}, which is similar to the terminal fall velocity of dendritic ice crystals at these conditions. A ridge of clouds grew in the middle of the channel near point 2, probably due to the latent heat of freezing that caused the air in the glaciated cloud to rise despite the glaciated cloud top lowering due to the relatively fast fall speed of the ice particles.

Satellites can also observe the effect of IN on convective clouds, as demonstrated by glaciogenic cloud seeding. The glaciation of convective clouds in response to heavy AgI seeding for hail prevention in Alberta, Canada, is shown in Fig. 14 (Rosenfeld and Woodley, 2003). Storm 2 was natural, whereas storm 3 was seeded. The $r_e(T)$ of the natural storm increased gradually with decreasing T down to nearly −40°C. The $r_e(T)$ in the seeded storm behaved more erratically, showing a mixture of glaciated and supercooled water cloud at $T > -28$°C, where full glaciation occurred. This is evident also in the erratic changes between the yellow (supercooled water) and red (ice) colors of the cloud, and probably reflect the inhomogeneous spreading of the IN particles within the cloud.

Natural aerosols that serve as IN are mainly desert dust and soil particles. Lidar observations showed that heavy concentrations of desert dust glaciate clouds near −20°C (Ansmann et al., 2008). Satellite observations of similar sloping layer clouds showed that clouds in heavy dust glaciate when their tops reach −21°C (area A in Fig. 15), whereas clouds above the dust layer glaciate at −35°C (area B in Fig. 15).

While IN obviously received much attention with respect to the ice forming processes in clouds, the most effective way for glaciating clouds at high temperatures is not necessarily with high IN concentrations, but rather by inducing scarcity of any type of aerosols. Smaller concentrations of CCN lead to a larger cloud drop size, which freeze faster and at

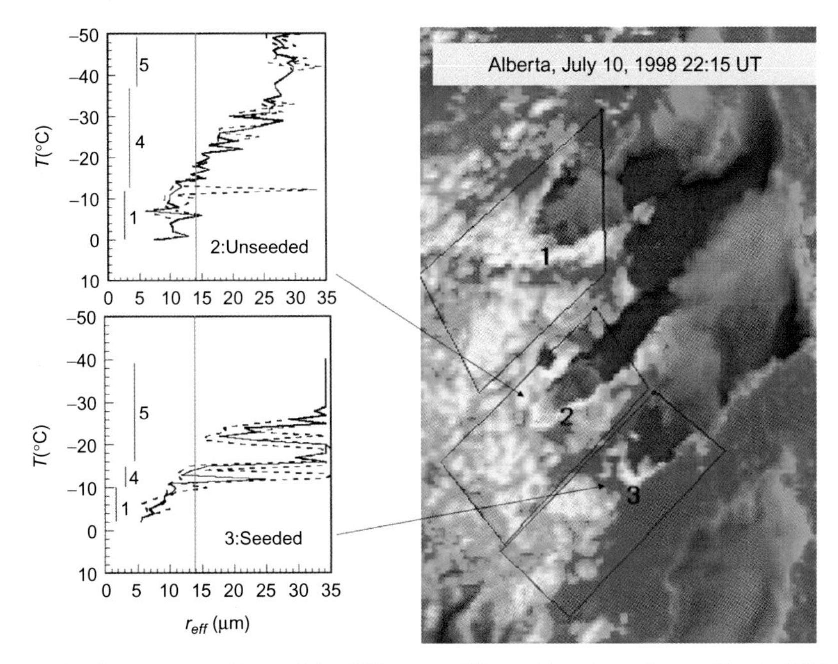

FIG. 14 Hailstorm clouds in an area of about 150 × 150 km over Alberta, Canada, as observed by the NOAA/AVHRR overpass of July 10, 1998 at 22:15 GMT, rendered as the microphysical RGB color scheme. storm 3 is seeded. The colors and the $r_e(T)$ graphs show accelerated glaciation at higher temperatures compared to storm 2. *Modified from Rosenfeld, D., Woodley, W.L., 2003. Closing the 50-year circle: from cloud seeding to space and back to climate change through precipitation physics. In Wei-Kuo Tao, Robert Adler (Eds.), Cloud Systems, Hurricanes, and the Tropical Rainfall Measuring Mission (TRMM). 234p., pp. 59–80, Meteorological Monographs 51, AMS (Chapter 6). © American Meteorological Society. Used with permission.*

higher temperatures. Furthermore, secondary ice formation (i.e., formation of ice particles without the mediation of IN) occurs when cloud droplets become larger than r_e of ~12 μm at a temperature range of −3°C to −8°C in a process known as "ice multiplication" (Hallett and Mossop, 1974). Satellite observations can detect glaciation temperature (T_g) in deep convective clouds. In this regard, T_g is defined as the temperature in which the top of a growing convective tower glaciates, as determined by the satellite-retrieved phase of the cloud top being ice. Such satellite observations show a steep increase in T_g of growing convective clouds with larger r_e at −5°C, as shown in Fig. 16 (Rosenfeld et al., 2011). Practically, clouds with droplets at the −5°C isotherm that are sufficiently large for the occurrence of ice multiplication, glaciate at $T_g > -20$°C, whereas clouds with smaller droplets can glaciate at temperatures down to the homogeneous ice nucleation temperature of −37.5°C (Rosenfeld and Woodley, 2000). This is supported by in situ aircraft observations in the Amazon, which observed ice initiation at a colder temperature with smaller re, as shown in Fig. 2 (Braga et al., 2017).

High T_g in clouds with small droplets is induced by high concentrations of IN such as desert dust ($r_e < 10$ μm in the case shown in Fig. 15), where desert dust can glaciate clouds at T_g near −20°C, which is still colder than T_g of pristine clouds with scarce IN.

Smaller droplets are induced by larger aerosol concentrations, but at the same time this

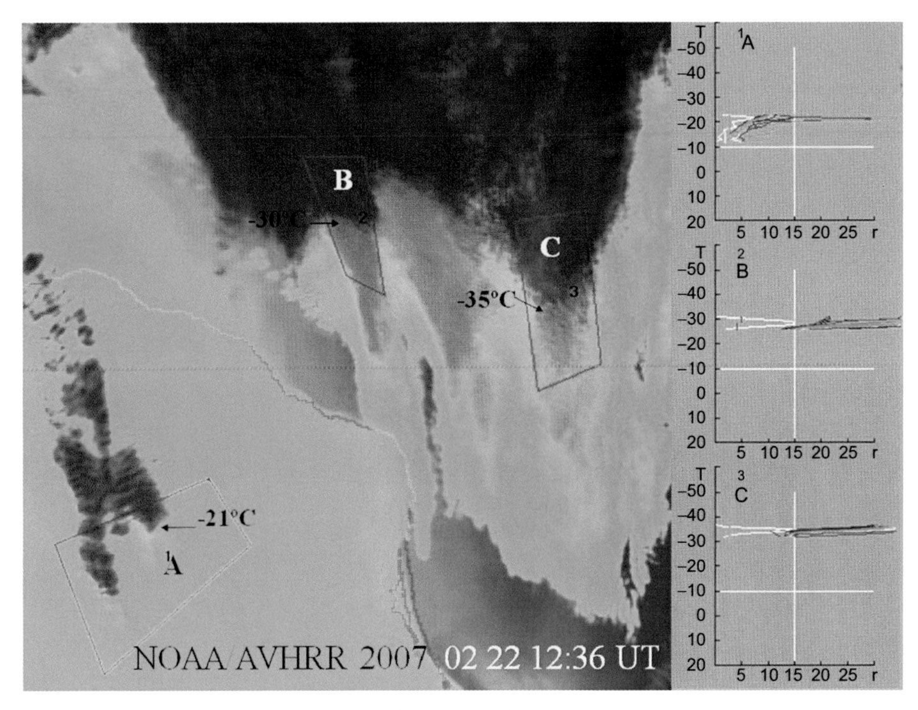

FIG. 15 Desert dust glaciating layer clouds over northern Libya. The microphysical color scale shows the glaciation as the transition from yellow water to red ice clouds. The cloud tops are sloping and getting colder northward. The clouds in area A are within the heavy desert dust and glaciate at −21°C. The clouds in area C are in the clean air above and glaciate at −35°C. Area B is an intermediate case.

often also means more IN concentrations. The smaller droplets would induce colder T_g, whereas the additional IN would induce warmer T_g. This issue was addressed by satellite observations of T_g as a function of aerosol type and optical depth (AOD), while excluding clouds with r_e >12 μm at −5°C where ice multiplication can be very active, as shown in Fig. 16 (Rosenfeld et al., 2011). The aerosol type was obtained from the Cloud-Aerosol Lidar and Infrared Pathfinder Satellite Observation (CALIPSO). The AOD was obtained from NASA's Moderate Resolution Imaging Spectroradiometer (MODIS) aerosol products. As expected, Fig. 17 shows increasing T_g for greater AOD, and higher T_g for desert dust compared to other aerosols for a given AOD. However, T_g of Eastern Asian air pollution was not much smaller. Clouds developing in smoke from Siberian forest fires far from any anthropogenic aerosol sources had by far the coldest T_g, reaching −35°C. Very small T_g was found also in clouds developing in the smoke of tropical forest fires (see the *orange line* in Fig. 11). This implies that air pollution aerosols, at least over East Asia, are a rich source of IN, and that particles other than those from biomass burning constitute most of the IN. Similar relationships were found between T_g and aerosol types that were scavenged in rain water (Zipori et al., 2015).

The nucleation of ice crystals can initiate mixed phase precipitation in clouds that

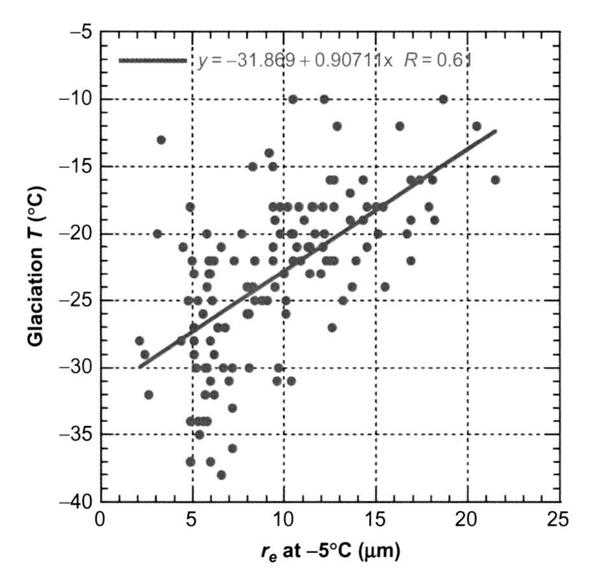

FIG. 16 The dependence of satellite-observed glaciation temperature of growing convective clouds on cloud drop effective radius at −5°C. *From Rosenfeld, D., Yu, X., Liu, G., Xu, X., Zhu, Y., Yue, Z., Dai, J., Dong, Z., Dong, Y., Peng, Y., 2011. Glaciation temperatures of convective clouds ingesting desert dust, air pollution and smoke from forest fires. Geophys. Res. Lett. 38, L21804, doi:10.1029/2011GL049423.*

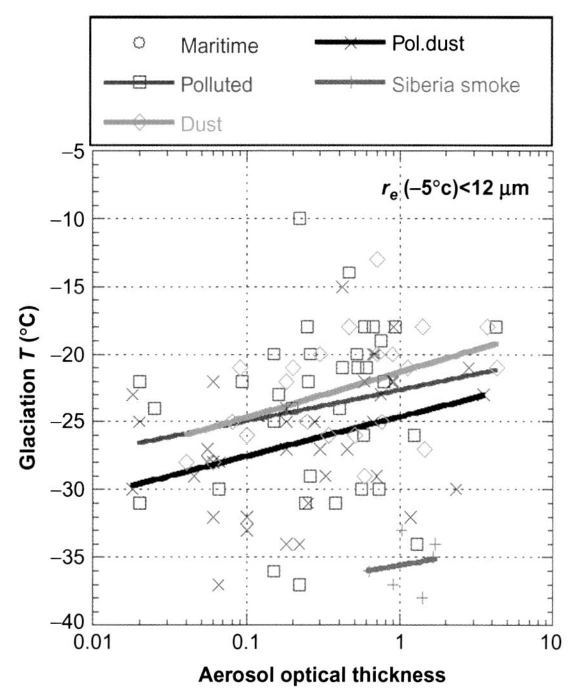

FIG. 17 The dependence of satellite-observed glaciation temperature of growing convective clouds on aerosol types and optical depth, for clouds with little secondary ice formation, as determined by drop effective radius at −5°C < 12 μm. Aerosol types were obtained by CALIPSO, and AOD was obtained from MODIS. *From Rosenfeld, D., Yu, X., Liu, G., Xu, X., Zhu, Y., Yue, Z., Dai, J., Dong, Z., Dong, Y., Peng, Y., 2011. Glaciation temperatures of convective clouds ingesting desert dust, air pollution and smoke from forest fires. Geophys. Res. Lett. 38, L21804, doi:10.1029/2011GL049423.*

otherwise would not precipitate due to suppressed coalescence, as indicated by satellite retrieved r_e < 14 μm. Conversely, adding CCN can suppress mixed phase precipitation in clouds that are deficient in IN when the reduction in r_e changes the conditions from favorable to unfavorable for ice multiplication. This was documented to be the case in shallow supercooled water clouds in Australia, where the Tropical Rainfall Measuring Mission (TRMM) documented suppressed rain and snow in such clouds, as detected by the TRMM precipitation radar. The vertical profile of the radar echo in the unpolluted clouds had a distinct signature of bright band at the melting level, which demonstrated that the precipitation was in the form of snow, which was absent in the polluted clouds (Rosenfeld, 2000).

9 VERTICAL MICROPHYSICAL PROFILES OF SEVERE CONVECTIVE STORMS

The updraft speed does not change the depth of the condensation-dominated growth zone for a given N_{db}, although it can increase N_{db} for a given Nccn. However, satellite observations showed that stronger updrafts aloft can delay the development of mixed phase and glaciation to greater heights and colder temperatures

(Rosenfeld et al., 2008b). Ultimately, in severe convective storms with very strong updrafts aloft (larger than 30 ms^{-1}), the $r_e(T)$ profiles becomes linear and does not show the usual increase above the height of initiation of precipitation, simply because there is no time for the precipitation growth in the strong updraft. For similar considerations, cloud glaciation is delayed to the homogeneous ice nucleation temperature of nearly −38°C, as shown in Fig. 18 and as documented by in situ aircraft measurements (Rosenfeld et al., 2006). Therefore, satellite retrieved linear $r_e(T)$ with small r_e at very cold T, and with T_g −38°C were shown to be indicators of severe convective storms producing large hail and tornados (Rosenfeld et al., 2008b). In fact, an experiment conducted at the National Severe Storms Laboratory (NSSL) showed that convective storms with such vertical profiles of $r_e(T)$ had predictive value for severe convective storms with a lead time of up to 2 h, because the precursor cloud towers already revealed the highly unstable nature of the atmosphere (Woodley et al., 2008).

10 APPLICATIONS OF VERTICAL PROFILES OF $R_E(T)$ TO RETRIEVE N_{DB} AND CCN(S)

The robust relationships between $r_e(T)$ and N_{db}, as reviewed in Section 3, can be utilized for obtaining N_{db} from satellite retrievals of $r_e(T)$. N_{db} can be retrieved under the assumption of adiabatic re if adiabatic LWC (LWC_a) is also known for the same T, because N_{db} becomes LWC_a divided by the mass of a mean-volume cloud drop in a unit volume or mass of the cloud. To calculate LWC_a one must retrieve cloud base temperature (T_b) and pressure (P_b). This can be done with the high resolution Imager data on board the Suomi/NPP satellite. The T_b is simply taken as the warmest cloudy pixel at the field of view, as illustrated in Fig. 8 (Zhu et al., 2014). According to Eq. (1), S_{max} can be calculated if W_b and N_{db} are known. Zheng and Rosenfeld (2015) have shown that W_b of the convective cloud can be approximated from its base height above the surface. With all components being measurable, it is now

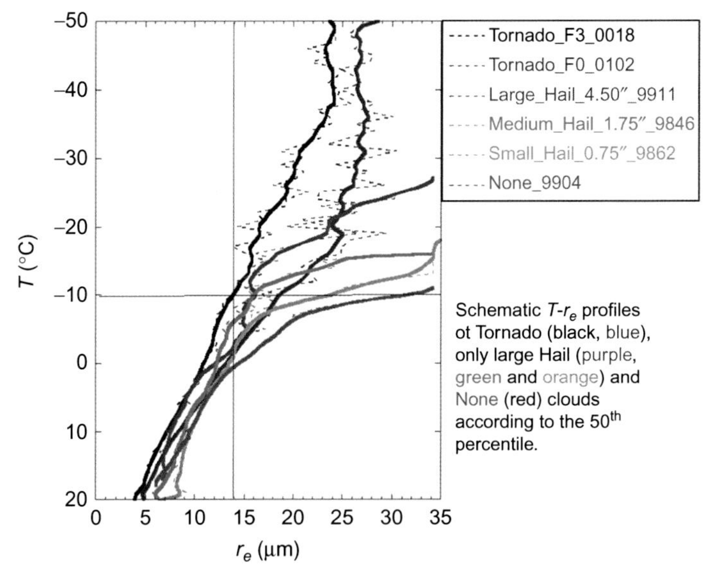

FIG. 18 $r_e(T)$ profiles for benign and severe convective storms, as measured in convective clouds by the Advanced Very High Resolution Radiometer (AVHRR, 1000 m resolution) over the Midwest of United States. The legend shows the tornado Fujita scale and hail size of each case. Note that as the storms become more severe, the $r_e(T)$ profiles become more linear and T_g occurs at lower temperatures. Furthermore, for the most severe storms r_e becomes smaller even above the isotherm of homogeneous ice nucleation temperature. The $r_e(T)$ curves are compiled from cases published in Rosenfeld et al. (2008b).

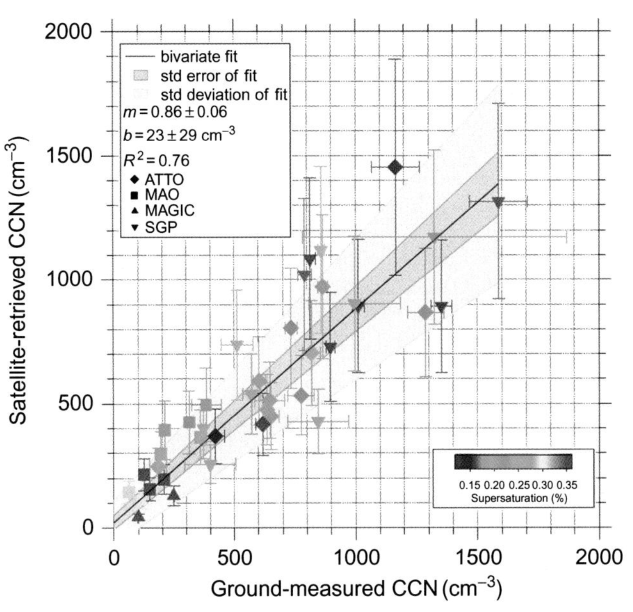

FIG. 19 The relationship between satellite-retrieved CCN concentrations and S_{max} at cloud base, and the ground-based instrument measurements of CCN at the same S_{max}. The slope and intercept of the best fit line are given in the key by *m* and *b*, respectively. The validation data are collected from the DOE/ASR sites on the SGP in Oklahoma and GOAmazon near Manaus, and over the northeast Pacific (MAGIC). In addition, data are obtained from the ATTO. The location is denoted by the marker shape, and S_{max} is shown by the color. *From Rosenfeld, D., Zheng, Y., Hashimshoni, E., Pöhlker, M. L., Jefferson, A., Pöhlker, C., Yu, X., Zhu, Y., Liu, G., Yue, Z., Fischman, B., Li, Z., Giguzin, D., Goren, T., Artaxoi, P., Barbosai, H.M.J., Pöschl, U., Andreae, M.O., 2016. Satellite retrieval of cloud condensation nuclei concentrations by using clouds as CCN chambers. Proc. Nat. Acad. Sci. doi:10.1073/pnas.1514044113.*

possible to use the retrieved cloud vertical profile of $r_e(T)$ to obtain S_{max}. Furthermore, N_{db} is by definition the CCN concentration that is nucleated at S_{max}. Therefore, it is now possible to use the cloud vertical profiles of $r_e(T)$ for retrieving CCN(S). This was done and validated over the DOE/ARM site of the Southern Great Plains, over the Amazon and on board a ship cruising between Los Angeles and Honolulu, as shown in Fig. 19 (Rosenfeld et al., 2016).

11 CONCLUSIONS

Cloud-penetrating radiation, i.e., passive and active microwave, can retrieve vertical profiles of hydrometeors, but not of cloud particles. Satellite retrieved vertical profiles of a cloud particle effective radius is possible only by composing cloud elements at different heights, under the assumption that they represent the composition of other taller clouds at the same height. This is a fair assumption for growing, non precipitating convective clouds.

Another assumption that simplifies greatly the interpretation of the observed vertical profiles is that cloud mixing with ambient air is close to inhomogeneous. It follows that the vertical profile of r_e then behaves nearly like the adiabatic cloud, while the water content of the same cloud can be far from adiabatic. An important outcome is a tight relationship between N_{db} and $r_e(T)$, which in turn determines the height for rain initiation when r_e reaches ~14 μm. Because aerosols often have a dominant role in determining N_{db}, they also determine the height for rain initiation, and even the height for cloud glaciation, which are detectable by satellite. This principle can also be used for reverse calculations of cloud base S_{max} and CCN(S) from the satellite-retrieved $r_e(T)$. To do this with acceptable accuracy, it is necessary to use the high resolution (375 m) of the Suomi/NPP Imager. Therefore, the most readily observable and most revealing properties of cloud microphysical vertical profiles are obtained from $r_e(T)$. These shapes led to important insights, including:

- Clouds do not produce significant warm rain when r_e at their tops does not exceed ~14 μm. This threshold has little sensitivity to cloud depth, and is applicable to both stratocumulus and deep convective clouds.
- Anthropogenic aerosols such as urban air pollution and smoke from forest fires are evident in decreasing r_e and increasing the height that is required to exceed r_e of 14 μm, thus suppressing precipitation from shallow clouds.
- Large hygroscopic aerosols are observed to have the opposite effect of air pollution, incurring larger r_e and restoring precipitation from polluted clouds. This accelerates the eventual washing down of the aerosols and cleansing the air mass.
- Convective clouds glaciate at a higher temperature when r_e near the −5°C isotherm is larger, probably because of the faster freezing of larger drops.
- Aerosols that act as good ice nuclei are observed to glaciate clouds at higher temperatures. The aerosols can be natural, like desert dust, or anthropogenic, like AgI for cloud seeding.
- Strong updrafts are evident in delaying the increase of r_e with height due to drop coalescence and the cloud glaciation to greater heights and colder T. This can be used for detecting and nowcasting severe convective storms.
- The satellite observed simple relationships between $r_e(T)$, CCN(S), and cloud base updrafts can serve as an evidence to the validity of applicability of these relationships to parameterization of the microphysical processes of convective clouds in large scale models.

References

Andreae, M.O., Rosenfeld, D., Artaxo, P., Costa, A.A., Frank, G.P., Longo, K.M., Silva-Dias, M.A.F., 2004. Smoking rain clouds over the Amazon. Science 303, 1337–1342.

Ansmann, A., Tesche, M., Althausen, D., Müller, D., Seifert, P., Freudenthaler, V., Heese, B., Wiegner, M., Pisani, G., Knippertz, P., Dubovik, O., 2008. Influence of Saharan dust on cloud glaciation in southern Morocco during the Saharan mineral dust experiment. J. Geophys. Res. Atmos. 113 (D4).

Beals, M.J., Fugal, J.P., Shaw, R.A., Lu, J., Spuler, S.M., Stith, J.L., 2015. Holographic measurements of inhomogeneous cloud mixing at the centimeter scale. Science 350 (6256), 87–90.

Braga, R.C., Rosenfeld, D., Weigel, R., Jurkat, T., Andreae, M.O., Wendisch, M., Pöschl, U., Voigt, C., Mahnke, C., Borrmann, S., Albrecht, R.I., Molleker, S., Vila, D.A., Machado, L.A.T., 2017. What determines the height for warm rain initiation and cloud glaciation in convective clouds over the Amazon basin?. Submitted to ACP.

Bruintjes, R.T., 1999. A review of cloud seeding experiments to enhance precipitation and some new prospects. Bull. Am. Meteorol. Soc. 80 (5), 805.

Burnet, F., Brenguier, J.L., 2007. Observational study of the entrainment-mixing process in warm convective clouds. J. Atmos. Sci. 64, 1995–2011.

Chang, F.L., Li, Z., 2002. Estimating the vertical variation of cloud droplet effective radius using multispectral near-infrared satellite measurements. J. Geophys. Res. Atmos. 107 (D15).

Freud, E., Rosenfeld, D., 2012. Linear relation between convective cloud drop number concentration and depth for rain initiation. J. Geophys. Res. 117, D02207. https://doi.org/10.1029/2011JD016457.

Freud, E., Rosenfeld, D., Andreae, M.O., Costa, A.A., Artaxo, P., 2008. Robust relations between CCN and the vertical evolution of cloud drop size distribution in deep convective clouds. Atmos. Chem. Phys. 8 (6), 1661–1675.

Freud, E., Rosenfeld, D., Kulkarni, J.R., 2011. Resolving both entrainment-mixing and number of activated CCN in deep convective clouds. Atmos. Chem. Phys. 11 (24), 12887–12900.

Gerber, H., 1996. Microphysics of marine stratocumulus clouds with two drizzlemodes. J. Atmos. Sci. 53 (12), 1649–1662. https://doi.org/10.1175/1520-0469.

Goren, T., Rosenfeld, D., 2015. Extensive closed marine stratocumulus downwind of Europe—a large cloud radiative effect or forcing? J. Geophys. Res. Atmos. 120 (12), 6098–6116.

Hallett, J., Mossop, S.C., 1974. Production of secondary ice particles during the riming process. Nature 249, 26–28. https://doi.org/10.1038/249026a0.

Han, Q., Rossow, W.B., Chou, J., Welch, R.M., 1998. Global variation of column droplet concentration in low-level clouds. Geophys. Res. Lett. 25, 1419–1422. https://doi.org/10.1029/98GL01095.

Johnson, D.B., 1982. The role of giant and ultragiant aerosol particles in warm rain initiation. J. Atmos. Sci. 39, 448–460.

Khain, A., Prabha, T.V., Benmoshe, N., Pandithurai, G., Ovchinnikov, M., 2013. The mechanism of first raindrops

formation in deep convective clouds. J. Geophys. Res. Atmos 118 (16), 9123–9140.
Khain, A., Ovtchinnikov, M., Pinsky, M., Pokrovsky, A., Krugliak, H., 2000. Notes on the state-of-the-art numerical modeling of cloud microphysics. Atmos. Res. 55 (3), 159–224.
King, N.J., Vaughan, G., 2012. Using passive remote sensing to retrieve the vertical variation of cloud droplet size in marine stratocumulus: An assessment of information content and the potential for improved retrievals from hyperspectral measurements. J. Geophys. Res. Atmos. 117 (D15).
Köhler, H., 1936. The nucleus in and the growth of hygroscopic droplets. Trans. Faraday Soc. 32, 1151–1161.
Konwar, M., Maheskumar, R.S., Kulkarni, J.R., Freud, E., Goswami, B.N., Rosenfeld, D., 2012. Aerosol control on depth of warm rain in convective clouds. J. Geophys. Res. 117, D13204. https://doi.org/10.1029/2012JD017585.
L'Ecuyer, T.S., Berg, W., Haynes, J., Lebsock, M., Takemura, T., 2009. Global observations of aerosol impacts on precipitation occurrence in warm maritime clouds. J. Geophys. Res. Atmos. 114(D9).
Lensky, I.M., Rosenfeld, D., 2006. The time-space exchangeability of satellite retrieved relations between cloud top temperature and particle effective radius. Atmos. Chem. Phys. 6, 2887–2894.
Merk, D., Deneke, H., Pospichal, B., Seifert, P., 2016. Investigation of the adiabatic assumption for estimating cloud micro-and macrophysical properties from satellite and ground observations. Atmos. Chem. Phys. 16 (2), 933–952.
Nakajima, T., King, M.D., 1990. Determination of the optical thickness and effective particle radius of clouds from reflected solar radiation measurements. Part I: theory. J. Atmos. Sci. 47 (15), 1878–1893.
Painemal, D., Zuidema, P., 2011. Assessment of MODIS cloud effective radius and optical thickness retrievals over the Southeast Pacific with VOCALS-REx in situ measurements. J. Geophys. Res. Atmos. 116 (D24).
Pawlowska, H., Brenguier, J.L., Burnet, F., 2000. Microphysical properties of stratocumulus clouds. Atmos. Res. 55, 15–33.
Pinsky, M., Khain, A., Mazin, I., Korolev, A., 2012. Analytical estimation of droplet concentrationat cloud base. J. Geophys. Res 117 (D18), D18211.
Prabha, T.V., Khain, A., Maheshkumar, R.S., Pandithurai, G., Kulkarni, J.R., Konwar, M., Goswami, B.N., 2011. Microphysics of premonsoon and monsoon clouds as seen from in situ measurements during the Cloud Aerosol Interaction and Precipitation Enhancement Experiment (CAIPEEX). J. Atmos. Sci. 68 (9), 1882–1901.
Rangno, A.L., Hobbs, P.V., 2005. Microstructures and precipitation development in cumulus and small cumulonimbus clouds over the warm pool of the tropical Pacific Ocean, Q. J. R. Q. J. R. Meteorol. Soc. 131, 639–673. https://doi.org/10.1256/qj.04.13.
Rosenfeld, D., 1999. TRMM observed first direct evidence of smoke from forest fires inhibiting rainfall. Geophys. Res. Lett. 26 (20), 3105–3108.
Rosenfeld, D., 2000. Suppression of rain and snow by urban and industrial air pollution. Science 287 (5459), 1793–1796.
Rosenfeld, D., Gutman, G., 1994. Retrieving microphysical properties near the tops of potential rain clouds by multispectral analysis of AVHRR data. Atmos. Res. 34 (1–4), 259–283. https://doi.org/10.1016/0169-8095(94)90096-5 .
Rosenfeld, D., Lensky, I.M., 1998. Satellite-based insights into precipitation formation processes in continental and maritime convective clouds. Bull. Am. Meteorol. Soc. 79, 2457–2476.
Rosenfeld, D., Woodley, W.L., 2000. Deep convective clouds with sustained supercooled liquid water down to—37.5°C. Nature 405, 440–442.
Rosenfeld, D., Woodley, W.L., 2003. Closing the 50-year circle: from cloud seeding to space and back to climate change through precipitation physics. In: Tao, Wei-Kuo, Adler, Robert (Eds.), Cloud Systems, Hurricanes, and the Tropical Rainfall Measuring Mission (TRMM). AMS, Boston. 234 p, pp. 59–80, Meteorological Monographs 51 (Chapter 6).
Rosenfeld, D., Lahav, R., Khain, A., Pinsky, M., 2002. The role of sea spray in cleansing air pollution over ocean via cloud processes. Science 297 (5587), 1667–1670.
Rosenfeld, D., Yu, X., Dai, J., 2005. Satellite retrieved microstructure of AgI seeding tracks in supercooled layer clouds. J. Appl. Meteorol. 44, 760–767.
Rosenfeld, D., Woodley, W.L., Krauss, T.W., Makitov, V., 2006. Aircraft microphysical documentation from cloud base to anvils of hailstorm feeder clouds in Argentina. J. Appl. Meteorol. 45, 1261–1281.
Rosenfeld, D., Fromm, M., Trentmann, J., Luderer, G., Andreae, M.O., Servranckx, R., 2007. The Chisholm firestorm: observed microstructure, precipitation and lightning activity of a pyro-cumulonimbus. Atmos. Chem. Phys. 7, 645–659.
Rosenfeld, D., Lohmann, U., Raga, G.B., O'Dowd, C.D., Kulmala, M., Fuzzi, S., Reissell, A., Andreae, M.O., 2008a. Flood or drought: how do aerosols affect precipitation? Science 321, 1309–1313.
Rosenfeld, D., Woodley, W.L., Lerner, A., Kelman, G., Lindsey, D.T., 2008b. Satellite detection of severe convective storms by their retrieved vertical profiles of cloud particle effective radius and thermodynamic phase. J. Geophys. Res. Atmos. 113 (D4).
Rosenfeld, D., Yu, X., Liu, G., Xu, X., Zhu, Y., Yue, Z., Dai, J., Dong, Z., Dong, Y., Peng, Y., 2011. Glaciation temperatures of convective clouds ingesting desert dust, air pollution and smoke from forest fires. Geophys. Res. Lett. 38,(L21804). https://doi.org/10.1029/2011GL049423.

Rosenfeld, D., Wang, H., Rasch, P.J., 2012a. The roles of cloud drop effective radius and LWP in determining rain properties in marine stratocumulus. Geophys. Res. Lett. 39, L13801. https://doi.org/10.1029/2012GL052028.

Rosenfeld, D., Woodley, W.L., Khain, A., Cotton, W.R., Carrió, G., Ginis, I., Golden, J.H., 2012b. Aerosol effects on microstructure and intensity of tropical cyclones. Bull. Am. Meteorol. Soc. 93 (2012), 987–1001.

Rosenfeld, D., Chemke, R., Prather, K., Suski, K., Comstock, J.M., Schmid, B., Tomlinson, J., Jonsson, H., 2014a. Polluting of winter convective clouds upon transition from ocean inland over central California: contrasting case studies. Atmos. Res. 135, 112–127.

Rosenfeld, D., Liu, G., Yu, X., Zhu, Y., Dai, J., Xu, X., Yue, Z., 2014b. High resolution (375 m) cloud microstructure as seen from the NPP/VIIRS Satellite imager. Atmos. Chem. Phys. 14, 2479–2496. https://doi.org/10.5194/acp-14-2479-2014.

Rosenfeld, D., Zheng, Y., Hashimshoni, E., Pöhlker, M.L., Jefferson, A., Pöhlker, C., Yu, X., Zhu, Y., Liu, G., Yue, Z., Fischman, B., Li, Z., Giguzin, D., Goren, T., Artaxoi, P., Barbosai, H.M.J., Pöschl, U., Andreae, M.O., 2016. Satellite retrieval of cloud condensation nuclei concentrations by using clouds as CCN chambers. Proc. Natl. Acad. Sci. https://doi.org/10.1073/pnas.1514044113.

Rudich, Y., Rosenfeld, D., Khersonsky, O., 2002. Treating clouds with a grain of salt. Geophys. Res. Lett. 29 (22) https://doi.org/10.1029/2002GL016055.

Simpson, J., Adler, R.F., North, G.R., 1988. A proposed tropical rainfall measuring mission (TRMM) satellite. Bull. Am. Meteorol. Soc. 69 (3), 278–295.

Suzuki, K., Nakajima, T.Y., Stephens, G.L., 2010. Particle growth and drop collection efficiency of warm clouds as inferred from joint CloudSat and MODIS observations. J. Atmos. Sci. 67 (9), 3019–3032.

Suzuki, K., Stephens, G.L., van den Heever, S.C., Nakajima, T.Y., 2011. Diagnosis of the warm rain process in cloud-resolvinig models using joint CloudSat and MODIS observations. J. Atmos. Sci. 68, 2655–2670. https://doi.org/10.1175/JAS-D-10-05026.1.

Szczodrak, M., Austin, P.H., Krummel, P.B., 2001. Variability of optical depth and effective radius in marine stratocumulus clouds. J. Atmos. Sci. 58, 2912–2926.

Twomey, S., 1959. The nuclei of natural cloud formation part II: the supersaturation in natural clouds and the variation of cloud droplet concentration. Geofisica pura e applicata 43 (1), 243–249.

Wendisch, M., Poschl, U., Andreae, M., Machado, L., Albrecht, R., Schlager, H., Rosenfeld, D., Martin, S., Abdelmonem, A., Afchine, A., Araujo, A., Artaxo, P., Aufmhoff, H., Barbosa, H., Borrmann, S., Braga, R., Buchholz, B., Cecchini, M., Costa, A., Curtius, J., Dollner, M., Dorf, M., Dreiling, V., Ebert, V., Ehrlich, A., Ewald, F., Fisch, G., Fix, A., Frank, F., Fuetterer, D., Heckl, C., Heidelberg, F., Hueneke, T., Jaekel, E., Jaervinen, E., Jurkat, T., Kanter, S., Kaestner, U., Kenntner, M., Kesselmeier, J., Klimach, T., Knecht, M., Kohl, R., Koelling, T., Kraemer, M., Krueger, M., Krisna, T., Lavric, J., Longo, K., Mahnke, C., Manzi, A., Mayer, B., Mertes, S., Minikin, A., Molleker, S., Muench, S., Nillius, B., Pfeilsticker, K., Poehlker, C., Roiger, A., Rose, D., Rosenow, D., Sauer, D., Schnaiter, M., Schneider, J., Schulz, C., de Souza, R., Spanu, A., Stock, P., Vila, D., Voigt, C., Walser, A., Walter, D., Weigel, R., Weinzierl, B., Werner, F., Yamasoe, M., Ziereis, H., Zinner, T., Zoeger, M., 2016. The ACRIDICON-CHUVA campaign: studying tropical deep convective clouds and precipitation over Amazonia using the new German research aircraft HALO. Bull. Am. Meteorol. Soc. 97 (10), 1885–1908. https://doi.org/10.1175/BAMS-D-14-00255.

Wood, R., 2005. Drizzle in stratiform boundary layer clouds. Part I: vertical and horizontal structure. J. Atmos. Sci. 62 (9), 3011–3033.

Woodley, W.L., Rosenfeld, D., Kelman, G., Golden, J.H., 2008. Short-term forecasting of severe convective storms using quantitative multi-spectral, satellite imagery: results of the early alert project. In: 24th Conference on Severe Local Storms, 27–31 October 2008, Savannah, Georgia.

Yuan, T., Li, Z., 2010. General macro-and microphysical properties of deep convective clouds as observed by MODIS. J. Climate 23 (13), 3457–3473.

Zheng, Y., Rosenfeld, D., 2015. Linear relation between convective cloud base height and updrafts and application to satellite retrievals. Geophys. Res. Lett.

Zhu, Y., Rosenfeld, D., Yu, X., Liu, G., Dai, J., Xu, X., 2014. Satellite retrieval of convective cloud base temperature based on the NPP/VIIRS Imager. Geophys. Res. Lett. 41. https://doi.org/10.1002/2013GL058970.

Zipori, A., Rosenfeld, D., Tirosh, O., Teutsch, N., Erel, Y., 2015. Effects of aerosol sources and chemical compositions on cloud drop sizes and glaciation temperatures. J. Geophys. Res. Atmos.

Further Reading

Martins, J.V., Marshak, A., Remer, L.A., Rosenfeld, D., Kaufman, Y.J., Fernandez-Borda, R., Koren, I., Correia, A.L., Zubko, V., Artaxo, P., 2011. Remote sensing the vertical profile of cloud droplet effective radius, thermodynamic phase, and temperature. Atmos. Chem. Phys. 11, 9485–9501. https://doi.org/10.5194/acp-11-9485-2011. http://www.atmos-chem-phys.net/11/9485/2011/acp-11-9485-2011.pdf.

CHAPTER

7

Polarimetric Technique for Satellite Remote Sensing of Superthin Clouds

Wenbo Sun[*,†], *Rosemary R. Baize*[†], *Gorden Videen*[‡,§], *Yongxiang Hu*[†]

[*]Science Systems and Applications Inc, Hampton, VA, United States
[†]NASA Langley Research Center, Hampton, VA, United States
[‡]Space Science Institute, Boulder, CO, United States
[§]US Army Research Laboratory, Adelphi, MD, United States

1 INTRODUCTION

Superthin clouds of optical depths smaller than ~0.3 cover ~50% of the globe (McFarquhar et al., 2000; Sun et al., 2011b, 2014) and play an important role in the radiation energy balance of the Earth (Dessler and Yang, 2003; Lee et al., 2009; Sun et al., 2011a,b), as well as in the remote sensing of aerosols (Sun et al., 2011b; Omar et al., 2013) and surface temperature (Sun et al., 2011a). For example, the aerosol optical depth (AOD) from NASA's Moderate Resolution Imaging Spectroradiometer (MODIS) aerosol product (MOD04) could be overestimated by ~100% when these clouds exist (Sun et al., 2011b). This is shown in Fig. 1 for the 1-year zonal mean aerosol optical depth at 0.55 μm from the MOD04 for daytime oceans in 2007. The aerosol optical depth of superthin-cloud-overcast sky identified by NASA's Cloud-Aerosol LIDAR and Infrared Pathfinder Satellite Observation (CALIPSO) (Winker et al., 2007) is systematically larger than that of clear sky, but smaller than ~0.3. This is because the superthin clouds are misclassified as aerosols in the MODIS aerosol product but their optical depth obviously deviates from that of aerosols. Fig. 1 shows that the impact of superthin clouds on the global aerosol product is very significant (Sun et al., 2011b). Also, due to failing to detect these clouds, the sea-surface temperature (SST) retrieved from NASA's Atmospheric Infrared Sounder (AIRS) (Chahine et al., 2006) satellite data is ~5–10 K lower at tropical and midlatitude regions, as shown in Fig. 2, where these clouds frequently occur (Sun et al., 2011a). Superthin clouds also can significantly affect the polarization state of reflected solar light from the Earth-atmosphere system (Sun and Lukashin, 2013). In order to use the highly accurate data from NASA's future Climate Absolute Radiance and Refractivity Observatory (CLARREO) mission (Wielicki

Remote Sensing of Aerosols, Clouds, and Precipitation
https://doi.org/10.1016/B978-0-12-810437-8.00007-1

© 2018 Elsevier Inc. All rights reserved.

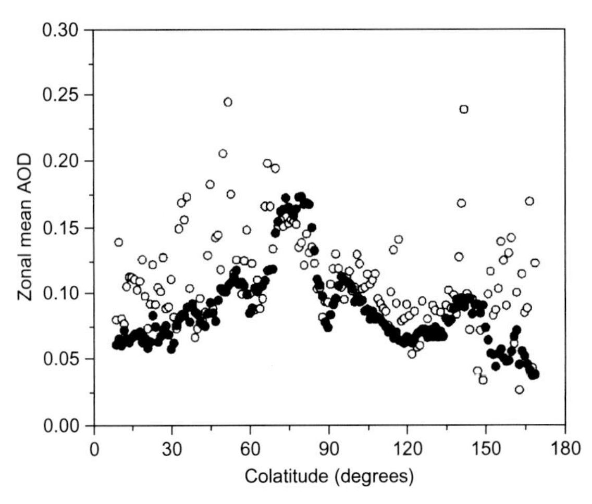

FIG. 1 One-year (2007) zonal mean MOD04 aerosol optical depth at 0.55 μm for clear *(filled circle)* and superthin clouds *(open circle)* over oceans identified by the CALIPSO LIDAR. These are all classified as clear-sky aerosols by the MOD04 product. *Reproduced from Sun, W., Videen, G., Kato, S., Lin, B., Lukashin, C., Hu, Y., 2011. A study of subvisual clouds and their radiation effect with a synergy of CERES, MODIS, CALIPSO, and AIRS data. J. Geophys. Res. 116, D22207. https://doi.org/10.1029/2011JD016422.*

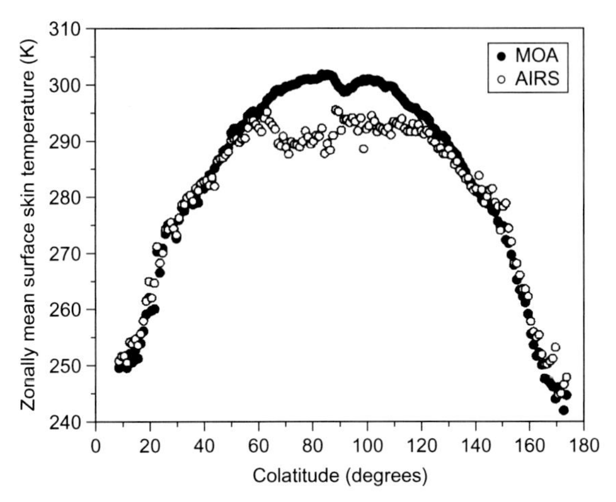

FIG. 2 One-year (2007) zonal mean surface skin temperatures from the AIRS and the Meteorological, Ozone, and Aerosol Data (MOA) for nighttime oceans. The temperature difference is evidently due to the AIRS cloud-screening algorithm, which fails to detect superthin clouds that mostly exist over tropical and subtropical areas. *Reproduced from Sun, W., Lin, B., Hu, Y., Lukashin, C., Kato, S., Liu, Z., 2011a. On the consistency of CERES longwave flux and AIRS temperature and humidity profiles. J. Geophys. Res. 116, D17101. https://doi.org/10.1029/2011JD016153.*

et al., 2013) to intercalibrate solar imagers like the MODIS, its follow-on Visible Infrared Imaging Radiometer Suite (VIIRS), or geostationary imagers, the polarization state of the reflected solar light must be known with sufficient accuracy. Therefore, a better knowledge of superthin clouds is also important for the CLARREO intercalibration project (Sun and Lukashin, 2013).

Data from the Central Equatorial Pacific Experiment (CEPEX) and the Tropical Ocean Global Atmosphere Coupled Ocean-Atmosphere Response Experiment (TOGA COARE) frequently show a geometrically thin layer of cirrus near the tropopause (Spinhirne et al., 1996). Prabhakara et al. (1993), using *Nimbus-4* Infrared Interferometer Spectrometer (IRIS) measurements, and Wang et al. (1994), using Stratospheric Aerosol Gas Experiment (SAGE) II measurements, both observed optically thin cirrus near the tropopause more than 50% of the time in warm pool regions. Using ground-based LIDAR, Uthe and Russell (1976) saw a high frequency of occurrence of superthin cirrus. More recently, Nee et al. (1998) found subvisual cirrus occurring approximately 50% of the time using LIDAR observations over Chang-Li, Taiwan, at 25°N latitude. Platt et al. (1998) also detected superthin cirrus with visible and infrared optical depths as low as 0.01 in Kavieng, Papua, New Guinea, at 3°S in 1993 using LIDAR. From near-global observations of optically thin cirrus during the LIDAR In-space Technology Experiment (LITE), Winker and Trepte (1998) found layers of cirrus occurring in thin sheets near the tropical tropopause with thicknesses between a few hundred meters and one kilometer that were unusually horizontally homogeneous. Even a sky overhead that looks very clear and blue can still have "blue" clouds at over 34,000 ft altitude (Packer and Lock, 1951).

Unfortunately, the variation in surface background reflection means superthin clouds generally cannot be detected by passive satellite instruments, like the Orbiting Carbon Observatory 2 (OCO-2) (Crisp et al., 2004), the Moderate Resolution Imaging Spectroradiometer (MODIS) (King et al., 1992), and the Advanced Very High Resolution Radiometer (AVHRR) (Brest and Rossow, 1992), which only measure the total radiance of the reflected solar light (Minnis et al., 2002; Mace et al., 2005; Ackerman et al., 2008). The resulting data products of many satellite and ground measurements are biased by these undetected clouds (Sun et al., 2011a,b; Omar et al., 2013). Using a strong-water-vapor-absorption channel such as 1.38 μm to exclude surface and low-layer effects can be effective on high cirrus (Gao and Kaufman, 1995), but may encounter difficulties for atmospheres with low water vapor (Ackerman et al., 1998). The reliability of the 1.38 μm channel method is also questionable if cloud optical depth is smaller than ~0.5, when the backscattered intensity is low (Roskovensky and Liou, 2003). In addition, superthin clouds may also exist in the lower layers of the atmosphere where there is ample water vapor. For example, as shown in Fig. 3, ground-based LIDAR measurements over Lanzhou, China frequently detect superthin water cloud layers at about 1–2 km in altitude, with backscatter signals similar to sunlight noise during daytime. In the lower layers of the atmosphere with ample water vapor, the sensitivity of the 1.38 μm channel is weak, hampering cloud detection capabilities; thus, low superthin water clouds cannot be detected by it. Furthermore, NOAA's polar orbiting High Resolution Infrared Radiation Sounder (HIRS) multispectral infrared data are usually used with the CO_2-slicing method for detecting thin cirrus clouds (Wylie et al., 1995; Wylie and Menzel, 1999). However, for superthin clouds, this requires the radiance of the background atmosphere and surface to be very close to that of the reference clear-sky environment, which can be difficult since the terrestrial background changes on spatial and temporal scales. In addition, the CO_2-slicing method is problematic when the difference between clear-sky and cloudy radiance for a spectral band is less than the instrument noise (Wylie and Menzel, 1999), as is the case for superthin clouds.

Since superthin clouds are hard to detect using space-borne instruments, their existence also complicates the retrieval of atmospheric constituents (Christi and Stephens, 2004). For example, the NASA Atmospheric CO_2 Observations from Space (ACOS) XCO_2 retrieval algorithm (O'Dell et al., 2012) defines clear-sky scenes for CO_2

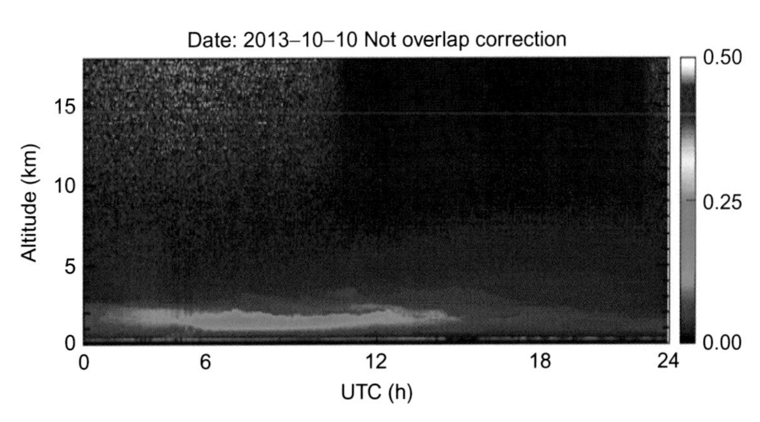

FIG. 3 Ground-based LIDAR measurements over Lanzhou, China frequently detect superthin water cloud layers at about 1–2 km altitude, with their backscatter signals near that of sunlight noise during daytime (Qiang Fu, personal communications).

retrieval for cases with atmospheric optical depth ≤ 0.3, which may still have superthin cloud contamination. When undetected, superthin clouds can introduce significant bias errors in the atmospheric carbon data measured by the ACOS and by the OCO-2 due to the scattering of incident light by these clouds. This scattering introduces uncertainties in the optical pathlength and thus the light absorption by CO_2, from which the CO_2 amount is retrieved.

Although many methods have been developed for detecting clouds (Gao and Kaufman, 1995; Wylie et al., 1995; Ackerman et al., 1998; Wylie and Menzel, 1999; Roskovensky and Liou, 2003), most superthin clouds are still missing constituents of the atmosphere in satellite data. LIDARs on NASA's Cloud-Aerosol LIDAR and Infrared Pathfinder Satellite Observation (CALIPSO) (Winker et al., 2007) and Cloud-Aerosol Transport System (CATS) missions are the only instruments in orbit that can detect superthin clouds; however, these can only cover small portions of the atmosphere. Long-term global surveys of superthin clouds using space-borne LIDARs are limited by their large operational cost and narrow field of view. Also, noise in the LIDAR instantaneous measurements can be significant, due to its relatively low transmitted power, ranging length, narrow field of view, and contamination by sunlight. Although sunlight contamination is not an issue for LIDAR in nighttime, limited photons from the narrow field of view received by LIDAR sensors still constitute errors for detection of superthin atmospheric constituents. To identify optically thin atmospheric components such as superthin clouds, LIDAR data must be averaged over a large spatial area to increase the number of photons measured and reduce the overall noise level. The data noise and spatial averaging process can cause difficulties in detection of many superthin clouds, resulting in large uncertainty in the daytime thin cloud and aerosol measurement. The High Spectral Resolution LIDAR (HSRL) technique (e.g., Rogers et al., 2011), which takes advantage of the spectral distribution of the LIDAR returns to discriminate aerosol and molecular signals, and thereby measures aerosol extinction and backscatter independently, would represent an advancement over the CALIPSO and the CATS measurements, but will also be limited in spatial coverage. Also, the signal-to-noise levels associated with LIDAR measurements can also limit the frequency with which superthin clouds can be detected by the HSRL. Therefore, improving the space LIDAR systems and developing an inexpensive passive-remote-sensing method with greater spatial coverage for reliable detection of superthin clouds have become critical issues for atmospheric remote-sensing practice. In this chapter, the method of using a passive polarimetric instrument such as the aerosol polarimetric sensor (APS) (Mishchenko et al., 2007) to detect superthin clouds is reviewed.

2 POLARIZATION SIGNATURE OF LIGHT BACKSCATTERED BY CLOUDS

A space-borne polarimetric instrument can measure the Stokes parameters I, Q, U, and V of the scattered light (Mishchenko et al., 2007). Since the circularly polarized component of radiance reflected by the ocean-atmosphere system is negligible (i.e., $V \asymp 0$) (Coulson, 1988), the degree of polarization (DOP) and angle of linear polarization (AOLP) can be defined in terms of Stokes parameters as (Sun and Lukashin, 2013):

$$\mathrm{DOP} = \frac{\sqrt{Q^2 + U^2}}{I} \tag{1}$$

and

$$\mathrm{AOLP} = \frac{1}{2}\tan^{-1}\left(\frac{U}{Q}\right) + \alpha_0 \tag{2}$$

where $\alpha_0 = 0$ degree if $Q > 0$ and $U \geq 0$; $\alpha_0 = 180$ degrees if $Q > 0$ and $U < 0$;

$\alpha_0 = 90$ degrees if $Q \leq 0$. In a right-handed Cartesian coordinate system, as shown in Fig. 4, the AOLP of the reflected light is the angle between the local meridian line in the direction of $\mathbf{e}_\theta$ and the electric vector **E** of the linearly polarized part of the reflected light, measured anticlockwise when viewed in the reverse direction of the reflected radiance (Sun and Lukashin, 2013).

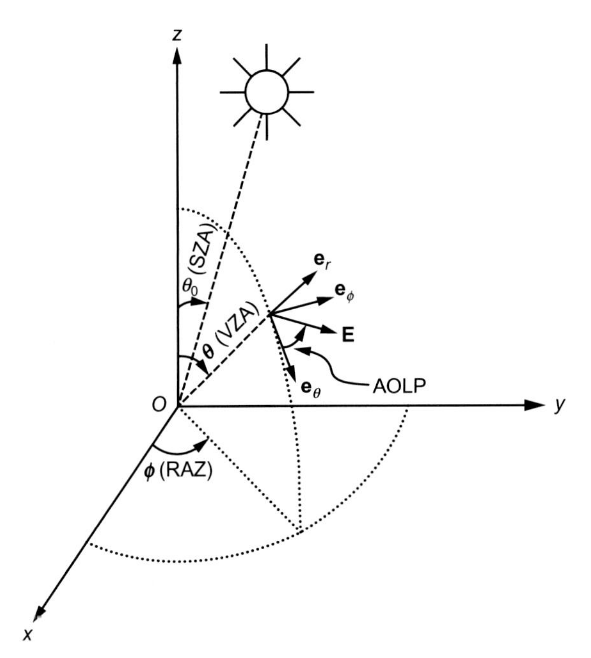

FIG. 4 The geometry of the system showing the scattered light from a surface located in the *x*-*y* plane. The Sun is located on the principal plane (*xoz*) and over the negative *x*-axis (i.e., the solar azimuth angle is 180 degrees). The direction of the reflected light is specified by the unit vector $\mathbf{e}_r$, and θ and ϕ denote the viewing zenith angle (VZA) and relative azimuth angle (RAZ), respectively. In the local right-handed orthonormal coordinate system $\mathbf{e}_r = \mathbf{e}_\theta \times \mathbf{e}_\phi$, where $\mathbf{e}_\theta$ lies in the meridian plane of the reflected light beam. The AOLP of the reflected beam in the direction of $\mathbf{e}_r$ is the angle between the local meridian line and the electric vector **E** of the linearly polarized light, counted anticlockwise when viewing in the reverse direction of the reflected beam. The electric field **E** of the linearly polarized part of the reflected light oscillates in the plane of $\mathbf{e}_\theta$ and $\mathbf{e}_\phi$. *Reproduced from Sun, W., Videen, G., Mishchenko, M.I., 2014. Detecting super-thin clouds with polarized sunlight. Geophys. Res. Lett. 41, 688–693. https://doi.org/10.1002/2013GL058840.*

The electric field **E** of the linearly polarized part of the reflected light oscillates in the plane of $\mathbf{e}_\theta$ and $\mathbf{e}_\phi$ in Fig. 4.

Natural solar radiation can be polarized by surface reflections as well as by scattering from atmospheric molecules and particles. When sunlight propagates through the clear atmosphere and is scattered back toward the Sun, the resulting signal is nearly unpolarized when the solar zenith angle (SZA) is not larger than ~40 degrees (Sun and Lukashin, 2013). The residual polarization in this direction is due to asymmetries in the system, due, for instance, to preferentially oriented ocean waves or nonzero angles between incidence and observation. By considering a longer solar wavelength, such as 670 nm, the contribution of molecular scattering is reduced. Moreover, unlike total radiance *I*, the DOP and AOLP are insensitive to surface roughness and absorption by atmospheric water vapor and other gases (Sun and Lukashin, 2013), which makes the polarization measurement robust in different environmental conditions, even when the detected components are within the lower layers of the atmosphere. In brief, remote sensing atmospheric particulates using polarization measurements in the backscattering region can minimize surface, molecule, and absorbing gas interferences, thus increasing the sensitivity to atmospheric particulates, like superthin clouds.

Because of the variations in surface reflections and atmospheric profiles, using total reflection intensity to detect superthin clouds is generally difficult. Also, although thin clouds can cause changes in the degree of polarization (DOP) of the reflected light (Sun and Lukashin, 2013), the dynamic range of the DOP change is mostly insufficient to unambiguously identify superthin clouds if the background polarized reflection is uncertain. However, another fundamental characteristic of light is its angle of linear polarization (AOLP). Fig. 5 shows the AOLPs at a wavelength $\lambda = 670$ nm (A) taken from the mean of 24-day

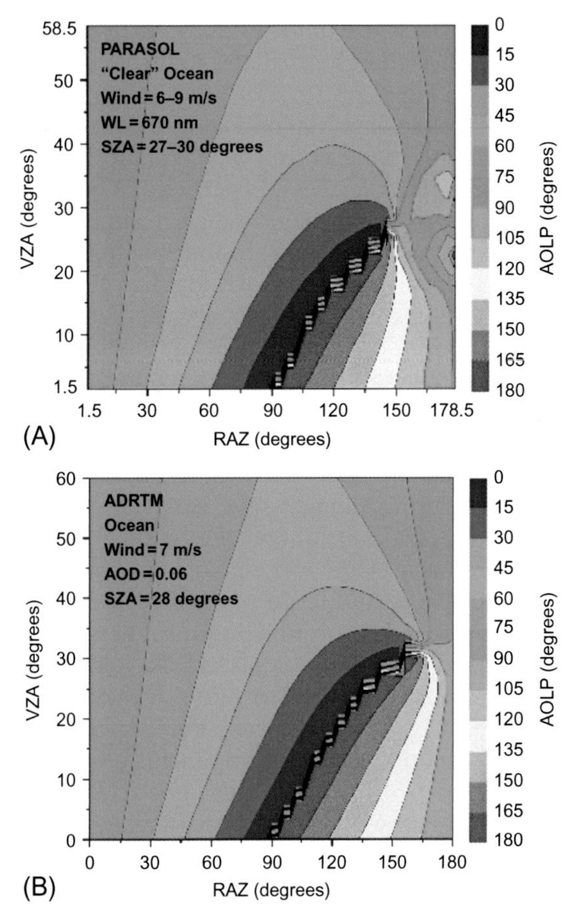

FIG. 5 The AOLP at 670 nm as a function of VZA and RAZ from (A) the mean of 24-day PARASOL measurements for clear-sky oceans identified by the PARASOL product, and (B) the ADRTM for clear-sky oceans. The 24 days of PARASOL data are taken from the first 2 days of each month across 2006. Data are binned in SZA from 27 to 30 degrees and in wind speed from 6 to 9 m/s, and averaged in 3 degrees bins in VZA and RAZ. In the modeling, the SZA is 28 degrees, the ocean wind speed is 7 m/s, the sea-salt AOD is 0.06, and the US standard atmosphere is used. *Reproduced from Sun, W., Videen, G., Mishchenko, M.I., 2014. Detecting super-thin clouds with polarized sunlight. Geophys. Res. Lett. 41, 688–693. https://doi.org/10.1002/2013GL058840.*

measurements for clear-sky oceans from the Polarization and Anisotropy of Reflectances for Atmospheric Science coupled with Observations from a LIDAR (PARASOL) product (Deschamps et al., 1994), and (B) calculated using the adding-doubling radiative transfer model (ADRTM) (Sun and Lukashin, 2013) for clear-sky oceans. The 24 days of PARASOL data are taken from the first 2 days of each month across 2006 and collected in a SZA bin of 27–30 degrees and a wind speed bin of 6–9 m/s, and averaged in 3 degrees bins in viewing zenith angle (VZA) and relative azimuth angle (RAZ). In the modeling, the SZA is 28 degrees, the ocean wind speed is 7 m/s, the sea-salt aerosols are of irregular shapes with an AOD of 0.06, and the US standard atmosphere [*National Oceanic and Atmospheric Administration, National Aeronautics and Space Administration, and United States Air Force*, 1976] is used. Note here that we only display the AOLPs over the RAZ range of 0–180 degrees. The AOLPs over the RAZ range of 180–360 degrees are symmetrically supplementary angles of the values over the RAZ range of 0–180 degrees (Sun and Lukashin, 2013). We find that the AOLPs from the PARASOL data and the model are similar except that at ~8 degrees off the exact-backscatter direction (RAZ = ~180 degrees and VZA = ~20 and ~36 degrees) the PARASOL result shows two distinct features in the AOLP of the reflected solar light. At these viewing angles, the dominant backscattered electric field in the PARASOL data for clear-sky oceans rotates from parallel to perpendicular to the Earth's surface; whereas, the modeled AOLPs do not show these features. We performed sensitivity studies with the ADRTM and found that ocean surface roughness due to winds and atmospheric gas or water vapor absorptions do not produce these features in the model outputs. However, this feature does appear when we incorporate a thin layer of cirrus into the atmosphere. The ADRTM simulation results of Fig. 6A and B demonstrate that the presence of a superthin layer of cirrus clouds can reproduce the AOLP features in Fig. 5A. In the simulation of Fig. 6A, we incorporate the solid-column ice-crystal shapes for the cirrus cloud, which may

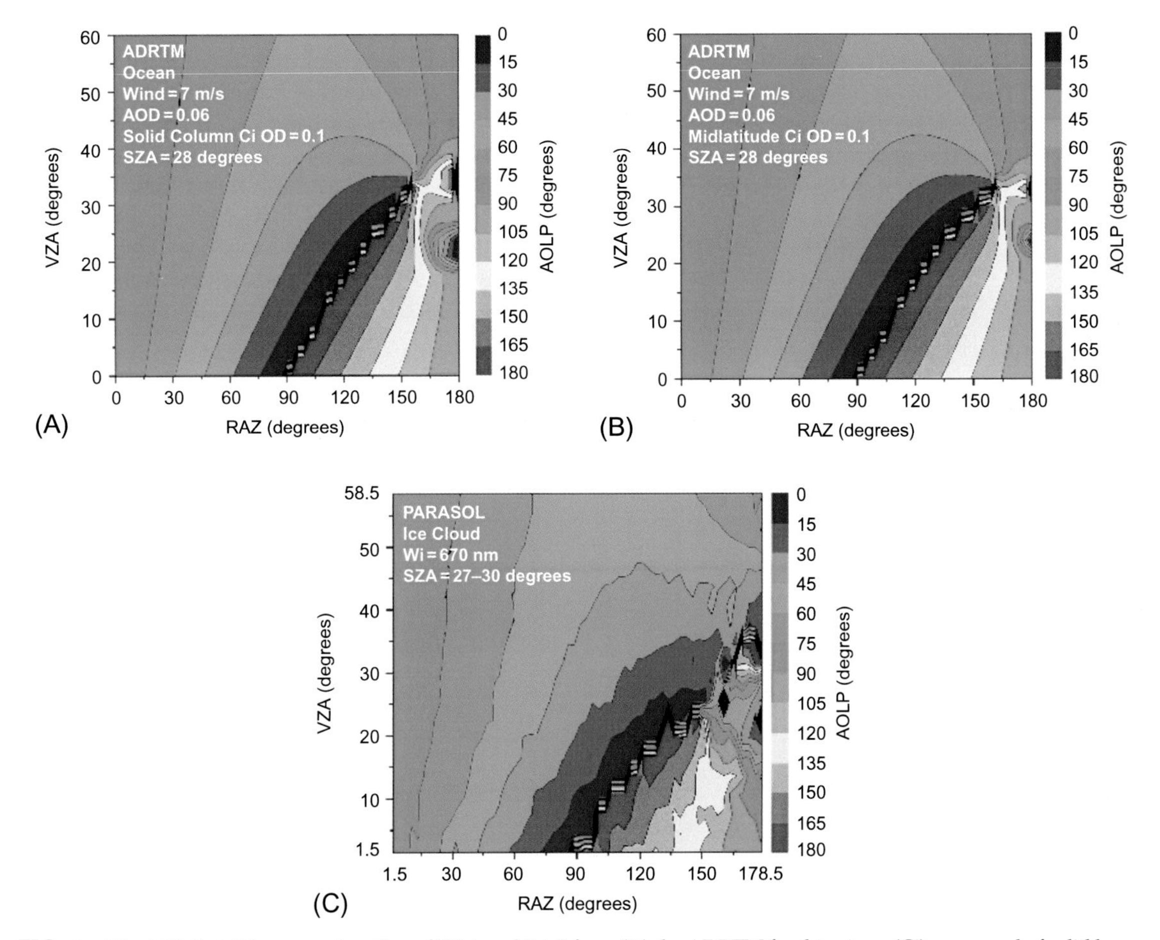

FIG. 6 The AOLP at 670 nm as a function of VZA and RAZ from (A) the ADRTM for thin cirrus (Ci) composed of solid hexagonal ice columns, (B) the ADRTM for thin cirrus composed of a mixture of typical midlatitude ice particle shapes, and (C) the mean of 24-day PARASOL measurements for ice clouds over both ocean and land identified by the PARASOL product. In the modeling, the optical depth (OD) of the cirrus clouds is 0.1 at 670 nm, the wind speed is 7 m/s, the sea-salt AOD is 0.06, the SZA is 28 degrees, and the US standard atmosphere is used. *Reproduced from Sun, W., Videen, G., Mishchenko, M.I., 2014. Detecting super-thin clouds with polarized sunlight. Geophys. Res. Lett. 41, 688–693. https://doi.org/10.1002/2013GL058840.*

not be the ubiquitous ice-crystal shapes in the atmosphere. Thin cirrus clouds are generally a mixture of various ice-crystal shapes such as hexagonal columns, plates, droxtals, bullet rosettes, and aggregates, etc. In the simulation of Fig. 6B, we incorporate a typical size distribution and particle-shape mixture of midlatitude cirrus clouds (Heymsfield and Platt, 1984). While not as sharp as the results of Fig. 6A, we can see in Fig. 6B that the distinct polarization-rotation features at the near-backscattering angles are still significant for clouds with an optical depth (OD) of only 0.1. These features remain visible even down to ODs of ~0.06. Our sensitivity studies on typical tropical cirrus produce similar results. Further evidence of the polarization-rotation features of ice clouds can be seen in the PARASOL data. Fig. 6C shows the AOLPs from the mean of the 24-day PARASOL data for ice clouds over both ocean and land identified by the PARASOL product. These data are particularly noisy because of limited sampling of PARASOL at several viewing-angles for ice clouds at the specific SZA. However, we can still see the polarization rotations above and below the exact-backscatter angle.

One mechanism that can produce such polarization-rotation features is the optical glory. The glory is an angular region of high intensity that may extend several degrees from the exact-backscatter direction and is accompanied by a strong p-polarization component (parallel with $\mathbf{e}_\theta$ within the meridian plane in Fig. 4) a few degrees from the exact-backscatter direction. This prominent p-polarization feature appears to be due to the internal field transmitted into the particle. Spherical droplets in water clouds can produce an especially strong glory. Fig. 7A shows the AOLPs of the averaged 24-day PARASOL data of water clouds over ocean.

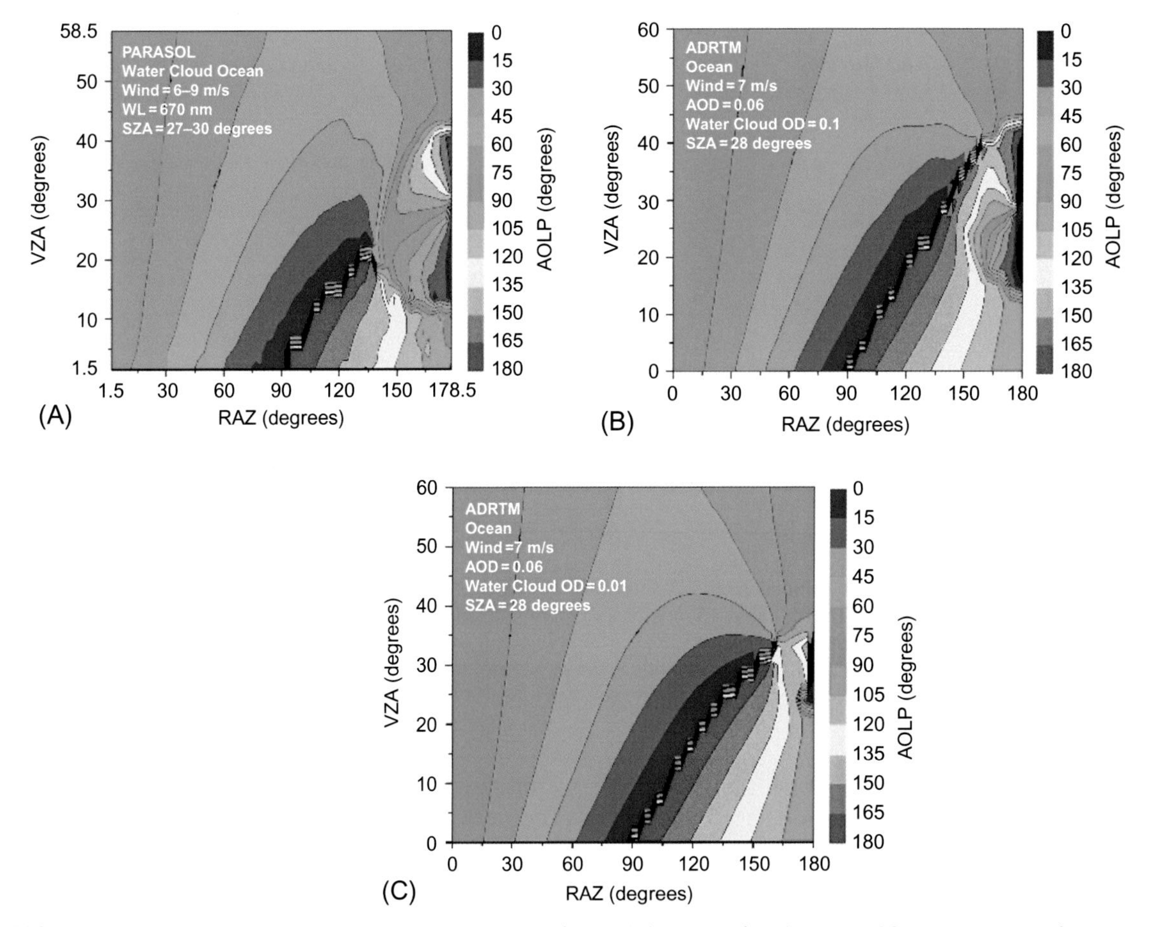

FIG. 7 The AOLP at 670 nm as a function of VZA and RAZ from (A) the mean of 24-day PARASOL measurements for water clouds over oceans identified by the PARASOL product, (B) the ADRTM for water clouds with an optical depth (OD) of 0.1, and (C) the ADRTM for water clouds with an OD of 0.01 at 670 nm over oceans. In the modeling, the wind speed is 7 m/s, the sea-salt AOD is 0.06, the SZA is 28 degrees, and the US standard atmosphere is used. *Reproduced from Sun, W., Videen, G., Mishchenko, M.I., 2014. Detecting super-thin clouds with polarized sunlight. Geophys. Res. Lett. 41, 688–693. https://doi.org/10.1002/2013GL058840.*

It is clear in this figure that the glory feature is a p-polarized band that extends ~20 degrees from the exact-backscatter direction (VZA ~28 degrees, RAZ = 180 degrees). Fig. 7B shows ADRTM results for water clouds with an OD of 0.1, the same OD as the cirrus clouds in the simulations of Fig. 6. Since the mean OD (~7.5) of water clouds from PARASOL data is much larger than 0.1, the p-polarized band in Fig. 7A extends larger than that in Fig. 7B. Also, model results in Fig. 7C demonstrate that subvisual water clouds having an OD of only 0.01 still display a prominent polarization feature. This polarization feature can be suppressed when particle absorption increases and morphology becomes more irregular (Muinonen et al., 2011; Volten et al., 2001). While cirrus clouds composed solely of irregularly shaped aggregated ice particles may not portray this feature (Sun and Lukashin, 2013), thin cirrus clouds generally contain significant quantities of simple particle shapes such as hollow or solid columns, plates, droxtals, bullet rosettes, etc. (Heymsfield and Platt, 1984), especially when close to the cloud top or in cold and dry regions. Additionally, since the glory prominence is dependent on particle morphology and absorption, our ADRTM simulations incorporating a dust cloud (Volten et al., 2001) do not result in a noticeable glory feature. Such results are consistent with Hubble Space Telescope images taken during the 2003 Mars opposition that also showed p-polarization features, which were suggested to result from thin clouds of nucleating ice crystals. These crystals were seen at the forward edge of a prominent dust storm and were visible against a desert background, but the dust storm itself did not display such p-polarization features (Shkuratov et al., 2005).

To further explain the physics behind the optical phenomenon, we refer to well-known aspects of the reflection theory in geometric optics as illustrated in Fig. 8. When natural light interacts with a water surface, the reflected field tends to be polarized parallel to the surface (s-polarized) and the transmitted field tends to be polarized within the plane of incidence (p-polarized). This occurs in ray-tracing from large water droplets

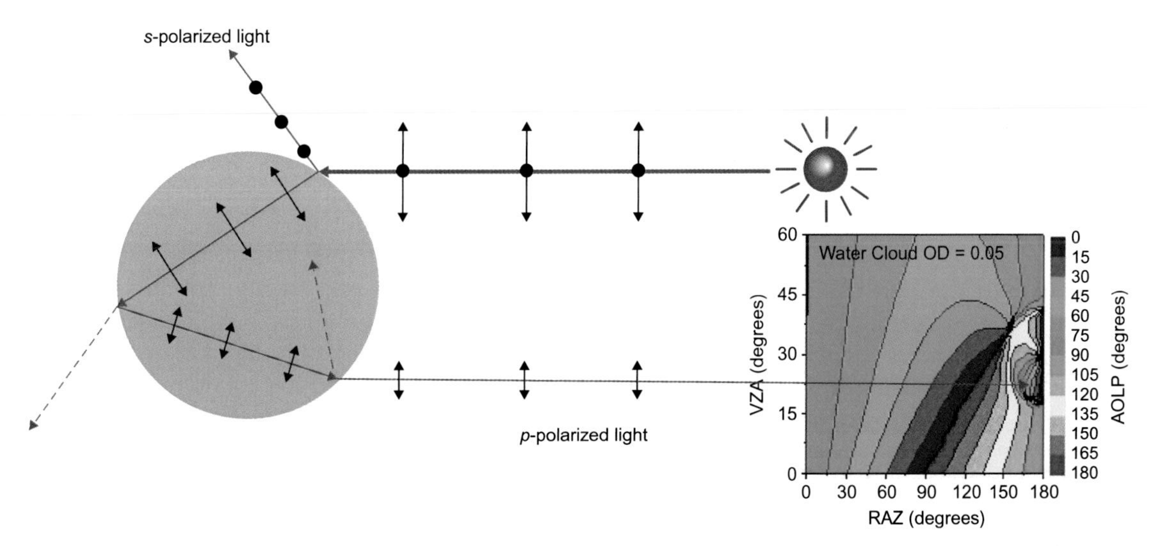

FIG. 8 Illustration of the physics for the near-backscatter p-polarization feature of the reflected light from clouds. *Reproduced from Sun, W., Baize, R.R., Videen, G., Hu, Y., Fu, Q., 2015. A method to retrieve super-thin cloud optical depth over ocean background with polarized sunlight. Atmos. Chem. Phys. 15, 11909–11918. https://doi.org/10.5194/acp-15-11909-2015.*

also where the internal fields have a tendency to favor the *p*-polarization state. One manifestation of this can be seen in the glory, which occurs at near-backscattered angles and is *p*-polarized. If the particle strongly absorbs light, like some aerosols, the refracted light cannot emerge from the particle, and the *p*-polarization feature is not observed. In addition, if the particle shape/surface is complex, like ice particle aggregates, the refracted light has weaker constructive interference, resulting in a weaker *p*-polarization feature. Furthermore, if the background surface can be characterized by single-scattering facets, like a water surface, the dominant reflected electric field from the surface at the near-backscatter direction is parallel to the surface (i.e., *s*-polarized to the meridian plane of reflected light) (Sun et al., 2014); thus, the *p*-polarization feature of the reflected light unambiguously indicates the presence of clouds. However, if the ground surface is not single-scattering-dominated, such as needle-leaf trees and grass lands, the electric field of reflected sunlight directly from the surface may not always be parallel to the surface, where the *p*-polarized light from the multiple-scattering surface introduces uncertainties in the method for the detection of superthin clouds over land areas.

In summary, the angle of linear polarization (AOLP) of scattered sunlight observed in two distinct angular regions near the exact-backscattering direction rotates from an angle parallel to the Earth's surface to an angle perpendicular to the Earth's surface when cloud particles are present in the atmosphere. A sensitivity study demonstrates that the polarization-angle feature of aerosols is not prominent enough to overcome the background polarization; however, this feature is detectable in water clouds having an optical depth of only ~0.01 and in ice clouds having an optical depth of only ~0.06, when the solar zenith angle (SZA) is not much larger than ~40 degrees over a strong polarization background, like oceans. Since most superthin clouds are over the tropical and midlatitude regions (Sun et al., 2011a,b), this limit of SZA will not affect the detection of most superthin clouds using the polarization features.

3 METHOD FOR RETRIEVING SUPERTHIN CLOUD OPTICAL DEPTH

In the last section, we showed that the dominant backscattered electric field from the clear-sky Earth-atmosphere system is nearly parallel-to the ocean surface (i.e., AOLP ~90 degrees); however, when clouds are present, this electric field can rotate in the perpendicular direction. This polarization feature of backscattered light from clouds can be used to detect all clouds including the superthin clouds. In this section, we will show how to retrieve the optical depth (OD) of superthin clouds with the polarized intensity of light backscattered from them. Note that this algorithm is developed as a means to remotely sense clouds that cannot be detected by other passive remote-sensing techniques. Any cases that include thick clouds which can be measured by conventional ways are out of the scope of this study.

To find a way to retrieve the optical depth (OD) of superthin clouds, we further modeled the angle of linear polarization (AOLP) of reflected sunlight from clouds of different thermodynamic phases, particle shapes, and optical depth over oceans. In the modeling, the atmosphere, including the cloud and aerosol layers, is assumed to be plane-parallel, and the cloud is assumed to be a homogeneous single layer over the ocean surface. The atmospheric profiles are from the US Standard Atmosphere (1976). The *p*-polarization feature of the reflected sunlight from clouds is our focus in this study. Note that *p*-polarization means ~0/180 degrees in terms of the AOLP in this work.

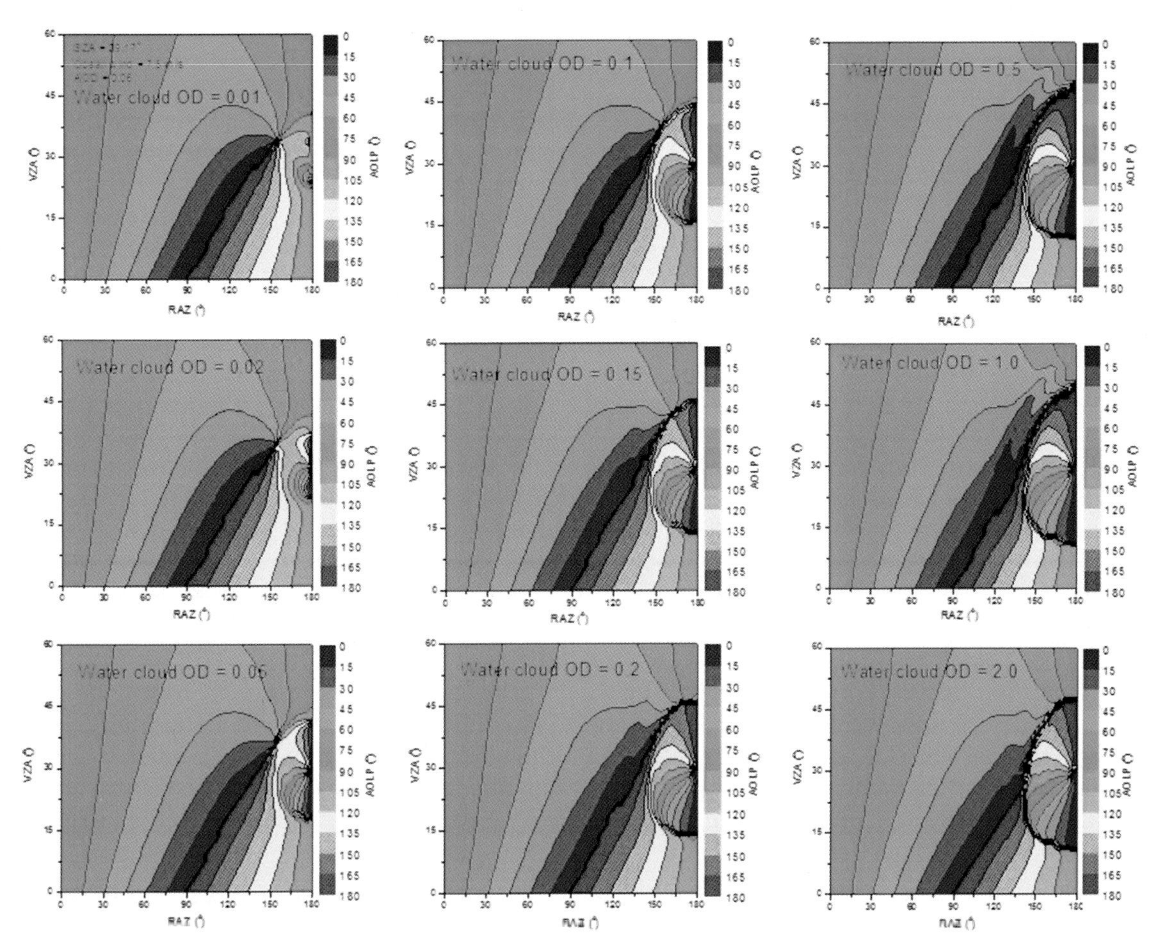

FIG. 9 The angle of linear polarization (AOLP) of reflected sunlight at 670 nm from water clouds showing that even clouds with optical depths (OD) of 0.01 exhibit the near-backscatter p-polarization feature. In the ADRTM modeling, the clouds' ODs are set from 0.01 to 2.0, the ocean wind speed is assumed to be 7.5 m/s, the solar zenith angle (SZA) is 29.17 degrees, and the aerosol optical depth (AOD) is 0.06. The C1 size distribution (Deirmendjian, 1969) is used for water cloud droplets. *Reproduced from Sun, W., Baize, R.R., Videen, G., Hu, Y., Fu, Q., 2015. A method to retrieve super-thin cloud optical depth over ocean background with polarized sunlight. Atmos. Chem. Phys. 15, 11909–11918. https://doi.org/10.5194/acp-15-11909-2015.*

Fig. 9 shows the modeled AOLP of reflected sunlight as a function of viewing zenith angle (VZA) and relative azimuth angle (RAZ) at a wavelength of 670 nm from water clouds with a different optical depth (OD) over the ocean. In the ADRTM modeling, the ocean wind speed is assumed to be 7.5 m/s, the solar zenith angle (SZA) is 29.17 degrees, and the aerosol optical depth (AOD) is 0.06. The modified gamma (MG) particle size distribution (PSD) is assumed for water cloud droplets;

$$dN/dR = N_0 R^{\nu} \exp\left(-\nu \frac{R}{R_0}\right) \tag{3}$$

where R denotes the droplet radius, R_0 is the modal radius, ν defines the shape of the distribution, and:

$$N_0 = \frac{\nu^{\nu+1}}{\Gamma(\nu+1)R_0^{\nu+1}} N_{tot} \quad (4)$$

is a constant with $\Gamma(\nu+1)$ as the gamma function and N_{tot} as the total number of particles per unit volume (Petty and Huang, 2011). The commonly used C1 size distribution (Deirmendjian, 1969), which is defined by Eq. (3) with $R_0 = 4\ \mu\text{m}$ and $\nu = 6$, is applied in this study. The water cloud is within an altitude range of 2–3 km. We can see that the near-backscatter p-polarization feature of the reflected light is evident even when the cloud OD is as small as 0.01. With the increase of the cloud OD, this pattern becomes stronger, and when cloud OD $> \sim 0.5$, it becomes saturated. Therefore, using the near-backscatter p-polarization feature of the reflected light, we can detect any water clouds, including subvisible ones with OD < 0.03, over oceans.

Fig. 10 shows the modeled AOLP of reflected sunlight as a function of VZA and RAZ at a wavelength of 670 nm from cirrus clouds. In the ADRTM modeling, the clouds' ODs are set from 0.01 to 2.0, the ocean wind speed is assumed to be 7.5 m/s, the SZA is 29.17 degrees, and the AOD is 0.06. The cirrus cloud is within an altitude range of 7–8 km. The size distribution of the ice particles in the cirrus clouds is from Heymsfield and Platt (1984) for the cloud temperature of −20°C to −25°C. The cirrus clouds are assumed to be composed of solid hexagonal column ice crystals with aspect ratios as given in Fu et al. (1998). The calculation of the single-scattering properties of the ice crystals is described in Baum et al. (2000). We can see that, similar to water clouds, cirrus clouds composed of hexagonal column ice crystals reflect sunlight with a significant near-backscatter p-polarization feature even when their OD is only ~0.01. This means that if ice cloud particle shapes are not complex, the clouds can be detected by this feature even if it is invisible with respect to standard passive remote-sensing techniques. It is known that superthin cirrus clouds more often appear in the tropical tropopause layer from 14.5 to 18.5 km (Fu et al., 2007; Virts et al., 2010), where the shapes of ice particles are more regular hexagonal columns, since they form in situ due to the large-scale slow uplift. Our results in Fig. 10 shows that these clouds could be well detected by the near-backscatter p-polarization feature of the reflected light. However, when cloud particle shapes are complex, such as a mixture of irregular particle shapes for tropical cirrus clouds as described in Meyer et al. (2004), Fig. 11 demonstrates that the p-polarization feature can hardly be seen for cloud ODs $< \sim 0.02$. Despite this limitation, the approach should be able to detect ice clouds with a mixture of complex particle shapes when their OD $> \sim 0.06$ (Sun et al., 2014).

The AOLP results reported here and in the last section demonstrate that superthin clouds can be detected reliably using the p-polarization feature of reflected sunlight from them. However, because the AOLP of reflected light, which can be derived from the ratio of polarized intensities Q and U, is not very sensitive to cloud optical depth, we must develop a new algorithm to retrieve the superthin cloud optical depth. In this section, we will show that the optical depth of superthin clouds can be retrieved at viewing angles in the neighborhood of the backscattering direction. Following the previous discussion (Sun et al., 2014), we refer to these regions as the "glory angles" that are within $\pm \sim 8$ degrees around the backscattering direction and include the blue and yellow spots in Fig. 12B. To exclude the effect of background reflection, we will use only the polarized component of the backscattered light parallel to the meridian plane of the reflected light (i.e., p-polarized light) for the OD retrieval. Since the clear-sky surface background reflection is only perpendicular to the meridian plane of the reflected light

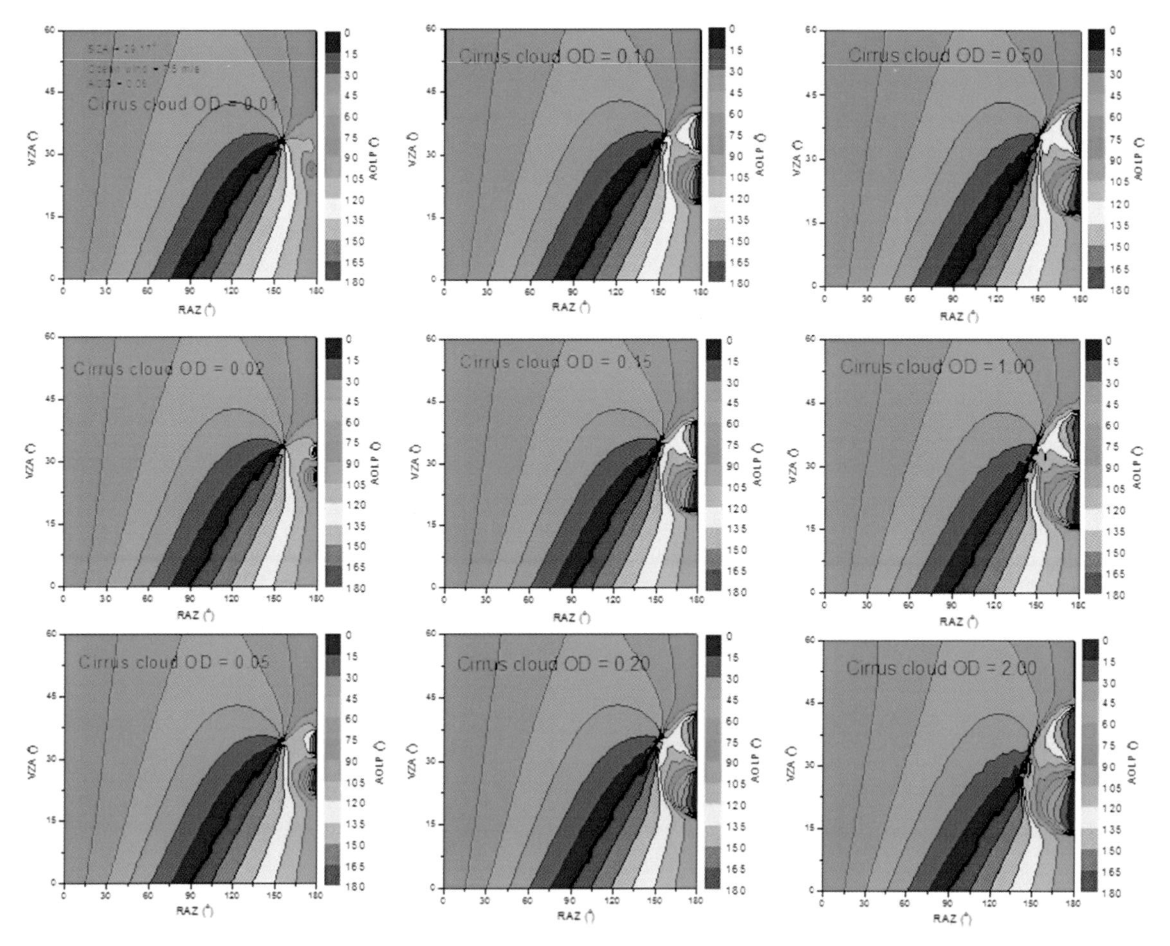

FIG. 10 The modeled AOLP of reflected sunlight at a wavelength of 670 nm from cirrus clouds *(hexagonal column particle shapes)*. In the ADRTM modeling, the clouds' optical depths (OD) are set from 0.01 to 2.0, the ocean wind speed is assumed to be 7.5 m/s, the solar zenith angle (SZA) is 29.17 degrees, and the aerosol optical depth (AOD) is 0.06. The size distribution of the ice particles in the cirrus clouds is from Heymsfield and Platt (1984) for the cloud temperature of −20°C to −25°C. The cirrus clouds are assumed to be composed of solid hexagonal column ice crystals with aspect ratios as given in Fu et al. (1998). Similar to water clouds, cirrus clouds composed of hexagonal column ice crystals exhibit the *p*-polarization feature even when their OD is ~0.01. *Reproduced from Sun, W., Baize, R.R., Videen, G., Hu, Y., Fu, Q., 2015. A method to retrieve super-thin cloud optical depth over ocean background with polarized sunlight. Atmos. Chem. Phys. 15, 11909–11918. https://doi.org/10.5194/acp-15-11909-2015.*

(*s*-polarized) in the neighborhood of the backscattering direction (see Fig. 12A), the *p*-polarized component of the backscattered light is caused by superthin clouds (Sun et al., 2014). Assuming that the linearly polarized electric field of the reflected light from the Earth-atmosphere system is $\mathbf{E_p}$, we can express its *p*-polarized component as (see Fig. 4),

$$E_p^{\perp} = E_p \cos(\text{AOLP}) \tag{5}$$

Therefore, the *p*-polarized reflectance is in the form:

$$I_p^{\perp} = I_p \cos^2(\text{AOLP}) \tag{6}$$

where $I_p = \sqrt{Q^2 + U^2}$ denotes the polarized reflectance. For a clear-sky system in which

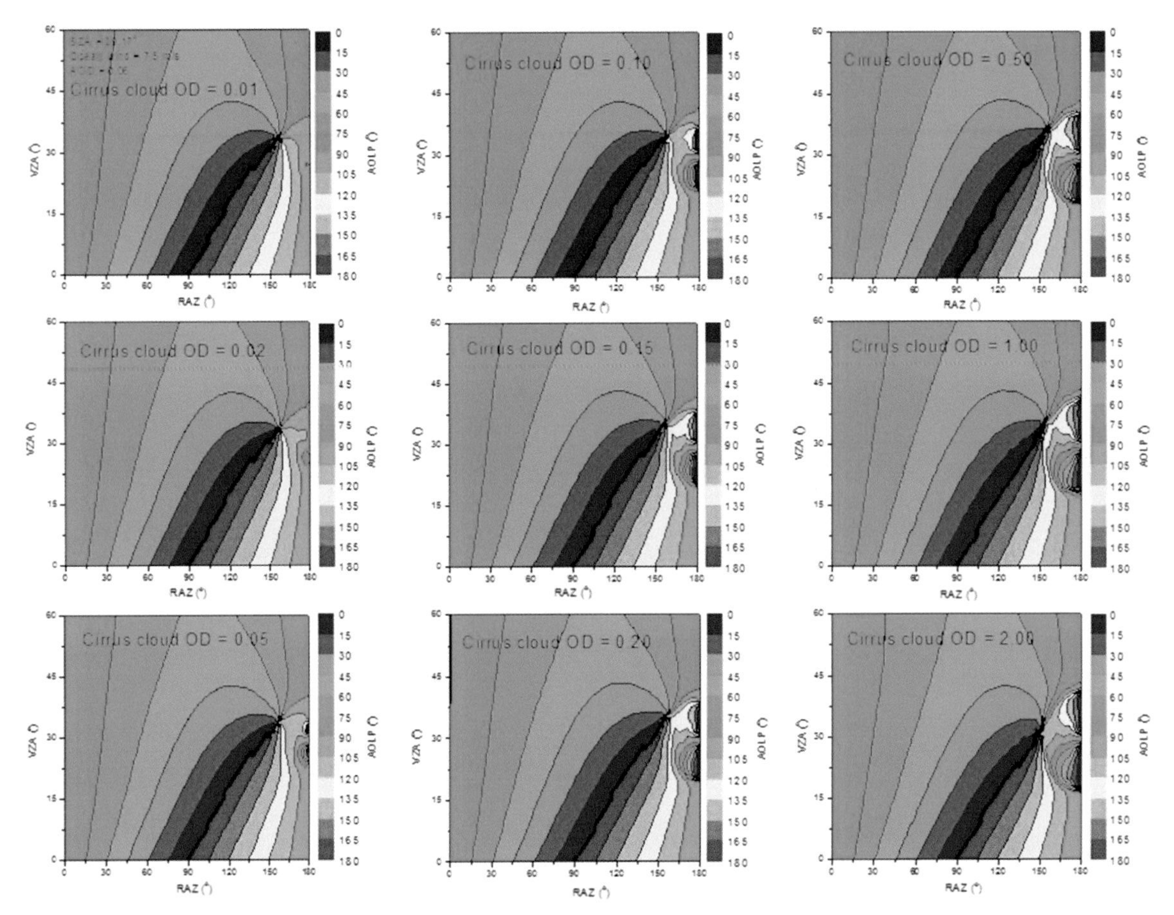

FIG. 11 The modeled AOLP of reflected sunlight at a wavelength of 670 nm from cirrus clouds *(complex particle shapes)*. In the ADRTM modeling, the clouds' optical depths (OD) are set from 0.01 to 2.0, the ocean wind speed is assumed to be 7.5 m/s, the solar zenith angle (SZA) is 29.17 degrees, and the aerosol optical depth (AOD) is 0.06. The size distribution of the ice particles in the cirrus clouds is from Heymsfield and Platt (1984) for the cloud temperature of −20°C to −25°C. The cirrus clouds are assumed to be composed of a mixture of complex particle shapes for tropical cirrus clouds as described in Meyer et al. (2004). Cirrus clouds composed of complex particle shapes crystals exhibit the *p*-polarization feature when their OD is >0.06. *Reproduced from Sun, W., Baize, R.R., Videen, G., Hu, Y., Fu, Q., 2015. A method to retrieve super-thin cloud optical depth over ocean background with polarized sunlight. Atmos. Chem. Phys. 15, 11909–11918. https://doi.org/10.5194/acp-15-11909-2015.*

the AOLP ~ 90 degrees at our specific observation direction, $I_p \cos^2(\text{AOLP}) \sim 0$. In a system containing superthin clouds or heavy aerosols in which the AOLP ~ 0/180 degrees in the observation direction, $I_p \cos^2(\text{AOLP}) \sim I_p$. At this specific observation direction, the OD can be retrieved from the polarized reflectance and the AOLP as:

$$\text{OD} = f^{-1}\left[I_p \cos^2(\text{AOLP})\right] \quad (7)$$

where $f^{-1}[]$ denotes a function that will be determined by the OD and $I_p\cos^2(\text{AOLP})$ correlation curve from modeling results for various superthin clouds.

Fig. 13 shows the modeled *p*-polarized reflectance at a wavelength of 670 nm as a function of

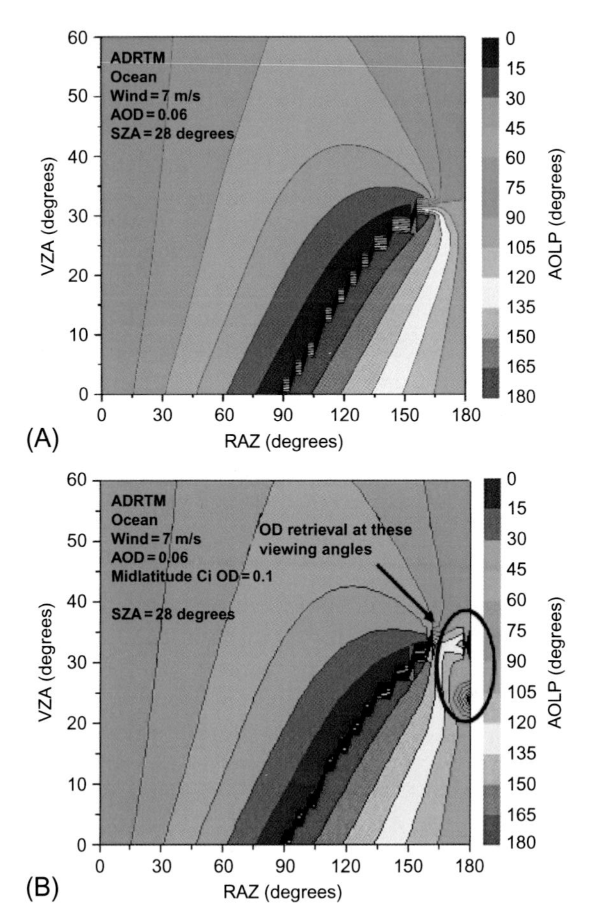

FIG. 12 The modeled AOLP at a wavelength of 670 nm as a function of viewing zenith angle (VZA) and relative azimuth angle (RAZ) from (A) the ADRTM for clear-sky oceans and (B) the ADRTM for oceans with a layer of superthin clouds. In the modeling, the solar zenith angle (SZA) is 28 degrees, ocean wind speed is 7 m/s, the sea-salt aerosol optical depth (AOD) is 0.06, and a layer of midlatitude cirrus cloud with an optical depth of 0.1 is assumed with an altitude range of 7–8 km. The difference between (A) and (B) shows the effect of superthin clouds on the reflected sunlight's AOLP. *Reproduced from Sun, W., Videen, G., Mishchenko, M.I., 2014. Detecting super-thin clouds with polarized sunlight. Geophys. Res. Lett. 41, 688–693. https://doi.org/10.1002/2013GL058840.*

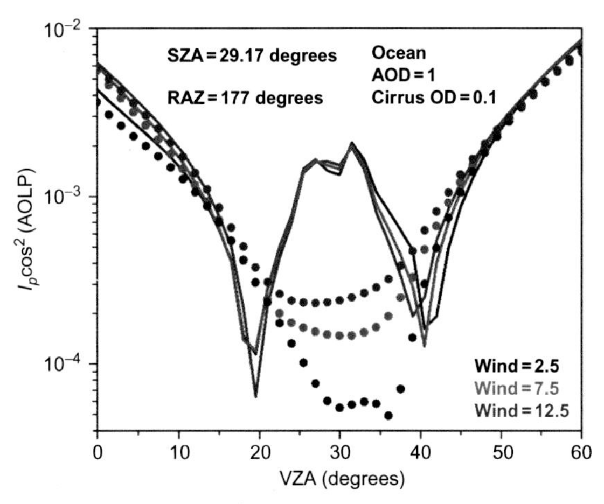

FIG. 13 Modeled perpendicularly polarized reflectance at a wavelength of 670 nm as a function of viewing zenith angle (VZA) and at a relative azimuth angle (RAZ) of 177 degrees for clear sky *(dots)* and superthin clouds *(solid lines)* having an optical depth (OD) of 0.1 over oceans with different wind speeds. In the modeling, the solar zenith angle (SZA) is 29.17 degrees, and the aerosol optical depth (AOD) is 0.1. The size distribution of the ice particles in the cirrus clouds is from Heymsfield and Platt (1984) for the cloud temperature of −20°C to −25°C. The cirrus clouds are assumed to be composed of hexagonal ice columns with aspect ratios as given in Fu et al. (1998). *Reproduced from Sun, W., Baize, R.R., Videen, G., Hu, Y., Fu, Q., 2015. A method to retrieve super-thin cloud optical depth over ocean background with polarized sunlight. Atmos. Chem. Phys. 15, 11909–11918. https://doi.org/10.5194/acp-15-11909-2015.*

VZA and at a RAZ of 177 degrees for a clear and superthin cloud scene over oceans with different wind speeds. In the modeling, the SZA is 29.17 degrees and the AOD is 0.1. The size distribution of the ice particles in the cirrus clouds is from Heymsfield and Platt (1984) for the cloud temperature of −20°C to −25°C. The cirrus cloud is assumed to be composed of hexagonal ice columns and is contained within an altitude range of 7–8 km. Results for clear oceans *(dotted curves)* and oceans with a cirrus layer with an OD of 0.1 *(solid curves)* are shown. Different colors represent different wind speeds: 2.5 *(black)*, 7.5 *(red)*, and 12.5 m/s *(blue)*. We can see that for clear-sky oceans, $I_p\cos^2$(AOLP) varies with the ocean surface roughness (wind speed) and its magnitude is low. However, when there is a layer of thin cloud over the oceans, at the glory angles, $I_p\cos^2$(AOLP) is about one order of magnitude larger than the clear-sky values and has a very small dependence on the ocean surface conditions. This is very important, since surface conditions can vary widely, and a significant dependence would complicate the OD retrievals. Fig. 13 demonstrates that $I_p\cos^2$(AOLP) is a robust quantity to use to retrieve the OD of optically thin clouds regardless of ocean surface conditions.

Additionally, Fig. 14 shows the modeled *p*-polarized reflectance at a wavelength of 670 nm as a function of VZA and at a RAZ of 177 degrees for superthin cirrus clouds over oceans with different ODs *(solid curves)*. Also shown in the figure is the result for clear oceans *(black dots)*. In the modeling, the SZA is 29.17 degrees, the wind speed is 7.5 m/s, and the aerosol optical depth (AOD) is 0.1. The size distribution of the ice particles in the cirrus clouds is from Heymsfield and Platt (1984) for the cloud temperature of −20°C to −25°C. The midlatitude cirrus cloud is assumed to be composed of a mixture of complex particle shapes as described in Baum et al. (2000) and to be within an altitude range of 7–8 km. We can see that with the increase of cloud OD, $I_p\cos^2$(AOLP) systematically increases at the glory angle region. When cloud OD approaches ~0.6, $I_p\cos^2$(AOLP) becomes saturated and it is difficult to differentiate the OD of the respective clouds. Therefore, this OD retrieval method may only work well for thin clouds that have OD < ~0.6. From the same calculations of Fig. 14, the modeled *p*-polarized reflectance at 670 nm as a function of cloud optical depth (OD) at a VZA of 28.5 degrees and a RAZ of 177 degrees for superthin cirrus clouds over oceans is displayed in Fig. 15. We can see that $I_p\cos^2$(AOLP) is nearly linearly related to cloud OD when the cloud OD < ~0.6. The OD and $I_p\cos^2$(AOLP) correlation curve in Fig. 15 is just an example of the OD and $I_p\cos^2$(AOLP) function in Eq. (7).

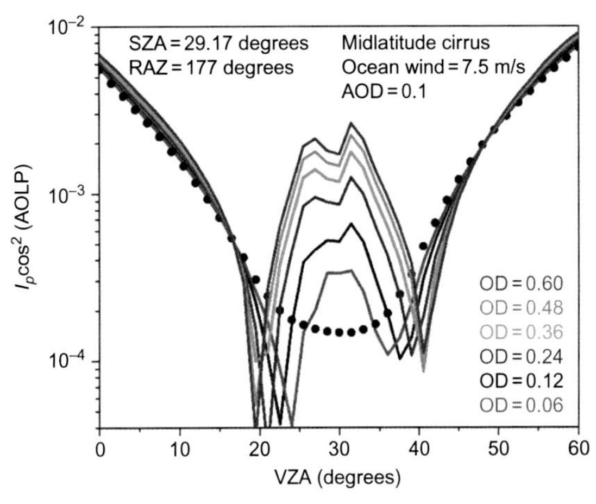

FIG. 14 Modeled perpendicularly polarized reflectance at a wavelength of 670 nm as a function of viewing zenith angle (VZA) and at a relative azimuth angle (RAZ) of 177 degrees for superthin cirrus clouds over oceans with different optical depth (OD) *(solid curves)*. Also shown in the figure is the result for clear sky over oceans *(black dots)*. In the modeling, the solar zenith angle (SZA) is 29.17 degrees, the wind speed is 7.5 m/s, and the aerosol optical depth (AOD) is 0.1. The size distribution of the ice particles in the cirrus clouds is from Heymsfield and Platt (1984) for the cloud temperature of −20°C to −25°C. The midlatitude cirrus clouds are composed of a mixture of complex particle shapes as described in Baum et al. (2000). These results suggest that the OD retrieval method may only work well for thin clouds that have OD < ~0.6. *Reproduced from Sun, W., Baize, R.R., Videen, G., Hu, Y., Fu, Q., 2015. A method to retrieve super-thin cloud optical depth over ocean background with polarized sunlight. Atmos. Chem. Phys. 15, 11909–11918. https://doi.org/10.5194/acp-15-11909-2015.*

Since this algorithm uses near-backscatter polarized reflectance for retrieval of cloud optical depth, it is sensitive to cloud thermodynamic phase or particle size and shape if the cloud is a cirrus or mixed-phase cloud. Thus, $I_p\cos^2$(AOLP) must be some function of ice-cloud particle size and shape, or liquid water cloud size distribution. Reliably detecting the thermodynamic phase of the clouds is a prerequisite for a good retrieval using this method. Using the oxygen A-band (759–770 nm) (Min et al., 2014) $I_p\cos^2$(AOLP) to estimate the altitude of the clouds could help to determine the cloud thermodynamic phase. On the other hand, a

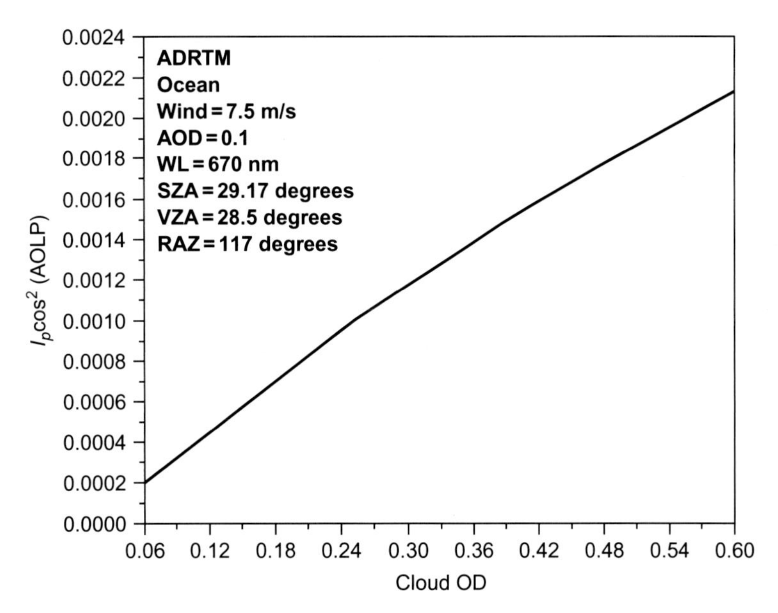

FIG. 15 Modeled perpendicularly polarized reflectance at a wavelength of 670 nm as a function of cloud optical depth (OD) at a viewing zenith angle (VZA) of 28.5 degrees and a relative azimuth angle (RAZ) of 177 degrees for superthin cirrus clouds over oceans. The solar zenith angle (SZA) is 29.17 degrees, the wind speed is 7.5 m/s, and the aerosol optical depth (AOD) is 0.1. The size distribution of the ice particles in the cirrus clouds is from Heymsfield and Platt (1984) for the cloud temperature of −20°C to −25°C. The midlatitude cirrus clouds are composed of a mixture of complex particle shapes as described in Baum et al. (2000). This figure shows the nearly linearly relationship between cloud OD and AOLP for OD < 0.6. *Reproduced from Sun, W., Baize, R.R., Videen, G., Hu, Y., Fu, Q., 2015. A method to retrieve super-thin cloud optical depth over ocean background with polarized sunlight. Atmos. Chem. Phys. 15, 11909–11918. https://doi.org/10.5194/acp-15-11909-2015.*

detailed study of the effects of particle size and shape of the clouds on the polarized reflectance from them is also necessary in the application of this method. As an example, Fig. 16 shows the modeled *p*-polarized reflectance at a wavelength of 670 nm as a function of VZA and at a RAZ of 177 degrees for superthin cirrus clouds over oceans with different ODs. In the modeling, the SZA is 29.17 degrees, the wind speed is 7.5 m/s, and the AOD is 0.06. The size distribution of the ice particles in the cirrus clouds is from Heymsfield and Platt (1984) for the cloud temperature of −20°C to −25°C. The cirrus clouds are assumed to be within an altitude range of 7–8 km and composed of mixtures of complex particle shapes for typical midlatitude cirrus clouds as described in Baum et al. (2000) *(open circles)* and tropical cirrus clouds as described in Meyer et al. (2004) *(solid curves)*, respectively. It can be seen that different mixtures of ice cloud particle shapes can result in different $I_p\cos^2$(AOLP). This can cause an uncertainty of ~0.05 in the retrieved cloud OD.

It is worth noting here that in Figs. 13, 14, and 16, for each case of the *p*-polarized reflectance curves of clouds, at the boundary of the glory angle region, $I_p\cos^2$(AOLP) drastically dips to very small values and our angular discretization in the modeling sometimes misses capturing the lowest point. A plausible explanation for this phenomenon is that it is the transitional region between backscattered *p*-polarized rays from the particles and the *s*-polarized rays from the surface, where surface background *s*-polarized reflectance cancels *p*-polarized reflectance from clouds, thus, *p*-polarized light is nearly zero

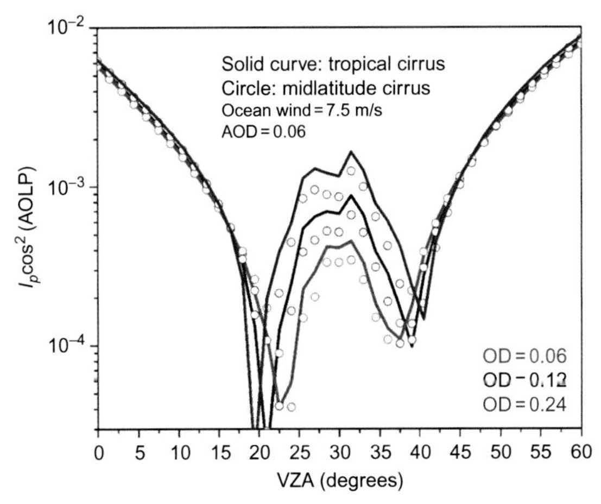

FIG. 16 Modeled perpendicularly polarized reflectance at a wavelength of 670 nm as a function of viewing zenith angle (VZA) and at a relative azimuth angle (RAZ) of 177 degrees for superthin cirrus clouds over oceans with different optical depth (OD). In the modeling, the solar zenith angle (SZA) is 29.17 degrees, the wind speed is 7.5 m/s, and the aerosol optical depth (AOD) is 0.06. The size distribution of the ice particles in the cirrus clouds is from Heymsfield and Platt (1984) for the cloud temperature of −20°C to −25°C. The clouds are composed of mixtures of complex particle shapes for typical midlatitude cirrus clouds as described in Baum et al. (2000) *(open circles)* and tropical cirrus clouds as described in Meyer et al. (2004) *(solid curves)*, respectively. This illustrates that inadequate knowledge of cloud particle shapes can result in uncertainties of up to ~0.05 in the retrieved cloud OD. *Reproduced from Sun, W., Baize, R.R., Videen, G., Hu, Y., Fu, Q., 2015. A method to retrieve super-thin cloud optical depth over ocean background with polarized sunlight. Atmos. Chem. Phys. 15, 11909–11918. https://doi.org/10.5194/acp-15-11909-2015.*

and AOLP is close to 90 degrees, so $I_p\cos^2$(AOLP) can be very small.

In summary, we show that the *p*-polarized reflectance $I_p\cos^2$(AOLP) at near-backscatter viewing angles can be used for the retrieval of the optical depth of superthin clouds, with little affect from ocean surface conditions. Our sensitivity study shows that for polarization intensity measurements with ~10% calibration error such as those from the PARASOL 670 nm channel (Fougnie et al., 2007), this algorithm can have ~0.006 uncertainty in the retrieved superthin cloud OD. This is a robust algorithm, which could be used to conduct inexpensive surveys for superthin clouds over midlatitude and tropical areas, where most superthin clouds exist (Sun et al., 2011b).

4 SUMMARY AND CONCLUSION

Although they are hard to be detected by passive or even active satellite instrument sensors, superthin clouds with optical depths smaller than ~0.3 exist globally and can affect the remote sensing of aerosols, surface temperature, and atmospheric composition significantly. Sea-surface temperature retrieved from NASA's Atmospheric Infrared Sounder (AIRS) satellite data is generally ~5–10 K lower than true values at tropical and midlatitude regions, where these clouds mostly exist. Using space-borne LIDAR for long-term global measurement of these clouds are limited by the large operational cost and narrow field of view of LIDAR systems. However, superthin clouds can be well detected by a polarimetric imager facing toward the back-scattering direction of sunlight, exploiting a distinct, characterizing feature of the angle of linear polarization of the scattered solar radiation. This superthin-cloud-detection technology will greatly improve the detection of superthin clouds and significantly impact the remote sensing of aerosols, surface temperature, and atmospheric composition, and thus, significantly improve data for climate modeling. Undeniably, as an algorithm based on a low-cost passive instrument measuring reflected solar light, it has difficulties in detecting superthin clouds over thick clouds, since the thick clouds' glory pattern is much stronger than that of superthin clouds. For those multilayer cases, this method can only tell there are thick clouds in the scene. Moreover, since the glory pattern is a special optical phenomenon of transparent cloud droplets or ice crystals, this algorithm is not sensitive to background aerosols that coexist with superthin clouds. However, heavy aerosols (OD > 0.2) can cause the electric field on the principal plane not parallel to the ocean surface at viewing zenith angles smaller than the backscattering angle, which will result in some ambiguities for retrieval, but they cannot produce the full glory pattern as shown in Figs. 5–7 and 9–11, i.e., electric fields in all radial directions around the backscattering direction. Also, as a method based on measurements of reflected sunlight, this algorithm obviously cannot work at night. It is important to note that previous studies suggest that, based on Raman LIDAR data, superthin clouds differ little between day and night, because of their insignificant absorption to solar radiation. Active instruments such as space-borne LIDARs can work during both day and night and can measure cloud/aerosol altitude, but their swath width is narrow while producing vertical profiles along the satellite track. In addition, the limited number of photons acquired from space-borne LIDARs reduces signal-to-noise level and introduces errors in their measured data. Thus, both passive and active instrument techniques have their advantages and disadvantages. Exploring innovative algorithms to make greater use of existing passive remote-sensing instruments and to complement active remote-sensing instruments is an obvious benefit.

Acknowledgments

This work is partly supported by NASA's CLARREO mission. The authors thank Bruce A. Wielicki for this support. Wenbo Sun and Gorden Videen acknowledge support through NASA Glory fund 09-GLORY09-0027, and thank Hal B. Maring and Michael I. Mishchenko for their input in this work.

References

Ackerman, S.A., Strabala, K.I., Menzel, W.P., Frey, R.A., Moeller, C.C., Gumley, L.E., 1998. Discriminating clear sky from clouds with MODIS. J. Geophys. Res. 103, 32141–32157. https://doi.org/10.1029/1998JD200032.

Ackerman, S.A., Holz, R.E., Frey, R., Eloranta, E.W., Maddux, B.C., McGill, M., 2008. Cloud detection with MODIS. Part II: validation. J. Atmos. Ocean. Technol. 25, 1073–1086. https://doi.org/10.1175/2007JTECHA1053.1.

Baum, B.A., Kratz, D.P., Yang, P., Ou, S.C., Hu, Y., Soulen, P.F., Tsay, S.C., 2000. Remote sensing of cloud properties using MODIS airborne simulator imagery during SUCCESS. I. Data and models. J. Geophys. Res. 105 (11), 767–780.

Brest, C.L., Rossow, W.B., 1992. Radiometric calibration and monitoring of NOAA AVHRR data for ISCCP. Int. J. Remote Sens. 13, 235–273.

Chahine, M.T., Pagano, T.S., Aumann, H.H., Atlas, R., Barnet, C., Blaisdell, J., Chen, L., Divakarla, M., Fetzer, E.J., Goldberg, M., Gautier, C., Granger, S., Hannon, S., Irion, F.W., Kakar, R., Kalnay, E., Lambrigtsen, B.H., Lee, S.-Y., Le Marshall, J., McMillan, W.W., McMillin, L., Olsen, E.T., Revercomb, H., Rosenkranz, P., Smith, W.L., Staelin, D., Strow, L.L., Susskind, J., Tobin, D., Wolf, W., Zhou, L.H., 2006. AIRS: Improving weather forecasting and providing new data on greenhouse gases. Bull. Am. Meteorol. Soc. 87, 911–926.

Christi, M.J., Stephens, G.L., 2004. Retrieving profiles of atmospheric CO2 in clear sky and in the presence of thin cloud using spectroscopy from the near and thermal infrared: A preliminary case study. J. Geophys. Res. 109. D04316. https://doi.org/10.1029/2003JD004058.

Coulson, K.L., 1988. Polarization and intensity of light in the atmosphere. A. Deepak, Hampton, VA.

Crisp, D., Atlas, R.M., Breon, F.-M., Brown, L.R., Burrows, J.P., Ciais, P., Connor, B.J., Doney, S.C., Fung, I.Y., Jacob, D.J., Miller, C.E., O'Brien, D., Pawson, S., Randerson, J.T., Rayner, P., Salawitch, R.J., Sander, S.P., Sen, B., Stephens, G.L., Tans, P.P., Toon, G.C., Wennberg, P.O., Wofsy, S.C., Yung, Y.L., Kuang, Z., Chudasama, B., Sprague, G., Weiss, B., Pollock, R., Kenyon, D., Schroll, S., 2004. The orbiting carbon observatory (OCO) mission. Adv. Space Res. 34, 700–709.

Deirmendjian, D., 1969. Electromagnetic scattering on spherical polydispersions. American Elsevier Publishing Company, Inc., New York, NY.

Deschamps, P.Y., Breon, F.-M., Leroy, M., Podaire, A., Bricaud, A., Buriez, J.-C., Seze, G., 1994. The POLDER mission: Instrument characteristics and scientific objectives. IEEE Trans. Geosci. Remote Sens. 32, 598–615.

Dessler, A.E., Yang, P., 2003. The distribution of tropical thin cirrus clouds inferred from Terra MODIS data. J. Clim. 16, 1241–1247.

Fougnie, B., Bracco, G., Lafrance, B., Ruffel, C., Hagolle, O., Tinel, C., 2007. PARASOL in-flight calibration and performance. Appl. Opt. 46, 5435–5451.

Fu, Q., Yang, P., Sun, W., 1998. An accurate parameterization of the infrared radiative properties of cirrus clouds for climate models. J. Clim. 11, 2223–2237.

Fu, Q., Hu, Y.X., Yang, Q., 2007. Identifying the top of the tropical tropospause layer from vertical mass flux analysis and CALIPSO lidar cloud observations. Geophys. Res. Lett. 34. L14813. https://doi.org/10.1029/2007GL030099.

Gao, B.C., Kaufman, Y.J., 1995. Selection of 1.375-μm MODIS channel for remote sensing of cirrus clouds and stratospheric aerosols from space. J. Atmos. Sci. 52, 4231–4237.

Heymsfield, A.J., Platt, C.M.R., 1984. A parameterization of the particle size spectrum of ice clouds in terms of the ambient temperature and the ice water content. J. Atmos. Sci. 41, 846–855.

King, M.D., Kaufman, Y.J., Menzel, W.P., Tanre, D., 1992. Remote-sensing of cloud, aerosol, and water-vapor properties from the Moderate Resolution Imaging Spectrometer (MODIS). IEEE Trans. Geosci. Remote Sens. 30, 2–27.

Lee, J., Yang, P., Dessler, A.E., Gao, B.C., Platnick, S., 2009. Distribution and radiative forcing of tropical thin cirrus clouds. J. Atmos. Sci. 66, 3721–3731.

Mace, G.G., Zhang, Y., Platnick, S., King, M.D., Minnis, P., Yang, P., 2005. Evaluation of cirrus cloud properties from MODIS radiances using cloud properties derived from ground-based data collected at the ARM SGP site. J. Appl. Meteorol. 44, 221–240.

McFarquhar, G.M., Heymsfield, A.J., Spinhirne, J., Hart, B., 2000. Thin and subvisual tropopause tropical cirrus: observations and radiative impacts. J. Atmos. Sci. 57, 1841–1853.

Meyer, K., Yang, P., Gao, B.C., 2004. Optical thickness of tropical cirrus clouds derived from the MODIS 0.66- and 1.375-μm channels. IEEE Trans. Geosci. Remote Sens. 42, 833–841.

Min, Q., Yin, B., Li, S., Berndt, J., Harrison, L., Joseph, E., Duan, M., Kiedron, P., 2014. A high-resolution oxygen A-band spectrometer (HABS) and its radiation closure. Atmos. Meas. Tech. 7, 1711–1722. https://doi.org/10.5194/amt-7-1711-2014.

Minnis, P., Young, D.F., Weilicki, B.A., Sun-Mack, S., Trepte, Q.Z., Chen, Y., Heck, P.W., Dong, X., 2002. A global cloud database from VIRS and MODIS for CERES. In: Proc. SPIE 3rd Intl. Asia-Pacific Environ. Remote Sensing Symp. 2002: Remote Sens. of Atmosphere, Ocean, Environment, and Space, Hangzhou, China, October 23–27, 4891, pp. 115–126.

Mishchenko, M.I., Cairns, B., Kopp, G., Schueler, C.F., Fafaul, B.A., Hansen, J.E., Hooker, R.J., Itchkawich, T., Maring, H.B., Travis, L.D., 2007. Accurate monitoring of terrestrial aerosols and total solar irradiance: introducing the glory mission. Bull. Am. Meteorol. Soc. 88, 677–691.

Muinonen, K., Tyynella, J., Zubko, E., Lindqvist, H., Penttilla, A., Videen, G., 2011. Polarization of light backscattered by small particles. J. Quant. Spectrosc. Radiat. Transf. 112, 2193–2212.

National Oceanic and Atmospheric Administration, National Aeronautics and Space Administration, and United States Air Force, 1976. U.S. Standard Atmosphere, NOAA-S/T 76-1562.

Nee, J.B., Len, C.N., Chen, W.N., Lin, C.I., 1998. Lidar observation of the cirrus cloud in the tropopause at Chung-Li (25 degrees N, 121 degrees E). J. Atmos. Sci. 55, 2249–2257.

O'Dell, C.W., Connor, B., Bosch, H., O'Brien, D., Frankenberg, C., Castano, R., Christi, M., Eldering, D., Fisher, B., Gunson, M., McDuffie, J., Miller, C.E., Natraj, V., Oyafuso, F., Polonsky, I., Smyth, M., Taylor, T., Toon, G.C., Wennberg, P.O., Wunch, D., 2012. The ACOS CO2 retrieval algorithm – Part 1: description and validation against synthetic observations. Atmos. Meas. Tech. 5, 99–121. https://doi.org/10.5194/amt-5-99-2012. http://www.atmos-meas-tech.net/5/99/2012.

Omar, A.H., Winker, D.M., Tackett, J.L., Giles, D.M., Kar, J., Liu, Z., Vaughan, M.A., Powell, K.A., Trepte, C.R., 2013. CALIOP and AERONET aerosol optical depth comparisons: one size fits none. J. Geophys. Res. 118, 4748–4766. https://doi.org/10.1002/jgrd.50330.

Packer, D.M., Lock, C., 1951. The brightness and polarization of the daylight sky at altitudes of 18,000 to 38,000 feet above sea level. J. Opt. Soc. Am. 41, 473–478.

Petty, G.W., Huang, W., 2011. The modified gamma size distribution applied to inhomogeneous and nonspherical particles: key relationships and conversions. J. Atmos. Sci. 68, 1460–1473.

Platt, C.M.R., Young, S.A., Manson, P.J., Patterson, G.R., Marsden, S.C., Austin, R.T., 1998. The optical properties of equatorial cirrus from observations in the ARM pilot radiation observation experiment. J. Atmos. Sci. 55, 1977–1996.

Prabhakara, C., Kratz, D.P., Yoo, J.-M., Dalu, G., Vernekar, A., 1993. Optically thin cirrus clouds: radiative impact on the warm pool. J. Quant. Spectrosc. Radiat. Transf. 49, 467–483.

Rogers, R.R., Hostetler, C.A., Hair, J.W., Ferrare, R.A., Liu, Z., Obland, M.D., Harper, D.B., Cook, A.L., Powell, K.A., Vaughan, M.A., Winker, D.M., 2011. Assessment of the CALIPSO lidar 532 nm attenuated backscatter calibration using the NASA LaRC airborne high spectral resolution lidar. Atmos. Chem. Phys. 11, 1295–1311. https://doi.org/10.5194/acp-11-1295-2011.

Roskovensky, J.K., Liou, K.N., 2003. Detection of thin cirrus from 1.38 μm/0.65 μm reflectance ratio combined with 8.6–11 μm brightness temperature difference. Geophys. Res. Lett. 30, 1985. https://doi.org/10.1029/2003GL018135.

Shkuratov, Y., Kreslavsky, M., Kaydash, V., Videen, G., Bell, J., Wolff, M., Hubbard, M., Noll, K., Lubenow, A., 2005. Hubble space telescope imaging polarimetry of Mars during the 2003 opposition. Icarus 176, 1–11. https://doi.org/10.1016/j.icarus.2005.01.009.

Spinhirne, J.D., Hart, W.D., Hlavka, D.L., 1996. Cirrus infrared parameters and shortwave reflectance relations from observations. J. Atmos. Sci. 53, 1438–1458.

Sun, W., Lukashin, C., 2013. Modeling polarized solar radiation from ocean-atmosphere system for CLARREO inter-calibration applications. Atmos. Chem. Phys. 13, 10303–10324. https://doi.org/10.5194/acp-13-10303-2103.

Sun, W., Lin, B., Hu, Y., Lukashin, C., Kato, S., Liu, Z., 2011a. On the consistency of CERES longwave flux and AIRS temperature and humidity profiles. J. Geophys. Res. 116. D17101. https://doi.org/10.1029/2011JD016153.

Sun, W., Videen, G., Kato, S., Lin, B., Lukashin, C., Hu, Y., 2011b. A study of subvisual clouds and their radiation effect with a synergy of CERES, MODIS, CALIPSO, and AIRS data. J. Geophys. Res. 116. D22207. https://doi.org/10.1029/2011JD016422.

Sun, W., Videen, G., Mishchenko, M.I., 2014. Detecting super-thin clouds with polarized sunlight. Geophys. Res. Lett. 41, 688–693. https://doi.org/10.1002/2013GL058840.

Sun, W., Baize, R.R., Videen, G., Hu, Y., Fu, Q., 2015. A method to retrieve super-thin cloud optical depth over ocean background with polarized sunlight. Atmos. Chem. Phys. 15, 11909–11918. https://doi.org/10.5194/acp-15-11909-2015.

Uthe, E.E., Russell, P.B., 1976. Lidar observations of tropical high-altitude cirrus clouds. In: Bolle, H.J. (Ed.), Radiation in the Atmosphere. Science Press, Enfield, NH, USA, pp. 242–244.

Virts, K.S., Wallace, J.M., Fu, Q., Ackerman, T.P., 2010. Tropical tropopause transition layer cirrus as represented by CALIPSO lidar observations. J. Atmos. Sci. 67, 3113–3129.

Volten, H., Muñoz, O., Rol, E., de Haan, J.F., Vassen, W., Hovenier, J.W., Muinonen, K., Nousiainen, T., 2001. Scattering matrices of mineral aerosol particles at 441.6 and 632.8 nm. J. Geophys. Res. 106, 17375–17401.

Wang, P.H., McCormick, M.P., Poole, L.R., Chu, W.P., Yue, G.K., Kent, G.S., Skeens, K.M., 1994. Tropical high cloud characteristics derived from SAGE II extinction measurements. Atmos. Res. 34, 53–83. https://doi.org/10.1016/0169-8095(94)90081-7.

Wielicki, B.A., et al., 2013. Climate absolute radiance and refractivity observatory (CLARREO): achieving climate change absolute accuracy in orbit. Bull. Am. Meteorol. Soc. 94, 1519–1539. https://doi.org/10.1175/BAMS-D-12-00149.1.

Winker, D.M., Trepte, C.R., 1998. Laminar cirrus observed near the tropical tropopause by LITE. Geophys. Res. Lett. 25, 3351–3354.

Winker, D.M., Hunt, W.H., McGill, M.J., 2007. Initial performance assessment of CALIOP. Geophys. Res. Lett. 34. L19803. https://doi.org/10.1029/2007GL030135.

Wylie, D.P., Menzel, W.P., 1999. Eight years of high cloud statistics using HIRS. J. Clim. 12, 170–184.

Wylie, D.P., Piironen, P., Wolf, W., Eloranta, E., 1995. Understanding satellite cirrus cloud climatologies with calibrated lidar optical depths. J. Atmos. Sci. 52, 4327–4343.

Wenbo Sun is a senior research scientist of Science Systems and Applications Inc. (SSAI) in Virginia, United States. He works in NASA Langley Research Center for climate-related satellite remote-sensing missions. He gained his B. Sc. degree in atmospheric physics from Peking University in 1988; he received his M.Sc. degree in ocean remote sensing from the State Ocean Administration (SOA) of China in 1991. He obtained his Ph.D. degree in atmospheric sciences from Dalhousie University, Canada in 2000. Wenbo Sun's research interests involve light scattering by arbitrarily shaped atmospheric particles, radiation transfer in clouds and aerosol layers, and satellite remote sensing of Earth-atmosphere system.

Rosemary R. Baize currently serves as the Deputy Director of Science Directorate at NASA Langley Research Center. She also serves as the Project Scientist for the Climate Absolute Radiance and Refractivity Observatory (CLARREO) Decadal Survey mission. She has a M.B.A. degree from the College of William and Mary, a M.Sc. degree in aerospace engineering from the University of Michigan, and a B.Sc. degree in aeronautical engineering from Purdue University. At the onset of her career, Ms. Baize tested advanced, high speed aircraft concepts, applying new measurement techniques in the wind tunnel and comparing the results to computational models. More recently, she has supported the formulation of new instruments and space mission concepts, including lower cost access to space options utilizing hosted payloads. Ms. Baize also worked in the Technology Commercialization Office at Kennedy Space Center where she managed projects focused on improving launch infrastructure and operations.

Gorden Videen is a physicist with the Army Research Laboratory and the Space Science Institute. He received his B.Sc. degree from the Physics Department of University of Arizona in 1986, his M.Sc. degree from the Optical Sciences Center of University of Arizona in 1991, and his Ph.D. degree from the Optical Sciences Center of University of Arizona in 1992. His primary focus recently has been on characterizing irregularly shaped aerosols using light-scattering techniques. Other research topics include analyzing surface structures and remote sensing.

Yongxiang Hu is a senior research scientist in the Science Directorate of NASA Langley Research Center. He graduated with a B.Sc. degree in Meteorology from Peking University in 1985, with a M.Sc. degree from the National Satellite Meteorological Center of China in 1988, and with a Ph.D. degree in atmospheric sciences from the University of Alaska in 1994. His expertise is in radiative transfer, lidar remote sensing, and climate modeling.

CHAPTER

8

Cloud Screening and Property Retrieval for Hyper-Spectral Thermal Infrared Sounders

Yu Someya, Ryoichi Imasu

Atmosphere and Ocean Research Institute, The University of Tokyo, Chiba, Japan

1 INTRODUCTION

Measurements of thermal infrared (TIR) radiation from space offer many benefits for monitoring of the atmospheric environment, such as gas concentrations and temperature profiles. To observe these targets, cloud contamination must be avoided using appropriate means. In addition, cloud properties elucidated from TIR data constitute important knowledge supporting atmospheric science.

Most satellites that have infrared sounders also have visible (VIS) and near infrared (NIR) imagers on board. Imaging is used for cloud detection in the field of view (FOV) of the sounder because of their high horizontal resolutions. The pixel size of imagers is generally less than that of TIR sounders. Therefore, imagers can detect horizontally small clouds. Moreover, VIS and NIR can detect low-level clouds. Detecting such clouds with TIR measurements is more difficult because of the small temperature contrast between clouds and surfaces. However, TIR measurements have the advantages for optically thin clouds, for clouds over high reflectance surfaces, and for nighttime observations. The most commonly used cloud climatological data analyses, International Satellite Cloud Climatology Project (ISCCP), uses both VIS and TIR imagers (Rossow and Garder, 1993). In most techniques used for data processing of ISCCP, clouds are discriminated from clear scenes through comparisons between observed and estimated clear sky radiances.

In the early phase of meteorological satellite sensor development, TIR data were used for cloud observations from space. Fritz and Winston (1962) discriminated clouds from the Earth surface using brightness temperature differences observed at a spectral channel in the atmospheric window region. Because this principle using temperature contrast between the Earth's surface and clouds is very simple and useful, it has been used as an important cloud detection algorithm for many TIR sounders. After launching multi-channel sounders, new cloud detection algorithms were developed, such as the CO_2 slicing method (Smith and

Remote Sensing of Aerosols, Clouds, and Precipitation
https://doi.org/10.1016/B978-0-12-810437-8.00008-6

© 2018 Elsevier Inc. All rights reserved.

Platt, 1978; Menzel et al., 1983) and the split window method (Inoue, 1985). Even when observed with simple filter type sounders in its infancy, TIR data can be used to detect optically thin clouds in the upper troposphere. High resolution Infrared Radiation Sounder (HIRS) has been used to observe them for approximately three decades (Wylie et al., 2005).

Recently, polar orbiting satellites have been equipped with hyper-spectral TIR sounders. The first demonstration of a hyper-spectral TIR sounder was made using the Interferometric Monitor for Greenhouse gases (IMG; Shimoda and Ogawa, 2000) launched on the Advanced Earth Observing Satellite (ADEOS) in 1996. Unfortunately, ADEOS ceased operations eight months after its launch because of hardware problems related to the solar paddle, but IMG recorded very fine and well calibrated spectra. After that precursor, several hyper-spectral sounders have been launched into polar orbits, such as Atmospheric Infrared Sounder (AIRS; Aumann et al., 2003), Tropospheric Emission Spectrometer (TES; Beer, 2006), Infrared Atmospheric Sounding Interferometer (IASI; Siméoni et al., 1997), Thermal and Near-infrared Spectrometer for Observation—Fourier Transform Spectrometer (TANSO-FTS; Kuze et al., 2009), and Cross-track Infrared Sounder (CrIS; Bloom, 2001). They are still in operation.

This chapter presents a review of cloud screening algorithms and cloud property retrieval methods using TIR sounder data for the five sounders described above.

2 SENSOR CHARACTERISTICS

The specifications of the five sensors are presented in Table 1. AIRS is a grating sensor. The others are Fourier Transform Spectrometers (FTS). The field of view (FOV) of TES is square. The FOV of the others is circular. The footprint geometry of each sounder is presented in Fig. 1.

TABLE 1 Sensor Specifications

	AIRS	TES	IASI	TANSO-FTS	CrIS
Spectrometer type	Grating	FTS	FTS	FTS	FTS
Spectral range (cm^{-1})	649–1135 1217–1613 2169–2674	650–3050	645–1210 1210–2000 2000–2760	12900–13200 5800–6400 4800–5200 700–1800	650–1095 1210–1750 2155–2550
Spectral resolution (cm^{-1})	$\frac{v}{\Delta v}=1200$	0.1 (nadir looking)	0.5	0.2	0.625 1.25 2.5
Size of IFOV at nadir	13.5 km circular	0.53 × 5.3 km rectangular	12 km circular	10.5 km circular	14 km circular
Orbit (km)	705	705	819	666	824
Platform	Aqua	Aura	Metop	GOSAT	Suomi-NPP
Launch date	May 4, 2002	Jul. 15, 2004	Oct. 19, 2006	Jan. 23, 2009	Oct. 28, 2011

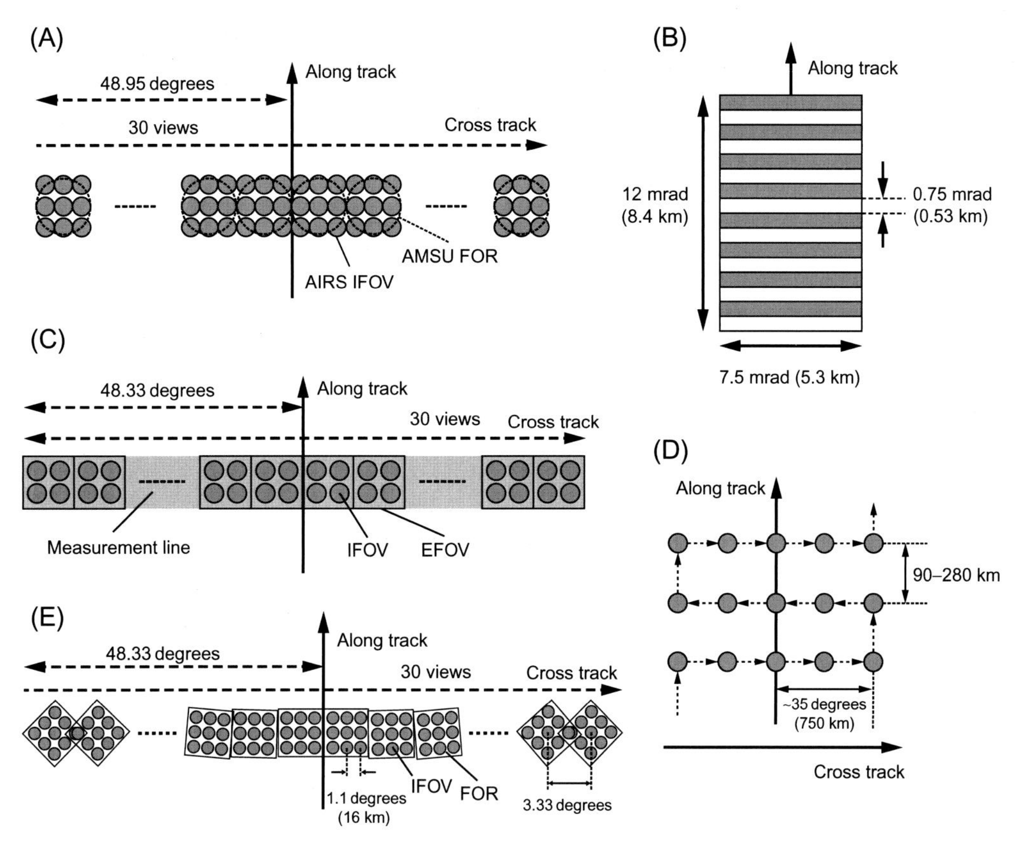

FIG. 1 Footprint geometry of the sounders, (A) AIRS, (B) TES, (C) IASI, (D) GOSAT, and (E) CrIS. AIRS and CrIS scan in the cross-track direction with a 3×3 array of footprints, which corresponds to a field of regard (FOR) of the microwave sensor. Similar to those footprints, IASI has a 2×2 IFOV array in an elementary field of view (EFOV). *Panel (A) modified from Aumann, H.H., Chahine, M.T., Gautier, C., Goldberg, M.D., Kalnay, E., McMillin, L.M., … Staelin, D.H., 2003. AIRS/AMSU/HSB on the Aqua mission: design, science objectives, data products, and processing systems. IEEE Trans. Geosci. Remote Sens. 41(2), 253–264, (B) modified from Beer, R., Glavich, T.A., Rider, D.M., 2001. Tropospheric emission spectrometer for the Earth Observing System's Aura satellite. Appl. Opt. 40(15), 2356–2367, (C) modified from Hilton, F., Armante, R., August, T., Barnet, C., Bouchard, A., Camy-Peyret, C., … Coheur, P.-F., 2012. Hyperspectral earth observation from IASI. Bull. Am. Meteorol. Soc. 93(3), 347, (D) modified from Kuze, A., Suto, H., Nakajima, M., Hamazaki, T., 2009. Thermal and near infrared sensor for carbon observation Fourier-transform spectrometer on the Greenhouse Gases Observing Satellite for greenhouse gases monitoring. Appl. Opt. 48(35), 6716–6733 and (E) modified from Han, Y., Revercomb, H., Cromp, M., Gu, D., Johnson, D., Mooney, D., … Borg, L., 2013. Suomi NPP CrIS measurements, sensor data record algorithm, calibration and validation activities, and record data quality. J. Geophys. Res. Atmos. 118(22).*

3 TYPICAL CLOUD DETECTION TECHNIQUES

Widely used cloud screening techniques using TIR data are presented in this section. Most techniques are used as fundamental techniques for operational processing of the hyper-spectral sounders.

3.1 Basic Threshold Method

One basic method, the threshold method, is the simplest and most popular method for cloud detection. Its principle is based on the facts that most of the surface reflectance is lower than those of clouds in the VIS region, or the surface temperatures are higher than those of clouds in the TIR region both over the land and over the ocean. Accordingly, it is difficult to distinguish clouds from highly reflective surfaces such as snow/ice and sun-glint in the VIS region, and from very cold surfaces as in high latitudes in the TIR region. In order to raise the detection accuracy, it is important to prepare precise databases such as land cover type, minimum reflectivity at every location, infrared emissivity, and the surface temperature. Most simply, only one value of reflectance in VIS and radiance (or brightness temperature) in TIR is set as the threshold. The cloudiness is judged as just clear or cloudy. However, in many cases, two thresholds are set and are connected linearly to judge the cloudiness by the confidence level, taking a value of 0%–100%.

3.2 CO_2 Slicing Method

This technique utilizes the difference of absorption strengths between the channel pair near a 15-μm CO_2 absorption band to obtain cloud top pressure (CTP). The concept of this technique is represented by the following equation:

$$\frac{R_{\lambda_1} - R_{\lambda_1}^{clr}}{R_{\lambda_2} - R_{\lambda_2}^{clr}} = \frac{\alpha_1 \epsilon_{\lambda_1} \int_{p_s}^{p_c} t_{\lambda_1}(p) dB_{\lambda_1}}{\alpha_2 \epsilon_{\lambda_2} \int_{p_s}^{p_c} t_{\lambda_2}(p) dB_{\lambda_2}}. \quad (1)$$

Here, λ stands for wavelength, R signifies the observed radiance, R^{clr} represents the clear sky radiance, α denotes the cloud fraction in the IFOV, ϵ represents the cloud emissivity, t is transmittance, p is pressure, B is the Planck function, p_s stands for the surface pressure, and p_c is the CTP. Also, $\alpha\epsilon$ is called the effective cloud amount (ECA). If λ_1 and λ_2 is sufficiently close, it can be regarded as $\alpha_1 \epsilon_{\lambda_1} = \alpha_2 \epsilon_{\lambda_2}$. p_c can be determined as the level at which the difference of left and right-hand sides of the equation is minimum. Subsequently, the cloud top temperature (CTT) and cloud top height (CTH) can be converted from CTP. Once CTP is determined, ECA is obtainable using the atmospheric window channel from

$$\alpha\epsilon_\lambda = \frac{R_\lambda - R_\lambda^{clr}}{R_\lambda^{bcd} - R_\lambda^{clr}}, \quad (2)$$

where R_{bcd} is the radiance assuming a dense cloud with determined CTP. If we assume that clouds exist homogeneously in the IFOV, i.e. $\alpha = 1$, the effective optical thickness, τ is calculated from

$$\tau = -\cos\theta \ln(1 - \epsilon), \quad (3)$$

where θ is the zenith angle of the observation.

This technique was used for both cloud screening and retrieval to obtain CTP and ECA.

3.3 Split Window Method

The existence of optically thin clouds generates characteristic spectra in the atmospheric window region in thermal infrared because their emissivity is strongly dependent on the wavelength (Inoue, 1985). The brightness temperature difference between two channels at the window region differs with cloud optical thickness and effective radius as a function of the brightness temperature at a reference channel. The magnitudes of the differences differ according to the thermodynamic phase of cloud particles, i.e., ice, water, or mixed. An example of this relation was simulated using a multiple-scattering radiative transfer code, the Polarization System for Transfer of Atmospheric Radiation version 3 (Pstar3; Ota et al., 2010), which is presented in Fig. 2. From this relation, if one of the cloud temperature, cloud optical thickness, and the effective radius is known, the other parameters can be estimated with previously calculated data tables. This method has been used widely to detect optically thin clouds from TIR observations (e.g. Inoue, 1987).

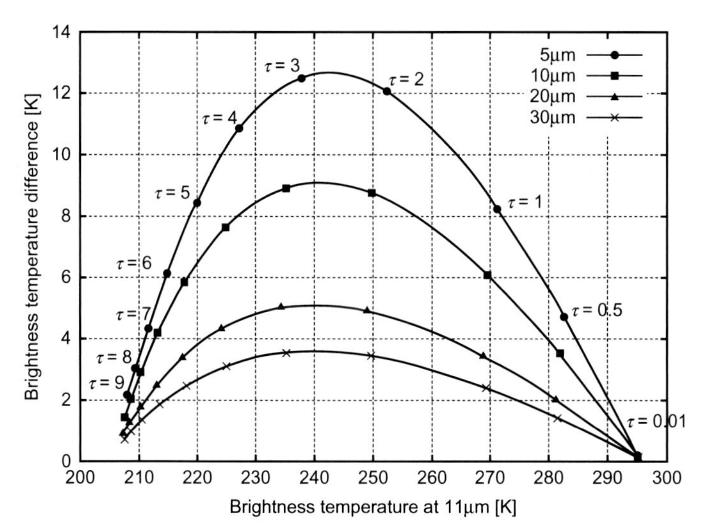

FIG. 2 Example of the brightness temperature difference between 11 and 12 μm as a function of the brightness temperature at 11 μm assuming the ice cloud with the mode radius of 5, 10, 20, and 30 μm and the optical thickness of 0.01–9.0 simulated by Pstar3. The tropical atmospheric profile is assumed. The cloud top was set at 14 km.

4 CLOUD DETECTION STRATEGY FOR THE SOUNDERS

This section presents a description of the strategies of cloud screening and cloud parameter retrieval for each sounder.

4.1 AIRS

The Atmospheric Infrared Sounder (AIRS) is one instrument onboard NASA's Aqua satellite. Data from the other instruments such as the Advanced Microwave Sounding Unit (AMSU) are also used jointly for processing. AIRS scans in a cross-track direction with 90 footprints per scan and a 48.95 degrees swath. An array of detector consists of the 3×3 IFOVs and its total space corresponds to one AMSU field of regard (FOR) as shown in Fig. 1.

In the AIRS processing, the cloud-cleared radiance which would be observed if the scene is clear is calculated to eliminate cloud contamination (Susskind et al., 2003; Aumann et al., 2007). AMSU data are used in this procedure because atmospheric profiles can be estimated using microwave sounders, even under cloudy conditions without precipitation. It is computed iteratively with improvement of parameters such as temperature profiles by combining AIRS data within the 3×3 array using those obtained from AMSU data.

4.1.1 Cloud Screening

This section introduces the cloud detection algorithm for AIRS Near-Real-Time (NRT) products (Goldberg et al., 2003). This algorithm was developed separately for use over ocean and over land.

Cloud detection over ocean consists of the following five tests. The FOV that has passed all the tests is identified as clear.

$$T_{b965}^{obs} > 270\text{K}$$

$$T_{b2616}^{obs} > \text{SST}$$

$$T_{b2616}^{obs} - T_{b2616}^{8\mu\text{m}} < T_1$$

$$T_{b2616}^{obs} - T_{b2616}^{11\mu m} < T_1$$

$$T_{b2616}^{8,11\mu m} > \text{SST}$$

In those expressions, T_{b965}^{obs} and T_{b2616}^{obs}, respectively, stand for the observed brightness temperature at 965.43 and 2616.095 cm^{-1}. Therein, the sea surface temperature (SST) is obtained from the National Centers for Environmental Prediction (NCEP). In addition, $T_{b2616}^{8\mu m}$ and $T_{b2616}^{11\mu m}$, respectively, denote the brightness temperatures at 2616 cm^{-1} converted from those at 8 and 11 μm channels. T_1 is the certain threshold value. The tests using the channel at 2616 cm^{-1} are conducted only for nighttime data because of the solar contamination.

Cloud detection over land uses AMSU data and the variability of AIRS data within the 3×3 array. This algorithm also consists of the following five tests.

$$T_{b2390}^{obs} - T_{b2390}^{AMSU} < T_2$$

$$\left|T_{b2558}^{obs} - T_{b937}^{obs}\right| < 10\text{K}$$

$$STD_{2390} < 3 \times \epsilon_{meas}$$

$$T_{b2445}^{obs} - T_{b2445}^{8\mu m} < T_1$$

$$T_{b2616}^{8,11\mu m} - \text{LST} < T_3$$

In those expressions, T_{b2390}^{obs}, T_{b2558}^{obs}, T_{b937}^{obs}, and T_{b2445}^{obs}, respectively, represent the observed brightness temperature at 2390, 2558.23, 937.81, and 2445.92 cm^{-1}. Therein, T_{b2390}^{AMSU} is the brightness temperature at 2390 cm^{-1} converted from AMSU channels 4, 5, and 6 (52.8, 53.594, and 54.4 GHz). T_2 and T_3 are thresholds. STD_{2390} represents the standard deviation of the brightness temperature at the 3×3 array using 2390 cm^{-1} channel. ϵ_{meas} denotes the measurement noise. $T_{b2445}^{8\mu m}$ is the converted brightness temperature at 2445.92 cm^{-1} from the 8 μm channel. The land surface skin temperature (LST) is also from NCEP.

4.1.2 Cloud Parameter Retrieval

The retrievable parameters processed from AIRS spectra are CTP and ECA. The 38 channels from 672.10 to 1133.91 cm^{-1} are used in this processing. The procedure assumes that the clouds in the FOR have two cloud tops and corresponding ECAs. Therefore, the theoretical radiance from kth IFOV ($k = 1, 2, \ldots 9$) at ith channel $R_{i,k}$ is calculated as

$$R_{i,k} = \left[1 - (\alpha\epsilon)_{1,k} - (\alpha\epsilon)_{2,k}\right] R_i^{clr} + (\alpha\epsilon)_{1,k} R_i^{bcd}(p_{c1}) + (\alpha\epsilon)_{2,k} R_i^{bcd}(p_{c2}), \quad (4)$$

where the definitions of variables are the same as those presented in Eq. (1). Clear sky radiances are calculated using the surface temperature, surface emissivity, and vertical profiles of temperature, water vapor, and ozone, which are derived from cloud-cleared radiances and AMSU observations. The unknown parameters in this equation, which are 18 ECAs and two CTPs, are estimated from iterative calculations as the values which minimize the difference between the calculated and the observed radiances. The current version, version 6, provides the CTP, CTH, CTT, and ECA for each FOV from the equation above (Olsen et al., 2013; Kahn et al., 2014).

The cloud thermodynamic phase was identified from brightness temperature differences in the atmospheric window region. The ice cloud test score is defined as the number that passes the following tests.

$$T_{b960} > 235$$

$$T_{b1231} - T_{b960} > 0$$

$$T_{b1231} - T_{b930} > 1.75$$

$$T_{b1227} - T_{b960} > -0.5$$

Here, T_b signifies brightness temperature. Subscript numbers are wavenumbers (cm^{-1}). The water cloud test score is obtained from the following tests.

$$T_{b1231} - T_{b960} < -1.0$$

$$T_{b1231} - T_{b930} < -0.6$$

If the value of the ice cloud test score minus the water cloud score is positive, the cloud is identified as ice cloud. The cloud is a water cloud if the value is negative. For values of 0, the phase is unknown.

Furthermore, the cloud is identified as ice cloud, the optical thickness, the effective diameter of cloud particles, and the temperature are estimated from the optimal estimation, as they are for the TES algorithm (Bowman et al., 2006) which is explained in the next section. The a priori values are set as 3.0 optical thickness, 30 μm diameter, and from the AIRS Level 2 product in temperature.

4.2 TES

The Tropospheric Emission Spectrometer (TES) is one of four sensors onboard NASA's Aura satellite. Although TES observes with both nadir and limb viewing, herein we describe the nadir viewing observation. As presented in Fig. 1, TES uses a rectangular footprint of approximately 5.3×8.4 km at the surface, with a 1×16 linear detector array.

4.2.1 *Cloud Screening*

Each scan of TES observations is categorized into the case of clear, broken, or inhomogeneous clouds, optically thick clouds, and optically thin clouds (Beer et al., 2002). The primal classification is based on the threshold technique using the brightness temperature converted from the averaged radiance between 867 and 900 cm^{-1}, except for water vapor absorption lines ($T_{b_{11}}$). If the $T_{b_{11}}$ values are close to the surface temperature within about 0.5 K and the values of pixels in the footprint are not so varied, the scene is identified as clear. The scene is categorized as broken or highly inhomogeneous cloud case if $T_{b_{11}}$ of pixels in the footprint have high variability. Scenes that have low variability of $T_{b_{11}}$ is categorized into two cases. A scene is identified as an optically thick cloud case if the $T_{b_{11}}$ values are significantly lower than the surface skin temperature for all pixels. If the $T_{b_{11}}$ values are slightly lower than the surface temperature, the scene is identified as an optically thin cloud case and the CO_2 slicing method is performed to estimate cloud top height.

4.2.2 *Cloud Parameter Retrieval*

Cloud effective optical thickness and cloud top pressure are retrieved simultaneously with surface temperature, surface emissivity, atmospheric temperature, water vapor, and ozone (Kulawik et al., 2006; Eldering et al., 2008) using the nonlinear optimal estimation algorithm (Bowman et al., 2006; Bowman et al., 2002; Worden et al., 2004). This algorithm obtains state vector $\boldsymbol{x}$ from minimizing the cost function:

$$C = ||\boldsymbol{y} - \boldsymbol{F}(\boldsymbol{x})||^2_{\boldsymbol{S}_\epsilon^{-1}} + ||\boldsymbol{x} - \boldsymbol{x}_a||^2_{\boldsymbol{S}_a^{-1}}, \quad (5)$$

where $\boldsymbol{y}$ signifies the measured radiance vector, $\boldsymbol{F}(\boldsymbol{x})$ denotes the calculated radiance vector, $\boldsymbol{S}_\epsilon$ represents the measurement noise covariance matrix, $\boldsymbol{x}_a$ is the a priori state vector, and $\boldsymbol{S}_a$ stands for the a priori error covariance matrix. The cloud characteristics are parameterized as

$$\tau(\nu, p) = \kappa(\nu) e^{-\beta(\ln p - \ln p_c)^2} \Delta s, \quad (6)$$

where τ represents effective optical thickness of the layer, ν is the frequency, p stands for pressure, κ signifies the effective extinction coefficient of the layer, β denotes the Gaussian with a parameter, p_c represents the cloud pressure, and s is the layer geometrical thickness. The a priori of cloud effective extinction coefficient κ is selected based on the brightness temperature difference (BTD) at 11 μm between the observed and the calculated clear sky radiance using the Global Modeling and Assimilation Office (GMAO) water vapor and temperature profiles as shown below.

$$BTD > 20K; \ \kappa = 4$$

$$10 < BTD \le 20K; \ \kappa = 1.3$$

$$6 < BTD \le 10K; \ \kappa = 0.8$$

$$0 < BTD \le 6K; \ \kappa = 0.015$$

The cloud pressure height is initially set as 500 hPa. The covariance matrix of the cloud effective extinction coefficient is set where the diagonal elements are 10 for the covariance of the log of the cloud extinction and the off diagonal elements are 0.9×diagonal. The covariance matrix of cloud pressure is 1 for the log scale of cloud pressure in the 1 sigma range of 183–1300 hPa. The radiative forward calculation model of this algorithm assumes the non-scattering cloud. Kulawik et al. (2006) investigates the error from this assumption by comparing the calculated radiance with and without scattering effects. They showed that the error from this assumption is negligibly small for high-level ice clouds, despite the relatively large error for low-level clouds.

4.3 IASI

The Infrared Atmospheric Sounding Interferometer (IASI) is one of eight instruments onboard the Meteorological Operational satellite programme (MetOp) satellites of the European Space Agency (ESA), which is a series of three polar-orbit satellites. MetOp-A and MetOp-B were launched respectively in 2006 and 2012. MetOp-C is scheduled for launch in 2018. Some other instruments such as the Advanced Very High Resolution Radiometer/3 (AVHRR/3) onboard the satellite is combined to detect clouds. IASI scans 30 EFOVs in a measurement line for cross-track direction with the maximum swath of 48.95 degrees. The EFOV consists of 2×2 IFOVs. The IASI footprint geometry is shown in Fig. 1.

4.3.1 *Cloud Screening*

For the recent version of IASI products, three cloud tests, i.e. the numerical weather prediction (NWP) test, AVHRR test, and artificial neural networks (ANN) test, are performed (EUMETSAT, 2014; August et al., 2012).

NWP TEST

The NWP cloud test is a type of threshold technique described in the previous section. Clear sky radiance is calculated using Radiative Transfer for TOVS (RTTOV) with NWP products from the European Centre for Medium-Range Weather Forecasts (ECMWF). The selected spectral channels for the test are at 832.5 and 900.25 cm^{-1}. The threshold value is set as 1 K.

AVHRR TEST

The second cloud test uses collocated AVHRR data. AVHRR, an imager also onboard the MetOp satellite, observes the horizontal resolution of approximately 1 km at nadir. AVHRR has six channels, designated respectively as 1, 2, 3a, 3b, 4, and 5, at the center wavelengths of 0.630, 0.865, 1.61, 3.74, 10.8, and 12.0 μm. This test consists of several individual tests using the threshold of VIS and NIR reflectance, the brightness temperature, and the BTD in the split window region. The scene is identified as cloudy if the integrated cloud fraction in the IASI footprint is higher than 2%.

ANN TEST

The third test is the artificial neural network (ANN) cloud detection. This test uses both IASI and AVHRR data as input parameters to train the neuron and returns the four levels of cloudiness: 1, clear; 2, partially cloudy (small cover); 3, partially cloudy (high cover); and 4, fully cloudy. For daytime observations, the 10 values of the mean plus/minus standard deviation of reflectance or brightness temperature at the channel 1, 2, 3a, 4, and 5 of AVHRR are used. For nighttime observations, the six values from the TIR channels of AVHRR are used. For IASI, the 18 channels between 680.250 and 2702.75 cm^{-1} are selected as inputs. These AVHRR and IASI data compose the input vector. The number of elements is 28 for daytime and 24 for nighttime.

The ANN model used in this method is a multilayer perceptron using backpropagation training. The numbers of hidden layers are two for daytime observations and three for nighttime. The activation function is the sigmoid function as $f(x)=1/(1+e^{-x})$. The model minimizes the error between the output and the training data with iterative calculations. These training data of cloudiness have been obtained from 24923 IASI IFOVs, which are categorized into the four levels using collocated AVHRR data.

4.3.2 Cloud Parameter Retrieval

If cloudy scenes are identified by the cloud tests, the two techniques, the CO_2 slicing method and the χ^2 method (Stubenrauch et al., 1999), are performed to estimate CTP, CTT, and ECA, in IFOV (EUMETSAT, 2014; August et al., 2012). The atmospheric information required for the calculations comes from NWP.

One channel of the channel pair used in the slicing calculation is fixed at 796.75 cm^{-1} as the reference channel. The other channel is selected from 41 channels between 707.5 and 756 cm^{-1}. Therefore, the slicing calculations are performed 41 times in maximum. The CTP is determined from the weighted mean of the obtained CTPs from these calculations. Once the CTP is determined, the ECA is estimated using a window channel at 900.50 cm^{-1} from the relation of Eq. (2).

The χ^2 method also uses the CO_2 absorption band near 15 μm. The CTP and ECA can be estimated as the pressure level and cloud amount which minimize differences between the simulated clear sky radiance and the measured one defined as

$$\chi^2(p, \alpha\epsilon) = \sum_{i=1}^{N} \left[\left\{ R_{\lambda_i}^{clr} - R_{\lambda_i}^{bcd}(p) \right\} \alpha\epsilon - \left\{ R_{\lambda_i}^{clr} - R_{\lambda_i}^{obs} \right\} \right]^2, \quad (7)$$

where N is the number of channels and the others are as shown in Eq. (1). An advantage of this method is that it always has solutions, even in the case of no solutions from the CO_2 slicing method.

Cloud thermodynamic phase is retrieved based on the method described by Strabala et al. (1994). This method determines the phase using combinations of the brightness temperature at 8 μm (T_{b_8}), 11 μm ($T_{b_{11}}$), and 12 μm ($T_{b_{12}}$) as shown below.

ice: $t_1 \leq (T_{b_8} - T_{b_{11}}) - (T_{b_{11}} - T_{b_{12}})$ or $T_{b_{11}} \leq 233$K
mix: $t_2 \leq (T_{b_8} - T_{b_{11}}) - (T_{b_{11}} - T_{b_{12}}) \leq t_1$

In other cases, the cloud phase is classified as liquid water. The temperature thresholds t_1 and t_2 are prepared as functions of latitude and month.

4.4 TANSO-FTS

The instrument onboard GOSAT, Thermal and Near-infrared Sensor for Observation (TANSO), is comprised of the main sensor, Fourier Transform Spectrometer (FTS), and Cloud and Aerosol Imager (CAI). TANSO-FTS has four spectral bands: one in near infrared for estimating the surface pressure, two in short-wavelength infrared for the measurements of columnar gas concentrations, and the last one in TIR of 5.56–14.3 μm for upper air gas concentration retrievals. CAI observes images at four spectral regions from ultraviolet to near-infrared. The TANSO-FTS scanning system uses a two-axis pointing mechanism. The scanning image of TANSO-FTS is presented in Fig. 1.

4.4.1 Cloud Screening

During daytime, cloud-contaminated scenes of FTS are distinguished from CAI images using the Cloud and Aerosol Unbiased Detection Intellectual Algorithm (CLOUDIA; Ishida and Nakajima, 2009) system. However, cloud screening must be performed using TIR data during nighttime because CAI is operated only in the daytime.

TIR THRESHOLD TECHNIQUE

One cloud screening method that uses TIR data is based on the threshold method applied to the brightness temperature level in the atmospheric window region (Imasu et al., 2010). As described in an earlier section, emissivity in this region is very important for the threshold method, as is the surface temperature, to ascertain an appropriate threshold value. That is particularly true over land that has spectral dependency of emissivity that varies widely among surface materials. As the first step of the procedure, the land cover type in each IFOV is determined, referring to the MODIS product (MCD12Q1). Then the type is related to spectral emissivity data of the ASTER spectral library (Baldridge et al., 2009) based on the assignment method proposed by Wilber et al. (1999). Although the vegetation activity and snow/ice coverage are not considered in the original assignment method, the normalized difference vegetation index (NDVI) retrieved from the CAI data and the snow/ice coverage derived from MODIS product (MOD10_L2) are introduced as assigned parameters for GOSAT data analyses. The surface temperature, which is the most important parameter, is estimated from the surface-level temperature over the land and from SST over the ocean. Then, the brightness temperature for clear sky, T_b^{clear}, is calculated theoretically using a radiative transfer code with the surface emissivity and temperature evaluated in advance. Once T_b^{clear} is evaluated, the threshold level for cloud screening can be set to be optimally lower than T_b^{clear} considering the measurement error level.

CO_2 SLICING TECHNIQUE

An improved CO_2 slicing technique (Someya et al., 2016) is planned to be applied to the operational data processing of the successor to GOSAT. Improvements from the original technique are channel reconstruction and their optimization. Channel reconstructions are performed to decrease the effects of measurement errors. First, the original channels are sorted based on their weighting function peak height. The pseudo-channels are redefined as sets of original channels of which the weighting function peaks are in the same height range consisting of 0.5 km height bins. The pairs of the pseudo-channels are optimized for various typical temperature profiles corresponding to patterns of latitude and atmospheric temperature at 500 hPa, which are used as indicators. The optimizations are done based on the spectral radiances simulated for cloudy conditions using Pstar3. For GOSAT data analysis, the channel pairs are determined by referring to the latitude and the temperature at 500 hPa in each IFOV of the sensor evaluated through the temporal-spatial interpolation of meteorological gridded data. This technique is also applied to the cloud screening for the sun-glint observation because CAI distinguishes clouds only slightly from high reflectance surfaces over the ocean.

4.5 CrIS

The Cross-Track Infrared Sounder (CrIS) is one component of the Cross Track Infrared and Microwave Sounder Suite (CrIMSS) onboard the Suomi National Polar-orbiting Partnership (Suomi NPP) spacecraft with other instruments including a microwave sounder. Similar to AIRS and IASI, the 3×3 detector array of IFOV, which is defined as the field of regard (FOR), sequentially scans forward to the cross track direction (Han et al., 2013). A measurement line comprises 30 FORs.

Cloud retrievals are not performed in the operational processing of CrIS data. Therefore, only cloud classifications for cloud clearing (JPSS Configuration Management Office, 2013) are described herein.

4.5.1 Cloudy Scene Classification

Just as it is with AIRS, the cloud cleared radiance is generated from CrIS data to avoid cloud contamination in cloudy scenes. The number of cloud formations is required for cloud clearing (Chahine, 1977). The meanings of formation include the expansion, the number of layers, and their heights. In CrIS data processing, "clear" is defined as the case in which the cloud fraction in the FOR is less than 50%. "Cloudy" is defined the case in which the cloud fraction is 50%–100%. "Cloud-free" means that the cloud fraction is 0%. These scene identifications for CrIS measurements also include several procedures (JPSS Configuration Management Office, 2013) as described below.

The first is based on the threshold technique. Atmospheric parameters and surface skin temperature are obtained from microwave observations. Because the surface reflectivity or emissivity is not obtainable from microwave observation, they are referred from a climatological database over land. The scene is identified as clear if the root mean square (RMS) difference of radiance in the FOVs within a FOR is smaller than the threshold.

The second procedure is the estimation of the number of cloud formations in the FOR for the cloud clearing. Principal Component Analysis (PCA) is performed on the dataset composed of the radiance spectra between 709.5 and 746 cm^{-1} of the nine FOVs within a FOR. The principal components (PCs) with large eigenvalues are associated with signatures of each cloud formation. Those with small eigenvalues are from measurement noise. Therefore, the number of the cloud formations is obtainable from the number of the significant principal components. The number of PCs is estimated from the two tests. The first test calculates the Residual Standard Deviation (RSD), defined as

$$RSD = \left[\frac{1}{N_{channel} \times (N_{FOV} - n)} \sum_{k=n+1}^{N_{FOV}} \lambda_k\right]^{1/2}, \quad (8)$$

where $N_{channel}$ is the number of spectral channels, N_{FOV} is the number of IFOV, and λ is the eigenvalue. The number of PCs is determined as the smallest n for which RSD is smaller than the measurement noise. The second test is the normalized χ^2 test. χ^2 is defined as

$$\chi^2 = \sum_{i=1}^{N_{channel}} \sum_{k=1}^{N_{FOV}} \frac{\left[R_{i,k} - R'_{i,k}(n)\right]^2}{\sigma_{i,k}^2}, \quad (9)$$

where $R_{i,k}$ represents the measured radiance, $R'_{i,k}(n)$ is the reconstructed radiance using the first n components, and $\sigma_{i,k}$ is the noise standard deviation. The number of PCs is determined as the smallest n for which χ^2 is smaller than $(N_{channel} - n)(N_{FOV} - n)$. The larger one is used if the determined numbers of PCs from the two tests differ. Based on the number of cloud formations, the FOVs within the FOR are grouped as clusters. They are labeled as either clear, partly cloudy, overcast, or no retrieval from the several tests using the number of cloud formations, the surface type, and thermal contrast. From these results, the cloud cleared radiance is retrieved with the combined use of the microwave observations.

5 CONCLUSION

This chapter introduced the cloud screening and cloud property retrieval methods which are used operationally for five hyperspectral sounders. These methods are based on various techniques that leverage the benefits of hyperspectral data. However, the performance of each technique has not been well investigated through comparison with more reliable cloud data such as those obtained by active sensors, i.-e., space lidar, because outputs from the cloud screening procedure are not products in themselves for the satellite project.

At most major space agencies of the world, higher-performance sounders are being developed to observe atmospheric environments more precisely. Some are scheduled to be launched soon. To use these sounder data at their maximum performance, cloud screening and clearing methods are extremely important. Both are expected to be optimized through intercomparisons of results obtained using various techniques.

References

August, T., Klaes, D., Schlussel, P., Hultberg, T., Crapeau, M., Arriaga, A., Calbet, X., 2012. IASI on Metop-A: operational level 2 retrievals after five years in orbit. J. Quant. Spectrosc. Radiat. Transf. 113 (11), 1340–1371.

Aumann, H.H., Chahine, M.T., Gautier, C., Goldberg, M.D., Kalnay, E., McMillin, L.M., Staelin, D.H., 2003. AIRS/AMSU/HSB on the Aqua mission: design, science objectives, data products, and processing systems. IEEE Trans. Geosci. Remote Sens. 41 (2), 253–264.

Aumann, H.H., Chahine, M.T., Barnet, C., Manning, E., Rosenkranz, P., Strow, L., Susskind, J., 2007. AIRS-Team Retrieval for Core Products and Geophysical Parameters Level 2, Version 4.0, JPL D-17006. Jet Propulsion Laboratory.

Baldridge, A., Hook, S., Grove, C., Rivera, G., 2009. The ASTER spectral library version 2.0. Remote Sens. Environ. 113 (4), 711–715.

Beer, R., 2006. TES on the Aura mission: scientific objectives, measurements, and analysis overview. IEEE Trans. Geosci. Remote Sens. 44 (5), 1102–1105.

Beer, R., Bowman, K.W., Brown, P.D., Clough, S.A., Eldering, A., Goldman, A., Jacob, D.J., Lampal, M., Logan, J.A., Luo, M., Murcray, F.J., Osterman, G.B., Rider, D.M., Rinsland, C.P., Rodgers, C.D., Sander, S.P., Shephard, M., Sund, S., Ustinov, E., Worden, H.M., Worden, J., 2002. TES Level 2 Algorithm theoretical basis document, version 1.15, JPL D-16474. Jet Propulsion Laboratory.

Bloom, H.J., 2001. The Cross-track Infrared Sounder (CrIS): a sensor for operational meteorological remote sensing. In: Paper presented at the Geoscience and Remote Sensing Symposium, 2001. IGARSS'01, IEEE 2001 International.

Bowman, K.W., Steck, T., Worden, H.M., Worden, J., Clough, S., Rodgers, C., 2002. Capturing time and vertical variability of tropospheric ozone: a study using TES nadir retrievals. J. Geophys. Res. Atmos. 107 (D23).

Bowman, K.W., Rodgers, C.D., Kulawik, S.S., Worden, J., Sarkissian, E., Osterman, G., Beer, R., 2006. Tropospheric emission spectrometer: retrieval method and error analysis. IEEE Trans. Geosci. Remote Sens. 44 (5), 1297–1307.

Chahine, M.T., 1977. Remote sounding of cloudy atmospheres. II. Multiple cloud formations. J. Atmos. Sci. 34 (5), 744–757.

Eldering, A., Kulawik, S.S., Worden, J., Bowman, K., Osterman, G., 2008. Implementation of cloud retrievals for TES atmospheric retrievals: 2. Characterization of cloud top pressure and effective optical depth retrievals. J. Geophys. Res. Atmos. 113 (D16).

EUMETSAT, 2014. IASI Level 2: Product Generation Specification, EPS.SYS.SPE.990013, v8A. EUMETSAT.

Fritz, S., Winston, J.S., 1962. Synoptic use of radiation measurements from satellite TIROS II. Mon. Weather Rev. 90 (1), 1–9.

Goldberg, M.D., Qu, Y., McMillin, L.M., Wolf, W., Zhou, L., Divakarla, M., 2003. AIRS near-real-time products and algorithms in support of operational numerical weather prediction. IEEE Trans. Geosci. Remote Sens. 41 (2), 379–389.

Han, Y., Revercomb, H., Cromp, M., Gu, D., Johnson, D., Mooney, D., Borg, L., 2013. Suomi NPP CrIS measurements, sensor data record algorithm, calibration and validation activities, and record data quality. J. Geophys. Res. Atmos.. 118(22).

Imasu, R., Hayashi, Y., Inagoya, A., Saitoh, N., Shiomi, K., 2010. Retrieval of minor constituents from thermal infrared spectra observed by GOSAT TANSO-FTS sensor. In: Proceedings of SPIE, p. 7857.

Inoue, T., 1985. On the temperature and effective emissivity determination of semi-transparent cirrus clouds by bi-spectral measurements in the 10 MU-M window region. J. Meteorol. Soc. Jpn. 63 (1), 88–99.

Inoue, T., 1987. A cloud type classification with NOAA 7 split—window measurements. J. Geophys. Res. Atmos. 92 (D4), 3991–4000.

Ishida, H., Nakajima, T.Y., 2009. Development of an unbiased cloud detection algorithm for a spaceborne multispectral imager. J. Geophys. Res. Atmos. 114.

JPSS Configuration Management Office. 2013. Joint Polar Satellite System (JPSS) Algorithm Theoretical Basis Document for the Cross Track Infrared Sounder (CrIS) Volume II, Environmental Data Records (EDR), Revision B, JPSS document code 474: 474-00056, JPSS office, 2013.

Kahn, B.H., Irion, F.W., Dang, V.T., Manning, E.M., Nasiri, S.L., Naud, C.M., Worden, J.R., 2014. The Atmospheric Infrared Sounder version 6 cloud products. Atmos. Chem. Phys. 14 (1), 399–426.

Kulawik, S.S., Worden, J., Eldering, A., Bowman, K., Gunson, M., Osterman, G.B., Beer, R., 2006. Implementation of cloud retrievals for Tropospheric Emission Spectrometer (TES) atmospheric retrievals: part 1. Description and characterization of errors on trace gas retrievals. J. Geophys. Res. Atmos. 111 (D24).

Kuze, A., Suto, H., Nakajima, M., Hamazaki, T., 2009. Thermal and near infrared sensor for carbon observation Fourier-transform spectrometer on the Greenhouse Gases Observing Satellite for greenhouse gases monitoring. Appl. Opt. 48 (35), 6716–6733.

Menzel, W.P., Smith, W.L., Stewart, T.R., 1983. Improved cloud motion wind vector and altitude assignment using VAS. J. Clim. Appl. Meteorol. 22 (3), 377–384.

Olsen, E.T., Fetzer, E., Hulley, G., Lambrigtsen, B., Manning, E., Blaisdell, J., Iredell, L., Susskind, J., Warner, J., Wei, Z., Blackwell, W., 2013. AIRS/AMSU/HSB version 6 changes from version 5, Version 1.0. Jet Propulsion Laboratory.

Ota, Y., Higurashi, A., Nakajima, T., Yokota, T., 2010. Matrix formulations of radiative transfer including the

polarization effect in a coupled atmosphere-ocean system. J. Quant. Spectrosc. Radiat. Transf. 111 (6), 878–894.
Rossow, W.B., Garder, L.C., 1993. Cloud detection using satellite measurements of infrared and visible radiances for ISCCP. J. Clim. 6 (12), 2341–2369.
Shimoda, H., Ogawa, T., 2000. Interferometric monitor for greenhouse gases (IMG). Adv. Space Res. 25 (5), 937–946.
Siméoni, D., Singer, C., Chalon, G., 1997. Infrared atmospheric sounding interferometer. Acta Astronaut. 40 (2), 113–118.
Smith, W.L., Platt, C.M.R., 1978. Comparison of satellite-deduced cloud heights with indications from radiosonde and ground-based laser measurements. J. Appl. Meteorol. 17 (12), 1796–1802.
Someya, Y., Imasu, R., Saitoh, N., Ota, Y., Shiomi, K., 2016. A development of cloud top height retrieval using thermal infrared spectra observed with GOSAT and comparison with CALIPSO data. Atmos. Meas. Tech. 9 (5), 1981–1992.
Strabala, K.I., Ackerman, S.A., Menzel, W.P., 1994. Cloud Properties inferred from 8–12-um Data. J. Appl. Meteorol. 33 (2), 212–229.
Stubenrauch, C.J., Rossow, W.B., Scott, N.A., Chedin, A., 1999. Clouds as seen by satellite sounders (3I) and imagers (ISCCP). Part III: spatial heterogeneity and radiative effects. J. Clim. 12 (12), 3419–3442.
Susskind, J., Barnet, C.D., Blaisdell, J.M., 2003. Retrieval of atmospheric and surface parameters from AIRS/AMSU/HSB data in the presence of clouds. IEEE Trans. Geosci. Remote Sens. 41 (2), 390–409.
Wilber, A.C., Kratz, D.P., Gupta, S.K., 1999. Surface emissivity maps for use in satellite retrievals of longwave radiation, NASA/TP-1999-209362.
Worden, J., Kulawik, S.S., Shephard, M.W., Clough, S.A., Worden, H., Bowman, K., Goldman, A., 2004. Predicted errors of tropospheric emission spectrometer nadir retrievals from spectral window selection. J. Geophys. Res. Atmos. 109 (D9).
Wylie, D., Jackson, D.L., Menzel, W.P., Bates, J.J., 2005. Trends in global cloud cover in two decades of HIRS observations. J. Clim. 18 (15), 3021–3031.

Further Reading

Beer, R., Glavich, T.A., Rider, D.M., 2001. Tropospheric emission spectrometer for the Earth Observing System's Aura satellite. Appl. Opt. 40 (15), 2356–2367.
Hilton, F., Armante, R., August, T., Barnet, C., Bouchard, A., Camy-Peyret, C., Coheur, P.-F., 2012. Hyperspectral earth observation from IASI. Bull. Am. Meteorol. Soc. 93 (3), 347.

CHAPTER

9

Surface Remote Sensing of Liquid Water Cloud Properties

*Christine Knist**, *Herman Russchenberg*[†]

*Deutsche Wetterdienst, Offenbach am Main, Germany [†]Delft University of Technology, Delft, The Netherlands

1 INTRODUCTION

Remote sensing of low-level liquid water clouds requires high instrumental sensitivity and vertical resolution since these clouds are composed of highly concentrated and inhomogeneously distributed small liquid water droplets. The key instrument measuring vertical profiles of nonprecipitating boundary layer clouds is the millimeter-wavelength Doppler radar because of its ability to penetrate clouds.

Millimeter-wavelength cloud radars (MMCRs) have been used in atmospheric cloud research over the last 30 years and their observations are indispensable for studying the microphysical properties of nonprecipitating boundary layer clouds (Kollias et al., 2007a).

Because of their short wavelengths, they are sensitive to small cloud droplets and ice crystals so that they detect all types of nonprecipitating clouds well before large hydrometeors are formed. Cloud radars are mainly operating at 35 GHz (8.7 mm wavelength, K_a-band) or 95 GHz (3.16 mm wavelength, W-band) because the atmospheric attenuation related to water vapor and oxygen reaches a local minimum at these frequencies. The capability of radars to detect the clouds' hydrometeors depends, in part, on the concentration and size of the particles present in the radar sample volume, the liquid or ice water content, and the radar wavelength.

Collocated observations of laser devices, such as the LIDAR or ceilometer, are more reliable for the estimation of the cloud base height. These instruments operate at wavelengths ranging from ultraviolet (UV) to the near infrared (NIR), which means that the sizes of particles they are sensitive to are different compared to millimeter-wavelength radars.

In addition to cloud radars and LIDARs, microwave radiometer (MWR) observations are widely used in atmospheric research since they provide vertically integrated amounts of cloud liquid water (*LWP*) and water vapor (*IWV*) in the atmosphere. The standard MWR is a dual-channel microwave receiver that measures the emission of microwave radiation from the atmosphere at two frequencies: 31.4 GHz and 23.8 GHz. In Section 2 we discuss

Remote Sensing of Aerosols, Clouds, and Precipitation
https://doi.org/10.1016/B978-0-12-810437-8.00009-8

© 2018 Elsevier Inc. All rights reserved.

methodologies and the underlying assumptions, followed by a case study in which we use ground-based radiation measurements as a means for validation (Section 3).

2 PRINCIPLE METHOD AND BASIC ASSUMPTIONS

The liquid water content (*LWC*), the droplet concentration, the effective radius, the visible optical extinction, and optical depth can be obtained from coincident cloud radar, LIDAR, and microwave radiometer observations. The used instruments and observables are presented in Table 1. To invert the observations in order to obtain the liquid water cloud properties of interest, certain fundamental assumptions about the cloud droplet size distribution (DSD) and the in-cloud structure are necessary.

2.1 Droplet Size Distribution

Underlying all the important microphysical and radiative properties is the droplet size distribution of liquid water clouds. In Miles et al. (2000), a large data set of in situ-probed droplet-size distributions was analyzed. Many of these recorded distributions were adequately modeled by a single-mode gamma or lognormal size distribution with long tails at larger particle sizes. In this study, the gamma distribution described by Walko et al. (1995) is used to parameterize the drop size distribution:

TABLE 1 Instruments and the Used Measured Quantities

Instruments	Measured Quantities
Cloud radar	Cloud top height (h_{ct}), profiles of radar reflectivity (Z)
LIDAR or ceilometer	Cloud base height (h_{cb})
Microwave radiometer (MWR)	Liquid water path (*LWP*)

$$n(r) = \frac{N}{R_n \Gamma(\nu)} \left(\frac{r}{R_n}\right)^{\nu-1} e^{\left(-\frac{r}{R_n}\right)}, \quad r \geq 0, \qquad (1)$$

where the droplet radius r ranges form zero to infinity. $\Gamma(\nu)$ is the gamma function and serves as a normalization constant to make the integral of $n(r)$ equal to the total number concentration N. The shape parameter ν controls the relative amount of smaller vs. larger droplets in the distribution. If ν is any real number greater than or equal to one, the gamma distribution is monomodal. For $\nu = 1$, Eq. (1) reduces to the exponential distribution in which the modal radius (the radius where $n(r)$ has a maximum value) is zero.

The proportionality between the moments of DSD forms the basis of many methods of retrieving the microphysical and optical cloud properties from the surface remote-sensing observations; this is widely accepted in the meteorology community (White et al., 1991; Frisch et al., 1995; Sassen et al., 1999; Kato et al., 2001; Matrosov et al., 2004). We can write

$$\langle r^a \rangle = k_{ab_\nu} \langle r^b \rangle^{\frac{a}{b}}, \quad 1 \leq a \leq b, \qquad (2)$$

where the correlation is embodied in the constant coefficient k_{ab_ν}

$$k_{ab_\nu} = \frac{\langle r^a \rangle}{\langle r^b \rangle^{\frac{a}{b}}} = \frac{r_m^a k_{a_\nu}}{\left(r_m^b k_{b_\nu}\right)^{\frac{a}{b}}} = \frac{\prod_{i=1}^{a-1}\left(1+\frac{i}{\nu}\right)}{\left(\prod_{i=1}^{b-1}\left(1+\frac{i}{\nu}\right)\right)^{\frac{a}{b}}} = \frac{\prod_{i=1}^{a-1}\left(1+i\epsilon^2\right)}{\left(\prod_{i=1}^{b-1}\left(1+i\epsilon^2\right)\right)^{\frac{a}{b}}}, \qquad (3)$$

which is a function of the DSD shape parameter or relative dispersion.

The proportionality factor allows one to infer the lower moments of the DSD, such as the liquid water content $\left(\propto N\langle r^3 \rangle\right)$, the optical extinction $\left(\propto N\langle r^2 \rangle\right)$, and the effective radius, from the observed radar reflectivity $\left(\propto N\langle r^6 \rangle\right)$. Commonly used is the expression of liquid water content as a function of the radar reflectivity.

By calculating the different moments of the gamma DSD, one can derive expressions for the radar reflectivity and the cloud LWC:

$$Z = 2^6 N\langle r^6 \rangle = 2^6 N r_m^6 \prod_{i=1}^{5}\left(1+\frac{i}{\nu}\right), \quad (4)$$

$$LWC = \frac{4}{3}\pi\rho_w N\langle r^3 \rangle = \frac{4}{3}\pi\rho_w N r_m^3 \prod_{i=1}^{2}\left(1+\frac{i}{\nu}\right). \quad (5)$$

Using Eq. (2) to substitute the third by the sixth moment, it follows that:

$$\begin{aligned} LWC &= \frac{\pi\rho_w}{6} k_{36_\nu} N^{\frac{1}{2}} Z^{\frac{1}{2}} \\ &= \frac{\pi\rho_w}{6}\left(\frac{\nu(\nu+1)(\nu+2)}{(\nu+3)(\nu+4)(\nu+5)}\right)^{\frac{1}{2}} N^{\frac{1}{2}} Z^{\frac{1}{2}}. \end{aligned} \quad (6)$$

This formulation is essentially equivalent to the $LWC - Z$ relations considered by White et al. (1991), Frisch et al. (1995) and Mace and Sassen (2000).

2.2 The Vertical Cloud Models

This subsection introduces three different vertical cloud models to infer the liquid water cloud microphysical and optical properties from the surface observations. These vertical cloud models are based on assumptions that are commonly used in the literature and on vertical characteristics that appear to be typical of warm boundary layer clouds. Here we review the different cloud models and assess the impact of the assumed in-cloud vertical structure on the microphysical and optical retrieval computations.

The vertical structure of the cloud liquid water content, the effective radius, and the optical extinction depend on the characteristics of the DSD parameters N, ν, and r_m. These properties are mainly affected by processes such as cloud droplet nucleation activity at the cloud base; adiabatic growth of droplets above the cloud base; entrainment-mixing processes with environmental drier air at the cloud top; and collision–coalescence (Pawlowska et al., 2006). Due to the various meteorological processes involved, it is difficult to estimate the needed characteristics of the DSD with height from the relatively few ground-based remote-sensing observations. They only provide information about the cloud location, the height-integrated liquid water content (LWP), and the profile of the cloud radar reflectivity. Hence, it is necessary to predefine the in-cloud vertical structure of the DSD parameters so that any moment of the DSD can be computed.

2.2.1 The Vertically Uniform (VU) Cloud Model

The simplest parameterization is a vertically uniform microphysical structure of the cloud layer. This cloud model has been used in atmospheric cloud research over the last 36 years (Stephens, 1978) and leads to the following relationship between the cloud optical depth τ and the vertically integrated liquid water path LWP:

$$\tau = \frac{3LWP}{2\rho_w r_e}. \quad (7)$$

This formulation is often used in satellite and surface retrieval schemes to infer the cloud properties from remotely sensed radiative fields. Surface-based retrieval methods use narrowband and broadband diffuse measurements for overcast clouds to derive the cloud optical depth, and the LWP from the microwave radiometer (MWR) in order to obtain the effective radius (e.g., Min and Harrison, 1996; Dong et al., 1998). Satellite retrievals of optical depth and effective radius are derived from measurements of spectral radiances using look-up tables (LUTs), in which the radiative transfer is parameterized by a vertically uniform cloud model (e.g., Han et al., 1995; Nakajima and King, 1990).

The cloud properties of interest computed on the basis of this model will be hereafter referred to as the vertically uniform (VU) cloud model. To parameterize the extent to which mixing processes may affect the vertical distribution of the cloud properties compared to the VU cloud model, two more vertical cloud models are introduced.

2.2.2 The Scaled Adiabatic Stratified (SAS) Cloud Model

Entrainment of subsaturated air into the cloud causes the departure of the cloud microphysical properties from those expected of adiabatically ascending parcels. This is frequently observed in the vertical profiles of the *LWC* of continental and maritime clouds (Chin et al., 2000; Kim et al., 2005), which deviate from adiabatic profiles. The subadiabatic character of the cloud, or the degree of adiabaticity (e.g., Boers et al., 2000; Kim et al., 2008), which relates nonadiabatic to purely adiabatic conditions, is often used to characterize the entrainment-mixing process.

The total degree of adiabaticity of a cloud layer can be quantified from the surface observation of MWR *LWP* with respect to the *LWP* that is expected for an adiabatic layer with equivalent cloud base and top height (Slingo, 1989):

$$A_F = \frac{LWP_{obs}}{LWP_{\mathrm{ad}}}, \tag{8}$$

where the adiabatic *LWP* (subscript ad) is defined within the model of an adiabatic cloud for which the *LWC* increases linearly with height from cloud base to cloud top (Brenguier, 1991). It can be determined from the vertical profiles of thermodynamic structures (e.g., radio soundings ascents) and the cloud geometrical thickness (Albrecht et al., 1990). If the degree of adiabaticity A_F is larger than 0.9, the cloud layer is nearly adiabatic. It is diluted when $0.5 \geq A_F \leq 0.9$ and strongly diluted when $A_F < 0.5$. In this context, the total degree of adiabaticity can be used as a proxy for the impact of mixing processes on the vertical distribution of the cloud properties. For a cloud layer in which probable impacts of mixing are characterized by a constant reduction of the adiabatic *LWC* with the cloud height, the rate at which the liquid water content increases can be parameterized by (e.g., Boers et al., 2000; Wood, 2005):

$$LWC(h) = A_F LWC_{\mathrm{ad}}(h), \tag{9}$$

where the total degree of adiabaticity A_F is used to scale the vertical gradient of the adiabatic *LWC*. It is assumed that mixing affects the adiabatic *LWC* uniformly through the entire cloud layer by the possible reduction in the *LWP* denoted by A_F. The *LWC* parameterization originates from the adiabatic *LWC* profile, where the droplet concentration N and the DSD shape parameter ν are assumed as constants with the cloud height. This imposes a vertical gradient in the mean particle size r_m, because the *LWC* is proportional to r_m^3 (Eq. 5). Thus, the proposed cloud model allows one to obtain a vertical structure in the *LWC* and the cloud droplet radius, while it additionally contains the property that the cloud properties may deviate from adiabatic conditions through A_F.

The cloud properties of interest computed on the basis of this cloud model will be hereafter referred to as the scaled-adiabatic stratified (SAS) cloud model. This notation refers to the parameterization of the *LWC* (Eq. 9), which is estimated from the MWR *LWP* -scaled adiabatic assumption. On the basis of the SAS cloud model or similar *LWC* parameterizations, several retrieval algorithms have been developed (e.g., Boers and Mitchell, 1994; Boers et al., 2000; Boers and Rotstayn, 2001; Kim et al., 2003; Wood, 2005; Turner, 2005; Turner et al., 2007b; Illingworth et al., 2007; Brandau et al., 2010) to obtain the cloud microphysics from surface and satellite remote-sensing observations.

2.2.3 The Radar Reflectivity-Homogeneous Mixing (HM) Cloud Model

Mixing and entrainment are crucial processes of influencing the DSD parameters (N, ν, and r_m) with height. The vertical profiles of liquid water content often become quasi-adiabatic (close to being adiabatic) in the lower part and subadiabatic in the upper part of the cloud column due to dry air mixed into the cloud near the top (Brenguier et al., 2000a; Pawlowska et al., 2006, and others). Such details are not included in the SAS cloud model, in which the cloud properties (LWC and r_m) are uniformly impacted through the entire cloud layer.

However, as aforementioned, the characterization of the mixing process for a specific atmospheric event is difficult to determine on the basis of the few observations. To describe the impact of mixing on the vertical distribution of the DSD parameters, commonly two extreme mixing events are assumed, namely homogeneous and inhomogeneous mixing (Boers et al., 2006; Kim et al., 2008). These two mixing scenarios impact the DSD parameters differently. In the homogeneous mixing scenario, the mixing of dry air in the cloud volume uniformly reduces the droplet sizes through the whole droplet spectrum by evaporation. The mixing of the entrained air is completed before the cloud droplets are totally evaporated so that the total number of cloud droplets does not change. In the opposite case of inhomogeneous mixing, it is assumed that a constant amount of cloud droplets of all sizes is completely evaporated so that the mean radius is preserved while the total number of droplets changes (see Baker et al., 1980; Kim et al., 2003, 2008; Boers et al., 2006; Burnet and Brenguier, 2007; Chosson et al., 2007).

According to the model presented in Boers et al. (2006), the sensitivity of entrainment-mixing to the DSD parameters as a function of the cloud height is indirectly observed in the vertical profile of the cloud radar reflectivity, because it is a function of the droplet concentration, the particle size, and the DSD distribution shape parameter at every measured in-cloud radar gate ($Z \propto N\langle r^6 \rangle$).

Because of the strong sensitivity of Z to the mean cloud radius, the uncertainty about whether the observed variable profile in Z is due to height variations in N, ν, or r_m is confined by assuming homogeneous mixing in every measured in-cloud radar gate. Thus the observed variable profile in the radar reflectivity is attributed to uniform variations in the mean particle size through the whole droplet spectrum and the droplet concentration and DSD shape parameter are vertically constant. This assumption was widely used in earlier studies of cloud radar-based remote-sensing retrieval techniques (White et al., 1991; Frisch et al., 1995; Sassen et al., 1999; Kato et al., 2001; Matrosov et al., 2004). The correlation of the vertical variation of Z to the mean particle size implies that the property increases or decreases with height depending on the shape of the reflectivity profile.

The cloud properties obtained from this model will be hereafter referred to as the radar reflectivity-homogeneous mixing cloud model and for convenience simply abbreviated with HM model. Here it should be noted that, in a strict sense, the SAS cloud model also relies on the assumption of homogeneous mixing in that, compared to this model, r_m through the entire cloud is uniformly impacted while N and ν are constants.

3 CASE STUDY: USING RADIATION MEASUREMENTS FOR VALIDATION

Radiation closure experiments of cloudy conditions provide an opportunity to assess the retrievals of the cloud microphysical and optical properties when in situ observations are not available. These experiments involve the use of the cloud microphysical and optical retrievals in radiation transfer models to predict the resulting radiative quantities at the surface (SFC) and the top-of-atmosphere (TOA). A comparative analysis between the modeled and the observed radiation is expected to provide the knowledge regarding the quality of the cloud retrievals. This approach determines the validity of the cloud retrievals in terms of being able to reproduce the radiation budget.

3.1 Data and Methods

To evaluate the cloud property retrievals on the basis of a cloudy sky radiation closure experiment, information from various instruments is required for retrieving the cloud properties from the three algorithms. It is also needed for applying the broadband shortwave radiation transfer model in the EarthCare simulator ECSIM (Donovan et al., 2008) in order to compare the simulated and measured broadband SW fluxes at the SFC.

The data used in this study were collected during the COPS campaign (Wulfmeyer et al., 2011) by the Atmospheric Radiation Measurement (ARM) Mobile Facility (AMF). The third deployment of the ARM Mobile Facility took place in the Murg Valley of the Black Forest region of Germany from March 2007 until January 2008. The AMF in the Murg Valley was equipped with several active remote sensors and a suite of passive instrumentation. The data collected from this observational site are available through the public access interface (http://www.archive.arm.gov). Table 2 shows stepwise the procedures of the cloudy sky closure experiment and the observations and their source of instrumentation used for the application of the retrieval technique and the evaluation study.

In Step 2, the three cloud property retrieval algorithms are applied using as input the following data. The cloud boundaries and the radar reflectivity profiles are obtained from WACR-ARSCL VAP[2] (Table 2), which merges observations from the 95 GHz W-band ARM Cloud Radar (WACR), the micropulse LIDAR, and the ceilometer to produce, among others, the cloud boundaries and the time-height profiles of the radar reflectivity (Clothiaux et al., 2001). The temporal resolution is 5 s and the vertical resolution 42.856 m. The WACR-ARSCL radar reflectivity profiles are additionally corrected for propagation losses using the Cloudnet two-way radar attenuation data[1] to reduce the error in the reflectivity profiles due to attenuation by liquid water and atmospheric gases. Furthermore, the *LWP* from the MWR retrieval algorithm (MWRRET)[2] of Turner (2007) is used, because it provides more accurate *LWP* retrievals than the standard ARM *LWP* estimates. The temporal resolution is 20 s. The profiles of pressure and temperature additionally needed for the SAS cloud model are based on radio-sounding ascents, MWR, SFC meteorological instruments, and NWP model outputs to determine thermodynamic profiles with a temporal resolution of 30 s and a vertical resolution equal to that of the cloud radar (see also Ebell, 2010; Ebell et al., 2011). To retrieve the cloud properties from the listed observations, the 1-min averaged data are used, because the surface observations of the broadband SW fluxes[2] (Table 2) needed in Step 4 are based on 1-min averaged radiation data from the Pyranometer.

The liquid water cloud retrievals obtained from the input observations present the essential input parameters to create the cloudy atmospheric scene in the EarthCARE simulator and to apply the embedded radiative transfer model. Accordingly, in Step 3 the setup for the ECSIM application is performed. A scene describing the physical state of the atmosphere is defined with a vertical resolution (Δh) of 43 m up to approximately 15 km according to the cloud radar height resolution. The TOA is set at 100 km while increasing the vertical resolution to 1 km above 15 km. The horizontal resolution is roughly approximated by 420 m $\times$ 420 m for a 1-min interval, using the wind speed observed at the cloud height that was on average 7 m/s. The retrieved vertical profiles of the *LWC*, the effective radius, the gamma DSD, and shape as functions of the height for each 1-min interval

[1] We acknowledge the Cloudnet project (European Union contract EVK2-2000-00611) for providing the Instrument Synergy/Target Categorization data, which was produced by the University of Reading using measurements from AMF in the Murg Valley.

[2] These data were obtained from the Atmospheric Radiation Measurement (ARM) Program sponsored by the U.S. Department of Energy, Office of Science, Office of Biological and Environmental Research, Climate and Environmental Sciences Division.

TABLE 2 Procedure of the Closure Experiment and the Instruments and Used Quantities

Procedure	Instrument/Product	Used Quantities	Reference
Step 1: Selection of the liquid case study	Cloudnet products from 95 GHz W-band ARM Cloud Radar (WACR), ceilometer, LIDAR, MWR, rain gauge	Target classification scheme	Hogan and O'Connor (2004)
	Pyranometer	direct SW fluxes	
Step 2: Application of the VU, SAS, and HM retrieval algorithms to retrieve the cloud properties	WACR-ARSCL VAP (Active Remotely-Sensed Cloud Locations Value Added Product) from WACR, ceilometer, micropulse LIDAR	Reflectivity profiles, cloud boundaries	Clothiaux et al. (2001); Kollias et al. (2007b)
	Cloudnet products from ECMWF model temperature, pressure, and humidity profiles, MWR	Two-way radar attenuation due to atmospheric gases and liquid water	Hogan and O'Connor (2004)
	MWR	*LWP*	Turner (2007)
	Radiosonde soundings, MWR, SFC meteorological instruments, NWP model outputs	Temperature, pressure, and water vapor profiles	Ebell et al. (2011)
Step 3: ECSIM atmospheric scene creation	Cloud retrievals (see step 2)	*LWC* and effective radii profiles, gamma DSD, and shape parameter	
	Radiosonde soundings, MWR, SFC meteorological instruments, NWP model outputs	Temperature, pressure, water vapor profiles	Ebell et al. (2011)
	Pyranometer	Surface albedo	See above
	Multifilter rotating shadowband radiometer (MFRSR)	Aerosol optical depth	See above
Step 4: Comparison between simulated and observed fluxes	ECSIM-embedded DISORT (Discrete Ordinate Radiative Transfer) solver	Simulated broadband SW fluxes on VU, SAS, and HM created scenes	Donovan et al. (2008)
	Pyranometer	Downwelling broadband SW flux at SFC	Shi and Long (2001); Long and Shi (2008)

are then added sequentially to define the total cloud field in the atmospheric scene (details see Section 3.2). In addition to the cloud field, ECSIM aerosol field extinction profiles are implemented and scaled such that their aerosol optical depth values are 0.16 (at 550 nm) according to the study of Ebell (2010), who analyzed aerosol optical depth from the MFRSR over the 9-month observation period. For ozone, oxygen, and other gases, profiles of the US standard atmosphere are assumed. Carbon dioxide is assumed to have a constant concentration of 380 ppm. The atmospheric profiles in the atmospheric scene are defined up to a height of 100 km. The SW surface albedo is estimated from the ratios of upward and downward broadband SW fluxes at the surface. The average surface albedo during the selected observational period is 0.213 (Ebell, 2010; Ebell et al., 2011) and assumed to be invariant with the wavelength. In total, three atmospheric scenes were created depending upon the cloud properties retrieved from the VU, SAS, and HM models, which are used as input for the radiative transfer calculations. In Step 4, the simulated surface broadband SW fluxes for each scene are then compared with observed SW flux at the surface (Shi and Long, 2001; Long and Shi, 2008).

3.2 Liquid Water Cloud Case Study

3.2.1 *Observations*

The stratocumulus cloud layer studied was observed on Oct. 24, 2007, over the AMF site in the Murg Valley of the Black Forest region of Germany.

The synoptic situation on Oct. 24, 2007, was dominated by a high-pressure region over Scandinavia and a low-pressure region over Southeast Europe. Over Central Europe, advection of moist air with northeasterly flow corresponded to the widespread cloudiness shown in the Aqua satellite image of the synoptic-scale cloud cover on Oct. 24, 2007, at 12.35 UTC (Fig. 1). The red plus sign indicates the location of the AMF site in the Murg Valley.

The remote-sensing observations of the cloud layer over the Murg Valley on Oct. 24, 2007, are demonstrated in Fig. 2. Shown are height versus time images of the WACR reflectivity factor (A), the vertical Doppler velocity (B), the ceilometer backscatter coefficient (C), and the Cloudnet target classification field (D).

After the precipitating multi-layer mixed cloud in the very early morning hours, a closed and cellular-structured stratocumulus cloud

FIG. 1 Aqua satellite image of the synoptic-scale cloud cover on Oct. 24, 2007, at 12.35 UTC. The *red* plus sign indicates the location of the AMF site in the Murg Valley.

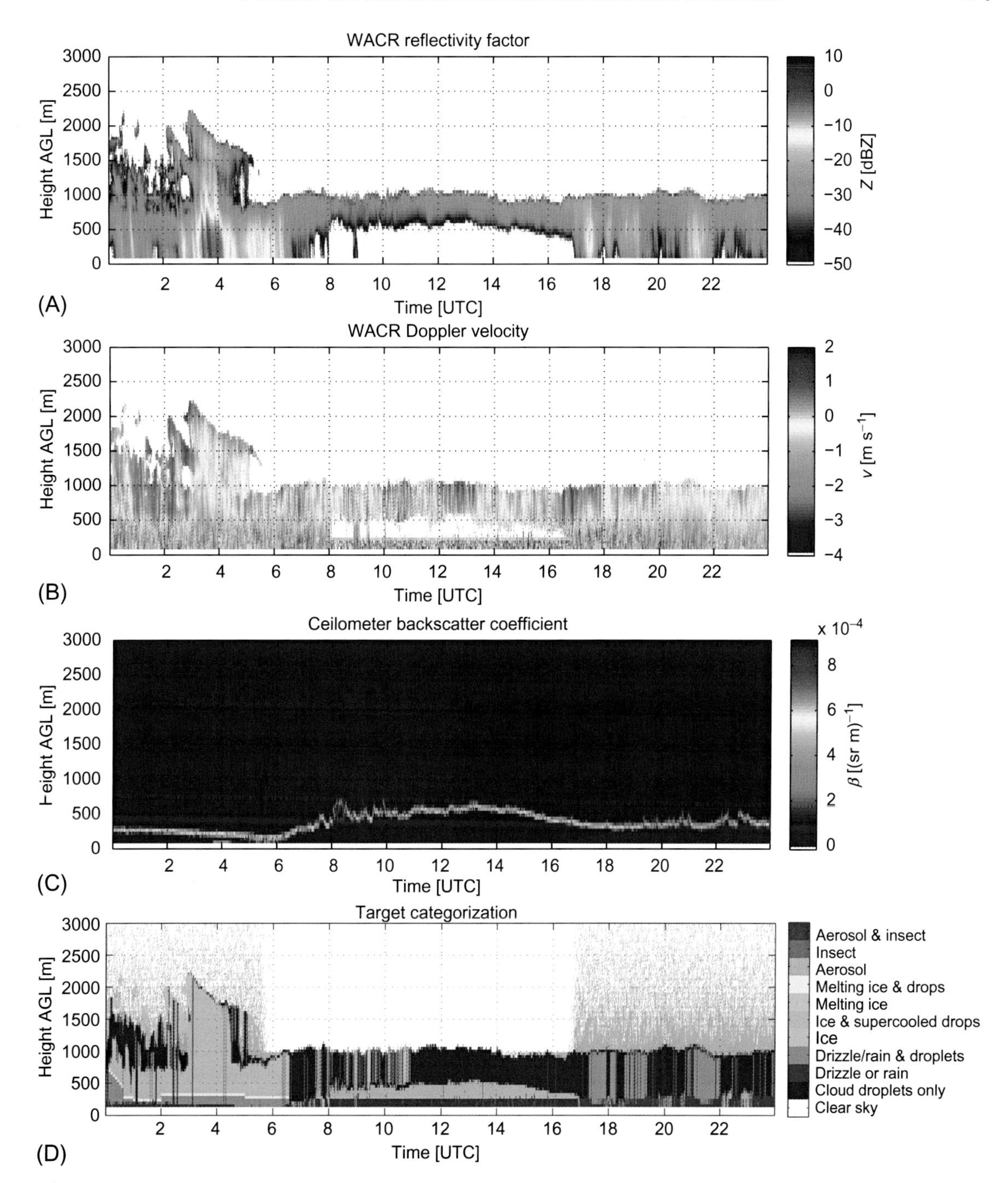

FIG. 2 (A) Height versus time images of WACR radar reflectivity factor, (B) Doppler velocity, (C) ceilometer backscatter coefficient, and (D) Cloudnet target classification on Oct. 24, 2007.

layer extending from approximately 0.6–1.1 km above ground was observed from around 8 UTC until late afternoon. Later in the evening and night, the stratocumulus cloud layer produced precipitation, mostly in the form of drizzle. The stratocumulus cloud was observed underneath a temperature inversion layer at about 1.1 km above ground, which suppressed vertical mixing and capped the lower part of the boundary layer (images of the vertical thermodynamic structure of the boundary layer are available under http://www.archive.arm.gov). During the radio-sounding ascent at 11.35 UTC, the boundary layer temperature below the inversion layer decreased from about 5°C to −3°C. Generally, the colder air triggered the formation of small liquid water droplets, which explains the relative low cloud base height at around 0.6 km above ground level observed from 8 to approximately 17 UTC.

Due to the strong optical attenuation by the water cloud, the ceilometer cloud penetration depth was limited to about 80 m (Fig. 2C). Furthermore, the intensity of the LIDAR backscatter coefficients were relatively high (partly larger than 5×10^{-5} s/rm) and the LIDAR depolarization signals were near zero (not shown), verifying the presence of rather large concentrations of small droplets at the cloud bottom. The upper part of the cloud layer was well detected by the cloud radar, which generally showed reflectivity values of around −30 dBZ (Fig. 2A). The cellular pattern in the cloud layer can be associated with the changes in the vertical motions observed in the Doppler-measured velocity (cf. Fig. 2B). For instance, the presence of regions dominated by the Doppler-measured updrafts (e.g., around 11 UTC or between 12.5 and 13 UTC in Fig. 2B) coincided with areas of increasing reflectivity observed after 11 UTC could be related to the growth of the cloud droplets with height until the delimiting temperature inversion at 1.1 km by containing water vapor rising air. Doppler-derived downdrafts indicate mixing with dry air above the cloud layer and thus cloud evaporation, which may explain that the decreasing cloud top height below 1 km above ground level shortly after 14 UTC coincided with a small region dominated by the Doppler-measured downdraft. Here it should be noted that generally an unambiguous identification of the organization of updrafts and downdrafts from the Doppler velocity field shown in Fig. 2B is not apparent.

During the nonprecipitating cloud layer period, the MWR estimated integrated water vapor (*IWV*) was generally low and increased from 1.1–1.3 cm and the MWR retrieved *LWP* varied between 40 and 150 g/m^2 (cf. Fig. 3). The relatively low amount of the *LWP* and the low values of radar reflectivity distributed over about 400–500 m cloud thickness indicate the presence of small droplets that are highly concentrated over the complete cloud layer as observed by the ceilometer at the cloud bottom.

Since the retrieval algorithms are only applicable for water clouds, it is necessary to identify observation periods that are not contaminated by ice or drizzle. As a rule of thumb, clouds that do not have a reflectivity higher than −17 dBZ are very likely composed of only small liquid droplets. Higher values of reflectivity suggest either the presence of drizzle droplets (Frisch et al., 1995) or, for subzero temperatures, the presence of larger particles such as ice. The observed stratocumulus cloud layer was slightly supercooled and among the liquid droplets, supercooled droplets and ice particles were diagnosed between 8 to about 11 UTC (Fig. 2D) by the Cloudnet target classification scheme. Here it should be noted that the target classification algorithm is unable to distinguish supercooled drizzle from ice. The algorithm indicates particles that have appreciable terminal velocities and such falling particles are likely to be composed of ice when they are present above wet bulb zero degree temperature isotherm. However, during midday when the cloud layer temperature slightly increased, the cloud layer did not produce detectable ice or

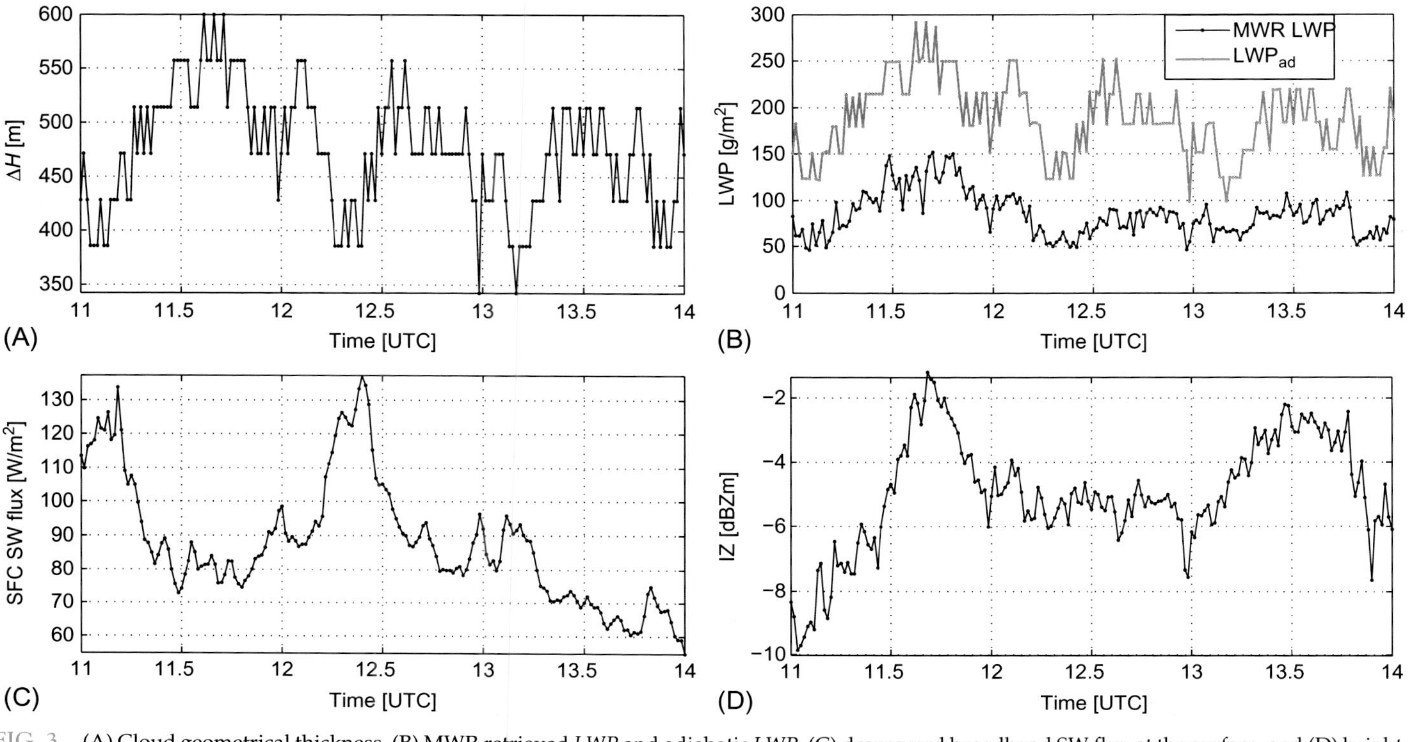

FIG. 3 (A) Cloud geometrical thickness, (B) MWR retrieved *LWP* and adiabatic *LWP*, (C) downward broadband SW flux at the surface, and (D) height-integrated radar reflectivity on Oct. 24, 2007, during 11–14 UTC.

drizzle particles anymore. The growth of ice or drizzle particles in the cloud layer is not unambiguously detectable from the radar reflectivity alone. However, combined with the fact that the LIDAR signal was extinguished, the absence of double peaks in the vertical structure of the reflectivity indicated a pure water cloud present from 11 UTC until the precipitation started at around 17 UTC.

Due to the valley location, it is also necessary to consider only time periods at which the solar zenith angle (SZA) is smaller than 73 degrees (cosine (SZA) > 0.3) to avoid topographical shadowing effects on the measured broadband SW fluxes at the surface (Ebell et al., 2011). At 11 UTC, the cosine of the SZA was 0.496 and decreased below 0.3 after 14 UTC. Therefore the application of the retrieval algorithms and the closure experiment are limited to the 3-h observation period from 11 to 14 UTC.

Fig. 3 shows the geometrical thickness (A), the MWR retrieved *LWP*, and the calculated adiabatic *LWP* ones (B), the observed SW fluxes at the surface (C), and the height-integrated reflectivity (D) as functions of the time. The curve of the cloud geometrical thickness (Fig. 3A) shows among the smaller fluctuations within ± 43 m (vertical resolution of the WACR) generally a wave-like propagation, ranging between 350 and 600 m, due to the cellular structure of the cloud layer associated with changes in the vertical motions observed in the Doppler-measured velocity (cf. Fig. 2B), that are responsible for the cloud structure. Concurrent with the alteration in the cloud thickness, the MWR *LWP* and the adiabatic *LWP* (Fig. 3B) exhibited similar fluctuations during the selected time period, such that as the geometrical increased (decreased), the *LWPs* similarly increased (decreased). The MWR *LWP* ranged from 45 to 151 g/m^2 with a mean value of 85 g/m^2 and standard deviation of 25 g/m^2, respectively. The observed *LWP* was always less than the calculated adiabatic values using the cloud radar tops, ceilometer cloud base heights, and the temperature profiles. For the geometrically thicker parts, the degree of adiabaticity was about 0.55; when the cloud layer thinned, it decreased to 0.4. Generally, the changes in the geometrical thickness lead to stronger fluctuations in the adiabatic *LWP* ($\pm 40\ g/m^2$). The broadband SW flux at the SFC inferred from the pyranometer (cf. Fig. 3C) was relatively low. Compared to the variations in the cloud depth and the *LWP*, it showed local maximums (minimums) when the geometrical thickness and the *LWP* decreased (increased). In turn, the variation of the vertically integrated reflectivity (Fig. 3D) mainly follows the variations of the cloud depth and the *LWP*. It ranged from −10 to −1.2 dBZm with a mean value of about −5.1 dBZm and standard deviation of 1.8 dBZm, respectively. The examination of the combined remote-sensing observations shows that the selected cloud layer is far from being an idealistic adiabatic liquid water cloud layer. This affects the cloud retrievals from the three vertical cloud models as will be discussed in the following. Here it should be noted that the valley location and the longwave cooling at cloud top helped to enhance the inversion layer with the result that the stratocumulus layer was persistent until Oct. 28, 2007, with a relatively sharp cloud top at about 1.1 km above ground.

3.2.2 *Cloud Property Retrievals*

In order to retrieve the liquid water cloud properties from the remote-sensing observations (Step 2 in Table 2), the shape of the gamma droplet size distribution ν was assumed to be 8.7 (Knist, 2014). This assumption is necessary since observations on the nature of the cloud-layer DSDs were not available.

Fig. 4A shows the height versus time images of the *LWC* retrieved from the HM model over the selected observation period. The HM model *LWC* profiles, which are dependent on the shapes of the reflectivity profiles, generally

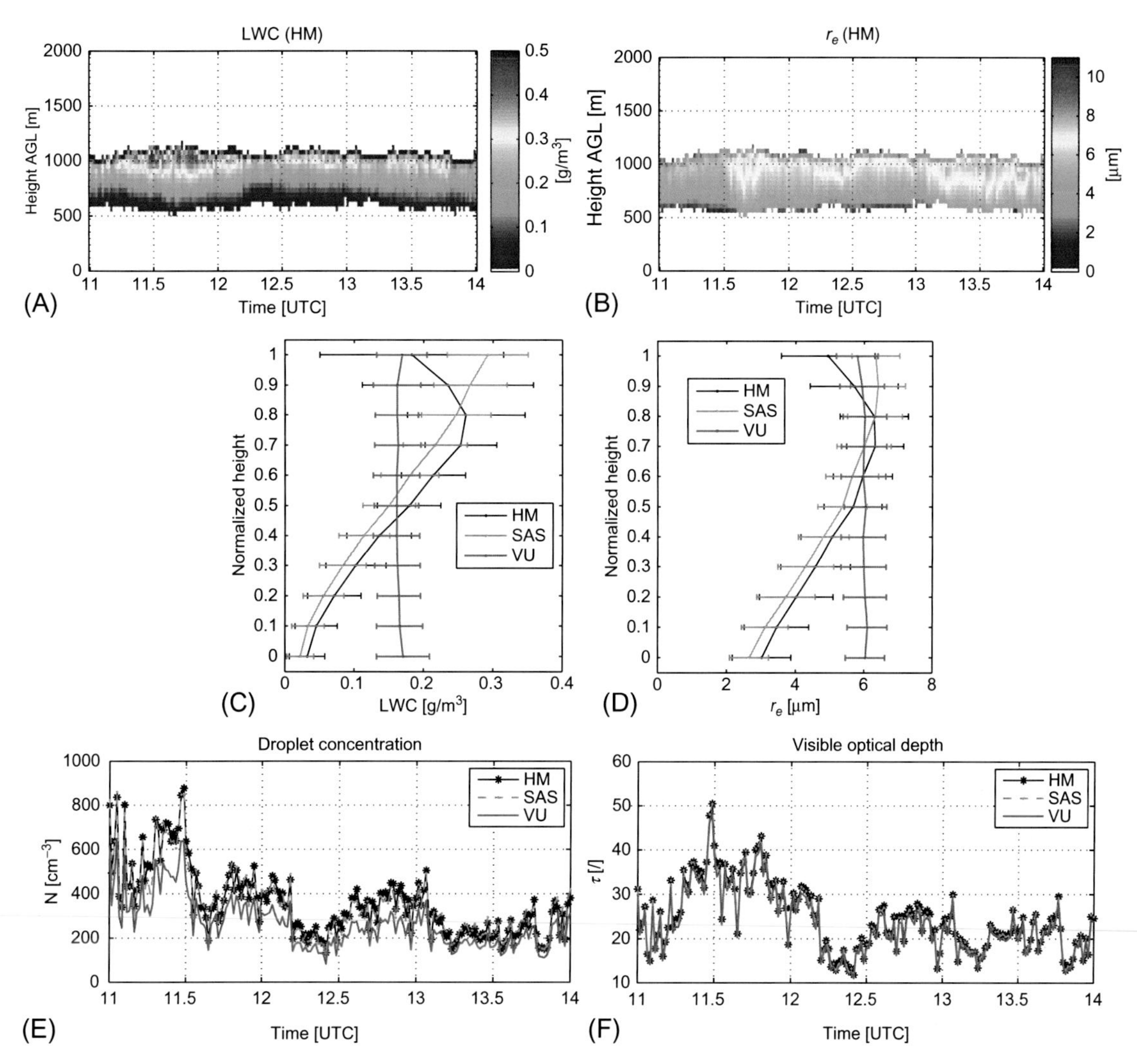

FIG. 4 (A) Time vs. height image of HM model *LWC* and (B) effective radius on Oct. 24, 2007, from 11 to 14 UTC. (C) 3-h average of VU, SAS, and HM model *LWC*, and (D) effective radius retrievals as functions of normalized height. The levels near $\hat{h} == 0, 0.1, 0.2 \ldots 1$ (=cloud top) of each measurement profile (in total 181 profiles) are used for calculating the statistics, which leads to slight deviations from strict vertical and linear profiles. (E) VU, SAS, and HM retrievals of droplet concentration and (F) visible optical depth.

exhibit an increase with the cloud height up to a certain height, at which the *LWC* values decrease to values close to zero until cloud top. As a result of the alteration observed in the cloud geometrical thickness and in the *LWP* (cf. Fig. 3B), the *LWC* reaches slightly higher values in the first hour of the observation period compared to the last 2 h where the observed *LWP* is lower for similar cloud depth.

The differences in the *LWC* vertical structures between the three models are demonstrated in Fig. 4C in terms of the 3-h average of the VU, SAS, and HM models *LWC* as functions of the cloud normalized height. Again, the height levels near $\hat{h} = 0$ (cloud base), 0.1, 0.2 ... 1 (cloud top) of each measurement profile (in total 181 profiles) are used for calculating the statistics. Here it should be noted that the variation in the geometrical thickness and the discretization to the selected height levels can lead to slight deviations from vertical and linear profiles in the averaged *LWC* profiles of the VU and SAS cloud models. Generally, the averaged *LWC* values in all the profiles are low because of the rather thick cloud compared to the relatively low observed *LWP*. The SAS model *LWC* is lower than the HM model *LWC* until the entrainment zone near cloud top at the height level 0.8 because of the low degree of adiabaticity. The standard deviation of the HM model *LWC* is larger than those of the VU and SAS models, especially in the upper part of the cloud layer, where the variability in the shape of reflectivity profiles is more pronounced. The internal pattern seen in the radar reflectivity observations of cloud layer can be only reflected in the HM model *LWC* retrievals (Fig. 4A), since the VU and SAS models retrieve either straight or linear *LWC* profiles.

The droplet concentrations retrieved from the three models are shown as functions of the time in Fig. 4E. Generally, the VU, SAS, and HM droplet concentrations vary over a wide range reflecting the alteration of the input observations of ΔH, Z, and the *LWP*. The HM model N ranges from 133 to 876/cm^3 with a mean value of about 360/cm^3 and standard deviation of 175/cm^3, respectively. While the SAS model retrieved N is very close to the one of the HM model, the concentration of the VU model is lower by a factor of 0.75.

The rather high values of N lead to low values of the effective radii. Fig. 4A shows the height versus time plots of the HM model r_e over the selected observation period and Fig. 4D shows the averaged r_e profiles of all models as functions of the cloud-normalized height. The retrieved effective radii from all models do not reach higher values than about 8 μm. This can show the impact of the assumption of the vertical constant N and DSD shape parameter as functions of the cloud height. This assumption implies that the droplet radii grow or evaporate uniformly through the whole droplet spectrum so that N and ν are kept constant as a function of height. This limits, for instance, the enhanced growth of some droplets at the expense of others so that fewer but bigger particles may occur. Furthermore, the fact that additional droplets may grow by unactivated cloud condensation nuclei (CCN) supplied from the subcloud layer through vertical motions is neglected. This raises the question whether such simple assumptions on the alteration of the DSD spectrum with cloud height are sufficient when non-idealistic behaviors are addressed. However, the impact of the vertical model assumptions on the shape of the effective radii profiles is similar to those in the *LWC* retrievals. The effective radii are very close to each other in the upper part of the cloud layer, except at cloud top, which is related to the differences appearing in the N retrievals between the SAS and VU cloud models.

The generally small values of r_e cause higher values of the visible optical depth, which are shown in Fig. 4F. This effect is especially seen between 11 and 12 UTC in which the values of the optical depth of all models are larger than 30. After 12 UTC, the optical depth is on average around 27 with a standard deviation of around 8. The optical depths derived using the different retrieval models are very close to each other. The relative difference between the VU and SAS model optical depth is 5%, because the maximum effective radius at cloud top of the SAS model is larger by a factor of 1.14. Here it should be noted that the correction procedure proposed in Kokhanovsky (2004) was applied on the retrievals of the extinction coefficient to take the variation of the extinction efficiency, especially for small droplets, into account.

Applying the retrieval algorithms on real remote-sensing observations of a certain cloud layer leads to uncertainties in the cloud properties that are dependent on many factors, including instrumental errors, various sources of errors in calculating the products of the instruments (e.g., reflectivity factor, cloud boundaries, *LWP*), or the validity of the fundamental algorithm assumptions. Especially the formation of drizzle or ice, which leads to an alteration of the assumed monomodal DSD to a bimodal distribution composed of particles with a mean radius of 20 μm or larger, can have a large impact. The filtering for periods that are not contaminated by drizzle or ice was accomplished by comparing the observations of different sensors and by using the Cloudnet target classification product. Although it cannot be predicted with certainty that the selected cloud layer really did not contain a second mode in the DSD, such a possibility is difficult to consider in estimating the retrieval uncertainty when in situ observations are not available. However, the algorithm responses to drizzle or ice particles in the cloud layer result in low amounts of droplet concentrations and large effective radii, which is not seen in the retrievals of the selected cloud layer.

The impact of the assumed constant droplet concentration and DSD shape parameter with height imposes further uncertainties on the cloud property retrievals. Considering the variability of the cloud observations, such as the cellular structure, the organization of the vertical motions, and the low degree of adiabaticity, it cannot be excluded that N and the shape of the distribution do not change with height since various mixing processes might be involved. They result in different responses of the DSD parameters of N, ν, and r_e, that impact the vertical distribution of the cloud microphysical and optical properties. The uncertainty in the retrievals can be evaluated from the uncertainty in the individual input observations.

The uncertainties in the cloud retrievals due to errors in the used products of the instruments are mainly driven by the MWR-retrieved *LWP* and the radar reflectivity. The radar reflectivity of the 95 GHz WACR is computed for a dielectric factor fixed at the temperature $T = 0°C$ ($|K_{94}|^2 = 0.71$). The midcloud temperature was about 0°C, so that errors in the reflectivity through the temperature dependence of the index of refraction of water are expected to be negligibly small. Furthermore, the WACR reflectivity was corrected for propagation losses using the two-way radar attenuation data from Cloudnet so that errors in the reflectivity profiles due to attenuation by liquid water and atmospheric gases can also be neglected. Potential errors in the reflectivity arise from uncertainties in the radar calibration term. Therefore, a calibration error of ±1.5 dB is assumed (Hogan et al., 2003). The cloud boundaries are obtained from the WACR-ARSCL VAP (see Table 2) that are based on the algorithm described in Clothiaux et al. (2001). It provides robust estimates of the cloud base height using all laser devices (micropulse LIDAR and ceilometer) and also of cloud top height from the cloud radar (Clothiaux et al., 2001), so that the error in the cloud top height is not expected to be larger than the vertical resolution of the WACR, 42.856 m. Such an uncertainty is negligibly small compared to the relatively thick cloud observed during the selected time period. Considering the uncertainty in the *LWP*, the MWRRET Value-Added Product provides an estimate of the one-sigma uncertainty in the cloud liquid water path retrieved using an advanced statistical and iterative approach (Turner, 2007). In the selected observation period, the *LWP* uncertainty is about $\pm 13\ g/m^2$ for *LWP* of $80\ g/m^2$ and increases to about $\pm 19\ g/m^2$ at $150\ g/m^2$. Since the shape of reflectivity profiles and the geometrical thickness can be considered as correct, the relative errors of the VU, SAS, and HM *LWC* retrievals are equivalent and depend only on the *LWP* uncertainty. Hence, the relative uncertainty of the *LWC* ranges from 12% to 20% over the observation period. The uncertainties in the droplet concentration, effective radii, and optical depth additionally require an estimate of the DSD shape parameter. The estimated DSD shape parameter of 8.7 can vary between 6.6 and 12.1 (Knist, 2014) With the estimated errors in Z, LWP, and ν, the total calculated relative uncertainty in the retrieval of N is between 48% and 56%. The uncertainty of r_e becomes approximately 14% and for τ it is between 21% and 29%. In the following section the liquid water cloud retrievals based on the three algorithms are used as input for the radiative transfer calculations to predict the radiative flux at the surface, which is then compared with the surface radiation observations.

3.3 Comparison of Simulated and Observed SW Flux

As described in Section 3.1, three atmospheric scenes are created depending on the cloud properties retrieved from the VU, SAS, and HM models and on the atmospheric state. Each scene includes 181 vertical profiles of the *LWC*, the effective radius, the gamma DSD, and the settings on the shape parameter. The optical properties, such as extinction coefficient and optical depth, are computed in ECSIM internally using the DSD coupled with scattering library files. The atmospheric state, such as temperature and humidity profiles, and the aerosol optical depth as well as the surface albedo are added.

Fig. 5 shows the observed and simulated SW fluxes at the surface depending upon the retrieved cloud properties of the VU (top), SAS (center), and HM (bottom) models as functions of the time. The error bars indicate the total uncertainty in the simulations due to the estimated uncertainties in the *LWC* and the effective radius. The results of the flux comparison are shown in Table 3. In general, the SW flux simulations of the retrievals from the VU, SAS, and HM models are in good agreement with the observations. The difference between the simulated SW flux of the vertically uniform cloud

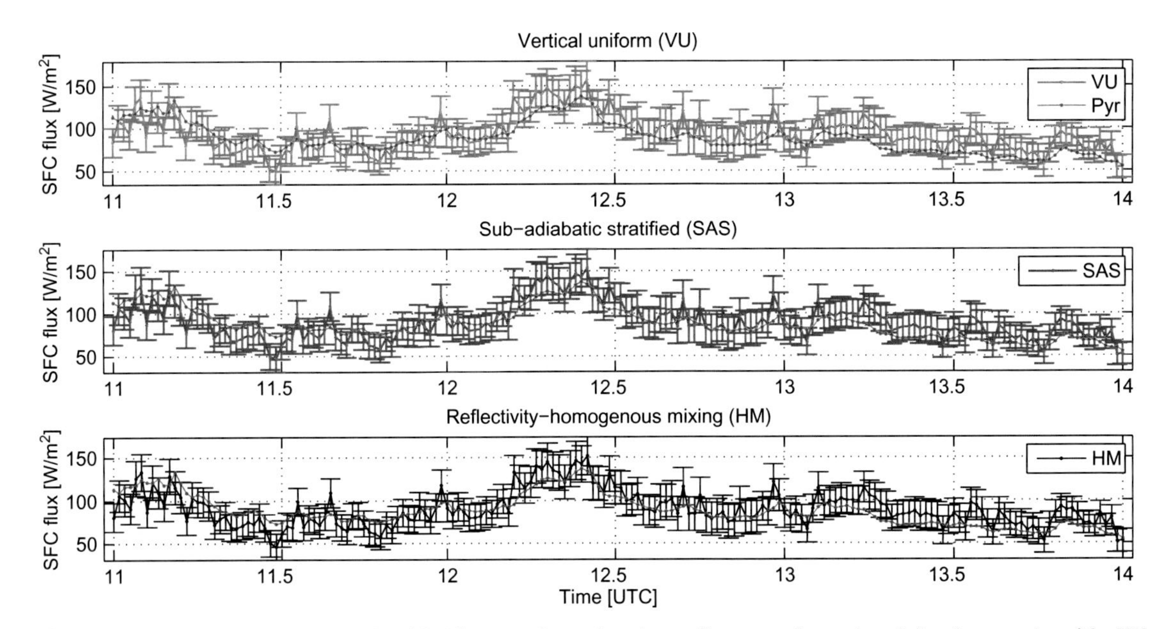

FIG. 5 Observed (*red* lines) and simulated SW fluxes at the surface depending upon the retrieved cloud properties of the VU (*top*), SAS (*center*), and HM (*bottom*) models as functions of the time. The error bars indicate the total uncertainty in the simulations due to the estimated uncertainties in the *LWC* and the effective radius.

TABLE 3 Mean Difference, RMS Difference, and Correlation Coefficient Between the VU, SAS, and HM Models Simulated and the Observed Broadband Shortwave Fluxes at the SFC

	VU Model	SAS Model	HM Model
Bias (W/m^2)	5.1105	2.0011	1.9179
(%)	6.4486	2.8076	2.7840
RMS difference (W/m^2)	13.9440	13.1652	13.1797
(%)	16.4794	15.3659	15.4861
Correlation coefficient	0.7941	0.7910	0.7858

and the measured SW flux is 5 W/m^2 with a standard deviation of about 13 W/m^2. The differences between the simulated fluxes of the SAS and HM cloud models and the observations are even lower, around 2 W/m^2 with a standard deviation of 13 W/m^2. The small differences in the comparison between the VU and SAS model simulations can be related to the small differences in the visible optical depth retrievals. The optical depth of the VU model is about 5% lower than the one of the SAS model, which may have led to the slight increase of the SW flux and thus to a slightly larger overestimation of the observed SW flux.

The simulated SW fluxes based on the retrievals obtained from the three models show stronger fluctuations compared to the SW flux observations, although the general trend and the local maximums and minimums are well represented (Fig. 5). Thus the correlation coefficients between the simulations and the observations during the observation period are around 0.79 (see also Fig. 6). The broader spread of the calculated SW fluxes is caused by the sensitivity of the cloud property retrievals (*LWC*, r_e, and thus τ) to the fluctuations observed in the input parameters, especially to the *LWP* and integrated reflectivity. They lead to a stronger variability in the cloud retrievals and thus in the simulated SW fluxes, which is not seen in the observation of the SW flux. The cloud layer variations of the *LWP* and the reflectivity are

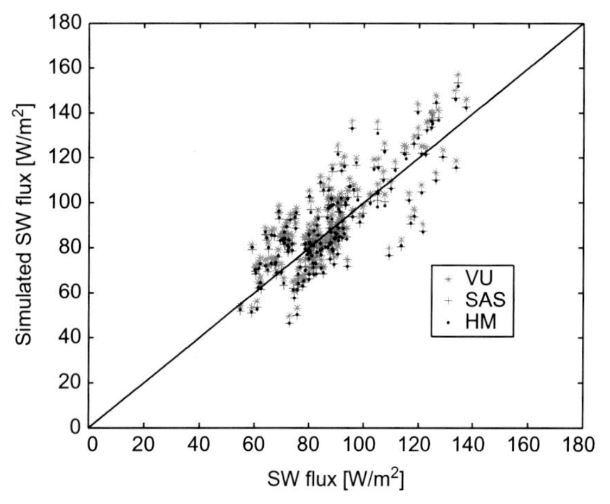

FIG. 6 Observed versus simulated SW flux at the SFC depending upon the retrieved cloud properties of the VU, SAS, and HM models.

obtained from sensors with narrow fields of view and the DISORT model in ECSIM computes the radiation fluxes column by column on the basis of the cloud radar and MWR-retrieved properties of the *LWC*, the effective radius, and the type of the DSD. These simulations are then compared with the surface observations from the pyranometer, which has a full atmospheric view and hence measures the averaged SW flux around the observational site. The cellular structure of the cloud layer might indicate that horizontal inhomogeneity in this cloud field exists. This means that the distributions of the observations of the *LWP* and the reflectivity and thus of the cloud property retrievals around the observational site are expected to be broader. This can explain the larger variability in the simulated fluxes, since the cloud property retrievals are obtained from observations, which characterize a smaller domain of the cloud layer.

However, the correlation between the calculated and observed fluxes is still acceptable.

The uncertainty in the comparisons and the small positive biases can be also related to the uncertainty in the retrievals due to errors in the observations and the assumption on the shape of the gamma DSD as indicated by the error bars in Fig. 5. Additionally, Fig. 7 shows the individual impact of the errors in the *LWP*, *Z*, and the gamma DSD shape parameter on the retrievals that propagate into estimates of the SW fluxes. Fig. 7A shows the mean observed and the mean simulated SW fluxes at the surface from the VU, SAS, and HM cloud models with the error bars indicating the impact of the uncertainty in the *LWC* and the effective radius caused by the *LWP* errors on the mean simulated fluxes. The estimated uncertainty in the *LWP* (Section 3.2) leads to a mean relative uncertainty in the *LWC* of 15% and in the effective radius of 5%. Such changes in the cloud properties lead, in this case, to changes in the mean simulated fluxes of 13–15 W/m^2. The uncertainty in the reflectivity only impacts the retrievals of effective radius, here by about 11%. This leads to changes in the mean fluxes of about 8 W/m^2.

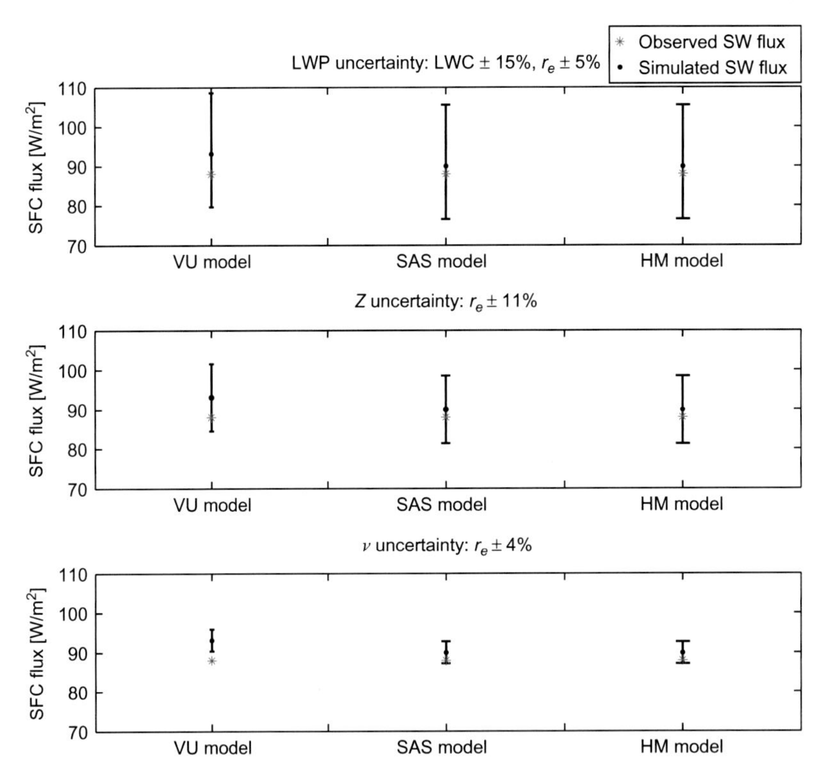

FIG. 7 Mean observed (*red*) and the mean simulated SW fluxes (black) at the surface from the VU, SAS, and HM cloud models with the error bars indicating the impact of the uncertainty in the *LWC* and effective radius due to *LWP* errors (*top*), due to reflectivity errors (*center*), and due to the assumed gamma DSD shape parameter (*bottom*) on the mean simulated fluxes.

The uncertainty in the DSD shape parameter has a small effect on the effective radius retrieval (4%), so that the changes in the mean fluxes are about 3 W/m^2. The uncertainties in the *LWC* and the effective radius retrievals can explain the differences between the simulated and observed SW fluxes of all three models that are also in the range of the uncertainty in the observed flux of $\pm$10 W/m^2 (Ebell et al., 2011). Other sources of errors, such as the effects of aerosol optical depth, water vapor, and the surface albedo may lead to changes in the simulated SW flux by a few W/m^2. However, the thermodynamic profiles, the surface albedo, and the aerosol optical depth values have been validated and confirmed to be suited for radiation studies at the AMF site in the Murg Valley (Ebell et al., 2011).

The results of the radiation closure experiment show that the retrieved cloud properties of the *LWC*, the effective radius, and thus the visible optical depth of all three models as inputs for the radiative transfer calculations reproduce reasonably the broadband SW flux at the surface for the selected case study. The impact of the differences in the vertical distribution of the *LWC* and the effective radius leads only to small differences in the visible optical depth and thus to small differences in the SW fluxes. Nevertheless, the simulated fluxes on the basis of the SAS and HM model retrievals show a slightly better performance than the results dependent on the VU model.

4 CONCLUSIONS

Ground-based remote sensing of clouds has seen huge development in recent decades. The inclusion of active instruments, such as radar and LIDAR, has enabled the retrieval of microphysical cloud properties as well the internal structures of clouds, with high spatial and temporal resolution. As in any indirect technique, models and parameterizations are used in the retrieval techniques. This limits the accuracy and representativeness of the retrievals. We have demonstrated that, under the right conditions, ground-based measurements of sky radiation can be used to validate the findings.

References

Albrecht, B.A., Fairall, C.W., Thomson, D.W., White, A.B., Snider, J.B., Schubert, W.H., 1990. Surface-based remote-sensing of the observed and the adiabatic liquid water-content of stratocumulus clouds. Geophys. Res. Lett. 17 (1), 89–92.

Baker, M.B., Corbin, R.G., Latham, J., 1980. The influence of entrainment on the evolution of cloud droplet spectra, I: a model of inhomogeneous mixing. Q. J. R. Meteorol. Soc. 106 (449), 581–598.

Boers, R., Mitchell, R.M., 1994. Absorption feedback in stratocumulus clouds—influence on cloud-top albedo. Tellus A 46 (3), 229–241.

Boers, R., Rotstayn, L.D., 2001. Possible links between cloud optical depth and effective radius in remote sensing observations. Q. J. R. Meteorol. Soc. 127 (577), 2367–2383.

Boers, R., Russchenberg, H., Erkelens, J., Venema, V., van Lammeren, A., Apituley, A., Jongen, S., 2000. Ground-based remote sensing of stratocumulus properties during CLARA, 1996. J. Appl. Meteorol. 39 (2), 169–181.

Boers, R., Acarreta, J.R., Gras, J.L., 2006. Satellite monitoring of the first indirect aerosol effect: retrieval of the droplet concentration of water clouds. J. Geophys. Res. Atmos. 111, D22208. https://doi.org/10.1029/2005JD006838.

Brandau, C.L., Russchenberg, H.W.J., Knap, W.H., 2010. Evaluation of ground-based remotely sensed liquid water cloud properties using shortwave radiation measurements. Atmos. Res. 96 (2-3), 366–377.

Brenguier, J.L., 1991. Parameterization of the condensation process—a theoretical approach. J. Atmos. Sci. 48 (2), 264–282.

Brenguier, J.L., Pawlowska, H., Schüller, L., Preusker, R., Fischer, J., Fouquart, Y., 2000a. Radiative properties of boundary layer clouds: droplet effective radius versus number concentration. J. Atmos. Sci. 57 (6), 803–821.

Burnet, F., Brenguier, J.-L., 2007. Observational study of the entrainment-mixing process in warm convective clouds. J. Atmos. Sci. 64 (6), 1995–2011.

Chin, H.N.S., Rodriguez, D.J., Cederwall, R.T., Chuang, C.C., Grossman, A.S., Yio, J.J., Fu, Q., Miller, M.A., 2000. A microphysical retrieval scheme for continental low-level stratiform clouds: impacts of the subadiabatic character on microphysical properties and radiation budgets. Mon. Weather Rev. 128 (7), 2511–2527.

Chosson, F., Brenguier, J.-L., Schueller, L., 2007. Entrainment-mixing and radiative transfer simulation in boundary layer clouds. J. Atmos. Sci. 64 (7), 2670–2682.

Clothiaux, E., Miller, M., Perez, R., Turner, D., Moran, K., Martner, B., Ackerman, T., Mace, G., Marchand, R., Widener, K., Rodriguez, D., Uttal, T., Mather, J., Flynn, C., Gaustad, K., Ermold, B., 2001. The ARM Millimeter Wave Cloud Radars (MMCRs) and the Active Remote Sensing of Clouds (ARSCL) Value Added Product (VAP). Technical Report DOE Tech. Memo. ARM VAP-002.1, supported by U.S. Department of Energy, Office of Science, Office of Biological and Environmental Research.

Dong, X.Q., Ackerman, T.P., Clothiaux, E.E., 1998. Parameterizations of the microphysical and shortwave radiative properties of boundary layer stratus from ground-based measurements. J. Geophys. Res.-Atmos. 103 (D24), 31681–31693.

Donovan, D.P., Voors, R.H., van Zadelhoff, G.J., Acarreta, J.-R., 2008. ECSIM model and algorithms document. Available at http://www.knmi.nl/zadelhof/.

Ebell, K., 2010. Characterization of clouds and their radiative effects using ground-based instrumentation at a low-mountain site. Dissertation, Universität Köln.

Ebell, K., Crewell, S., Loehnert, U., Turner, D.D., O'Connor, E.J., 2011. Cloud statistics and cloud radiative effect for a low-mountain site. Q. J. R. Meteorol. Soc. 137, 306–324.

Frisch, A.S., Fairall, C.W., Snider, J.B., 1995. Measurement of stratus cloud and drizzle parameters in ASTEX with a k-alpha-band doppler radar and a microwave radiometer. J. Atmos. Sci. 52 (16), 2788–2799.

Han, Q., Rossow, W., Welch, R., White, A., Chou, J., 1995. Validation of satellite retrievals of cloud microphysics and liquid water path using observations from FIRE. J. Atmos. Sci. 52 (23), 4183–4195.

Hogan, R.J., O'Connor, E.J., 2004. Facilitating cloud radar and lidar algorithms: the Cloudnet instrument synergy/target categorization product. Cloudnet documentation, Department of Meteorology, University of Reading, Reading, United Kingdom. Available at http://www.cloud-net.org.

Hogan, R.J., Bouniol, D., Ladd, D.N., O'Connor, E.J., Illingworth, A.J., 2003. Absolute calibration of 94/95-GHz radars using rain. J. Atmos. Ocean. Technol. 20 (4), 572–580.

Illingworth, A.J., Hogan, R.J., O'Connor, E.J., Bouniol, D., Brooks, M.E., Delanoe, J., Donovan, D.P., Eastment, J.D., Gaussiat, N., Goddard, J.W.F., Haeffelin, M., Baltink, H.K., Krasnov, O.A., Pelon, J., Piriou, J.M., Protat, A., Russchenberg, H.W.J., et al., 2007. Cloudnet—continuous evaluation of cloud profiles in seven operational models using ground-based observations. Bull. Am. Meteorol. Soc. 88 (6), 883–898.

Kato, S., Mace, G.G., Clothiaux, E.E., Liljegren, J.C., Austin, R.T., 2001. Doppler cloud radar derived drop size distributions in liquid water stratus clouds. J. Atmos. Sci. 58 (19), 2895–2911.

Kim, B.G., Schwartz, S.E., Miller, M.A., Min, Q.L., 2003. Effective radius of cloud droplets by ground-based remote sensing: relationship to aerosol. J. Geophys. Res.-Atmos. 108 (D23), 4740. https://doi.org/10.1029/2003JD003721.

Kim, B.G., Klein, S.A., Norris, J.R., 2005. Continental liquid water cloud variability and its parameterization using Atmospheric Radiation Measurement data. J. Geophys. Res. Atmos 110, D15S08. https://doi.org/10.1029/2004JD005122.

Kim, B.-G., Miller, M.A., Schwartz, S.E., Liu, Y., Min, Q., 2008. The role of adiabaticity in the aerosol first indirect effect. J. Geophys. Res.-Atmos. 113 (D5), D05210. https://doi.org/10.1029/2007JD008961.

Knist, C., 2014: Retrieval of liquid water cloud properties from ground-based remote sensing observations. Dissertation at Delft University of Technology, ISBN 978-94-6259-263-6.

Kokhanovsky, A., 2004. Optical properties of terrestrial clouds. Earth Sci. Rev. 64 (3-4), 189–241.

Kollias, P., Clothiaux, E.E., Miller, M.A., Albrecht, B.A., Stephens, G.L., Ackerman, T.P., 2007a. Millimeter-wavelength radars—new frontier in atmospheric cloud and precipitation research. Bull. Am. Meteorol. Soc. 88 (10), 1608–1624.

Kollias, P., Clothiaux, E.E., Miller, M.A., Luke, E.P., Johnson, K.L., Moran, K.P., Widener, K.B., Albrecht, B.A., 2007b. The atmospheric radiation measurement program cloud profiling radars: Second-generation sampling strategies, processing, and cloud data products. J. Atmos. Oceanic Technol. 24 (7), 1199–1214.

Long, C., Shi, Y., 2008. An automated quality assessment and control algorithm for surface radiation measurements. Open Atmos. Sci. J. 2. https://doi.org/10.2174/1874282300802010023.

Mace, G.G., Sassen, K., 2000. A constrained algorithm for retrieval of stratocumulus cloud properties using solar radiation, microwave radiometer, and millimeter cloud radar data. J. Geophys. Res.-Atmos. 105 (D23), 29099–29108.

Matrosov, S.Y., Uttal, T., Hazen, D.A., 2004. Evaluation of radar reflectivity-based estimates of water content in stratiform marine clouds. J. Appl. Meteorol. 43 (3), 405–419.

Miles, N.L., Verlinde, J., Clothiaux, E.E., 2000. Cloud droplet size distributions in low-level stratiform clouds. J. Atmos. Sci. 57 (2), 295–311.

Min, Q.L., Harrison, L.C., 1996. Cloud properties derived from surface MFRSR measurements and comparison with GOES results at the ARM SGP site. Geophys. Res. Lett. 23 (13), 1641–1644.

Nakajima, T., King, M.D., 1990. Determination of the optical-thickness and effective particle radius of clouds from reflected solar-radiation measurements, I. Theory. J. Atmos. Sci. 47 (15), 1878–1893.

Pawlowska, H., Grabowski, W.W., Brenguier, J.L., 2006. Observations of the width of cloud droplet spectra in stratocumulus. Geophys. Res. Lett. 33 (19), L19810. https://doi.org/10.1029/2006GL026841.

Sassen, K., Mace, G.G., Wang, Z., Poellot, M.R., 1999. Continental stratus clouds: a case study using coordinated remote sensing and aircraft measurements. J. Atmos. Sci. 56 (14), 2345–2358.

Shi, Y., Long, C.N., 2001. Best estimate radiation flux value-added procedure: Algorithm operational details and explanations. Technical Report DOE/SC-ARM/TR-008, DOE, ARM, Pacific Northwest National Laboratory Richland, Washington. Supported by U.S. Department of Energy, Office of Science, Office of Biological and Environmental Research.

Slingo, A., 1989. A GCM parameterization for the shortwave radiative properties of water clouds. J. Atmos. Sci. 46 (10), 1419–1427.

Stephens, G.L., 1978. Radiation profiles in extended water clouds, II: parameterization schemes. J. Atmos. Sci. 35 (11), 2123–2132.

Turner, D.D., 2005. Arctic mixed-phase cloud properties from AERI lidar observations: algorithm and results from SHEBA. J. Appl. Meteorol. 44 (4), 427–444.

Turner, D.D., 2007. Improved ground-based liquid water path retrievals using a combined infraredand microwave approach. J. Geophys. Res. Atmos. 112 (D15), D15204. https://doi.org/10.1029/2007JD008530.

Turner, D.D., Clough, S.A., Lijegren, J.C., Clothiaux, E.E., Cady-Pereira, K.E., Gaustad, K.L., 2007b. Retrieving liquid water path and precipitable water vapor from the Atmospheric Radiation Measurement (ARM) microwave radiometers. IEEE Trans. Geosci. Remote Sens. 45 (11), 3680–3690.

Walko, R.L., Cotton, W.R., Meyers, M.P., Harrington, J.Y., 1995. New Rams cloud microphysics parameterization, I: the single-moment scheme. Atmos. Res. 38 (1-4), 29–62.

White, A.B., Fairall, C.W., Thomson, D.W., 1991. Radar observations of humidity variability in and above the marine atmospheric boundary-layer. J. Atmos. Ocean. Technol. 8 (5), 639–658.

Wood, R., 2005. Drizzle in stratiform boundary layer clouds. Part I: vertical and horizontal structure. J. Atmos. Sci. 62 (9), 3011–3033.

Wulfmeyer, V., Behrendt, A., Kottmeier, C., Corsmeier, U., Barthlott, C., Craig, G.C., Hagen, M., Althausen, D., Aoshima, F., Arpagaus, M., Bauer, H.-S., Bennett, L., Blyth, A., Brandau, C., et al., 2011. The Convective and Orographically-induced Precipitation Study (COPS): the scientific strategy, the field phase, and research highlights. Q. J. R. Meteorol. Soc. 137, 3–30.

Further Reading

Barnard, J.C., Long, C.N., 2004. A simple empirical equation to calculate cloud optical thickness using shortwave broadband measurements. J. Appl. Meteorol. 43 (7), 1057–1066.

Ebell, K., Loehnert, U., Crewell, S., Turner, D.D., 2010. On characterizing the error in a remotely sensed liquid water content profile. Atmos. Res. 98 (1), 57–68.

CHAPTER

10

Measuring Precipitation From Space

Francisco J. Tapiador

University of Castilla-La Mancha (UCLM), Toledo, Spain

1 INTRODUCTION

Measuring hydrometeors (rain, snow, hail, graupel, etc.) from space is important for environmental studies as ground estimates do not provide the adequate temporal and spatial coverage nor provide information about the precipitation in 3D, an aspect which is critical to understanding the thermodynamics of latent heat release in the atmosphere (Tao et al., 2001). This topic, at its turn, is central to an improved understanding of the Earth System and to global warming research (Tao et al., 2006, 2016). In terms of economic activities, measuring precipitation from space is fundamental for activities such as hydropower planning, operations at dams, and fresh water availability management; not to mention to agriculture, forestry, or natural resources management.

Why do we need remote sensing estimates of precipitation if we have Numerical Weather Prediction (NWP) models that can not only estimate but also forecast the occurrence of hydrometeors? One reason is that below the basic-scale level, models still struggle to provide precise estimates of precipitation for hydrological applications. While the forecasts are improving fast, there are still many situations with a low degree of predictability, and that affects the overall consistency of the estimates for hydrological applications. Apart from that, there are at least three major reasons to still require satellite-derived estimates of precipitation.

The first one is because of the need of a reference metric to gauge model performance. Unless included with temperature, precipitation is not provided to the model either in the form of initial or boundary conditions. Rather, precipitation is generated by the model itself from other variables, so in a sense it is more independent from any input information the model could have. This is important in terms of ensuring statistical independence in the verification (cfr. Ebert et al., 2007) and validation.

The second reason is that models are heavily tuned to observed conditions (Voosen, 2016). Remote precipitation estimates are thus required to allow modelers to fine-tune the many empirical parameters all models embed. Also, in the weather forecasting realm, assimilation of precipitation into models is known to improve the predictions.

A third major reason to require remote estimate precipitation is to understand the physics behind the precipitation processes. Without observations, cumulus and microphysics parameterizations cannot be improved. Detailed observations in 4D (3D plus time) are critical for deriving parameters from first-principles, a must in order not to tie those parameters to particular atmospheric situations.

Remote Sensing of Aerosols, Clouds, and Precipitation
https://doi.org/10.1016/B978-0-12-810437-8.00010-4

© 2018 Elsevier Inc. All rights reserved.

Rain gauges and disdrometers can only give point estimates of precipitation and they present several limitations due to undercatch and the limited representativeness of the measurements of the surrounding area (Nešpor et al., 2000; Ciach, 2003; Duchon and Biddle, 2010; Duchon and Essenberg, 2001). In contrast to temperature or humidity, precipitation is a geophysical quantity highly variable in space and time, which makes it difficult to get its precise measurement. Even at small, below-kilometer scale, precipitation is highly variable (Tapiador et al., 2010). Thus, instantaneous estimates of rainfall from a single disdrometer used to calibrate radars can be more than 70% off (Tapiador et al., 2017).

To give an idea on the limitations of ground estimates of precipitation, the combined measuring area of all the rain gauges in the world is smaller than a standard football field (Kidd et al., 2017). Besides, gauges and disdrometers are limited to land and a few atolls. Over the oceans, the only means to estimate precipitation precisely and accurately is by using satellites. Ground radars can provide remotely-sensed estimates of precipitation, but they are limited to land and the estimates are difficult to homogenize to build comprehensive, gapless, regional products. Truly global measurements are simply nonexistent for other than satellite-based products (Tapiador et al., 2012).

Satellite estimates of precipitation are produced from infrared/visible and from microwave wavelengths. The rest of the electromagnetic (EM) spectrum is not useful as the radiation there is either blocked by the atmosphere or is insensitive to either the properties of the hydrometeors or to theirs effects on the immediate environment.

2 INFRARED AND VISIBLE FREQUENCIES METHODS

Methods based in the infrared/visible (IR/VIS) frequencies assume some sort of correlation between the cloud top brightness temperature and surface rainfall (Vicente et al., 1998). While it is known that such a relationship is only indirect (large cloud vertical development do indicate convection and thus large rain rates, but the exact location of surface rain is not exactly matched by coldest tops, see Fig. 1), and notwithstanding that the internal structure of precipitation systems is hidden in those wavelengths, IR methods have been widely used in the past before the advent of dedicated microwave (MW) sensors (e.g., Wilheit et al., 1977; Adler and Negri, 1988).

The major advantage of IR techniques is that the sensors can be placed at geostationary orbit, thus providing hemispheric view at high temporal resolution (tens of minutes as in 2017). The frequency of the IR EM radiation also allows high spatial resolution (a few kilometers). IR/VIS methods can provide estimates 24/7, if information from the VIS is critical, which is often the case as VIS enhances the ability to discriminate some microphysical processes.

IR-based methods compared reasonably well with observations when aggregated in space and time since averaging blurs the unprecise identification of where and when the rain have fallen over a particular point in space.

When several IR bands are available, the precipitation estimates improve. Thus, several EUMETSAT products from the METEOSAT series satellites have been proposed taking advantage of frequencies peaking at different heights (Fig. 2). However, the rationale is still the same, namely an indirect relationship between cloud temperature aloft and precipitation at ground. At the end, clouds are opaque to IR radiation and the IR is not sensitive to hydrometeor-size particles, making any estimate highly indirect.

3 MICROWAVE-BASED METHODS

More direct estimates of precipitation from space can be derived using microwave and millimeter-wave frequencies between Ku (10 GHz) and W (95 GHz) bands. Within this range, precipitation-sized particles absorb, emit,

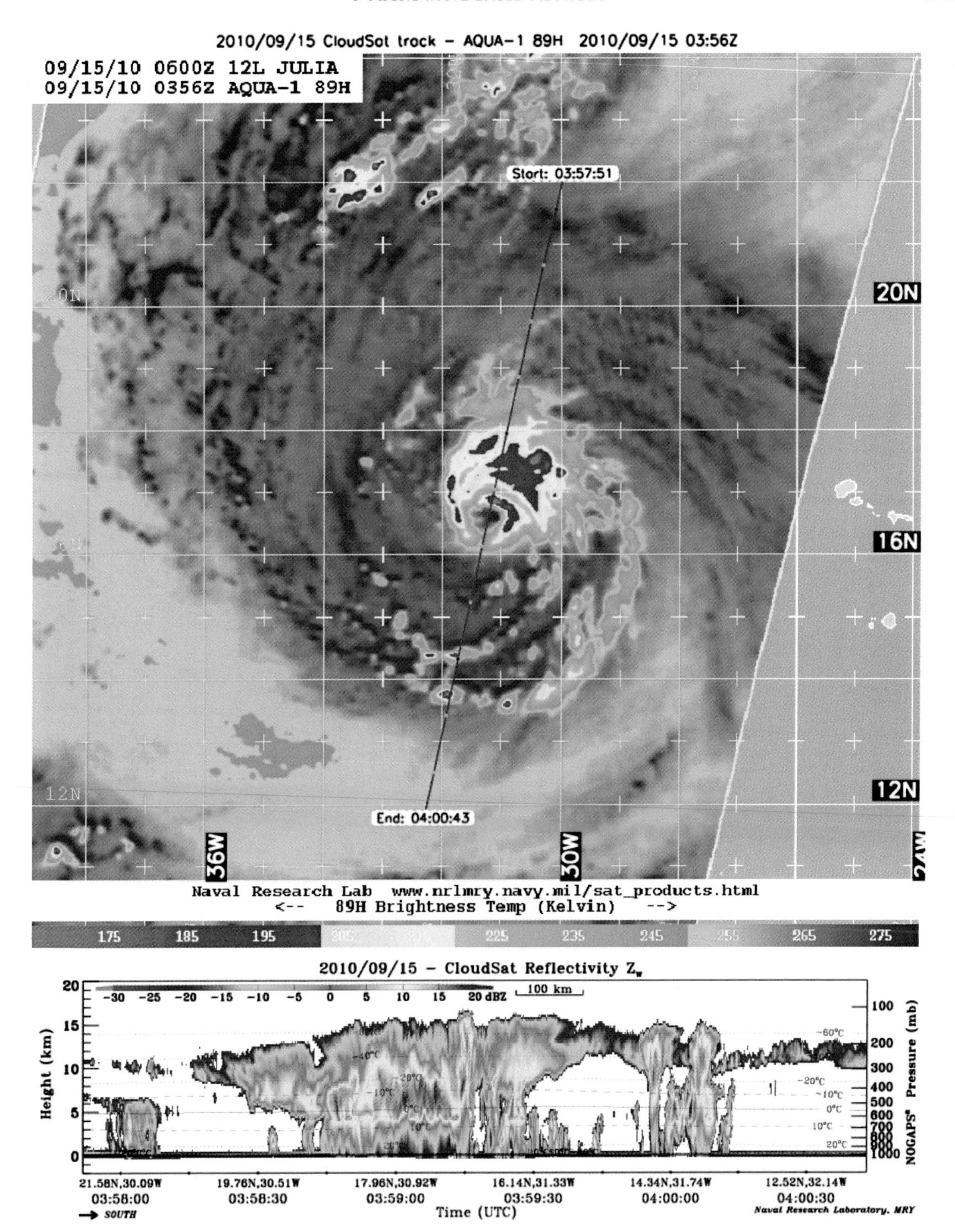

FIG. 1 An example of PMW retrieval over ocean (from MODIS-Aqua sensor, top), and CloudSat reflectivity (bottom, which is a good proxy for liquid precipitation below the 273 K isoline; Stephens et al., 2008). The figure illustrates how cold cloud tops do not correlate with surface rainfall. *Source: http://cloudsat.atmos.colostate.edu.*

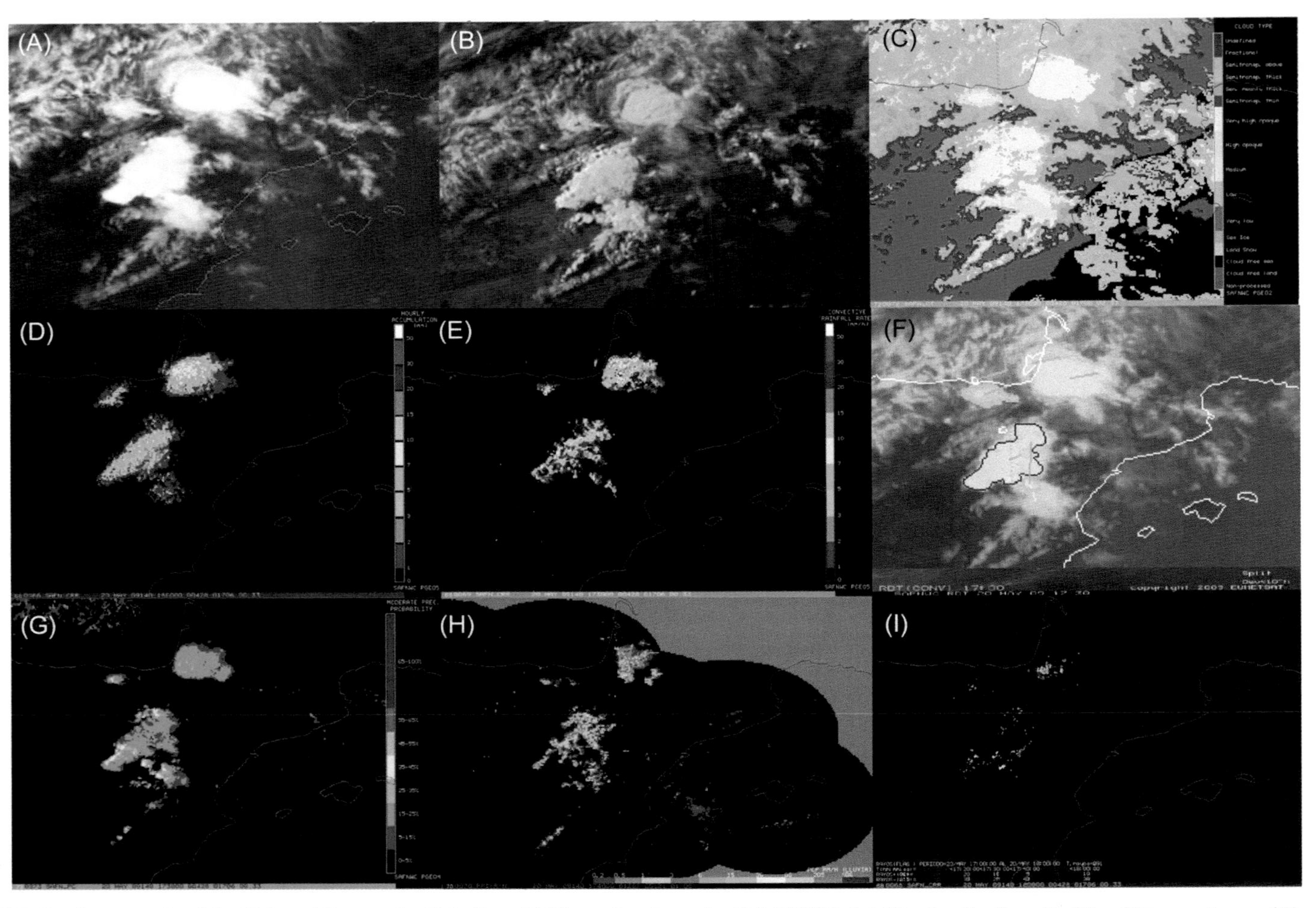

FIG. 2 An example of the IR-based Convective Rain Rate (CRR) product from the EUMESAT's Satellite Applications Facility of Nowcasting and Very Short Range Forecasting (NWC-SAF). The panel depicts a comparison between several remote sensing-derived products for a severe storm occurred over Spain on May 20, 2009. (A) Air mass RGB composite, (B) natural RGB composite, (C) Cloud Type (CT), (D) convective rainfall rate (CRR), hourly accumulations, (E) convective rainfall rate, instantaneous, (F) rapidly-developing thunderstorms (RDT), (G) precipitating clouds (PC), (H) radar (PPI), and (I) lightning. *Source: http://www.nwcsaf.org/web/guest/topical-images-gallery.*

and scatter electromagnetic radiation. Microwave satellite sensors can sense the thermal emission emanating from the top of the atmosphere, or measure the power backscattered from a series of pulses. The first approach is named "passive microwave methods," and the second "active microwave techniques."

3.1 Passive Microwave Methods

PMW techniques are based on the fact that microwave radiation emitted from the Earth's surface interacts with some atmospheric components such as water vapor, cloud drops, and hydrometeors. Depending on the wavelength of the MW radiation, scattering or emission dominates the signal measured by the radiometer. By using several frequencies, both effects can be considered.

Depending upon the surface emissivity, the relative contribution of the hydrometeor-affected signal to the signal received by the sensor can be small, and surface emission can dominate, especially over land. Thus, separate algorithmical strategies exist for land and ocean (cfr. Turk et al., 2000).

The easiest part is the ocean. Emission-based techniques are used there with frequencies below approx. 20 GHz. At these wavelengths, the ocean looks radiometrically cool and the existence of liquid phase precipitation features as a warmer area in the radiation captured by the radiometer. Emission methods are not used over land because it has high surface emissivity that masked the small increase due to precipitation, nor can it be employed for high rainfall rates as the relative contribution of hydrometeors to the emission signature saturates exponentially with increasing optical depth.

Over land, estimation is more complicated. An indirect scattering-based approach with higher frequencies is used there. Frozen precipitation scatters the upwelling microwave radiation resulting in the satellite perceiving a radiometric cooling in the precipitating areas. The technique works well for convective type precipitation: large vertical development clouds generated in such process are frozen above a certain height (the freezing layer; also known as the melting layer, or the "bright band"), and liquid precipitation at ground comes from the melting of those hydrometeors. Then, as MW radiation can penetrate clouds (in contrast with the IR), there is a good correlation between detecting more ice aloft and having more rainfall at ground. The method also works for the oceans.

The main limitation of MW sensors that affects estimation from space is that in order to maintain a reasonable antenna size, sensors onboard low orbit satellites have a coarse spatial resolution (from 5 to 60 km). Also, the limited field of view inherent to low Earth orbits reduces their temporal revisit. An associated problem is the beam-filling effect, which arises when only a fraction of the (large) Instantaneous Field of View (IFOV) of the sensor is filled with precipitation.

3.2 Active Microwave Methods

The principle of active microwave methods is the same as the meteorological ground radar. A beam of microwave-frequency EM radiation is emitted by the radar and the properties of the backscattered signal (Z) are measured and related with the concentration and phase (liquid, solid, mixed) of the hydrometeors.

The relationship between rainfall (R) and Z is a function of the particle size distribution (PSD), which is the number of particles per size in a volume. If the EM signal is polarized, then it is possible to better estimate the PSD, as large drops are more oblate than small ones. Using two frequencies, for instance the Ka and the Ku bands, allows to further discriminate the phase of the hydrometeors.

4 METHODS BASED ON IR+MW FUSION

In order to combine the best of both worlds, having the directness of MW frequencies and the high spatial and temporal resolutions of IR sensors, fused algorithms have been developed (Barrett et al., 1987; Adler et al., 1993). The idea behind these techniques is to propagate a precise, IR-interpolated MW-based estimate through the IR-derived atmospheric flow. The first part consists of finding some sort of relationship between many IR pixels and a single, lower resolution MW estimate of precipitation. The second part involves calculating atmospheric vectors to be used to advect the former (as in Bellerby, 2006), generally considering the dynamics of the clouds as seen in the IR (thus, if the pixel becomes colder, then vertical development is inferred and thus, more rain assigned). The details of the fusion, morphing (Joyce et al., 2004), or physically-based propagation (Tapiador, 2008) vary depending on the algorithm. For a review of these, the reader is referred to Tapiador et al. (2017) (Fig. 3).

5 THE TROPICAL RAINFALL MEASUREMENT MISSION (TRMM)

The TRMM mission (Simpson et al., 1988, 1996; Kummerow et al., 1998) is a paradigmatic example of a disruptive technology in atmospheric remote sensing. The time TRMM was active (1998–2015) can be dubbed as the "TRMM-era of remote sensing of precipitation from space." Aimed as a tool to investigate latent heat release in the Tropics (e.g., Grecu et al., 2009), and to advance precipitation science, the satellite tremendously increased our knowledge of the water

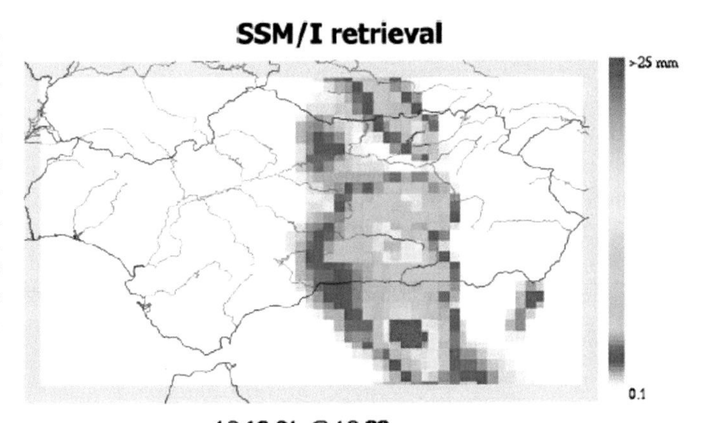

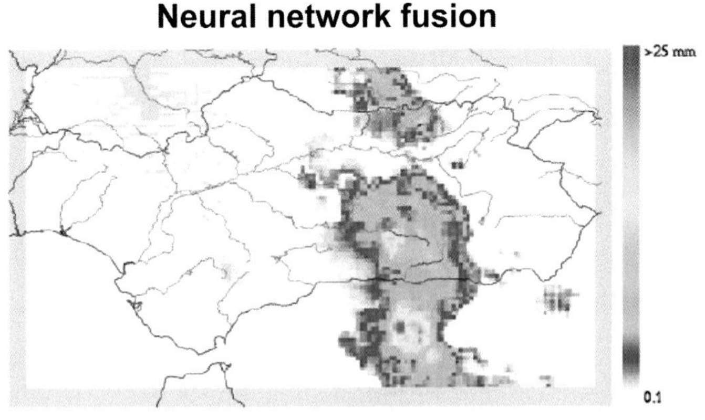

FIG. 3 An example of a neural network-based fusion of PMW estimates of precipitation from SSM/I sensor with IR data from the Meteosat satellite. The fusion improves both the spatial and the temporal resolution of the PMW estimates while taking advantage of the more direct estimate of precipitation provided by the cloud-penetrating MW radiation. *From Tapiador, F.J., Kidd, C., Levizzani, V., Marzano, F.S., 2004. A neural networks-based fusion technique to estimate half-hourly rainfall estimates at 0.1 degrees resolution from satellite passive microwave and infrared data. J. Appl. Metereol. 43 (4), pp. 576–594, ISSN 0894-8763.*

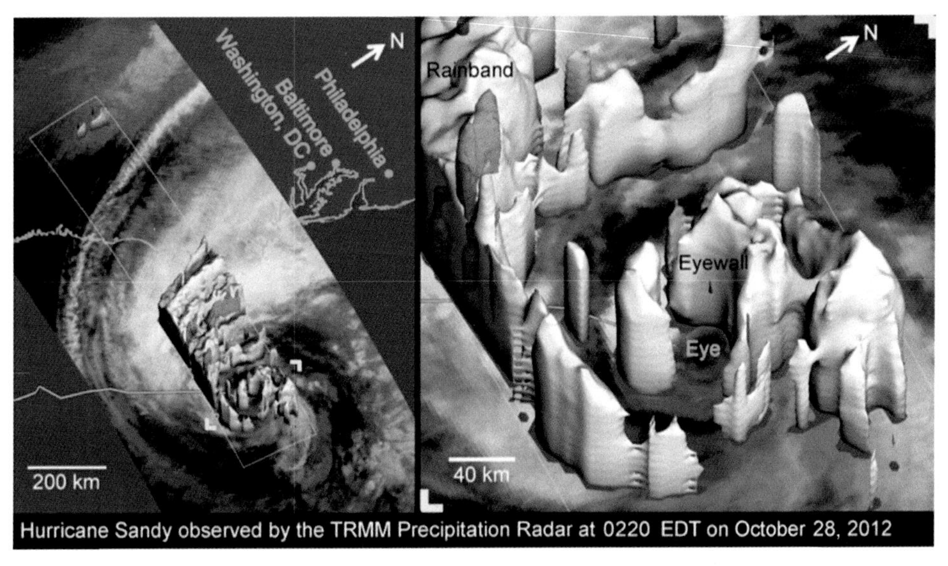

FIG. 4 An example of the potential of TRMM for investigating precipitation in 3D. Hurricane Sandy analyses. *Source: https://pmm.nasa.gov.*

cycle in the whole globe, in spite of the 35S–35N nominal coverage. Algorithms for both the radiometer (TMI) and the Precipitation Radar (PR), and the combination of both exist (Haddad et al., 1997; Iguchi et al., 2000; Kummerow et al., 2001). TRMM-based products such as the TMPA (Huffman et al., 2010) are key to understanding the global water cycle and to validate models and assumptions in satellite algorithms (Fig. 4).

6 THE GLOBAL PRECIPITATION MEASURING (GPM) MISSION

February 27, 2014 marked another milestone in the remote sensing of precipitation. For the first time, a two-band precipitation radar, the DPR, was set in orbit onboard the Core Observatory (CO) of the Global Precipitation Measuring (GPM) Mission (Hou et al., 2014; Skofronick-Jackson et al., 2017). The inclination of the orbit provided the GPM satellite with an almost global coverage, and the enhanced capabilities of the DPR compared with its predecessor, TRMM's PR, allowed a better understanding of solid precipitation (Fig. 5).

GPM-derived products such as the IMERG (Huffman et al., 2015) are becoming increasingly used for a variety of scientific applications. Future climatologies from GPM will be invaluable to characterize the effects of global warming on the biota and human activities. The combination under the Precipitation Measurement Missions (PMM) umbrella of the GPM-CO with satellites such as Megha-Tropiques, Suomi NPP, DMSP F17/F18, MetOp B/C, GCOM-W1, NOAA 18/19, and JPSS-1 can only benefit from the scientific understanding of the global water cycle (Fig. 6).

7 THE FUTURE OF SATELLITE ESTIMATES OF PRECIPITATION

There are many applications of remote estimates of precipitation: direct use in hydrological or crop models, or as a component of insurance and actuarial activities, definition of weather derivatives, in hydropower operations,

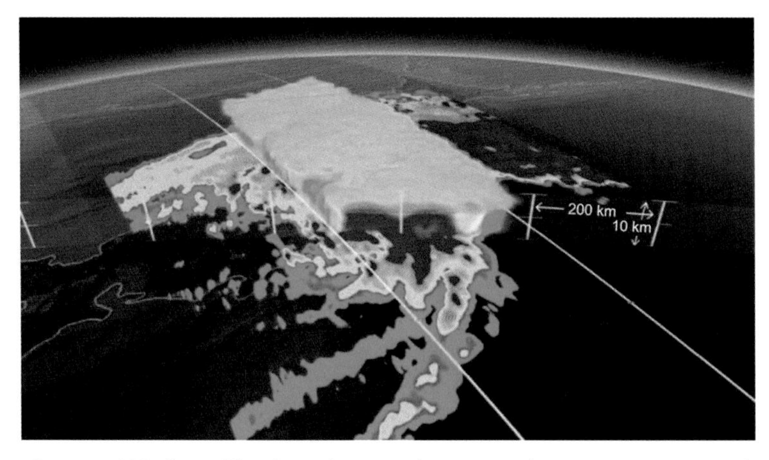

FIG. 5 GPM-derived view of Typhoon Phanfone showing the inner volumetric rain rates and the solid and liquid precipitation. *Source: https://pmm.nasa.gov.*

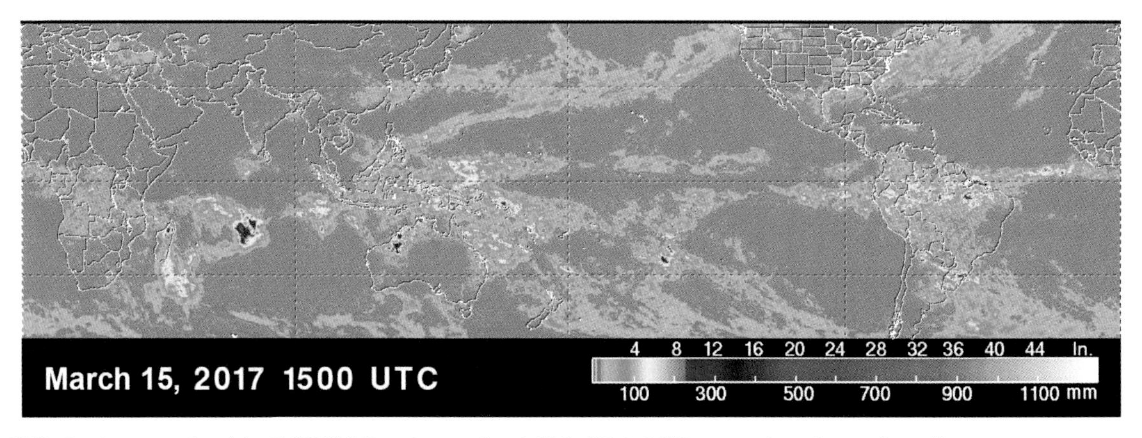

FIG. 6 An example of the IMERG 7 Day Accumulated Global Rainfall Data product. *Source: https://pmm.nasa.gov.*

verification and validation of climate and weather models, etc. (Tapiador et al., 2017). The need of increasingly precise estimates, especially for the solid phase of precipitation, is likely to grow in the next decades. The next years are also likely to witness the emergence of several other players. While the United States and Japan have been leading the field, new sensors from China and India are likely to add new potentialities to the existing constellation of radar and radiometers.

The future of measuring precipitation from space may involve some of the following strategies: Doppler radars, small CubeSats, enhanced GPM-type dual polarizations, multiwavelength radars, and the use of millimeter-wave radiances (Haddad et al., 2017). Each one of these choices presents pros and cons, but the ultimate choice, or the combination of choices, will more likely cohabit with GPM, which is extremely likely to become the ultimate reference for forthcoming missions. Geosounders (MW sensors on geostationary orbit) have also been proposed in Europe and China. Their role, if they are finally launched, are to complement precipitation estimates from radars and radiometers, with a focus on tracking hurricane phenomena.

TRMM was a mission with a nominal lifetime of 3 years that lasted for two decades and ended just because the satellite ran out of fuel. GPM has a lifetime of 5 years but enough fuel for two decades, so providing the hardware is as at least as reliable as TRMM's, we can expect GPM to hover over the field for many years from now (2017), thus increasing the invaluable knowledge base from TRMM with additional and more detailed information.

Thanks to GPM and the combination of new sensors with enhanced capabilities we now have a more detailed picture of precipitation on Earth to benefit agriculture, industry, and transportation sectors and to help boost economic development based on spatial technology.

Acknowledgments

Funding from projects CGL2013-48367-P, CGL2016-80609-R (Ministerio de Economía y Competitividad), UNCM08-1E-086 (Ministerio de Ciencia e Innovacion), and CYTEMA (UCLM) is gratefully acknowledged.

References

Adler, R.F., Negri, A.J., 1988. A satellite infrared technique to estimate tropical convective and stratiform rainfall. J. Appl. Meteorol. https://doi.org/10.1175/1520-0450(1988)027<0030:ASITTE>2.0.CO;2.

Adler, R.F., Negri, A.J., Keehn, P.R., Hakkarinen, I.M., 1993. Estimation of monthly rainfall over Japan and surrounding waters from a combination of low-orbit microwave and geosynchronous IR data. J. Appl. Meteorol. https://doi.org/10.1175/1520-0450(1993)032<0335:EOMROJ>2.0.CO;2.

Barrett, E.C., Kidd, C., Bailey, J.O., 1987. The use of SMMR data in support of the Bristol/NOAA interactive scheme (BIAS) for satellite improved rainfall monitoring. Annual Rep. U.S. Dept. of Commerce, Cooperative Agreement NA86AA-H-RA001, 77 pp.

Bellerby, T.J., 2006. High-resolution 2-D cloud-top advection from geostationary satellite imagery. IEEE Trans. Geosci. Remote Sens. 44, 3639–3648. https://doi.org/10.1109/TGRS.2006.881117.

Ciach, G.J., 2003. Local random errors in tipping-bucket rain gauge measurements. J. Atmos. Ocean. Technol. 20, 752–759. https://doi.org/10.1175/1520-0426(2003)20<752:LREITB>2.0.CO;2.

Duchon, C.E., Biddle, C.J., 2010. Undercatch of tipping-bucket gauges in high rain rate events. Adv. Geosci. 25, 11–15. https://doi.org/10.5194/adgeo-25-11-2010.

Duchon, C.E., Essenberg, G.R., 2001. Comparative rainfall observations from pit and aboveground rain gauges with and without wind shields. Water Resour. Res. 37, 3253–3263. https://doi.org/10.1029/2001WR000541.

Ebert, E.E., Janowiak, J.E., Kidd, C., 2007. Comparison of near-real-time precipitation estimates from satellite observations and numerical models. Bull. Am. Meteorol. Soc. 88, 47–64. https://doi.org/10.1175/BAMS-88-1-47.

Grecu, M., Olson, W.S., Shie, C.L., L'Ecuyer, T.S., Tao, W.K., 2009. Combining satellite microwave radiometer and radar observations to estimate atmospheric heating profiles. J. Clim. 22, 6356–6376. https://doi.org/10.1175/2009JCLI3020.1.

Haddad, Z.S., Smith, E.A., Kummerow, C.D., Iguchi, T., Farrar, M.R., Durden, S.L., Alves, M., Olson, W.S., 1997. The TRMM "Day-1" radar/radiometer combined rain-profiling algorithm. J. Meteorol. Soc. Jpn. Ser. II 75, 799–809.

Haddad, Z.S., Sawaya, R.C., Kacimi, S., Sy, O.O., Turk, F.J., Steward, J., 2017. Interpreting millimeter-wave radiances over tropical convective clouds. J. Geophys. Res. Atmos. 122, 1650–1664. https://doi.org/10.1002/2016JD025923.

Hou, A.Y., Kakar, R.K., Neeck, S., Azarbarzin, A.A., Kummerow, C.D., Kojima, M., Oki, R., Nakamura, K., Iguchi, T., 2014. The global precipitation measurement mission. Bull. Am. Meteorol. Soc. 95, 701–722. https://doi.org/10.1175/BAMS-D-13-00164.1.

Huffman, G.J., Adler, R.F., Bolvin, D.T., Nelkin, E.J., 2010. The TRMM multi-satellite precipitation analysis (TMPA). In: Hossain, F., Gebremichael, M. (Eds.), Satellite Rainfall Applications for Surface Hydrology. Springer Netherlands, Dordrecht, pp. 3–22. https://doi.org/10.1007/978-90-481-2915-7_1.

Huffman, G.J., Bolvin, D.T., Braithwaite, D., Hsu, K., Joyce, R., 2015. Algorithm Theoretical Basis Document (ATBD) Version 4.5. NASA Global Precipitation Measurement (GPM) Integrated Multi-satellitE Retrievals for GPM (IMERG). NASA, Greenbelt, MD, USA.

Iguchi, T., Kozu, T., Meneghini, R., Awaka, J., Okamoto, K., 2000. Rain-profiling algorithm for the TRMM precipitation radar. J. Appl. Meteorol. 39, 2038–2052. https://doi.org/10.1175/1520-0450(2001)040<2038:RPAFTT>2.0.CO;2.

Joyce, R.J., Janowiak, J.E., Arkin, P.A., Xie, P., 2004. CMORPH: a method that produces global precipitation estimates from passive microwave and infrared data at high spatial and temporal resolution. J. Hydrometeorol. 5, 487–503. https://doi.org/10.1175/1525-7541(2004)005<0487:CAMTPG>2.0.CO;2.

Kidd, C., Becker, A., Huffman, G.J., Muller, C.L., Joe, P., Skofronick-Jackson, G., Kirschbaum, D.B., Kidd, C., Becker, A., Huffman, G.J., Muller, C.L., Joe, P., Skofronick-Jackson, G., Kirschbaum, D.B., 2017. So, how much of the earth's surface is covered by rain gauges? Bull. Am. Meteorol. Soc. 98, 69–78. https://doi.org/10.1175/BAMS-D-14-00283.1.

Kummerow, C., Barnes, W., Kozu, T., Shiue, J., Simpson, J., 1998. The tropical rainfall measuring mission (TRMM) sensor package. J. Atmos. Ocean. Technol. 15, 809–817. https://doi.org/10.1175/1520-0426(1998)015<0809:TTRMMT>2.0.CO;2.

Kummerow, C., Hong, Y., Olson, W.S., Yang, S., Adler, R.F., McCollum, J., Ferraro, R., Petty, G., Shin, D.-B., Wilheit, T.T., 2001. The evolution of the Goddard profiling algorithm (GPROF) for rainfall estimation from passive microwave sensors. J. Appl. Meteorol. 40, 1801–1820. https://doi.org/10.1175/1520-0450(2001)040<1801:TEOTGP>2.0.CO;2.

Nešpor, V., Krajewski, W.F., Kruger, A., Nešpor, V., Krajewski, W.F., Kruger, A., 2000. Wind-induced error of raindrop size distribution measurement using a two-dimensional video disdrometer. J. Atmos. Ocean. Technol. 17, 1483–1492. https://doi.org/10.1175/1520-0426(2000)017<1483:WIEORS>2.0.CO;2.

Simpson, J., Adler, R.F., North, G.R., 1988. A proposed tropical rainfall measuring mission (TRMM) satellite. Bull. Am. Meteorol. Soc. 69, 278–295. https://doi.org/10.1175/1520-0477(1988)069<0278:APTRMM>2.0.CO;2.

Simpson, J., Kummerow, C., Tao, W.-K., Adler, R.F., 1996. On the tropical rainfall measuring mission (TRMM). Meteorol. Atmos. Phys. 60, 19–36. https://doi.org/10.1007/BF01029783.

Skofronick-Jackson, G., Petersen, W.A., Berg, W., Kidd, C., Stocker, E., Kirschbaum, D.B., Kakar, R., Braun, S.A., Huffman, G.J., Iguchi, T., Kirstetter, P.E., Kummerow, C., Meneghini, R., Oki, R., Olson, W.S., Takayabu, Y.N., Furukawa, K., Wilheit, T., 2017. The global precipitation measurement (GPM) mission for science and society. Bull. Am. Meteorol. Soc. https://doi.org/10.1175/BAMS-D-15-00306.1.

Stephens, G.L., Vane, D.G., Tanelli, S., Im, E., Durden, S., Rokey, M., Reike, D., Partain, P., Mace, G.G., Austin, R., L'Ecuyer, T., Haynes, J., Lebsock, M., Suzuki, K., Waliser, D., Wu, D., Kay, J., Gettelman, A., Wang, Z., Marchand, R., 2008. CloudSat mission: performance and early science after the first year in orbi. J. Geophys. Res. 113. https://doi.org/10.1029/2008JD009982. D00A18.

Tao, W.-K., Lang, S., Olson, W.S., Meneghini, R., Yang, S., Simpson, J., Kummerow, C., Smith, E., Halverson, J., 2001. Retrieved vertical profiles of latent heat release using TRMM rainfall products for February 1998. J. Appl. Meteorol. 40, 957–982. https://doi.org/10.1175/1520-0450(2001)040<0957:RVPOLH>2.0.CO;2.

Tao, W.-K., Smith, E.A., Adler, R.F., Haddad, Z.S., Hou, A.Y., Iguchi, T., Kakar, R., Krishnamurti, T.N., Kummerow, C.D., Lang, S., Meneghini, R., Nakamura, K., Nakazawa, T., Okamoto, K., Olson, W.S., Satoh, S., Shige, S., Simpson, J., Takayabu, Y., Tripoli, G.J., Yang, S., 2006. Retrieval of latent heating from TRMM measurements. Bull. Am. Meteorol. Soc. 87, 1555–1572. https://doi.org/10.1175/BAMS-87-11-1555.

Tao, W.-K., Takayabu, Y.N., Lang, S., Olson, W., Shige, S., Hou, A., Jackson, G., Jiang, X., Lau, W., Krishnamurti, T., Waliser, D., Zhang, C., Johnson, R., Houze, R., Ciesielski, P., Grecu, M., Hagos, S., Kakar, R., Nakamura, K., Braun, S., Bhardwaj, A., 2016. TRMM latent heating retrieval and comparison with field campaigns and large-scale analyses, AMS Meteorological Monographs—Multi-Scale Convection-Coupled Systems in the Tropics, 2.1–2.234 (Chapter 2). http://journals.ametsoc.org/toc/amsm/56.

Tapiador, F.J., 2008. A physically based satellite rainfall estimation method using fluid dynamics modelling. Int. J. Remote Sens. 29, 5851–5862. https://doi.org/10.1080/01431160802029677.

Tapiador, F.J., Checa, R., de Castro, M., 2010. An experiment to measure the spatial variability of rain drop size distribution using sixteen laser disdrometers. Geophys. Res. Lett. 37, 1–6. https://doi.org/10.1029/2010GL044120.

Tapiador, F.J., Turk, F.J., Petersen, W., Hou, A.Y., García-Ortega, E., Machado, L.A.T., Angelis, C.F., Salio, P., Kidd, C., Huffman, G.J., de Castro, M., 2012. Global precipitation measurement: methods, datasets and applications. Atmos. Res. 104, 70–97. https://doi.org/10.1016/j.atmosres.2011.10.021.

Tapiador, F.J., Navarro, A., Moreno, R., Jiménez-Alcázar, A., Marcos, C., Tokay, A., Durán, L., Bodoque, J.M., Martín, R., Petersen, W., de Castro, M., 2017. On the optimal measuring area for pointwise rainfall estimation: a dedicated experiment with fourteen laser disdrometers. J. Hydrometeorol. https://doi.org/10.1175/JHM-D-16-0127.1.

Turk, F.J., Rohaly, G., Hawkins, J., Smith, E.A., Marzano, F.S., Mugnai, A., Levizzani, V., 2000. Meteorological applications of precipitation estimation from combined SSM/I, TRMM and geostationary satellite data. In: Pampaloni, P., Paloscia, S. (Eds.), Microwave Radiometry and Remote

Sensing of the Earth's Surface and Atmosphere. VSP Int. Sci. Publisher, Utrecht, The Netherlands, pp. 353–363.

Vicente, G.A., Scofield, R.A., Menzel, W.P., 1998. The operational GOES infrared rainfall estimation technique. Bull. Am. Meteorol. Soc. 79, 1883–1893. https://doi.org/10.1175/1520-0477(1998)079<1883:TOGIRE>2.0.CO;2.

Voosen, P., 2016. Climate scientists open up their black boxes to scrutiny. Science. 354 (6311). https://doi.org/10.1126/science.354.6311.401.

Wilheit, T.T., Chang, A.T.C., Rao, M.S.V., Rodgers, E.B., Theon, J.S., 1977. A satellite technique for quantitatively mapping rainfall rates over the oceans. J. Appl. Meteorol. 16, 551–560.

Further Reading

Tapiador, F.J., Kidd, C., Levizzani, V., Marzano, F.S., 2004. A neural networks-based fusion technique to estimate half-hourly rainfall estimates at 0.1 degrees resolution from satellite passive microwave and infrared data. J. Appl. Meteorol. 0894-8763. 43 (4), 576–594.

Tapiador, F.J., Navarro, A., Levizzani, V., Cattani, E., García-Ortega, E., Huffman, G.J., Kidd, C., Kucera, P.A., Kummerow, C.D., Masunaga, H., Petersen, W.A., Roca, R., Sánchez, J.-L., Tao, W.-K., Turk, F.J., 2017. Global precipitation measurements for validating climate models. Atmos. Res. 197, 1–20.

CHAPTER

11

Measurement of Precipitation from Satellite Radiometers (Visible, Infrared, and Microwave): Physical Basis, Methods, and Limitations

Atul K. Varma

Space Applications Centre, Ahmedabad, India

1 BACKGROUND

Rainfall affects the lives and economies of a majority of the Earth's population. It brings both sorrow and celebrations. Millions of dollars of property and crops are damaged and thousands of lives are lost every year due to strong winds and flash floods associated with torrential rains from severe storms. These figures are far more devastating for densely populated countries like India with a population of nearly 1.3 billion. As of today, the flash floods associated with torrential rains are one the greatest challenges in weather/hydrological predictions. A deficiency of rainfall causes drought and failure of crops, that often results in unprecedented human miseries and a large-scale migration of humans and animals in search of water. In order to effectively safeguard property and lives from such fury of nature, it is of utmost importance to predict the rainfall accurately. Until numerical weather prediction (NWP) models accurately predict the rainfall field in different time scales, the rainfall measurements available in near real time may still be desired in many applications. The role of precipitation in the dynamics of the Earth's atmosphere is enormous. The latent heat released by precipitation is the largest source of diabatic heating in the tropical atmosphere and a significant heat source in the temperate latitudes. In order to make NWP models improve their predictability, a 3D distribution of rainfall measurements is desired for their initialization and verification, as well as assimilation in different time steps. Thus, irrespective of the development of NWP models and forecasting techniques, the necessity of precipitation estimation/measurement can never be substituted with model predictions alone. While the rainfall measurements are of paramount importance for atmospheric researchers, for the oceanographic community, their importance is felt through their effect on the density of the ocean

Remote Sensing of Aerosols, Clouds, and Precipitation
https://doi.org/10.1016/B978-0-12-810437-8.00011-6

© 2018 Elsevier Inc. All rights reserved.

surface waters. The rain is found to be isolated near the surface water layer from the thermocline, and reduce the vertical advection and entrainment (Godfrey and Lindstrom, 1989; Miller, 1976). In some of the oceanic regions, the salinity variation caused by rainfall generates horizontal density gradients, which in turn produce ocean currents, the same as those caused by thermal variations (Taft and Kessler, 1987; Cooper, 1988).

Precipitation is probably the longest observed and most widely recorded hydrological phenomenon, yet a lot still needs to be explored about its global distribution. Conventionally, rainfall over the ground is measured using rain gauges and radars. Rain gauges provide only point measurements and hence they poorly represent the spatial variability of the precipitation that varies from a few meters to several kilometers, making the sampling requirements for measuring its spatial distribution extremely stringent. The distribution of the rain gauges is far from adequate to present its meaningful distribution and variability for the study of various rain-induced events/processes, e.g., flash flood, dam failure, river catchments, etc. The rain gauge measurements are also not error-free; the most common cause of error in rain gauge measurements is due to winds, which some past studies show induces significant undercatch (e.g., Nespor and Sevruk, 1999). On the other hand, radars are better representative of the areal rain, but their coverage is limited only over land and are also affected by their high cost. The radar measurements often suffer due to poor calibration of radar reflectivity (Z) and also of the Z-R (reflectivity versus rain rate) relationship. These measurements also suffer from other uncertainties such as rain-path attenuation, contamination by ground return power, subpixel and vertical rain variability, and complex terrain effects (Krajewski et al., 2006). During severe weather conditions, ground observation networks often fail to respond due to various reasons, such as power failure, structural damage, flooding, human inaccessibility, etc. Thus, keeping in mind the difficulties and limitations of the ground measurements, the most reliable and convenient means to measure the precipitation over a large area is by using the satellite observations. The satellites have an advantage of providing frequent uniform coverage over a large area. They have the capability of continuous observation over an evolving storm (from geostationary platform) and provide near real time digital data for further analysis. However, the satellite measurements suffer from large errors. With the advent of imaging technology and superior spatiotemporal sampling from space-borne platforms, the satellite based rain estimations are increasingly becoming more and more accurate. For algorithm development and its fine tuning, and for the determination of the measurement accuracy, well calibrated surface observations representing satellite pixel size area are required. In order to cater to the requirement of such high quality surface measurements, a number of validation sites equipped with state-of-the-art instruments are developed. Most of these sites house a high-density network of rain gauges and a well calibrated radar. While the validation sites over the land are many, one of the few oceanic sites is located at Kwajalein Atoll in the Republic of the Marshall Islands in the western Pacific oceans (www.atmos.washington.edu/MG/KWAJ/GV.html), which is extensively being used for validating precipitation (e.g., Schumacher and Houze, 2000). An extensive validation plan drawn for Global Precipitation Measurement Mission is described by Petersen and Schwaller (2008) and Schwaller and Morris (2011).

In the next sections, the basis of rainfall estimation, procedures, limitations, and difficulties in rain validation are discussed.

2 SATELLITE RAINFALL ESTIMATION METHODS

The precipitation measurement from space-borne sensors is conventionally carried out in visible, thermal, infrared, and microwave regimes of the electromagnetic spectrum. Based on observation wavelength, the satellites assume rainfall estimations can be divided into three broad categories:

(1) Visible (Vis) and Thermal infrared (TIR) methods
(2) Microwave (MW) methods
(3) Merged rainfall methods

2.1 Visible (Vis) and Thermal Infrared (TIR) Methods

Visible (Vis) and infrared (IR) techniques are grouped together because they share common characteristics. Both provide indirect assessment of rainfall as Vis/IR radiances do not penetrate clouds and hence rain droplets inside the clouds are not directly sensed by them (Barrett and Martin, 1981; Bhandari and Varma, 1995). Because of the indirect relationship between the satellite measured Vis or IR radiances and rainfall, the techniques developed for one region and for a particular temporal and spatial scale do not necessarily work equally well in another region/scale.

The basic premise for rain measurement using visible band images is that a brighter cloud is a thicker cloud, which brings more rain. This is not always true as brightness is not a function of thickness alone. The factors governing cloud brightness are determined by the physics of optics that requires gaining insight of the scattering theory of electromagnetic radiations. The dimension of the cloud and its orientation with respect to incident beam is important for determining cloud brightness. The other factors that may affect the cloud brightness are phase (water/ice/mixed phase), size and density of cloud droplets, etc. (Barrett and Martin, 1981). Generally, the visible band data alone is rarely used for rain estimation because the relationship between cloud brightness and rainfall is weak, brightness of the Vis images are highly dependent on sun inclination angle and unavailability of visible band images during nighttime.

The basic physical premise for rain measurement using the TIR band satellite observations is that the cloud top temperature is an indicator of cloud top height, and that the higher and colder cloud tops resulting from stronger convection produce more rain. This assumption is not always true because often clouds that are colder are not always rainier. However, it is noticed that the relationship between rain rate and brightness temperature in TIR images is stronger compared to that between rain rate and cloud brightness in Vis images.

The prime advantage of Vis/IR observations for rain estimation is that such observations are available from a geostationary platform every 15–30 min. The other advantage is due to their finer spatial resolution compared to passive microwave observations. For over the last half a century, many techniques for rain estimation using Vis/IR observations were developed. In some of these techniques, using observations from a single channel (Vis or IR) different cloud types were identified that produced different rain rates (referred to as cloud indexing methods). Another set of techniques were developed that were based on the evolution of clouds in a sequence of images (referred to as life history methods). Some other techniques that took advantage of both Vis and IR observations to identify and assign higher rain rates to brighter and colder clouds were referred to as Bispectral techniques. There are some more advanced techniques that employed a conceptual model and other ancillary observations to estimate the rain rate (referred to as cloud model techniques).

The earlier important work on precipitation estimation can be traced to Woodley et al. (1972) and Griffith et al. (1978); these studies concluded that bright clouds in the Vis imagery produce more rainfall than darker clouds. Further, they found that the growing area of bright clouds in the Vis and cold tops in the TIR imagery produce more rainfall than those not expanding. In contrast, they found that with the same cloud characteristics, that is, area, brightness/brightness temperature (*Tb*) in a decaying stage produce little or no rainfall. The rainfall increases with the merging of cumulonimbus clouds, and the heaviest rain occurs where storms are most vigorous in the upwind portion of the clouds.

Despite of an already known complex relationship between cloud brightness/temperature in Vis/IR images and surface rain rate, a very simple scheme was proposed by Arkin (1979) which linearly related the fraction cloud cover within a large grid box ($>1^\circ \times 1^\circ$) to rain rate. Arkin carried out his study using observations collected during GARP (Global Atmospheric Research Program) Atlantic Tropical Experiment (GATE). He found that a pixel is rainy if its brightness temperature (*Tb*) measured in thermal infrared band (10.5–12.5 μm) <235 K. This scheme has many limitations. First, it is not able to account for rain coming from warm clouds and rain associated with orography. According to Liu and Zipser (2009), the warm clouds contribute up to a 20% area over oceans and 7.5% over land. Second, the Arkin method does not account for rain evaporated below the clouds when the clouds' air is unsaturated. Third, and the most severe limitation, is that it restricts maximum daily rain to 71.2 mm. Nevertheless, the Arkin method gained popularity due to its simple approach and for its ability to estimate large-scale rain with reasonable accuracy over tropical regions dominated by convective clouds. The Arkin method is successfully applied to GOES satellites and the distributed precipitation product is referred to as GOES Precipitation Index (GPI). The GPI products (Joyce and Arkin, 1997) in a $2.5^\circ \times 2.5^\circ$ grid in a daily, pentad, and monthly scale are routinely available as GPCP (Global Precipitation Climatology Project) products from the National Oceanic and Atmospheric Administration (NOAA) ftp site (ftp://ftp.cpc.ncep.noaa.gov/precip/gpi). Utilizing known diurnal variability of cloudiness over the India region Bhandari and Varma (1995) suggested a lesser number of satellite observations for precipitation estimation using Arkin's method. Arkin et al. (1989) found that their method is applicable to the India region also. They, however, suggested a threshold of 265–270 K to define a raining area over the orographic region along the west coast of India. The GPI referred to as Quantitative Precipitation Estimation (QPE) is also regularly produced in a $1^\circ \times 1^\circ$ grid every 3-h from the INSAT-3D satellite over the India region. Fig. 1 shows a typical example of INSAT-3D derived QPE for the time period from 0300 UTC to 0600 UTC on August 19, 2016.

Notwithstanding the shortcomings of GPI (Arkin's method) as described in the last paragraph, the Arkin method provided reasonable estimates of precipitation over large spatial and temporal scales. There, however, existed a persisting requirement for finer spatial and temporal scale estimates, especially for nowcasting and for undertaking relief and rescue operations during disasters associated with intense rain events. In order to address such requirements, Scofield (1987) provided a scheme for rainfall estimation and outlooks for 30 min based on a man-computer interactive scheme called IFFA (Interactive Flash Flood Analyzer). His scheme required a dedicated computer system with specialized software (referred to as Man-Computer Interactive Data Assessment System (MaCIDAS)) and an experienced meteorologist. The job of the meteorologist was to use his experience and, through available specialized image processing tools, detect the active convective systems in the cloud image and assign the appropriate rain amount to them. The active area in a storm is the area of strong updrafts usually in the upwind portion of the storm where heavy rain occurs. The wind referred to is shear wind between thunderstorm motion and anvil level wind. Scofield (1987) has provided a number of clues to identify the active area of the thunderstorm in the Vis and TIR observations. The assigned rainfall values to the active area are further modified for over-shooting cloud

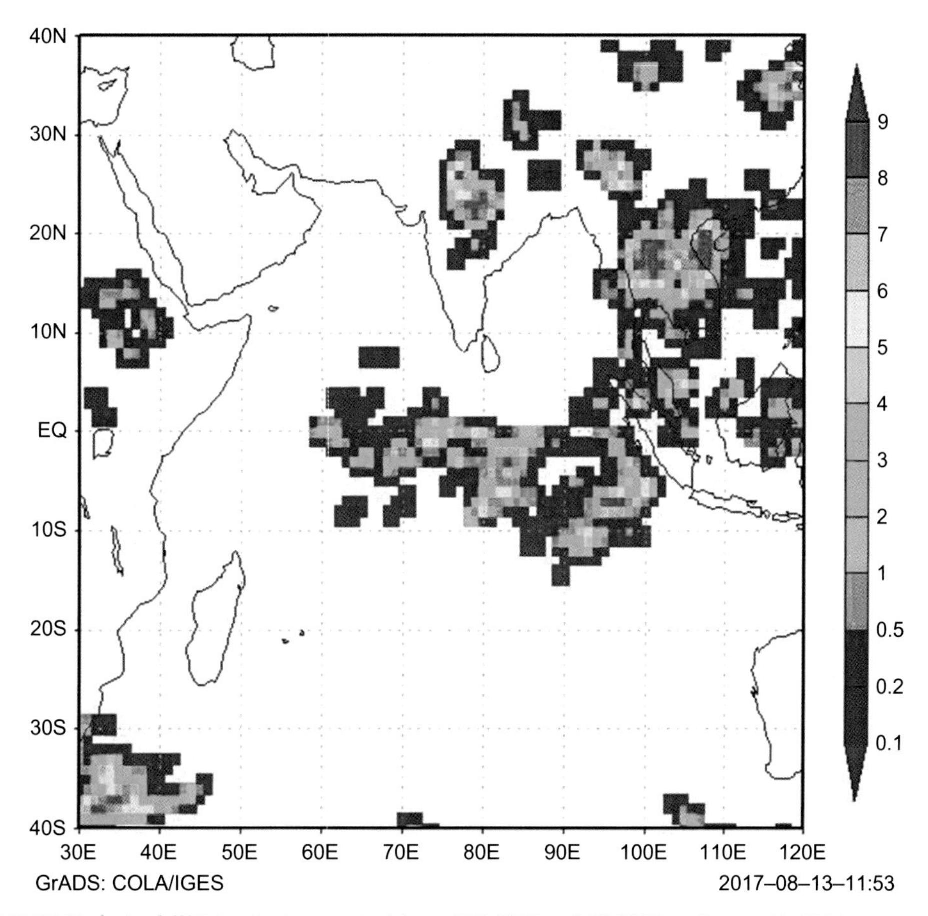

FIG. 1 INSAT-3D derived QPE for the time period from 0300 UTC to 0600 UTC on August 19, 2016.

tops, available moisture, and location of the equilibrium level, etc. The IFFA was a very successful method that provided a 30-min estimation of precipitation but needed considerable time for image processing, interpretation, and computation. A meteorologist was able to analyze only one precipitating system at a time. The Scofield (1987) method was very subjective and cumbersome and was highly dependent upon the skill of a trained meteorologist. The necessity of training is emphasized by Scofield himself in his paper wherein he asserts "this scheme requires a dedicated person(s) round the clock to run this scheme." In order to overcome the subjective nature of IFFA and eliminate continuous human intervention, and to reduce the processing time for rainfall estimation, especially when several storms are present in an image, it was found to be necessary to automatize the IFFA while retaining all its features.

In an attempt to automate the Scofield technique, the Auto-Estimator (A-E) technique was developed (Vicente et al., 1998). This scheme automated many of the features of IFFA and provided 15-min rain estimations at pixel-scale. The precipitation estimates were calibrated against radar measurements and an exponential relationship between rain rate and brightness temperature measured at 10.7 μm was developed. The adjustments were done for precipitable water (PW) and relative humidity (RH) (called PWxRH factor), the equilibrium level correction using environmental parameters from the Eta model, and orography correction using 850 mb u and v winds from the Eta model and elevation model. The A-E method required radar measurements to identify raining pixels. If radar rain is found to be zero, the A-E defines the pixel as nonraining. The PWxRH factor used precipitable water in inches and relative humidity in fraction. The probability of rain was found to be nil for PWxRH < 1 for GOES pixel of size (4×4 km^2). The PW and RH correction was carried out for increasing the precipitation values for wetter environment and reducing for a dryer environment. The equilibrium level (or level of neutral buoyancy) correction was applied to enhance the precipitation amount from warm clouds that is not supported by the observed brightness temperature values. The orographic correction was carried out to enhance the precipitation value along the windward side and to reduce in the leeward side of the slopes (Vicente et al., 2002). The orography correction gradient of slopes along the wind direction above the atmospheric boundary layer (i.e., usually at 850 mb level) is computed. The slopes are computed from the digital elevation model at pixel scale (for example, ETOPO2 with $2' \times 2'$ resolution from National Geophysical Data Centre, NOAA, United States). The correction is a function of wind speed and the gradient of slope. The orographic correction is applied to the brightness temperature of the pixel under consideration which results in reduction/enhancement of *Tb*. As the relationship between *Tb* and rain rate is nonlinear, this may significantly affect the rainfall amount at the pixel.

It is observed that A-E method often produces false rain from thin cirrus clouds. It also required radar observations in near real-time, which are not readily available in many parts of the world. This limited the use of A-E outside North America, and also over the vast oceans. In order to further improve the A-E method, Scofield and Kuligowaski (2003) (which is also described by Ramirez-Beltran et al., 2008) proposed further modifications to A-E referred to as the Hydro-Estimator (H-E). The H-E was made fully automated without the limitations associated with A-E.

In the H-E method, the cirrus clouds are avoided by analyzing the brightness temperature of the pixel under consideration with respect to its neighboring pixels. Rain (R) from core and noncore areas of a storm is identified and different R-Tb relationships are suggested. This allows higher precipitation rates for the convective cores. For the convective core, the relationship between Tb (K) and the core rain rate (R_c) (mm h^{-1}) is adopted from the A-E method and is given as:

$$R_c = ae^{-bTb^{1.2}}$$

The coefficients a and b are dynamically calculated for each pixel for the given value of precipitable water (PW) from the NWP model fields. The maximum possible rain value at any pixel is dependent upon the availability of PW. Thus, the maximum precipitation becomes a function of available moisture in the atmosphere.

For a noncore, the relationship between Tb and noncore rain rate (R_s) is given as:

$$R_s = (250 - Tb) \times \frac{R_{max}}{5} \text{ and } R_s < 12\text{mm/h}; \text{ and } R_s < R_c$$

In H-E, the precipitation at a pixel is considered the combination of both core and noncore precipitation. The fraction of core and noncore rain is worked out by analyzing an area of 101×101 pixels surrounding the pixel under consideration, and the mean (μ) and standard deviation (σ) of Tb in this area is determined. The μ and σ are used to determine Z-factor, where,

$$Z = \frac{\mu - Tb}{\sigma}$$

All pixels with $Z \geq 1.5$ are sufficiently colder from the mean to be considered as convective. Hence the value of Z is restricted to 1.5. If $Z < 0$; $R = 0$, i.e., the pixel is either cirrus or inactive convective,

Otherwise,

$$R = \frac{\left[R_c \times Z^2 + R_s \times (1.5 - Z)^2\right]}{\left[Z^2 - (1.5 - Z)^2\right]}$$

If $Z = 1.5$, the pixel rain rate R reduces to core-rain only. On the other hand, if $Z = 0$, the pixel rain rate R is determined by purely noncore rain. The above relation provides the initial precipitation amount that is further modified for the wetness/dryness of the atmosphere and for warm rain situations. The following modifications are applied in H-E rain:

- Correction for wet/dry environment
- Correction for equilibrium level/warm top
- Orography correction

NOAA regularly produces H-E products over different regions of the globe; these products can be accessed from the website (http://www.orbit.nesdis.noaa.gov/smcd/emb/ff/). The H-E technique is also applied to 30-min INSAT-3D observations over the India region. Fig. 2 shows a typical instantaneous rain on July 22, 2016 at 1200 UTC from INSAT-3D. The figure shows a strong monsoon activity over the India region. The H-E at NOAA is implemented on the MaCIDAS (Man Computer Interactive Data Assessment System) and uses environmental fields derived from the NAM model, whereas the H-E adopted for INSAT-3D is independently coded in FORTRAN so that runs on Linux and Windows and uses the NCEP NWP model forecast fields. The H-E scheme running for INSAT-3D, makes corrections to the NCEP model and derived relative humidity and water vapor over higher altitudes using a histogram equalization approach. This removes the observed dry bias in the model fields over higher altitudes and improves the precipitation estimation (Varma and Gairola, 2015; Kumar and Varma, 2016).

Apart from H-E, the other IR based operational algorithm at NOAA/NESDIS is referred to as GOES Multispectral Rainfall Algorithm (GMSRA) that makes use of all the 5 channels (visible: 0.65 μm, near infrared: 3.9 μm, middle infrared: 6.7 μm, and thermal infrared: 11 and 12 μm) of GOES imager for rain estimation (Ba and Gruber, 2001). These channels are used for identifying optically thick clouds and thin cirrus clouds, overshooting cold tops, determining cloud particle size, and cloud growth rate, etc. The GMSRA offers different relationships for determining the rain for different regions. For identifying thick cloud during daytime, the visible reflectance value >0.40 is used. Like H-E, nonprecipitating cirrus clouds are identified using a temperature gradient measured at 11 μm channel with respect to surrounding pixels as well as effective cloud radius near the cloud top derived using solar irradiance at 3.9 μm. The difference of brightness temperature measured at 11 and 6.7 μm (Inoue, 1997) and cloud top temperature at 11 μm is also used for rain identification from very deep convective cores. The rain rate is calculated using predetermined probability distributions of rain rates and a mean rain rate using concurrent observations of GOES and gauge-adjusted-radar rain rates. Further corrections in the rain rate are applied based on cloud growth in the two subsequent images and available moisture.

In the GMSRA method, the various cloud properties such as cloud thickness, cloud top temperature (CTT), CTT gradients, cloud growth and cloud particle size, etc. are utilized, which were used for rainfall identification and estimation by various researchers (e.g., Arkin and Meisner, 1987; Vicente et al., 1998; Adler and Negri, 1988; Vicente et al., 1998; Rosenfield and Gutman, 1994) in the past. The innovation in this technique is that it combines all the above

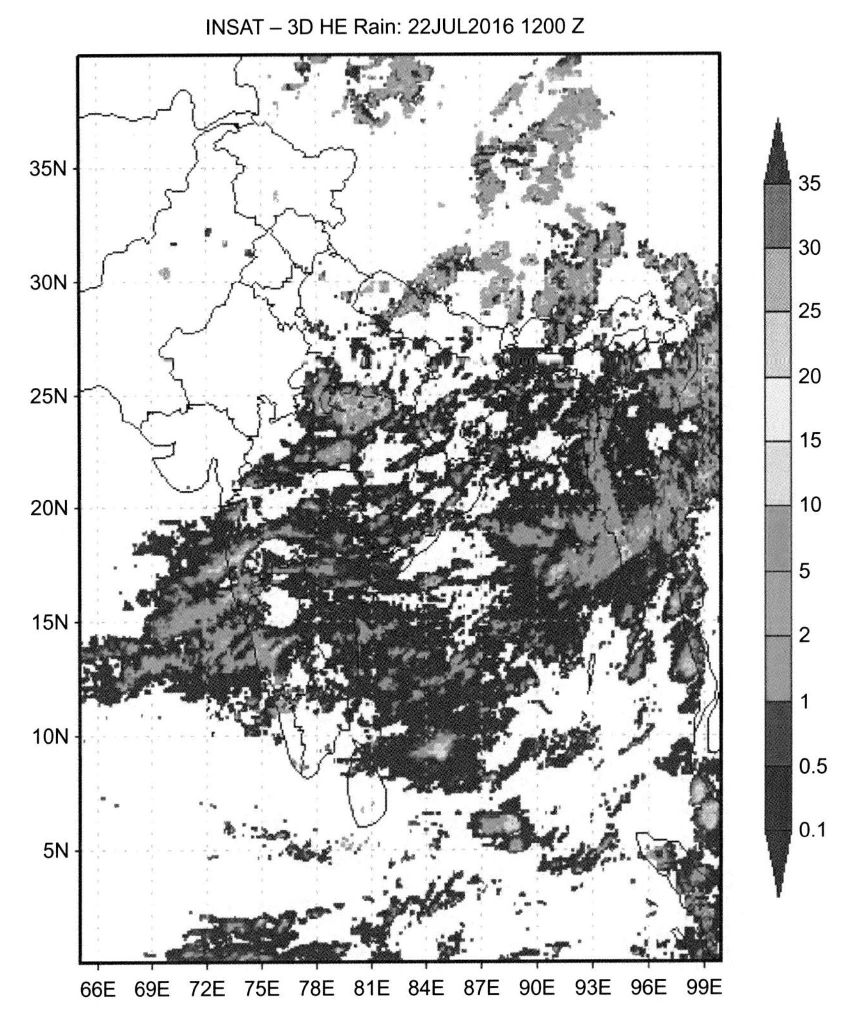

FIG. 2 Typical instantaneous rain on July 22, 2016 at 1200 UTC from INSAT-3D. Figure shows a strong monsoon activity over the India region.

cloud properties in a single and comprehensive rainfall estimation scheme.

2.2 Passive Microwave Methods

At microwave wavelengths, precipitation-size drops interact strongly whereas the cloud droplets interact weekly with the radiation, which allows the detection of precipitation by microwave radiometers. In the microwave regime of electromagnetic spectrum, the clouds are transparent, especially at higher wavelengths (e.g., <35 GHz). Due to direct interaction of microwave radiances with hydrometeors, the microwave based precipitation estimation techniques offer more direct measurement of rainfall than the Vis/IR techniques. The most critical disadvantage of passive microwave precipitation estimation is that they have poor spatial and temporal resolution due to their large wavelength and non-geostationary orbit of the spacecraft.

The physical basis of passive microwave rainfall estimation schemes is somewhat more complicated than Vis/IR techniques with intrusion of both emission and scattering processes at play. Spencer et al. (1989) have calculated the scattering and absorption properties of rain for the three main wavelengths that have been used to measure precipitation. They observed: (1) Ice scatters do not absorb the microwave radiation, (2) Liquid drops both absorb and scatter, but absorption dominates, and (3) Scattering and absorption both increase with frequency and with rain rate, and scattering by ice particles increases much more rapidly with frequency than scattering by liquid. At low microwave frequencies ($\sim < 22$ GHz), the wavelength is large with respect to water droplets that makes scattering coefficient very small, and hence in this regime the absorption dominates over scattering. On the contrary, at higher frequencies ($\sim > 60$ GHz), the trend is reversed and the scattering dominates over the absorption. Thus, at a very large wavelength, the rain column is almost transparent and the satellite-borne radiometer receives a significant amount of radiances from the surface. As the frequency increases (or wavelength decreases), the radiometer receives a major contribution from the different layers of the rain structure. At very high frequencies, microwave radiances are scattered by the ice on the cloud top and the radiometer does not directly sense the rain below. Thus, the low-frequency channels, for which the absorption dominates, interact more directly with hydrometeors compared to the high-frequency channels (e.g., 85/89/91 GHz). Hence the rain measurement using low frequency channels is more direct in nature. The passive microwave measurements are also effected by water vapor, cloud liquid, oxygen, surface temperature, and the surface emissivity, etc. This makes it very difficult to differentiate rain from the background, especially in the low rain regime.

For a given vertical profile of environmental parameters such as temperature, humidity, and hydrometeors, an appropriate radiative transfer model (e.g., Liu, 1998) can be used to calculate microwave brightness temperatures for different frequencies and polarizations at the top of the atmosphere as a function of rain rate (Wilheit et al., 1977). While it may not be possible to precisely know the profiles of hydrometeors, such model simulations provide insight into behavior of brightness temperatures in the presence of rain. Fig. 3 shows a typical behavior of brightness temperatures over oceans at 53° incidence angle measured at 10, 19, 37, and 85 GHz H-polarized frequencies as a function of rain rate. It can be observed that the brightness temperature at low frequency channels (10–37 GHz) first increases with rain rate (e.g., for 19 GHz upto ~15–20 mm h^{-1} of rain rate), and then decreases. The decrease in the brightness temperature is due to increasing droplet size and hence the scattering coefficient at higher rain rates. Fig. 3 also shows the variation of a horizontally polarized 85 GHz brightness temperature with rain rate for Marshall Palmer distribution of ice particles and a given freezing level. In this case, it can be observed that the brightness temperature decreases with increasing

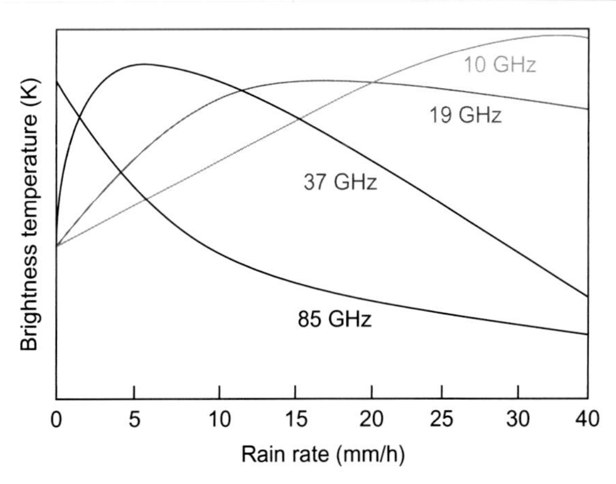

FIG. 3 A typical behavior of H-polarized brightness temperatures at 10, 19, 37, and 85 GHz with rain rate.

rain rate due to the increasing value of the scattering coefficient. It is obvious from Fig. 3 that over oceans, which is a cold background at microwave frequencies (emissivity ~0.4), rain rates can be measured at suitable frequencies with either emission or scattering based algorithms depending upon which process dominates at that frequency. The emission based algorithms are not able to differentiate rain over the land surface due to highly varying emissivity and warm background. However, the scattering based channels (e.g., 85/89/91 GHz) at which the brightness temperature decreases with rain rate, are employed for rainfall estimation over the land surface.

A number of satellite missions with onboard passive microwave radiometer/sounder are currently being used for rainfall estimation. Table 1 shows such radiometers and sounders and their frequency channels. Table 1 also shows channels available on radiometers as well as in or around the 22.235 water vapor absorption line. The table also shows sounders working in oxygen complex between 50 and 60 GHz for vertical temperature sounding and near water vapor absorption line at 183.31 GHz for vertical humidity sounding. While the most appropriate channels for precipitation retrieval are available on radiometers, sounder channels are also exploited for precipitation retrieval (e.g., Varma et al., 2016).

Rain measurement indices: Apart from brightness temperature (Tb), several Tb based indices are found to be more sensitive of precipitation than Tb itself. Some of the commonly used indices are described below:

Polarization Difference:

$$D = Tb(V) - Tb(H)$$

The difference between vertical and horizontal brightness temperature is sensitive to precipitation as difference between Vertical (V) and Horizontal (H) polarization decreases with scattering which increases with rain rate.

Scattering Index:

The scattering index as defined by Grody (1991) provides the depression in the 85GHz brightness temperature due to scattering by hydrometeors. The nonraining 85 GHz Tb is approximated by nonscattering lower frequency channels (19 and 22 GHz). It is expressed as:

$$SI = -451.9 - 0.44 \times Tb_{V19} - 1.775 \times Tb_{V22} + 0.00575 \times Tb_{V22}^{2} - Tb_{V85}$$

The value of SI >10 indicates the presence of rain (Grody, 1991; Varma et al., 2015). The

TABLE 1 The Presently Available Passive Microwave Radiometer/Sounder on Different Satellite Missions, and Available Frequency Channels

Satellite-Radiometer/Sounder	Channels With Frequency in GHz and Polarization (In Bracket)
GPM-GMI	10.65 (V/H), 18.7 (V/H), 23.8 (V), 36.64 (V/H), 89.0 (V/H), 165.6 (V/H), 183.31 ± 3 (V), and 183.31 ± 7 (V)
DMSP-SSMIS	19.35 (V/H), 22.235 (V), 37 (V/H), 50.3–63.28 (V/H), 91.65 (V/H), 150 (H), 183.31 ± 1 (H), 183.31 ± 3 (H), and 183.31 ± 7 (H)
METOP-B/NOAA 18 and 19/NPP (AMSU A1/A2/B)	23.8 (V), 31.4 (V), 50.3-63.28 (H), 89.0 (V), 157 (V), 183.31 ± 1 (H), 183.31 ± 5 (H), and 190.311 (V)
GCOM-W1	6.925 (V/H), 10.65 (V/H), 18.7 (V/H), 23.80 (V/H), 36.5 (V/H), and 89 (V/H)
MT-SAPHIR	183.31 ± 0.2 (H), 183.31 ± 1.1 (H), 183.31 ± 2.8 (H), 183.31 ± 4.2 (H), 183.31 ± 6.8 (H), and 183.31 ± 11 (H)

SI increases with rain rate. Grody (1991) developed SI for SSM/I channels, which can further be calibrated using radiative transfer simulations (Varma et al., 2015).

Polarization Corrected Temperature (*PCT*):

The PCT is computed from a weighted difference between the vertical and horizontal polarizations of the 85–89 GHz channels, which segregates the effect of attenuation from hydrometeors, irrespective of the surface temperatures. At 85/89 GHz, due to scattering from the cloud ice, very cold brightness temperatures (below 100 K) can be observed. Such low temperatures can also result from surface water bodies with low emissivity leading to ambiguity in the rain identification. Fortunately, using polarization diversity of the measurements, the effect of rain that produces unpolarized radiances and surface water that produces highly polarized radiances can be segregated. In order to remove the surface effects, PCT was suggested by Spencer et al. (1989). Unlike 85 GHz, PCT is monotonic when water vapor, cloud water, and precipitation increase (Liu and Curry, 1998). Spencer et al. (1989) showed that PCT for surface water is higher than that for cloud water, and PCT for cloud water is higher than that for precipitation. Kumar et al. (2009) also showed how PCT is able to identify and thus allow removal of false precipitation signatures from 85 GHz observations from SSM/I. PCT values vary inversely with precipitation, and

$$\mathrm{PCT} = \alpha TB_{89V} - \beta TB_{89H}$$

where, $\alpha = 1.818$ and $\beta = 0.818$ (Spencer et al., 1989).

The PCT for other channels is also attempted. They, however, offer a smaller dynamic range due to smaller value of single scattering albedo. Lee et al. (2002) have given PCT for 37 GHz channels of SSM/I with values of $\alpha = 2.18$ and $\beta = 1.18$.

2.2.1 Goddard Profiling Algorithm

The Goddard Profiling Algorithm (GPROF) developed by Kummerow et al. (1996, 2001) was intended to provide simultaneous measurements of surface precipitation and the vertical profiles of the hydrometeors. This is possible using a number of frequency channels that provide radiances from different layers of the precipitating cloud. The GPROF algorithm is an operational algorithm that is applied to radiometric measurements of brightness temperatures from Tropical Rainfall Measuring Missions (TRMM) Microwave Imager (TMI) and Global Precipitation Measurement (GPM) Microwave Imager (GMI). The algorithm is based on Bayes' Theorem. Bayesian theory is a general approach to solving inverse problems such as retrieving dependent geophysical parameters from a set of remote sensing observations. The Bayer's theorem of conditional probability is defined as

$$p_{\mathrm{post}}(a|b) = \frac{P_{\mathrm{f}}(b|\,a)P_{\mathrm{pr}}(a)}{\int P_{\mathrm{f}}(b|\,a)P_{\mathrm{pr}}(a)da}$$

where, "b" is the measurement vector (e.g., set of brightness temperatures) with associated respective measurement error. "a" is the set of state vector (e.g., surface precipitation and all hydrometeor profiles) that effects the "b". $P(a|b)$ is conditional probability of "a" for given "b". The P_{post} is posterior and P_{pr} is prior probability density function. The above equation relates the inverse problem to the forward problem. The $P_{\mathrm{pr}}(a)$ is known from prior measurements of a certain profile "a". The $P_{\mathrm{f}}(b|a)$ is known from the forward radiative transfer model. The denominator normalizes the integral. The $P_{\mathrm{post}}(a|b)$ is the probability distribution of the state vector given the measurement vector. The application of Bayer's theorem produces a probability distribution of hydrometeor profiles, which is integrated to calculate the mean value of a vector.

In practice, radiative transfer simulations using hydrometeors and other environmental profiles are used to generate a large a prior database of brightness temperatures and corresponding environmental and hydrological profiles to calculate $P_{pr}(a)$ and $P_f(b|a)$. Ideally, such a prior database must have the representation of all possible environmental conditions. The observations of the profile of hydrometeors, especially under severe weather conditions, are very limited. In order to meet the requirement of such a large database, the hydrometeors and other environmental profiles of different seasons and geographical regions generated by cloud resolving model (CRM) are used. Presently, high-resolution mesoscale numerical weather forecast models, such as Weather Research Forecast (WRF), can be used as CRM.

2.2.2 Rain Retrieval Using Humidity Sounder Channels Near 183.31 GHz

The humidity sounder, SPHIR, onboard Megha-Tropiques (MT) satellite operates in 6 H-polarized frequency channels around the water vapor absorption line at 183.31 GHz. The central frequency of SAPHIR channels are 183 ± 0.2, ± 1.1, ± 2.8, ± 4.2, ± 6.8, and ± 11.0 GHz. SAPHIR is a cross scanning instrument that sweeps $\pm 50°$ on both sides of the nadir over a swath of ~1700 km with an effective field of view ranging from ~10 km at nadir to ~22 km at two ends of the swath.

In order to reduce the false alarms, most of the passive microwave rain estimation schemes are in two steps. In the first step, the rain affected pixels are identified and in the second step, the quantitative rain estimation is made (e.g., Varma et al., 2003). Varma et al. (2016) developed a rain identification and estimation algorithm using channels close to 183.31 GHz SAPHIR-humidity sounder on Megha-Tropiques (MT). Based on observed brightness temperatures at all the six channels of the SAPHIR, they proposed a probabilistic rain identification algorithm. They showed that false alarms and the missing rain values were always low (<0.9 mm h^{-1}). Further they proposed a radiative transfer simulations supported rain retrieval algorithm. Fig. 4A shows the monthly rain rate from SAPHIR and TRMM 3B42V7 for the month of August in 2016. The TRMM 3B42-V7 is the most advanced rain algorithm that combines both IR and microwave measurements and then adjusts the rain values with surface observations. Fig. 4B shows in stantaneous rain from SAPHIR associated with hurricane Mathew in the east Atlantic over the Caribbean Islands on October 6, 2016 at ~0206 UTC.

2.2.3 Multichannel Merged Rainfall Algorithm

The IR measurements offer high spatial and temporal sampling, whereas microwave measurements are more direct and hence considered to be more accurate. Therefore, merging of IR and microwave observations has the potential of providing the most accurate measurements in smaller time scales. Several algorithms that merge IR and microwave and are also calibrated with surface observations are being used operationally to produce rainfall products in finer scales. For example, TRMM-3B42 is one such precipitation product that combines IR and microwave. The TRMM-3B42 algorithm, also referred to as TRMM Multisatellite Precipitation Analysis (TMPA), first calibrates available IR observations with Microwave Precipitation Measurements from Low Orbiting Satellites (LEO) on monthly scale. Resulting coefficients are then applied to IR observations from other geostationary satellites. The final adjustment in the measurements is done with surface observations (Huffman et al., 2007; Yong et al., 2015). In practice, TMPA estimates are produced in four stages—(1) all available microwave precipitation estimates are combined in a $0.25° \times 0.25°$ latitude-longitude grid after being calibrating by using probability matching as suggested by Miller (1972), (2) IR precipitation estimates are generated using calibrated microwave observations, (3) generation of merged microwave-IR

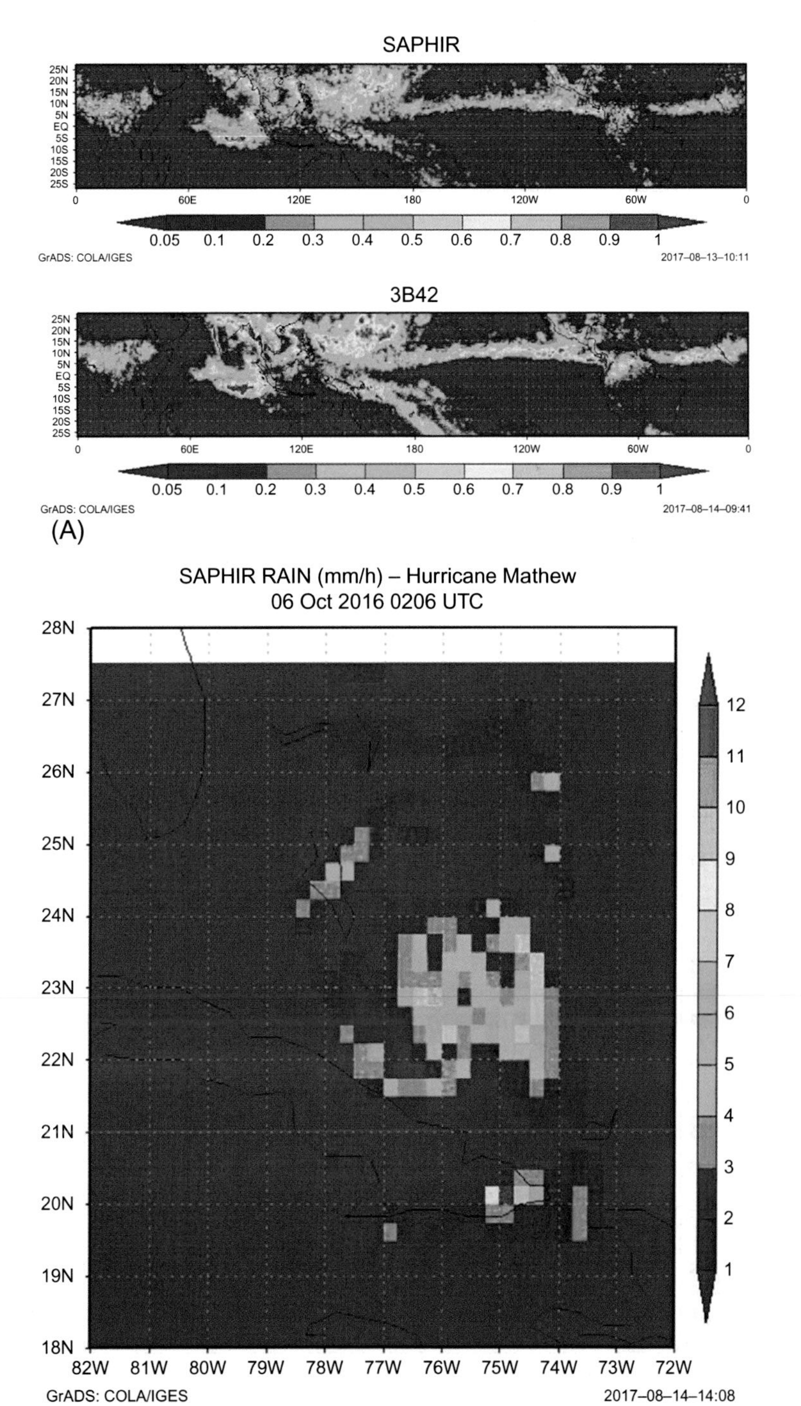

FIG. 4 (A) Monthly rain rate from SAPHIR and TRMM 3B42V7 for the month of August 2016 and (B) instantaneous rain averaged in $0.25° \times 0.25°$ grid from SAPHIR associated with hurricane Mathew in the east Atlantic over Caribbean Islands on October 6, 2016 at ~0206 UTC.

precipitation which is produced by filling the gaps in calibrated microwave measurements with microwave-calibrated IR measurements, and (4) rain gauge data based correction is applied by first computing the ratio of monthly gauge precipitation and monthly accumulated merged IR-microwave precipitation in a $0.25° \times 0.25°$ latitude-longitude grid and then applying it in 3-h microwave-IR merged accumulations. The TRMM-3B42 is available in 3-h accumulations in a $0.25° \times 0.25°$ grid over global land and oceans between 50°S and 50°N. The Integrated MultisatellitE Retrievals for GPM (IMERG) is another product from GPM constellation, which is similar to TMPA, available every 30-min in a $0.1° \times 0.1°$ grid (Huffman et al., 2015). Another product is from JAXA called GSMaP, which is available hourly in a $0.1° \times 0.1°$ grid (Aonashi et al., 2009).

The active measurement of precipitation is not under the scope of this chapter. However, several algorithms exist which utilize the availability of passive and active measurements from the same platform for precipitation estimation. One such algorithm was proposed by Ghosh et al. (2014) using passive and active measurements from a scatterometer onboard the Oceansat-2 satellite. Scatterometer is essentially a radar that measures the radar cross-section at larger incidence angles. The Oceansat-II scatterometer has a pencil beam antenna with dual feed assembly generating two rotating beams at two different incidence angles and polarizations. The scatterometer is also used for measuring apparent brightness temperature (*Tb*) at both V-polarization from outer beam and H-polarization from inner beam from instrument noise with an appropriate model. It may, however, be noted here that *Tb* measurements from scatterometer are relatively noisy due to the fact that it is not an optimal radiometer. The scatterometer is traditionally used for wind vector measurement over oceans. The basic Level-1 measurements the Oceansat-2 scatterometer provides is VV and HH polarized radar back scattering coefficient and V and H polarized brightness temperatures at Ku band (13.8 GHz). The Oceansat-2 scatterometer has a large swath of about 1400 and 1840 km for its inner and outer beams which, if exploited for precipitation estimation, can match spatial and temporal sampling of other operational precipitation measurement missions like SSMIS for global rain measurement. The scatterometer measurement of ocean surface wind vectors are severely affected by falling rain, which requires flagging and possible correction. Thus, it is necessary to have precipitation detection/measurement along with scatterometer measurements of wind vectors. Using a neural network based setup, Ghosh et al. (2014) provided procedures for rain detection and measurement over global oceans. They used the concurrent Oceansat-2 radiometer and scatterometer observations along with National Centre for Environmental Prediction (NCEP) NWP model derived wind speed and direction, total precipitable water (TPW) and relative humidity (RH), and Tropical Rainfall Measuring Mission (TRMM) and Advanced Microwave Scanning Radiometer (AMSR-E) for Earth Observation Satellite (EOS) derived rainfall values. They applied NN separately for rain identification and measurement and for five different geographical regions referred to as region I (25°N–25°S), region II (15°N–45°N), region III (35°N–70°N), region IV (15°S–45°S), and region V (35°S–70°S). On applying NN for rain identification, they found rain identification accuracy of about 93%, 87%, 90%, 79%, and 85%, and no-rain detection accuracy of about 97%, 87%, 86%, 84%, and 86% for these regions. In the second step, the NN is applied for rain estimation at the rain affected pixels, and the rms error of rain estimation for regions I to V (rain rates varying from >0 to approximately 45, 25, 25, 45, and 20 mm h^{-1}) is found to be 1.86, 0.69, 0.47, 0.56, 0.46 mm h^{-1}, respectively. Fig. 5 shows the

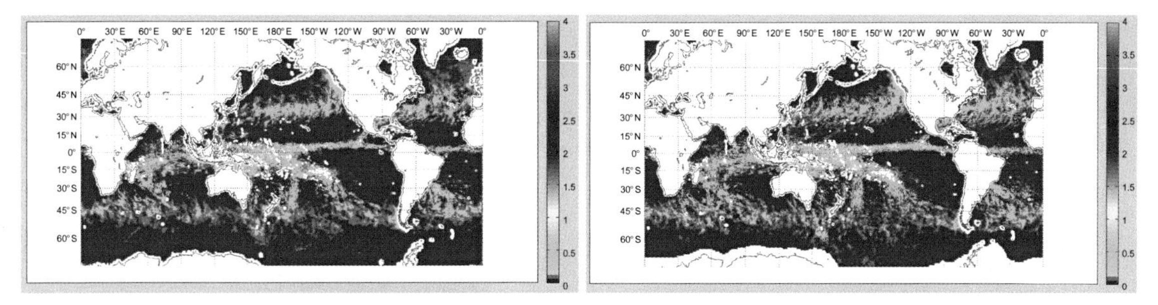

FIG. 5 Rain rate (mm h^{-1}) measurement from Oceansat-II scatterometer in the spatial resolution is 0.25° × 0.25°—monthly averaged for January 2010 (left panel) and corresponding monthly averaged rain from AMSR-E (right panel). *From Ghosh, A., Varma, A.K., Shah, S. Gohil, B.S., Pal, P.K., 2014. Rain identification and measurement using Oceansat-II Scatterometer observations. Remote Sens. Environ. 142, 20–32.*

monthly averaged rain from the month of January 2010 from the Oceansat-2 radiometer/scatterometer and AMSR-E.

Apart from the radiometer-scatterometer, the combined use of radiometer-altimeter can also be used for rain estimation. The altimeter is a nadir looking radar system (Chelton et al., 1989) that primarily provides very precise measurements of Sea Surface Height (SSH), Significant Wave Height (SWH), and Ocean surface Wind Speed (OWS). The SSH is derived by precisely measuring the travel time taken by the radar pulse from satellite to the surface and back. For the precise measurement of SSH, various corrections particularly those related to propagation of the radar pulse through atmospheric, tidal, and sea surface are determined and applied. The correction due to propagation of radar pulse is required due to variations in the refractive index along its propagation path through the atmosphere. One of the most important corrections desired is due to the presence of water vapor in the atmosphere. Due to the high variability of water vapor in space and time, its real-time monitoring is necessary, and it is traditionally carried out using concurrent observations from a passive microwave radiometer. Thus, most of the altimetric missions are accompanied with a nadir looking passive microwave radiometer. The presence of rain induces significant errors and biases in the retrieval of altimeter measurements of oceanic geophysical parameters like SSH, SWH, and OWS, which necessitates its detection and estimation to flag the rain corrupted observations, and to possibly develop correction methods. The effect of rain on measurements from a single frequency (Ku-band) radar altimeter was examined by a number of researchers, (e.g., Srokosz and Guymer, 1988; Guymer and Quartly, 1992; Walsh et al., 1984; Varma et al., 1994; Contreras et al., 2003). While most of the altimeter worked in a single frequency at Ku-band, the SARAL AltKa is working at Ka band. There are also a few dual frequency (C and Ku band) radar altimeter systems (e.g., TOPEX, JASON) for providing better ionosphric correction, which also provide opportunity for the detection and retrieval of rain using differential attenuation at two frequencies (Bhandari and Varma, 1996; Varma et al., 2001, 2006; Tournadre and Bhandari, 2009; Quilfen et al., 2006). The SARAL AltiKa provides measurements in Ka-band, which results in higher attenuation (as compared to Ku-band) from clouds and precipitation. For Ka-band, the two-way attenuation from rain as well as clouds was determined by Walsh et al. (1984). A theoretical simulation on effect of rain on Ka-band SARAL Altimeter measurements was provided by Tournadre et al. (2009a,b). The passive

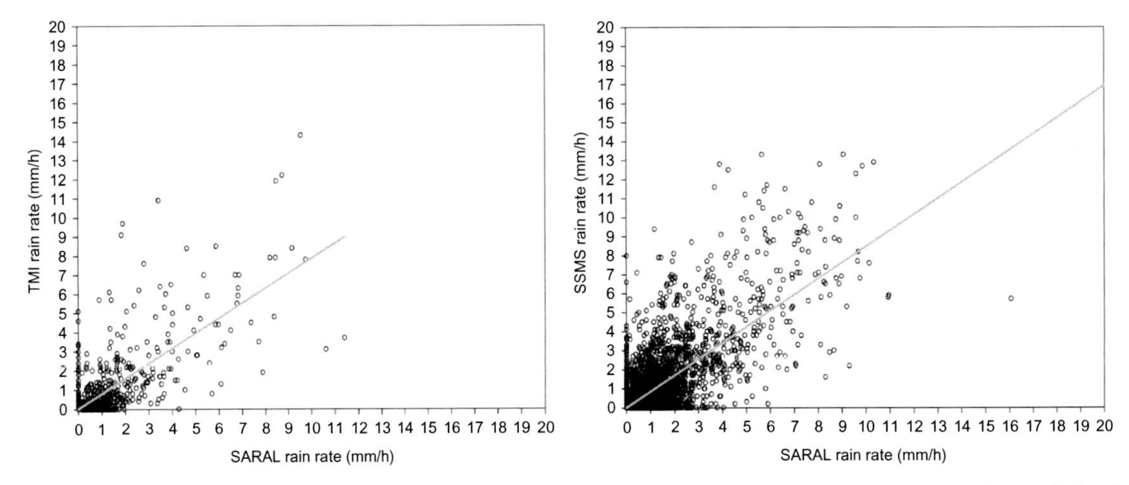

FIG. 6 Comparison of instantaneous rain rate from SARAL and TMI (left-panel) SARAL and SSMIS (right-panel), for the period from June to September 2014 and for regions within ±40° latitudes.

microwave radiometers accompanying altimeter missions are also found to be sensitive to precipitation (e.g., Wilheit et al., 1977; Varma et al., 2002; Gairola et al., 2004). Varma et al. (1999) demonstrated the use of the TOPEX Microwave Radiometer (TMR) onboard the TOPEX/Poseidon satellite for global rain estimation. As measurements from both the altimeter and radiometer are sensitive to rain, it is worthwhile to use them in a synergy for a more precise measurement of rain. One such synergic use of concurrent observations from TOPEX Altimeter and accompanying TMR for rain estimations is carried out by Gairola et al. (2005). The availability of Ka-band AltiKa measurements along with an accompanying passive microwave radiometer offers a unique opportunity for measuring precipitation.

The SARAL carried a nadir looking radar altimeter operated at Ka-band (35.75 GHz) and a nadir looking passive microwave radiometer operating at 23.8 and 37 GHz. Concurrent SARAL and TRMM Precipitation Radar (PR) observations are used to show that several SARAL measured parameters such as normalized radar backscattering coefficient (σ_0), Automatic Gain Control (AGC), peakiness, waveform, and brightness temperature at 23.8 and 37 GHz are sensitive to rain. Based on probability distribution of rain sensitive parameters, a probabilistic rain identification algorithm was developed, which was optimized to reduce false alarm cases. In the second step, a genetic algorithm based on a rain retrieval algorithm was developed. The algorithm was applied to an independent SARAL dataset and the resulting rain compared with TMI and SSMIS showed correlation of 0.77 and 0.78 and rmsd of 0.49 and 0.43, respectively. Fig. 6 shows the scatter plot of comparison of rain from SARAL with TMI (left panel) and SSMIS (right panel).

2.3 Problems Associated With Passive Microwave Measurement of Precipitation

The Vis and IR radiances do not penetrate the clouds and hence precipitation measurement using them is always indirect in nature. Furthermore, the brightness observed in the visible band images depends upon the Satellite-Cloud-Sun geometry and no observation is available during nighttime. On the other hand, microwave radiances directly interact with the hydrometeors and hence such measurements are more direct in nature. Nevertheless, the microwave measurement of precipitation suffers from large errors due to a variety of reasons (Varma and Pal, 2012). The major problems associated with passive microwave remote sensing of precipitation are primarily due to uncertainty in determining, (1) background fields for rain identification, (2) rain-type classification, (3) horizontal and vertical distribution of hydrometeors and associated beam filling error, (4) poor calibration and validation due to different shape/size of the measurement area and geometry, (5) shape (solid/liquid) and distribution of hydrometeors. Apart from such uncertainties, there are deficiencies in radiative transfer model assumptions and/or formulations, and instrument errors.

Fig. 7 shows the scatter plot of concurrent observations of brightness temperature over the Bay of Bengal during July 2003 at 19.3 GHz (V) from TRMM TMI and rain rate from TRMM precipitation radar (PR). The red line shows the average plot, whereas the green line shows the radiative transfer model simulated variation. The plot on the left-hand side (Fig. 7A) is for the convective rain and for the right-hand side (Fig. 7B) is for the startiform rain. These plots are taken from Varma et al. (2004). A large variation in the brightness temperature (~30 K) can be observed in the absence of rain (i.e., rain = 0). This large variation in the brightness temperature is due to variations in the surface and the atmospheric parameters such as SST, wind speed, water vapor, cloud liquid water, etc. There may also be some contribution to this variation coming from the possible error in collocating two observations, but that may not be too large an effect because both TMI and PR are on the same observing platform. Such huge variation in the brightness temperature can cause uncertainties in the rain identification, especially in the low rain regime.

From Fig. 7, it can also be observed that brightness temperature versus rain rate relationship is different for convective and stratiform rain types. Thus, rain type identification is necessary for accurate rain estimation. Varma and Liu (2010) and Hong et al. (1999) have provided algorithms for rain type classification. Their algorithms were essentially based on horizontal variability of the brightness temperature and/or identification of the cloud ice in high frequency

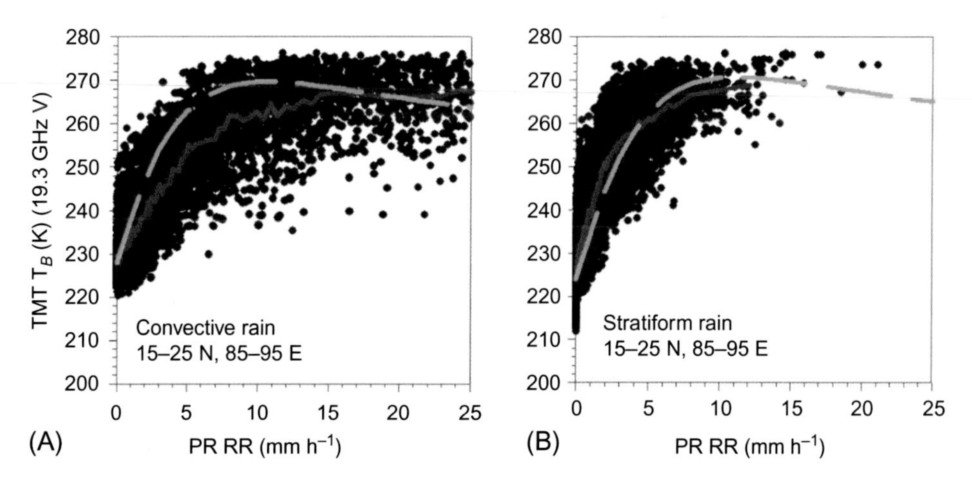

FIG. 7 Scatter plot of concurrent observations of brightness temperature over the Bay of Bengal during July 2003 at 19.3 GHz (V) from TRMM TMI and rain rate from TRMM precipitation radar (PR) for (A) convective and (B) stratiform rain. The red line shows the average plot, whereas the green line shows the radiative transfer model simulated variation. *From Varma, A.K., Liu, G., Noh, Y.J., 2004. Sub-pixel scale variability of rainfall and its application to mitigate the beam-filling problem. J. Geophys. Res. 109, D18210. https://doi.org/10.1029/2004JD004968.*

channels (e.g., 85 GHz V and H). The rain identification is never perfect and hence may lead to certain errors in rain estimation.

The passive microwave observations have a large instantaneous field of view (IFOV) and it is unlikely that such a large field of view is filled with uniform rain. Because the relationship between brightness temperature and rain rate (*Tb-R*) is nonlinear, the variability of rain over the IFOV may lead to error in the rain estimation. From Fig. 7, it can be observed that for convective rain, the radiative transfer simulated brightness temperatures are significantly higher than the average observed brightness temperatures. The overestimation can be several tens of kelvins and is largely due to horizontal inhomogeneity of the rain field within the satellite IFOV (Varma et al., 2004; Varma and Liu, 2010; Spencer et al., 1983). Often microwave observations are presented over an effective field of view (EFOV) that is based on integration time of the microwave energy by the radiometer. The footprint of different frequency channels is also different. Data convolution and deconvolution is a common technique to resample the measurements to a single resolution. Rapp et al. (2009) showed that resampling of the microwave measurements of brightness temperature does not necessarily result in reducing errors due to the partial beam filling in the precipitation retrieval vis-à-vis using native data. Apart from the horizontal variability of the precipitation there is also vertical variability over the IFOV. While horizontal variability is responsible for deviation of average brightness temperature and rain rate relationship from the radiative transfer simulated observations, the vertical variability is responsible for the large scatter of values around the mean relationship, as can be seen in Fig. 7. A thorough discussion on vertical variability of rain profiles on the *TB-R* relation was given by Liu and Fu (2001). As can be expected, given the same surface rain rate, a profile corresponding to a deeper rain column results in a higher brightness temperature in the emission channels. In this study, an observed mean rain profile was taken from Liu and Fu (2001). A second order difference in mean and simulated *Tb-R* relationship may possibly arise due to the difference in the vertical profiles as considered in the radiative transfer simulations of brightness temperatures and the actual profiles that have given rise to the observed brightness temperatures. The convergence of observed and simulated *Tb-R* relationship at higher rain rates is expected as such high pixel-averaged rain rates tend to have more uniform beam filling. A slightly higher brightness temperature at very high pixel-averaged rain rates in Fig. 7 is possibly due to a deeper rain column in radiative transfer simulations that actually existed.

There are several issues related to calibration and validation of the microwave measurement of precipitation that are to be discussed in the next section. As discussed in Section 2.2.1, for forward simulations of brightness temperatures, precise knowledge of the profile of the hydrometeors is desired. Such measurements, especially under severe weather conditions, are very limited. Thus, a cloud resolving model is often used that provides the desired profiles of the hydrometeors. There are also errors involved in the scattering computation from the ice particles as there are uncertainties in determining their precise shape, size, and distribution. Apart from that, there are uncertainties in the fall velocity and temperature of the hydrometeors. The radiative transfer models have their own errors due to formulations and approximations, especially more so in the fast models. The instrument errors resulting from instrument noise, blackbody calibrations, etc. may also lead to errors in the precipitation estimation.

2.4 Validation of Precipitation

The precipitation is known to vary in all possible spatial scales ranging from few meters to several kilometers (McCollum and Krajewski,

1998; Tustison et al., 2003) and that poses difficulty not only in the validation of the precipitation but also in its retrieval using satellite observations. Several researchers (e.g., Wilheit, 1986; Chiu, et al., 1990; Varma et al., 2004; Varma and Liu, 2006) described and also suggested a possible stochastic solution to the so-called beam filling problem resulting from the nonuniform distribution of precipitation within a satellite footprint that leads to the severe underestimation of the precipitation at microwave frequencies. Many researchers have examined the precipitation variability at a small scale and reported its spatial distribution as a lognormal distribution (e.g., Lopez, 1976; Kedem and Slud, 1994). Tustison et al. (2001, 2003) examined the scale dependency of the precipitation and the errors associated with them when interpolated from one scale to another. Gupta and Waymire (1993) examined the mesoscale variability of the precipitation and modeled the variability in short time intervals based on the concept of random cascades. Kundu and Bell (2003) tried to capture the spatial and temporal variability of the mesoscale rainfall in a stochastic model. Several investigators tried to examine the temporal variability of the precipitation through auto-correlation analysis. For example, Yule (1945) examined the autocorrelation in an annual time series of precipitation data of Greenwich for the years 1815–1924 and found that the autocorrelation coefficients in annual precipitation with lag 1–24 years varied from −0.12 to +0.19.

The satellite based precipitation products are generated in different scales ranging from instantaneous to a season, and from pixel-scale (~4 km × 4 km) to 100 s of kilometer (e.g., 2.5° × 2.5° latitude-longitude grid). The quantitative validation of high spatial and temporal resolution measurement of precipitation from satellite poses a great challenge due to the high variability of the precipitation observed in nature, which make rain statistics dependent upon spatial and temporal scales over which rain data is averaged (Kundu and Bell, 2003). Piyush et al. (2012) unraveled the variability of the precipitation at a pixel scale to explore and get an insight of the validation accuracy of precipitation measured from satellites at pixel scale. In their study, they considered pixels of size 4 and 8 km to understand the validation problem.

The study of Piyush et al. (2012) was focused on the validation problem associated with satellite based pixel scale instantaneous precipitation measurements that arises either due to misalignment in space and/or time of two mutually comparing measurements, or due to their different scales of measurement. They used TOGA-COARE (Tropical Ocean Global Atmosphere Coupled Ocean-Atmosphere Response Experiment) (Webster and Lukas, 1992) IOP (Intensive Observation Period) precipitation data collected by Massachusetts Institute of Technology (MIT) C-band Doppler radar installation on a ship over the Western Pacific warm pool (2°S, 156°E) from November 1992 to February 1993. The observations were available every 10 min in 2 km grid (Short et al., 1997). The TOGA-COARE observations utilized by Piyush et al. (2012) were obtained during two ship cruises from November 10 to December 10, 1992 and from December 15, 1992 to January 18, 1993. During cruise 1 and 2, the average rainfall reported was 3.4 and 6.6 mm day^{-1}, respectively, and the wettest periods during these cruises were November 23–24, 1992 and December 20–25, 1992, respectively (Short et al., 1997). They examined the uncertainty in the comparison of two precipitation products when they are measured in the same scale (4 and 8 km) but are slightly misplaced in space and time. This was done by analyzing the autocorrelation of temporal, spatial, and spatiotemporal series of precipitation at 4 and 8 km scales. They found that the autocorrelation function of precipitation measurements drops to 0.45 and 0.58 in 10-min time for 4 and 8 km pixel, respectively. Similarly, spatial auto-correlation drops to 0.64 and 0.61 for 4 km pixel with 4 km displacement and for 8 km pixel with 8 km displacement, respectively. Their result showed that the drop of autocorrelation in the spatiotemporal series is even more rapid. Further, they computed the root mean square fractional error (RMSFE) as:

$$\mathrm{RMSFE} = \sqrt{\frac{\sum_{i=1}^{n} \left(\frac{R(t) - R(t+\Delta t)}{R(t)} \right)^2}{n}}$$

where, $R(t)$ is rain measured at time t and $R(t+\Delta t)$ is rain measured after time lag Δt over the same area. They found RMSFE value of 0.63–0.90 and 0.71–0.94 for 10–60-min time lag for 4 and 8 km pixel, respectively. Thus, the results show that a few minutes or/and a few kilometers of difference in the alignment of two observations may have severe effects on the validation results. This may happen when we compare satellite derived precipitation with radar observations which may not always be perfectly aligned in space and time due to inaccuracies in their geolocation and time determination. Thus, proper care must be taken in geolocation and time determination to mitigate this problem.

Further, they examined the subpixel scale rain variability. Fig. 8 shows the distribution of rain at 2 km × 2 km pixel within 8 km × 8 km area. The different probability distributions (PDFs) in Fig. 8 are drawn for different 8 km ×8 km area averaged rain bins. The figure shows close to lognormal distribution of rain rates within a large area of 8 km × 8 km. The PDFs of Fig. 8 agree with similar distributions reported by Varma and Liu (2006). They examined the probability of finding 2 km × 2 km rain rate close to 8 km × 8 km averaged rain rate and found a probability of 0.1, 0.24, 0.41, 0.59, 0.87 for 2 km × 2 km rain rate matching with 8 km × 8 km window average rain rate (R) with percentage error of 10%, 25%, 50%, 75%, and 100% of R, respectively. It may be noted that a percentage error of 100% of R covers the entire lower range of precipitation (given by (8)), thus 13% of sub-window rain rates have $r > 2 \times \mathrm{R}$. They showed a Monte Carlo simulated validation of precipitation averaged over 8 km × 8 km pixel with 2 km × 2 km averaged concurrent rain (Fig. 9).

The Monte-Carlo simulated comparison in Fig. 9 shows a correlation (R) of 0.73 and an error of estimation of 4.3 mm h^{-1} for a satellite precipitation measurement range of 0–70 mm h^{-1}. The hypothetical ground based measurements show

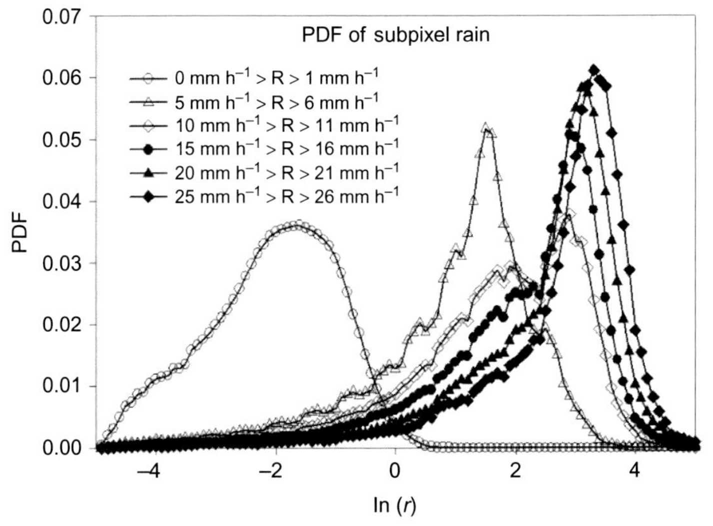

FIG. 8 Conditional Probability Distribution of rain rates (r_i) over 2 km × 2 km areas within 8 km × 8 km pixel with pixel averaged rain rates (R) bin of >0–1, 5–6, 10–11, 15–16, 20–25, 25–26 mm h^{-1}. *From Piyush, D.N., Varma, A.K., Pal, P.K., Liu, G., 2012. An analysis of rainfall measurements over different spatio-temporal scales and potential implications for uncertainty in satellite data validation. J. Meteorol. Soc. Jpn. 90(4), 439–448. https://doi.org/10.2151/JMSJ.2012-408.*

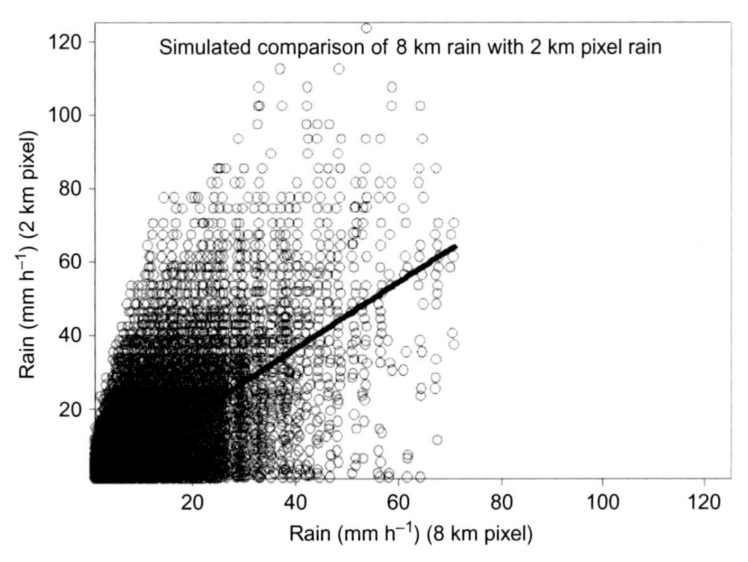

FIG. 9 A Monte Carlo Simulated comparison of rain over 8 km × 8 km pixel with 2 km × 2 km averaged rain using a prior probability distribution of rain (R) and Conditional Probability Distribution of rain rates (r_i) (as shown in Fig. 8). *From Piyush, D.N., Varma, A.K., Pal, P.K., Liu, G., 2012. An analysis of rainfall measurements over different spatio-temporal scales and potential implications for uncertainty in satellite data validation. J. Meteorol. Soc. Jpn. 90(4), 439–448. https://doi.org/10.2151/JMSJ.2012-408.*

a much larger dynamic range compared to hypothetical satellite measurements, which is due to the fact that higher rain rates are confined to smaller areas with steep gradients (Goldhirsh and Walsh, 1982).

The study by Tustison et al. (2001, 2003) and the above validation study by Piyush et al. (2012) show the difficulties in the validation of precipitation from satellitse. The comparison of precipitation as shown in Fig. 9 by Piyush et al. (2012) is carried out with the assumption of 2 km × 2 km of surface observations compared with 8 km × 8 km of satellite observations. Precipitation values are derived from the same radar observations, however, in practice, these are two observations from different sources and hence they have different accuracies. In such cases, the comparison statistics will become further degraded than what is presented in Fig. 9. Also in real practice, both measurements may have their own geolocation and/or temporal errors that may further deteriorate the comparison statistics. Piyush et al. assumed sub-pixel rain at 2 km × 2 km scale, which may not be always true. Rain measured at a smaller scale (or point measurements from gauge) will lead to more errors. In case of microwave measurement of precipitation, the pixel size is large (~25 km) and hence the validation error will also be larger. The study highlights the errors in the calibration and validation of precipitation from the satellite measurements which have the potential to be completely off from the ground measurements at different scale. In order to have insight of precipitation validation, it is worthwhile to view the validation results from different algorithms by NOAA at http://www.star.nesdis.noaa.gov/smcd/emb/ff/aboutProductValidation.php.

3 CONCLUSION

This chapter focuses on the retrieval of precipitation using visible/infrared and microwave observations from satellites. It is emphasized that both Vis/IR and microwave methods have their own advantages and disadvantages. In order to overcome the disadvantage of Vis/IR methods due to their indirect sensing of precipitation advanced precipitation schemes, such as H-E and GMSRA based on a conceptual model of precipitation process and many auxiliary information from NWP, model fields are developed. The H-E and GMSRA provide precipitation estimates at a very fine scale of a few kilometers (e.g., ~4 km × 4 km) at every 15–30 min from geostationary platforms. Thus, in effect, these precipitation products are capable of capturing the growth and decay cycle of even small convective cells.

On the other hand, the satellite microwave radiances directly interact with hydrometeors and thus, compared to Vis/IR, they are more suitable for precipitation measurement. However, the large footprint size of microwave observations and the associated partial beam filling problem results in large errors in rain estimation. This chapter discusses the beam filling problem and its possible mitigation. There are various microwave brightness temperatures based indices which are found to be more sensitive to precipitation than the corresponding brightness temperatures themselves. Also discussed in this chapter is a state-of-the-art algorithm, the Goddard profiling (GPROF) algorithm, which is currently being used for operational retrievals by many operational meteorological agencies. GPROF is based on Bayes' Theorem, which relates retrieval with prior knowledge of the atmospheric conditions and associated satellite measurements from radiative transfer simulations. Furthermore, probabilistic precipitation identification and a radiative transfer simulations supported retrieval scheme is discussed for retrieving precipitation using microwave sounder operating at frequencies close to a 183.31 GHz water vapor absorption line. In this chapter, there is also discussion of retrieval of precipitation using combined passive and active microwave measurements.

The IR measurements offer high spatial and temporal sampling and microwave measurements that more directly interact with hydrometeors. In order to exploit the positive aspect of both the measurements, it is imperative to combine them optimally to get the best estimates of the precipitation. Various algorithms such as TRMM 3B42 V7 and IMERG are discussed in this chapter that combine Vis and microwave measurements for the best possible estimation of precipitation in fine space and time domains.

Once the retrieval of the precipitation is accomplished, the next step is to specify its accuracy with respect to truth (New et al., 2001). The surface rain gauge stations are sparse and near nonexistent over vast oceans. The high variability of precipitation makes it difficult to directly compare gauge measurements with averaged satellite measurements over certain areas. In this chapter, the variability of precipitation at small spatial and temporal scales and its effect on validation is discussed. Usually, a gridded dataset of daily/monthly precipitation from surface gauges is used for validation of satellite observations. This gridded dataset is generally produced using inverse distance weighting schemes for rain gauges that are taken for consideration for producing rain in a grid box. The various gridded products may differ in the number of stations that are used for each estimate and the formulation of the distance weighting function. The gridding schemes interpolate over data-sparse areas, that, however, may lead to large errors when there are steep gradients in the spatial distribution of precipitation. Because there are different variants of the averaging schemes used for different gridded products (e.g., Rudolf et al., 1994; Xie et al., 1996), they lead to different estimates of gridded precipitations. For example, Xie et al. (1996) compared their precipitation estimates with GPCP (Global Precipitation Climatology Project) monthly estimates in 2.5° × 2.5° boxes and found that the mean absolute difference between the two estimates as high as about 15% at grid cells with five or more stations within a 2.5° box in both datasets; this difference increased to 40%–90% in grid cells where both datasets had only one station within a box. This is an example that shows that, due to the high variability of the precipitation, the gridded surface observations themselves do not perfectly match with each other even in very coarse scales. Thus, validation remains a very important issue for satellite precipitation measurements. For the assessment of the accuracy of the satellite measurements, it is necessary that data from the dense network of rain gauges and well calibrated radars observations be utilized. Such measurements are not available in all locations and thus, specifically developed validation sites, such as Kwajalein Atoll, must be utilized. Different climatic regions/seasons and associated different rain types and processes do exist, and the performance of the retrieval algorithms may also be different in these different situations. For example, the precipitation type and processes over land and ocean may differ; similarly, they may also differ in tropical, sub-tropical, and extra-tropical regions, and also over hilly terrains and coastal regions (Varma and Liu, 2010), etc. Hence, for the assessment of the quality of the satellite derived precipitation, there is a need for more validation sites to be developed to account for the variation in precipitation types/processes in different parts of the Earth.

The first satellite mission specifically designed for precipitation measurement over global tropics was TRMM. The TRMM provided valuable data from 1998 for nearly 17 years. As a follow-up mission to TRMM, the GPM with enhanced capabilities was launched in February 2014. Unlike TRMM (orbit inclination 35°), the GPM with 65° of orbit inclination allows it to cover and provide precipitation in higher latitudes as well. GPM provides measurements of rain, water, and snow with much better accuracies than TRMM. High quality observations from GPM have given impetus to understanding of the water cycle and its variability and interconnections with global energy, carbon, and other cycles. The measurements from GPM in conjunction with other satellites such as SMAP (Soil Moisture Active Passive), GRACE-FO (Gravity Recovery and Climate Experiment-Follow On), ICESat-2 (Ice, Cloud, and land Elevation Satellite) and SWOT (Surface Water Ocean Topography), etc., may help in understanding water cycles like never before.

References

Adler, R.F., Negri, A.J., 1988. A satellite infrared technique to estimate tropical convective and stratiform rainfall. J. Appl. Meteorol. 27, 30–51.

Aonashi, K., Awaka, J., Hirose, M., Kozu, T., Kubota, T., Liu, G., Shinge, S., Kida, S., Seto, S., Takahashi, N., Takayabu, Y.N., 2009. GSMaP passive microwave precipitation retrieval algorithm: algorithm description and validation. J. Meteorol. Soc. Jpn. 87A, 119–136.

Arkin, P.A., 1979. The relationship between fractional coverage of high cloud and rainfall accumulation during GATE over the B- scale array. Mon. Weather Rev. 107, 1382–1387.

Arkin, P.A., Meisner, B.N., 1987. Spatial and annual variation in the diurnal cycle of large scale tropical convective cloudiness and precipitation. Mon. Weather Rev. 115, 2009–2032.

Arkin, P.A., Krishna Rao, A.V.R., Kelker, R.R., 1989. Large scale precipitation and outgoing longwave radiation from INSAT-1B during the 1986 southwest monsoon season. J. Clim. 2, 619–628.

Ba, M.B., Gruber, A., 2001. GOES multispectral rainfall algorithm (GMSRA). J. Appl. Meteorol. 40, 1500–1514.

Barrett, E.C., Martin, D.W., 1981. The Use of Satellite Data in Rainfall Monitoring. Academic Press, London. 340 pp.

Bhandari, S.M., Varma, A.K., 1995. On estimation of large scale monthly rainfall estimation over the indian region using minimal INSAT-VHRR data. Int. J. Remote Sens. 16, 2023–2030.

Bhandari, S.M., Varma, A.K., 1996. Potential of simultaneous dual-frequency radar altimeter measurements from TOPEX for rainfall estimation over oceans. Remote Sens. Environ. 58, 13–20.

Chelton, D.B., Walsh, E.J., MacArthur, J.L., 1989. Pulse compression and sea level tracking in satellite altimetry. J. Atmos. Ocean. Technol. 6, 407–438.

Chiu, L.S., North, G.R., David, A.S., McContell, A., 1990. Rain estimation from satellites effects of finite field of view. J. Geophys. Res. 95, 2177–2185.

Contreras, R.F., Plant, W.J., Keller, W.C., Hayes, K., Nystuen, J., 2003. Effects of rain on Ku-band backscatter from the ocean. J. Geophys. Res. 108 (C5), 34-1–34-15.

Cooper, N.S., 1988. The effect of salinity on tropical ocean models. J. Phys. Oceanogr. 18, 697–707.

Gairola, R.M., Varma, A.K., Pokhrel, S., Agarwal, V.K., 2004. Integrated satellite microwave and infrared measurements of precipitation during a Bay of Bengal cyclone. Indian J. Radio Space Phys. 33, 115–124.

Gairola, R.M., Pokhrel, S., Varma, A.K., Agarwal, V.K., 2005. A combine passive-active microwave retrieval of quantitative rainfall from TOPEX/Poseidon altimeter and TMR. Int. J. Remote Sens. 26 (8), 1729–1753.

Ghosh, A., Varma, A.K., Shah, S., Gohil, B.S., Pal, P.K., 2014. Rain identification and measurement using Oceansat-II Scatterometer observations. Remote Sens. Environ. 142, 20–32.

Godfrey, J.S., Lindstrom, E.J., 1989. The heat budget of the equatorial western Pacific surface mixed-layer. J. Geophys. Res. 94 (C6), 8007–8017.

Goldhirsh, J., Walsh, E.J., 1982. Rain measurements from space using a modified Seasat-type radar altimeter. IEEE Trans. Antennas Propag. AP-30 (4), 726–733.

Griffith, C.G., Woodley, W.L., Grube, P.G., Martin, D.W., Stout, J., Sikdar, D.N., 1978. Rain estimation from geostationary satellite imagery- Visible and Infrared studies. Mon. Weather Rev. 106, 1153–1171.

Grody, N.C., 1991. Classification of snow cover and precipitation using special sensor microwave imager (SSM/I). J. Geophys. Res. 96, 7423–7435.

Gupta, V.K., Waymire, E., 1993. A statistical analysis of mesoscale rainfall as a random cascade. J. Appl. Meteorol. 32 (2), 251–267.

Guymer, T.H., Quartly, G.D., 1992. The effect of rain on ERS-1 altimeter data'. In: Proc. First ERS-1 Symp.—"Space at the Service of our Environment". Cannes, France, ESA SP -35, 445-450.

Hong, Y., Kummerow, C., Olson, W.S., 1999. Seperation of convective and stratiform precipitation using microwave brightness temperature. J. Appl. Meteorol. 38, 1195–1213.

Huffman, G.J., Adler, R.F., Bolvin, D.T., Gu, G., Nelkin, E.J., Bowman, K.P., Hong, Y., Stocker, E.F., Wolff, D.B., 2007. The TRMM multi-satellite precipitation analysis: quasi-global, multi-year, combined-sensor precipitation estimates at fine scale. J. Hydrometeorol. 8, 38–55. https://doi.org/10.1175/JHM560.1.

Huffman, G.J., Bolvin, D.A., Braithwaite, D., Hsu, K., Joyce, R., 2015. Algorithm theoretical document (ATBD) Ver 4.5, NASA global precipitation measurement (GPM) Integrated Multi-Satellite Retrieval for GPM (IMERG), NASA, Greenbelt, MD, USA (https://pmm.nasa.gov/sites/default/files/document_files/IMERG_ATBD_V4.5.pdf).

Inoue, T., 1997. Day-to-night cloudiness change to cloud type inferred from split window measurements onboard NOAA Polar Orbiting Satellites. J. Meteorol. Soc. Jpn. 75, 59–66.

Joyce, R., Arkin, P.A., 1997. Improved estimates of tropical and subtropical precipitation using the GOES precipitation index. J. Atmos. Oceanic Technol. 14, 997–1011.

Kedem, B., Slud, E., 1994. Partial likelihood analysis of logistics regression and autoregression. Stat. Sin. 4, 89–106.

Krajewski, W.F., Kitelekos, A., Goska, R., 2006. A GIS based methodology for the assessment of weather radar beam blockage in mountainous regions: two examples from U.S NEXRAD network. Comput. Geosci. 32 (3), 283–302.

Kumar, P., Varma, A.K., 2016. Assimilation of INSAT-3D hydro-estimator method retrieved rainfall on short range weather prediction. Q. J. R. Meteorol. Soc. 143, 384–394.

Kumar, R., Mishra, A., Gairola, R.M., Varma, A.K., Das, I., 2009. Evaluation of precipitation features in high frequency SSM/I measurements over Indian land and Oceanic regions. IEEE Geosci. Remote Sens. Lett. 6 (3), 373–377.

Kummerow, C., Olson, W.S., Giglio, L., 1996. A simplified scheme for obtaining precipitation and vertical hydrometer profiles from passive microwave sensors. IEEE Trans. Geosci. Remote Sens. 34, 1213–1232.

Kummerow, C., Hong, Y., Olson, W.S., Yang, S., Adler, R.F., Mccollum, J., Ferraro, R., Petty, G., Shin, D.-B., Wilheit, T.T., 2001. The evolution of the goddard profiling algorithm GPROF) for rainfall estimation from passive microwave sensors. J. Appl. Meteorol. 40, 1801–1820.

Kundu, P.K., Bell, T.L., 2003. A stochastic model of space-time variability of mesoscale rainfall: Statistics of spatial averages. Water Resour. Res. 39 (12), 1328 https://doi.org/10.1209.2002WR001802.

Lee, T.F., Turk, F.J., Hawkins, J., Richardson, K., 2002. Interpretation of TRMM TMI images of tropical cyclones. Earth Interact. 6, 1–17.

Liu, G., 1998. A fast and accurate model for microwave radiance calculations. J. Meteorol. Soc. Jpn. 76, 335–343.

Liu, G., Curry, J.A., 1998. An investigation of the relationship between emission and scattering signals in SSM/I data. J. Atmos. Sci. 55, 1628–1643.

Liu, G., Fu, Y., 2001. The characteristics of tropical precipitation profiles as inferred from satellite radar measurements. J. Meteorol. Soc. Jpn. 79 (1), 131–143.

Liu, C., Zipser, E.J., 2009. "Warm Rain" in the tropics: seasonal and regional distribution based on 9 years of TRMM data. J. Clim. 22, 767–779.

Lopez, R.E., 1976. Radar characteristics of the cloud populations of tropical disturbances in the northwest atlantic. Mon. Weather Rev. 104, 268–283.

McCollum, J.R., Krajewski, W.F., 1998. Investigations of error sources of the global precipitation climatology project emission algorithm. J. Geophys. Res. 103 (D22), 28,711–28,719. https://doi.org/10.1029/98JD02580.

Miller, J.R., 1972. A climatological Z-R relationship for convective storms in the northern great planes. In: Proceedings of 15th Conference on Radar Meteorology, Boston, AMS, pp. 153–154.

Miller, J.R., 1976. The salinity effect in a mixed-layer depth model. J. Phys. Oceanogr. 6, 29–35.

Nespor, V., Sevruk, B., 1999. Estimation of wind induced error or rainfall gauge measurements using a numerical simulation. J. Atmos. Ocean. Technol. 16, 450–464.

New, M., Todd, M., Hulme, M., Jones, P., 2001. Review: precipitation measurements and trends in the twentieth century. Int. J. Climatol. 21, 1899–1922.

Petersen, W.A., Schwaller, M.R., 2008. Global precipitation mission—ground validation: science implementation plan. Available from, https://pmm.nasa.gov/sites/default/files/document_files/GPM_GVS_imp_plan_Jul08.pdf.

Piyush, D.N., Varma, A.K., Pal, P.K., Liu, G., 2012. An analysis of rainfall measurements over different spatio-temporal scales and potential implications for uncertainty in satellite data validation. J. Meteorol. Soc. Jpn. 90 (4), 439–448. https://doi.org/10.2151/JMSJ.2012-408.

Quilfen, Y., Tournadre, J., Chapron, B., 2006. Altimeter dual frequency observations of surface winds, waves, and rain rate in tropical cyclone Isabel. J. Geophys. Res. 111 C01004.

Ramirez-Beltran, N.D., Kuligowaski, R.J., Harmsen, E.W., Castro, J.M., Cruz-Pol, S., Cardona, M.J., 2008. Rainfall estimation from convective storms using the hydro-estimator and NEXRAD. WSEAS Trans. Syst. 7 (10), 1016–1027.

Rapp, A.D., Lebsock, M., Kummerow, C., 2009. On the consequences of resampling of microwave radiometer observations for use in retrieval algorithms. J. Appl. Meteorol. Climatol. 48, 1981–1993.

Rosenfield, D., Gutman, G., 1994. Retrieving microphysical properties near the tops of potential rain clouds by multi spectral analysis of AVHRR data. Atmos. Res. 34, 259–283.

Rudolf, B., Hauschild, H., Rueth, W., Schneider, U., 1994. Terrestrial precipitation analysis: operational method and required density of point measurements. In: Desbois, M., Desalmand, F. (Eds.), Global Precipitation and Climate Change. Springer, Berlin, pp. 173–186.

Schumacher, C., Houze Jr., R.A., 2000. Comparison of radar data from TRMM satellite and Kwajalein oceanic validation site. J. Appl. Meteorol. 39, 2015–2164.

Schwaller, M.R., Morris, K.R., 2011. A ground validation network for the global precipitation measurement mission. J. Atmos. Oceanic Technol. 28, 301–319.

Scofield, R.A., 1987. The NESDIS operational convective precipitation estimation technique. Mon. Weather Rev. 115, 1773–1792.

Scofield, R.A., Kuligowaski, R.J., 2003. Status and outlook of operational satellite precipitation algorithms for extreme-precipitation events. Weather Forecast. 18, 1037–1051.

Short, sb_ma, Kucera, P.A., Ferrier, B.S., Gerlach, J.C., Rutledge, S.A., Thiele, O.W., 1997. Shipboard radar rainfall patterns within the TOGA COARE IFA. Bull. Am. Meteorol. Soc. 78, 2817–2836.

Spencer, R.W., Olson, W.S., Rongzhang, W., Martin, D.W., Weinman, J.A., Santek, D.A., 1983. Heavy thunderstorms observed over land by the Nimbus 7 scanning multichannel microwave radiometer. J. Clim. Appl. Meteorol. 22, 1041–1046.

Spencer, R.W., Goodman, H.M., Hood, R.E., 1989. Precipitation retrieval over land and ocean with SSM/I: identification and characteristics of the scattering signal. J. Atmos. Ocean. Technol. 6, 254–273.

Srokosz, M.A., Guymer, T.H., 1988. A study of the effect of rain on Seasat radar altimeter data. In: Proc. IGARSS '88 Symp., Edinburg, Scotland, 13-16 Sept, 1988, ESA SP-284, 651-654.

Taft, B.A., Kessler, W.S., 1987. On the effect of salinity on the dynamics of the tropical Pacific circulation. Trop. Ocean-Atmos. Newsl. 41, 810.

Tournadre, J., Bhandari, S., 2009. Analysis of short space-time scale variability of oceanic rain using Topex/Jason Tandem Mission measurements. J. Atmos. Ocean. Technol. 26 (1), 74–90.

Tournadre, J., Lambin, J., Steunou, N., 2009a. Cloud and rain effects on ALTIKA/SARAL Ka band radar altimeter. Part I: modeling and mean annual data availability. IEEE Trans. Geosci. Remote Sens. 47 (6), 1806–1817.

Tournadre, J., Lambin, J., Steunou, N., 2009b. Cloud and rain effects on ALTIKA/SARAL Ka band radar altimeter. Part II: definition of a rain/cloud flag. IEEE Trans. Geosci. Remote Sens. 47 (6), 1818–1826.

Tustison, B., Harris, D., Foufoula-Georgiou, E., 2001. Scale issues in verification of precipitation forecasts. J. Geophys. Res. 106 (11), 11775–11784.

Tustison, B., Foufoula-Georgiou, E., Harris, D., 2003. Scale-recursive estimation for multisensory quantitative precipitation forecast verification: a preliminary assessment. J. Geophys. Res. 107, 8377. https://doi.org/10.1029/2001JD001073.

Varma, A.K., Gairola, R.M., 2015. Algorithm Theoretical Basis Document (Modified): Hydro-Estimator, SAC/EPSA/AOSG/SR/04/2015, p. 29.

Varma, A.K., Liu, G., 2006. Small scale horizontal rain-rate variability observed by satellite. Mon. Weather Rev. 134, 2722–2733. https://doi.org/10.1175/MWR3185.1.

Varma, A.K., Liu, G., 2010. On classifying rain types using satellite microwave observations. J. Geophys. Res. 115. https://doi.org/10.1029/2009JD012058. D07204.

Varma, A.K., Pal, P.K., 2012. Use of TRMM precipitation radar to address the problem of rain detection in passive microwave measurements. Indian J. Radio Space Phys. 41, 411–420.

Varma, A.K., Prakash, W.J., Bhandari, S.M., 1994. Monitoring Indian summer monsoon rainfall using spaceborne radar altimeters. In: Proc. of National Symposium on Microwave Remote Sensing and Users' Meet (MRSUM), Space Applications Centre, Ahmedabad, India, Jan. 10-11, pp. 261–264.

Varma, A.K., Gairola, R.M., Kishtawal, C.M., Pandey, P.C., Singh, K.P., 1999. Rain rate estimation from nadir looking microwave radiometer (TMR) for correction of radar altimetric measurements. IEEE Trans. Geosci. Remote Sens. 35 (5), 2556–2568.

Varma, A.K., Gairola, R.M., Pandey, P.C., Singh, K.P., 2001. Use of TOPEX-altimeter for the study of diurnal and spatial distribution of south-west monsoon rainfall over the Bay of Bengal and the Arabian Sea. Remote Sens. Environ. 77 (1), 112–121.

Varma, A.K., Gairola, R.M., Pokhrel, S., Mathur, A.K., Gohil, B.S., Agarwal, V.K., 2002. Rain rate measurements over global oceans from IRS-P4 MSMR. Proc. Indian Acad. Sci. Earth Planet. Sci. 111 (3), 257–266.

Varma, A.K., Pokhrel, S., Gairola, R.M., Agarwal, V.K., 2003. An empirical algorithm for cloud liquid water from MSMR and its utilization in rain identification. IEEE Trans. Geosci. Remote Sens. 41 (8), 1853–1858.

Varma, A.K., Liu, G., Noh, Y.J., 2004. Sub-pixel scale variability of rainfall and its application to mitigate the beam-filling problem. J. Geophys. Res. 109. 10.1029/2004JD004968. D18210.

Varma, A.K., Piyush, D.N., Gohil, B.S., Basu, S.K., Pal, P.K., 2015. A radiative transfer algorithm for identification and retrieval of rain from Megha-Tropiques MADRAS. Adv. Space Res. 55, 1576–1589.

Varma, A.K., Piyush, D.N., Gohil, B.S., Pal, P.K., Srinivasan, J., 2016. Rain detection and measurement from Megha-Tropiques microwave sounder. J. Geophys. Res. Atmos. 121. https://doi.org/10.1002/2016JD024907.

Varma, A.K., Pokhrel, S., Gairola, R.M., Agarwal, V.K., 2006. Study of geophysical parameters associated with the Orissa super cyclone using active and passive microwave remote sensing measurements. Int. J. Remote Sens 27 (18), 3753–3765. https://doi.org/10.1080/01431160410001705060.

Vicente, G.A., Scofield, R.A., Menzel, W.P., 1998. The operational GOES infrared rainfall estimation technique. Bull. Am. Meteorol. Soc. 79, 1883–1898.

Vicente, G.A., Davenport, J.C., Scofield, R.A., 2002. The role of orographic and parallax corrections on real time high resolution satellite rainfall rate distribution. Int. J. Remote Sens. 23 (2), 221–230.

Walsh, E.J., Monaldo, F.M., Goldhirsh, J., 1984. Rain and cloud effects on a satellite dual frequency radar altimeter system operating at 13.5 and 35 GHz. IEEE Trans. Geosci. Remote Sens. GE 22, 615–622.

Webster, P.J., Lukas, R., 1992. TOGA COARE: the coupled ocean-atmosphere response experiment. Bull. Am. Meteorol. Soc. 73, 1377–1416.

Wilheit, T.T., 1986. Some comments on passive microwave measurement of rain. Bull. Am. Meteorol. Soc. 67, 1226–1232.

Wilheit, T., Cheng, A.T.C., Rao, M.S.V., Rodgers, E.B., Theon, J.S., 1977. A satellite technique for quantitatively mapping rainfall rates over the oceans. J. Appl. Meteorol. 16, 551–560.

Woodley, W.L., Sancho, B., Miller, A.H., 1972. Rainfall estimation from satellite cloud photographs. NOAA Tech. Memo. ERL OD-11, 43 pp. Available from: U.S. Department of Commerce/National Oceanic and Atmospheric Administration/National Environmental Satellite, Data, and Information Service, Washington, DC 20233.

Xie, P.P., Rudolf, B., Schneider, U., Arkin, P.A., 1996. Gauge-based monthly analysis of global land precipitation from 1971 to 1994. J. Geophys. Res. Atmos. 101, 19023–19034.

Yong, B., Liu, D., Gourley, J.J., Tian, Y., Huffman, G.J., Ren, L., Hong, Y., 2015. Global view of real-time TRMM multi-satellite precipitation analysis—implications for its successor global precipitation measurement mission. Bull. Am. Meteorol. Soc. 283–296. https://doi.org/10.1175/BAMS-D-14-00017.1.

Yule, G.U., 1945. On the method of studying time-series based on their internal correlations. J. R. Meteorol. Soc. 108, 208–225.

CHAPTER

12

Development of a Rain/No-Rain Classification Method Over Land for the Microwave Sounder Algorithm

Satoshi Kida, Takuji Kubota*, Shoichi Shige†, Tomoaki Mega‡*

***Earth Observation Research Center, Japan Aerospace Exploration Agency, Tsukuba, Japan**
†Kyoto University, Kyoto, Japan ‡Osaka University, Osaka, Japan

1 INTRODUCTION

The observation of global precipitation is important for both scientific research and operational purposes. Globally, there are currently more than 200,000 rain gauges actively measuring precipitation on land (New et al., 2001); however, their temporal and spatial distribution make it difficult to produce a consistent global dataset of precipitation for land hydrology uses. One of the best methods for the global acquisition of precipitation over land is to use data from passive microwave radiometers (MWRs) aboard low earth-orbiting (LEO) satellites.

Passive MWRs aboard LEO satellites are generally divided into two types: imagers and sounders. Microwave imagers (MWIs) include the Tropical Rainfall Measuring Mission (TRMM) Microwave Imager (TMI), the Advanced Microwave Scanning Radiometer for the Earth Observation System (AMSR-E) aboard the Aqua satellite, and the Special Sensor Microwave Imager (SSM/I) of the Defense Meteorological Satellite Program (DMSP). These imagers have window channels suitable for observing emission signatures from raindrops (e.g., the 10- and 19-GHz channels) and for observing the scattering signatures from ice particles (e.g., the 37- and 85-GHz channels). Microwave sounders (MWSs) aboard LEO satellites include the Advanced Microwave Sounding Unit (AMSU) aboard the National Oceanic and Atmospheric Administration (NOAA) satellites. They have been developed primarily for profiling atmospheric temperature and moisture using opaque spectral regions. To optimize sensor performance, 20 channels are divided among three separate total-power radiometers: AMSU-A1, AMSU-A2, and AMSU-B (note: NOAA-18 replaced AMSU-B with the similar Microwave Humidity Sounder, MHS). AMSU-A has window channels at 23.8, 31.4, 50.3, and 89 GHz

Remote Sensing of Aerosols, Clouds, and Precipitation
https://doi.org/10.1016/B978-0-12-810437-8.00012-8

© 2018 Elsevier Inc. All rights reserved.

and the 50- to 60-GHz oxygen band. AMSU-B/MHS has window channels at 89 and 150 GHz and around the 183-GHz water vapor line. Because there have been four AMSU instruments in orbit since the launch of NOAA18 in 2005, together with several MWIs, there have been more observations of rainfall in time and space, with swaths being ~2200-km wide. Actually, Kubota et al. (2009) demonstrated effective performance of the merger of the MWSs in addition to the MWIs by the ground-radar validation around Japan.

In addition to the NOAA operational Microwave Surface and Precipitation Products System (MSPPS) Day-2 algorithm (Ferraro et al., 2005), recently various techniques such as neural network approaches (Surussavadee and Staelin, 2008; Sano et al., 2015); the Water Vapor Strong Lines at 183GHz (183-WSL) technique (Laviola and Levizzani, 2011); a technique using canonical analysis (Casella et al., 2015); the Microwave Integrated Retrieval System (MIRS) (Boukabara et al., 2011); the Goddard Profiling algorithm (GPROF) (Kidd et al., 2016); Linear discriminant analysis (LDA) (You et al., 2016); and the Global Satellite Mapping of Precipitation (GSMaP) (Shige et al., 2009) have been developed for the MWS.

For a dataset merged from the MWIs and the MWSs, bias errors in the retrieved rain rates that vary between MWIs and MWSs are troublesome for many applications. The only way to try to overcome this problem is to develop a consensus algorithm applicable to both MWIs and MWSs based on the same physical principle (Shige et al., 2010). Therefore, Shige et al. (2009) developed an over-ocean rainfall retrieval algorithm for MWS (GSMaP_MWS) that shares at a maximum a common algorithm framework with the GSMaP algorithm for MWI (GSMaP_MWI; (Kubota et al., 2007; Aonashi et al., 2009)).

Although the MWS precipitation algorithms for retrievals over oceans employ emission signatures using lower-frequency channels such as the 23- and 31-GHz channels of AMSU-A (Shige et al., 2009; Vila et al., 2007), rain estimates from the MWS algorithms have lower accuracy than those from MWI algorithms (Lin and Hou, 2008; Kida et al., 2010). This is because the resolution of the emission channels of MWSs is coarser than MWIs, which contributes to their failure to detect warm rain that has small scale and little ice aloft.

Over land, the MWS and MWI algorithms employ only scattering signatures. The precipitation retrieval algorithm for MWRs, which use the channel at around 89 GHz, tends to miss precipitation systems that have few ice particles aloft because the precipitation retrieval algorithm over land employs only the scattering signature from ice particles aloft. However, because the MWSs use the 150-GHz channel, which is better than the 89-Ghz channel for detecting areas of light precipitation (Bennartz, 2002), there is the potential to detect precipitation systems that have few ice particles aloft. The MSSPPS Day-2 algorithm uses the difference of the brightness temperatures (Tbs) between 89 and 150 GHz for detecting precipitation.

Over coastal mountain ranges of the Asian monsoon region, heavy orographic rainfall is frequently associated with low storm heights (Shige and Kummerow, 2016). This leads to the conspicuous underestimation of rainfall using microwave radiometer algorithms that conventionally assume that heavy rainfall is associated with high storm height. Several previous studies demonstrated the underestimation in regions such as Japan (Kubota et al., 2009), Korea (Kwon et al., 2008), Taiwan (Taniguchi et al., 2013), and India (Shige et al., 2015). Utilization of 150 GHz in addition to 89 GHz channels can be helpful for detecting precipitation with low storm height and mitigating such underestimation in the orographic rainfall.

The development of the rain-estimation algorithms has been enhanced since the launch of the TRMM satellite in 1997. Developed under a United States–Japan joint mission, TRMM carries rain observation sensors such as the TMI and the Precipitation Radar (PR) (Kummerow et al., 1998). Accurate measurement of precipitation by the PR and simultaneous observations by the TMI and PR have greatly advanced algorithm development for the MWI. Moreover, the simultaneous observations by the PR and the MWS data can also lead to the enhancement of the MWS algorithms.

In this study, we investigated a new rain/no-rain classification (RNC) method for the precipitation retrieval algorithm for MWSs using high-frequency channels such as 150 and around 183 GHz. The proposed RNC method was applied to the GSMaP_MWS retrievals over land, and its performance was compared with that of the TRMM/PR observations and the original GSMaP_MWS. The revised GSMaP_MWS over land was also compared with the MSPPS Day-2 product because the dataset of the MSPPS Day-2 product is used by some global precipitation maps, such as the CMORPH (Joyce et al., 2004) and TMPA 3B42 (Huffman et al., 2007).

Section 2 presents a description of the data used in this study. The original RNC method is compared with the TRMM/PR observations, and the revised RNC method using the 150-GHz channel is proposed in Section 3. In Section 4, the performance of the revised RNC method is compared with that of the original RNC method and the PR observations, and conclusions are stated in Section 5.

2 DATA

2.1 Spatial and Temporal Matchup of AMSU Data Against PR Data

In this study, we used the Tbs measured by the AMSU on the NOAA-15, 16, 17, and 18 satellites. For comparison with the estimates from the AMSU, the precipitation product from the PR on the TRMM was used. Because the AMSU-B (MHS) is a cross-track scanner with an angular swath of ± 50 degrees, its effective field of view (FOV) varies from 16×16 km at nadir to 52×27 km on the limb. The PR is a nadir-looking instrument that measures backscatter from precipitation particles, observing the three-dimensional structure of precipitation with fine resolution ($\sim$5 km).

In this study, the PR level-2 standard product 2A25 version 6 (Iguchi et al., 2000; Iguchi et al., 2009) was used. Because the TRMM and NOAA satellites have different orbits, we matched PR 2A25 data and NOAA-15, 16, 17, and 18 AMSU products during 2007. In order to more directly compare the PR observation with the estimation from the AMSU, a weighted average of the PR observations was performed in the neighborhood of a given AMSU FOV with a method similar to Kida et al. (2009). The varying FOV of the AMSU was considered with the same methodology of Kida et al. (2010). Overall, 2141 rain events for matched PR–AMSU orbits were found within a 20-min window. Then, for each AMSU-B pixel, the rain rate from the PR, storm height (rain-top height) from the PR, and freezing-level height from the Japan Meteorological Agency Climate Data Assimilation System (JCDAS) were collocated. The storm height from the PR will be helpful for a diagnosis of shallow or tall precipitation systems. Because the PR has higher resolution than AMSU-B, rain estimates from the PR were averaged within the AMSU-B footprint.

2.2 MSPPS Algorithm

The MSPPS Day-2 rainfall algorithm for the AMSU was developed at NOAA. The algorithm originates from the work of Weng and Grody (Weng and Grody, 2000) and, subsequently, from Zhao and Weng (Zhao and Weng, 2002). A simultaneous retrieval of the ice water path (IWP) and ice particle effective diameter (De) from Tb89 and Tb150 was performed through two processes: simplifying the radiative transfer equation into a two-stream approximation and estimating the cloud- base and cloud-top Tbs through the use of Tb23 and Tb31. The rain rate was computed based on an IWP and rain-rate relation derived from the GPROF algorithm database, which contains the profiles of various hydrometeors generated from cloud-resolving models.

Over land, the IWP is computed only if there is no snow on the ground, as determined from AMSU; there is a valid IWP and De retrieval (between 0.0 and 3.0 kg m^2 and between 0.0 and 3.5 mm, respectively); and Tb89 – Tb150 is >3 K. The rain rate is computed based on an IWP and rain-rate relation only when the land surface temperature (Ts) exceeds 269 K and Ts – Tb190 is ≤10 K.

2.3 The GSMaP_MWS Algorithm

The GSMaP algorithm is a PR-consistent advanced MWR algorithm (Kubota et al., 2007; Aonashi et al., 2009). The basic concept of the GSMaP algorithm is to find the optimal rainfall that provides the radiative transfer model (RTM) FOV-averaged Tbs that best fit the observed Tbs. The GSMaP algorithm consists of two elements: a forward calculation part and a retrieval part. The forward calculation part calculates the look-up tables (LUTs) providing the relationship between the surface rainfall rates and Tbs with an RTM. The variations in path lengths through which the atmosphere is viewed by cross-track scanning should be taken into account. For the AMSU, the received polarization also varies with scan angle because of the rotating-reflector/fixed feed horn antenna design. Therefore, as described in Shige et al. (2009), the LUTs are produced for each scan angle from the RTM calculations with mixed vertical and horizontal polarizations as a function of scan angle (Grody et al., 2001).

In the retrieval part, the rainfall rates are estimated from the observed Tbs using the LUTs after the application of the RNC classification.

The GSMaP_MWS has been developed based on the GSMaP_MWI. The GSMaP_MWS shares the RTM of the forward calculation part with the GSMaP_MWI, except for some changes in frequencies, scan angles, and polarizations. Over oceans, the GSMaP_MWS combines an emission-based estimate from Tb23 and a scattering-based estimate from Tb89, depending on a scattering index (SI) computed from Tb89 and Tb150 (Shige et al., 2009).

Fig. 1 shows the original RNC method over land for the GSMaP_MWS. The original RNC

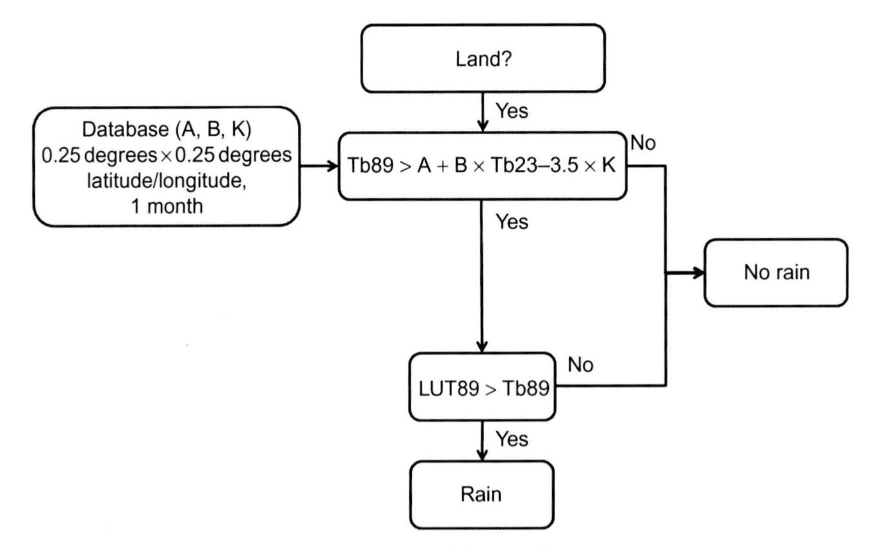

FIG. 1 Flowchart of the original RNC method over land in GSMaP_MWS.

method over land has two steps. The first step is the method described by Seto et al. (2008), which checks the depression of the Tb at 89 GHz owing to scattering from ice particles. The SI from Tb89 and Tb23 (hereafter, the SI is referred to as SI89) is calculated as

$$\mathrm{SI89} = \mathrm{Tb89}_c - \mathrm{Tb89} \tag{1}$$

and

$$\mathrm{Tb89}_c = a + b \times \mathrm{Tb23} + \sigma \times k, \tag{2}$$

where $\mathrm{Tb89}_c$ denotes the estimated 89-GHz Tb over a clear region. When SI89 in a target pixel is >0, the target pixel is classified as a rain pixel. According to M2d, as described by Seto et al. (2008), the coefficients *a*, *b*, and *k* are calculated in a 0.25 × 0.25 degrees latitude–longitude grid for each month. In this study, the coefficients were calculated from Tbs observed by the AMSU during a 5-year period (January 2006 to December 2010).

In the second step, the scattering signature at 89 GHz using the LUT is checked. The condition that rain exists is given by

$$\mathrm{LUT89} > \mathrm{Tb89}, \tag{3}$$

where LUT89 is the 89-GHz Tb at 0 mm h^{-1} in the LUT.

3 COMPARISON OF RAIN/NO-RAIN CLASSIFICATION (RNC) OVER LAND USING THE 89-GHz CHANNEL WITH PR AND MSPPS RAIN ESTIMATES

To evaluate the performance of the RNC, this study used the four scenarios presented in Table 1, which is a 2×2 contingency table. In Table 1, N_1 is the number of pixels where rain observed by the PR exists and the RNC method classifies rain. N_2 is the number of pixels where rain observed by the PR exists and the RNC method classifies no-rain. N_3 is the number of pixels where rain observed by the PR does not exist and the RNC method classifies rain. N_4 is the number of pixels where rain observed by the PR does not exist and the RNC method classifies no-rain. The probability of detection (POD) is the ratio of correct estimates of rain to the number of observed rain events. The POD is defined as $\mathrm{POD}=N_1/(N_1+N_2)$. This statistic, also known as the hit rate, is an index of rain detection ability. The false-alarm ratio (FAR) is the fraction of estimates of rain that were incorrect or an index of false rainfall. The FAR is defined as $\mathrm{FAR}=N_3/(N_1+N_3)$. The threat score (TS) is defined as $\mathrm{TS}=N_1/(N_1+N_2+N_3)$ and it provides the number of correct estimates of rain divided by the total number of occasions on which a rain event was observed by the PR or classified by the RNC method. The frequency bias (FB) is the ratio of the rain area classified by the RNC method to the PR-observed rain area. The FB is defined as $\mathrm{FB}=(N_1+N)/(N_1+N_2)$. Here, FB=1 reflects unbiased estimates, indicating that the number of rain areas classified by the RNC method was the same as the number of rain areas observed by the PR. A further description of these statistics can be found in Wilks (Wilks, 2006).

TABLE 1 A 2 × 2 Contingency Table for Evaluation

		The RNC Method	
		Rain	**No-Rain**
PR	Rain	N_1	N_2
	No-Rain	N_3	N_4

Elements N_1 through N_4 are assigned the observed event counts in each category.

Fig. 2 shows the case where the AMSU matched the PR on Jan. 2, 2007, (TRMM orbit number 52041, NOAA-15 orbit number 44915) in South America. The POD, FAR, TS, and FB of the MSPPS (Fig. 2B) with reference to the PR (Fig. 2A) are 0.733, 0.188, 0.626, and 0.903, respectively. The POD, FAR, TS, and FB of the original GSMaP_MWS (Fig. 2C) with reference

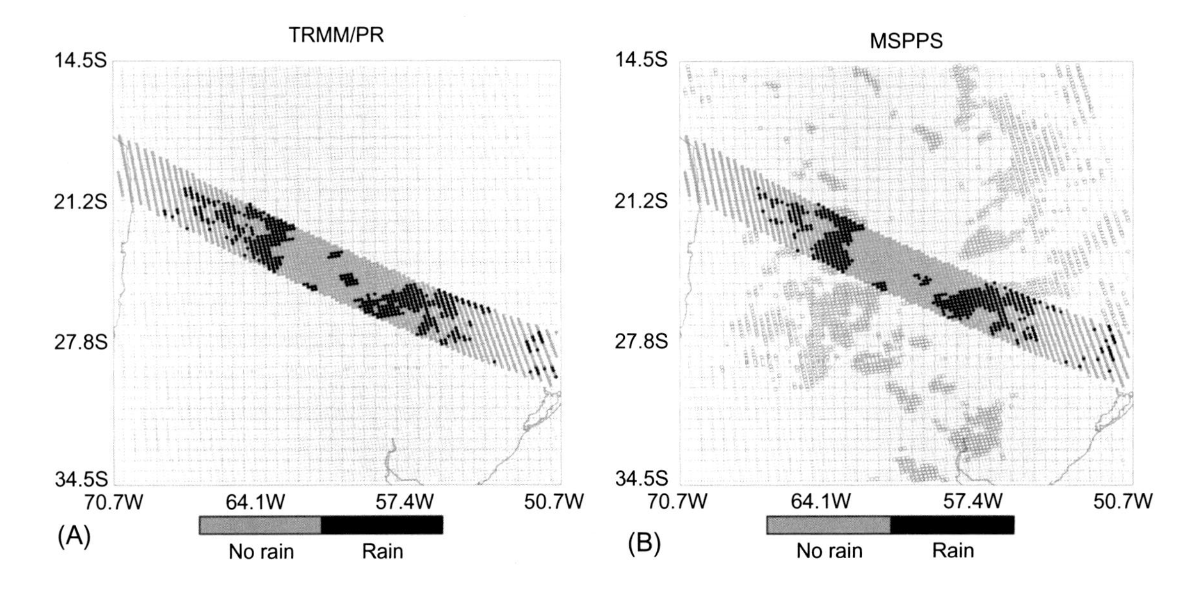

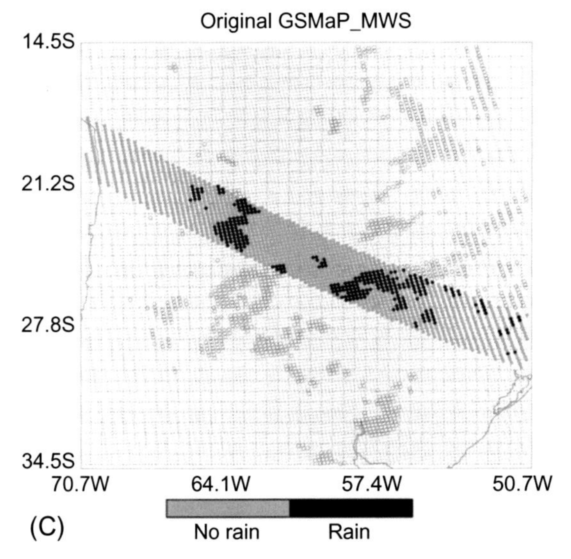

FIG. 2 Case in which the MWS observation matches the PR observation on Jan. 2, 2007, (TRMM orbit number 52041, NOAA-15 orbit number 44915). (A) PR, (B) MSPPS, and (C) GSMaP_MWS with the original RNC method. *Black* areas show rain and *gray* areas show no rain.

to the PR (Fig. 2A) are 0.563, 0.117, 0.524, and 0.639, respectively. While the rain area of the MSPPS is similar to that of the PR, the rain area of the original GSMaP_MWS is smaller. Specifically, in the region around the area of strong rain, the original GSMaP_MWS fails to detect the region of light rain. In the precipitation area detected by both the PR and the GSMaP_MWS, the averaged storm height is about 5.8 km and the conditional rain rate is 2.73 mm h^{-1}. In the rain area that was not detected by the GSMaP_MWS, the averaged storm height is about 3.4 km and the conditional rain rate is 0.55 mm h^{-1}. The average freezing height within this region in this case is about 5.0 km, and although the GSMaP_MWS can detect areas of strong precipitation (e.g., convective precipitation where there are scattering signatures from ice particles), it fails to detect areas of weak precipitation, such as shallow convective precipitation or stratiform precipitation that have little ice aloft.

In order to examine statistical differences between the GSMaP_MWS with the original method and the MSPPS, POD, FAR, and TS were evaluated in all AMSU–PR matchup data, described in Section 2.1. The skill scores were calculated for the MWS retrievals with reference to the PR, and then skill scores for the GSMaP and the MSPPS were compared in Fig. 3. The POD values of the original method are considerably lower than those of the MSPPS, although the FAR of the original method is better. The TS of the original method is also lower than those of the MSPPS. Because the TS takes into account both false alarms and missed events, this suggested that total performances of the GSMaP with the original method were worse than those of the MSPPS.

The reasons behind the poor results in the GSMaP_MWS were further examined using the AMSU–PR matchup data. Fig. 4 shows probability density functions (PDFs) of storm height in regions where the GSMaP_MWS detects and does not detect rain areas. While about 66% of the precipitation pixels detected by the GSMaP_MWS have storm heights >4.0 km, about 80% of the precipitation pixels missed by the GSMaP_MWS have storm heights <4.5 km.

Contiguous rain areas detected by the PR and GSMaP_MWS were identified as precipitation systems. The area of each precipitation system detected by the GSMaP_MWS was compared with the area of the corresponding precipitation system detected by the PR. Fig. 5 shows a PDF of the difference in rain areas over land between the PR and the GSMaP_MWS using the original RNC method. Most (about 75% of all cases) precipitation systems from the GSMaP_MWS have smaller areas than observed by the PR. The ratio of the total rain area from the GSMaP_MWS to that of the PR is about 52%.

Thus, the shallow precipitation was missed and rain areas detected by the original method were narrower. These can lead to lower values of the POD and TS of the original method.

4 PROPOSED RNC METHOD

4.1 Comparison of Scattering Signatures Between the 89- and 150-GHz Channels

As has been noted previously by several authors (Bennartz and Bauer, 2003; Chen and Staelin, 2003; Ferraro et al., 2000), the 150-GHz channel is more sensitive than the 89-Ghz channel to smaller-sized particles and, therefore, it is more useful for detecting areas of light precipitation (Bennartz, 2002). To compare the performances of rain detection using the 150- and 89-GHz channels, we investigated the 89- and 150-GHz scattering signatures (hereafter, referred to as SS89 and SS150, respectively). We can compute SS89 and SS150 as

$$SS89 = TB89_0 - TB89 \quad (4)$$

and

$$SS150 = TB150_0 - TB150, \quad (5)$$

where $TB89_0$ and $TB150_0$ are the Tbs of the 89- and 150-GHz channels in clear regions around rain systems, respectively, and TB89 and TB150 are the Tbs of the 89- and 150-GHz channels in rain systems, respectively. When SS89 and/or SS150 are >0 in a target pixel, the target pixel is defined as a rain pixel.

For each precipitation system observed by the PR, we calculated the rain area derived from the

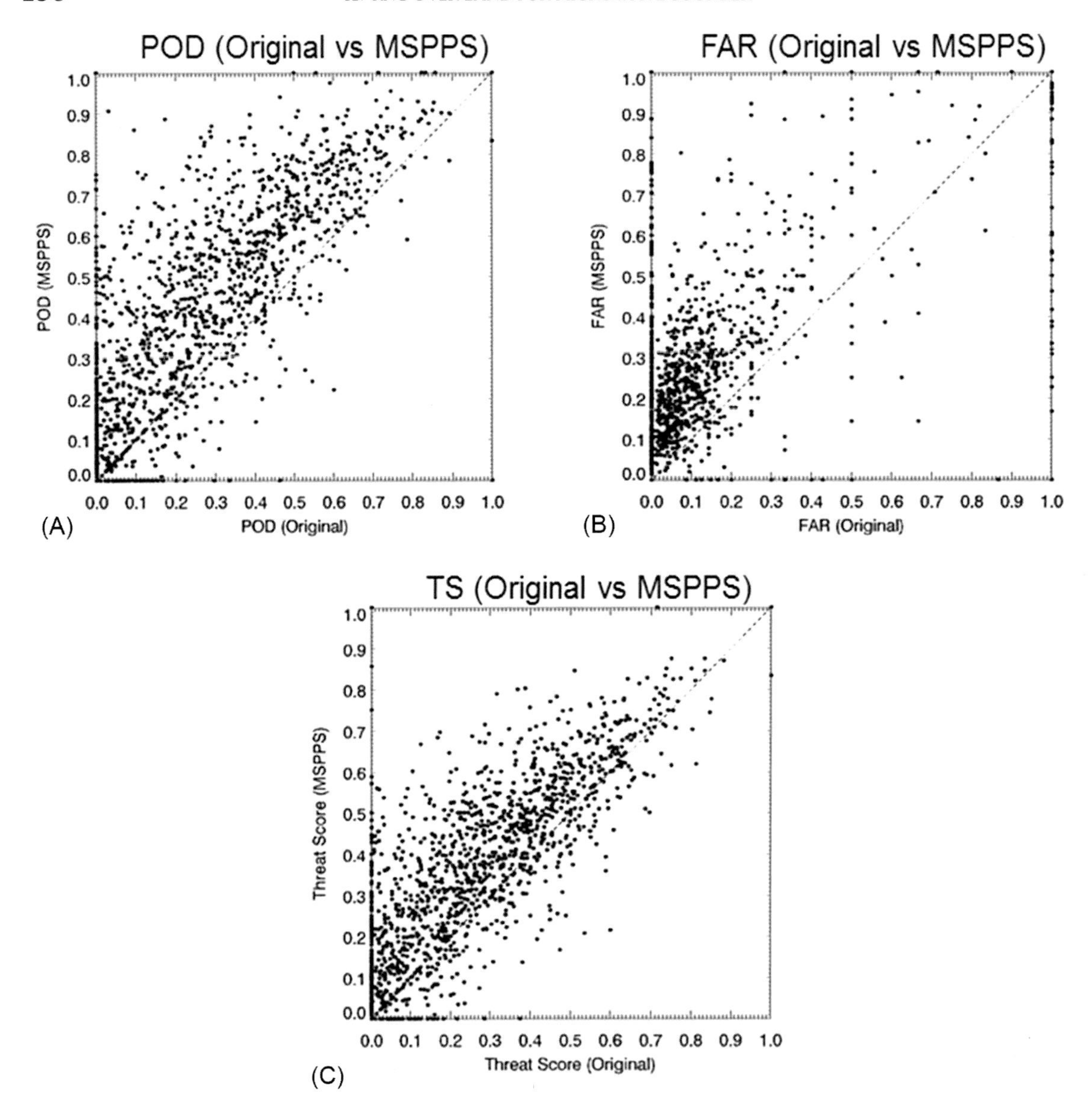

FIG. 3 Scatter diagrams of POD (A), FAR (B), and TS (C) of PR-AMSU matched-up cases between the GSMaP original RNC method and the MSPPS. Horizontal axis shows skills of the original RNC method and vertical axis shows skills of the MSPPS.

89- and 150-GHz channels (hereafter, the rain area is referred to as RA89 and RA150, respectively). Fig. 6 shows a PDF of the difference in the rain area between RA150 and RA89. The RA150 is larger than the RA89 in about 80% of the 2385 precipitation systems where the difference between RA150 and RA89 is >1 pixel. On average, RA150 is about 13% larger than RA89. Fig. 7 shows histograms of storm height as detected by the 89- and 150-GHz channels. Because the 150-GHz channel has higher sensitivity than the 89-GHz channel to smaller-sized

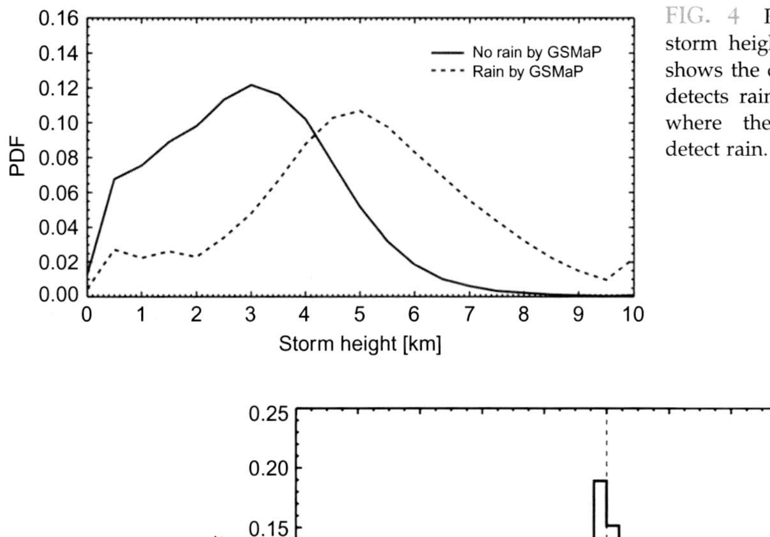

FIG. 4 Probability density function (PDF) of storm height observed by the PR. The solid line shows the case where the GSMaP original method detects rain, and the dashed line shows the case where the GSMaP original method fails to detect rain.

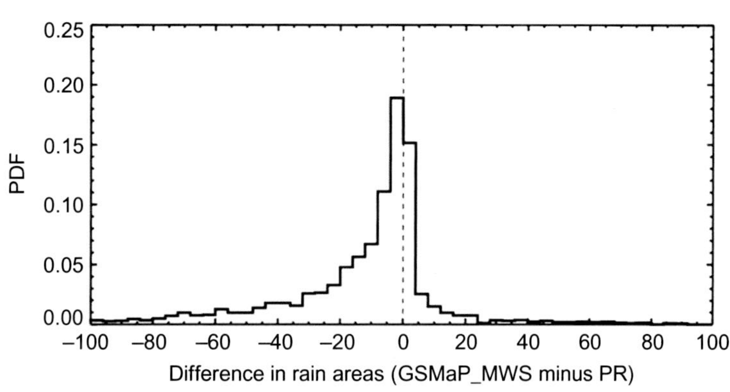

FIG. 5 Probability density function (PDF) of difference of rain areas between the original method and PR. The unit of the difference is a pixel.

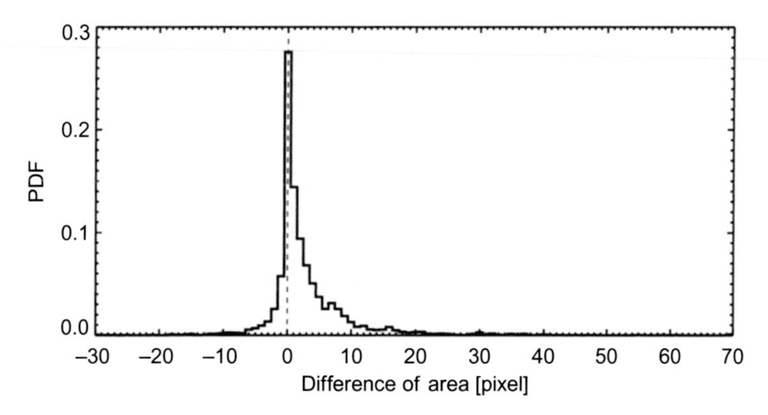

FIG. 6 Difference of precipitation area detected between 89- and 150-GHz brightness temperatures.

ice particles and raindrops (as found in shallow rain), shallow rain with a storm height of <4.5 km can be detected by the scattering signature at 150 GHz. Therefore, this result suggests that the use of the scattering signature at 150 GHz will lead to better detection of precipitation areas than using the scattering signature at 89 GHz.

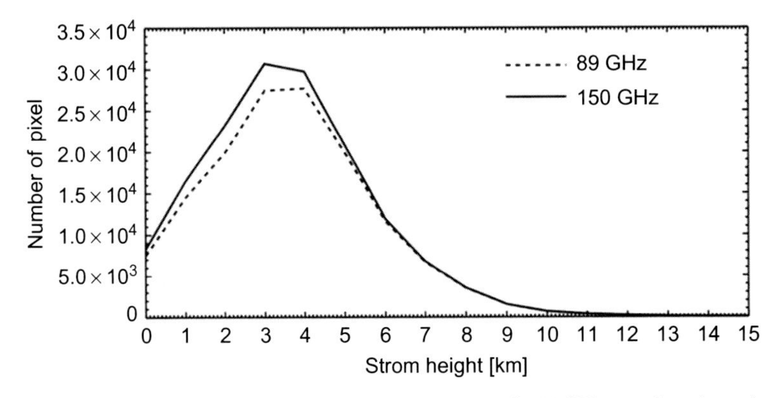

FIG. 7 Number of rain pixels detected by scattering signatures at 89 and 150 GHz as a function of storm height.

4.2 Proposed Method Using Scattering Index (SI)

In this study, we proposed a new RNC method that employs the SI, which is computed based on the difference between the Tbs at 89 and 150 GHz. The SI is similarly defined as for the GSMaP_MWS over the ocean (Shige et al., 2009):

$$SI = (Tb89 - LUT89) - (Tb150 - LUT150), \quad (6)$$

where LUT89 and LUT150 are the Tbs for $0\ \text{mm h}^{-1}$ as computed by the RTM calculation. Because the SI is increased by an increase in rain rate, it is used as the RNC threshold value. The RNC threshold value of the SI is used in those regions where the original GSMaP_MWS algorithm determines no-rain.

To establish the RNC threshold value of the SI, we used the TS. Fig. 8 shows the TS as a function of the RNC threshold value of the SI. Because the 150-GHz Tb has higher sensitivity than the 89-GHz TB to water vapor and cloud, if a very low RNC threshold value of the SI is employed, water vapor and clouds could be misclassified as rain, leading to low TS values. Conversely, if a high RNC threshold value of the SI is employed, the precipitation is no longer detected by the SI, resulting in the same TS as the original GSMaP_MWS (dashed line in Fig. 8).

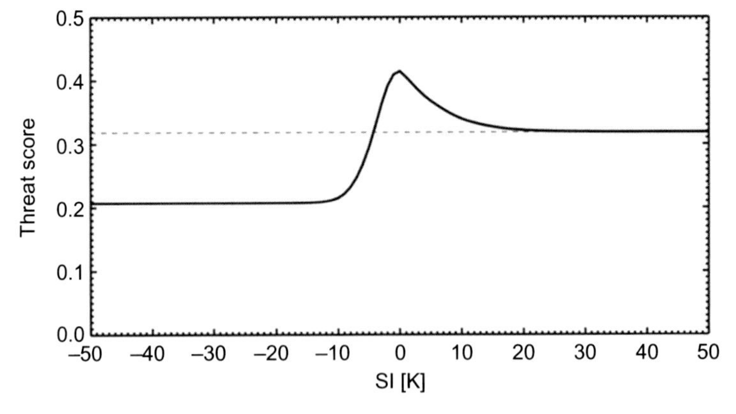

FIG. 8 Threat score as a function of an RNC threshold value for the Scattering Index (SI).

When the RNC threshold value of SI is zero, the TS becomes a maximal value; therefore, we used zero as the RNC threshold value of the SI.

Over snow-covered and desert areas, scattering signals caused by precipitation particles are often contaminated by snow and sand particles on the surface. When only the SI was applied to the RNC method of the GSMaP_MWS, we found that the scattering signatures caused by sand and surface snow could be misclassified as the scattering signature of precipitation particles in the Sahara Desert, and in mid-latitude and highly mountainous regions in winter, respectively. To screen out such scattering signatures, we used the channels at 89, 186, and 190 GHz. Over desert surfaces and snow cover, Tb89 is reduced because it is strongly affected by surface scattering. Conversely, both Tb186 and Tb190 are free from the effects of surface emission and are sensitive to different atmospheric levels. The difference between Tb186 and Tb190 increases as the quantity of precipitation particles at lower levels increases. Therefore, in the revised RNC method we employed the two Tbs: $Tb89 - Tb190 > 0$ and $Tb186 - Tb190 > -8$. Therefore, the combination of the original method, SI, and screening of the desert and the surface snow was implemented in the GSMaP_MWS, and Fig. 9 summarizes a flowchart of the revised RNC method over the land.

4.3 Results

Fig. 10 shows the results of the proposed method for the same case as in Fig. 2. The POD, FAR, TS, and FB of this case are 0.761, 0.219, 0.626, and 0.974, respectively. In the precipitation system observed by the PR, while the region of light rain is missed by the original method (Fig. 2C), it is detected by the proposed method. Although the FAR is changed from 0.117 to 0.219 because of the increased detection of areas of light rain, POD is improved from 0.563 to 0.761, TS is improved from 0.524 to

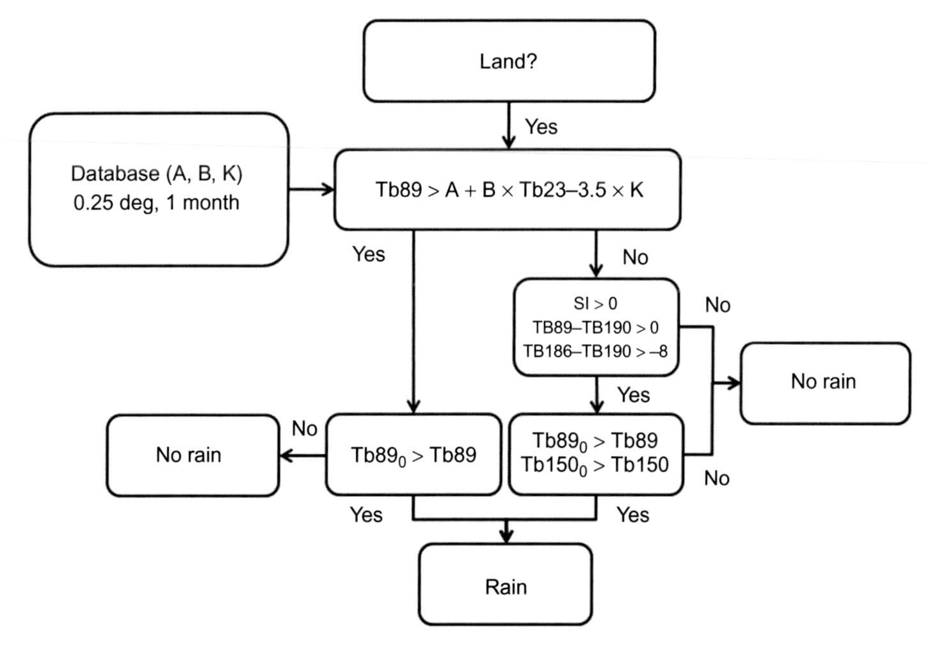

FIG. 9 Flowchart of the revised RNC method over land in GSMaP_MWS.

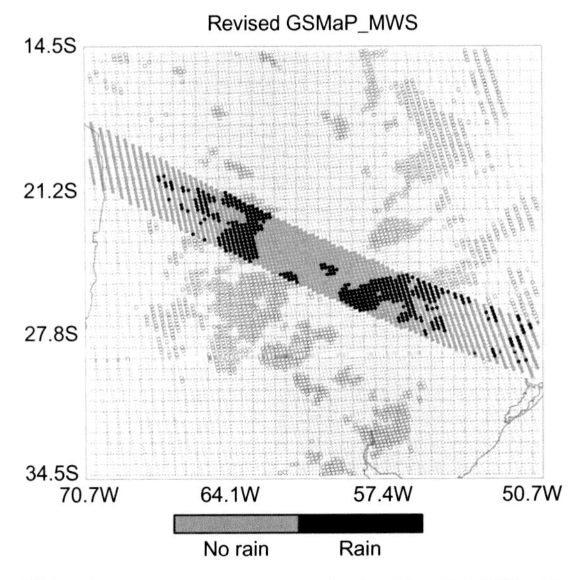

FIG. 10 Same as Fig. 2, except for the GSMaP_MWS with the revised RNC method.

0.626, and FB is improved from 0.639 to 0.974. The skill scores of the proposed method are improved and they are similar to those of the MSPPS. The rain amount is changed from 0.740 to 0.792 mm h^{-1}. While the rain area of the GSMaP_MWS is increased by about 52%, the rain amount has increased only by about 7%. This is because the proposed RNC method mainly detects weak precipitation that the original RNC method fails to detect.

Fig. 11 shows the PDF of storm height detected by the proposed RNC method in all matched-up cases. Because the SI is used for the revised RNC method, precipitation that has a mean storm height of 3.52 km can be detected. The number of detections of shallow precipitation systems with low cloud-top heights (~3 km) may be inadequate because most of the shallow precipitation systems are likely to be cumulus clouds with a horizontal scale that is usually smaller than that of the MWSs (i.e., AMSU-B has resolution of about 16 km at nadir).

Fig. 12 shows the change in the skill scores between the original and proposed methods in all matched-up cases, and Table 2 shows the change of average skill scores between the original and revised methods. The FB of the revised method is more than double that of the original method, although the rain area of the GSMaP_MWS is still smaller than that of the PR because of the failure to detect very shallow precipitation, such as rain from cumulus clouds. The POD of the revised method is improved from that of the original method because the revised method is designed to detect more rain pixels than the original method. Although the FAR of the revised method is higher than that

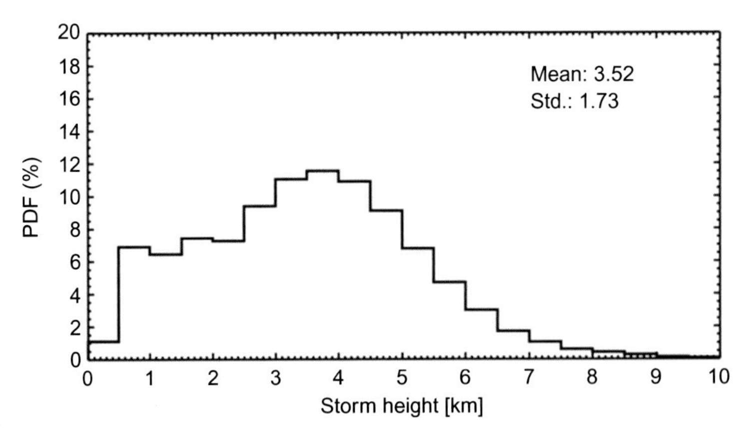

FIG. 11 Probability density function (PDF) of storm height in the rain pixel detected by the revised RNC method.

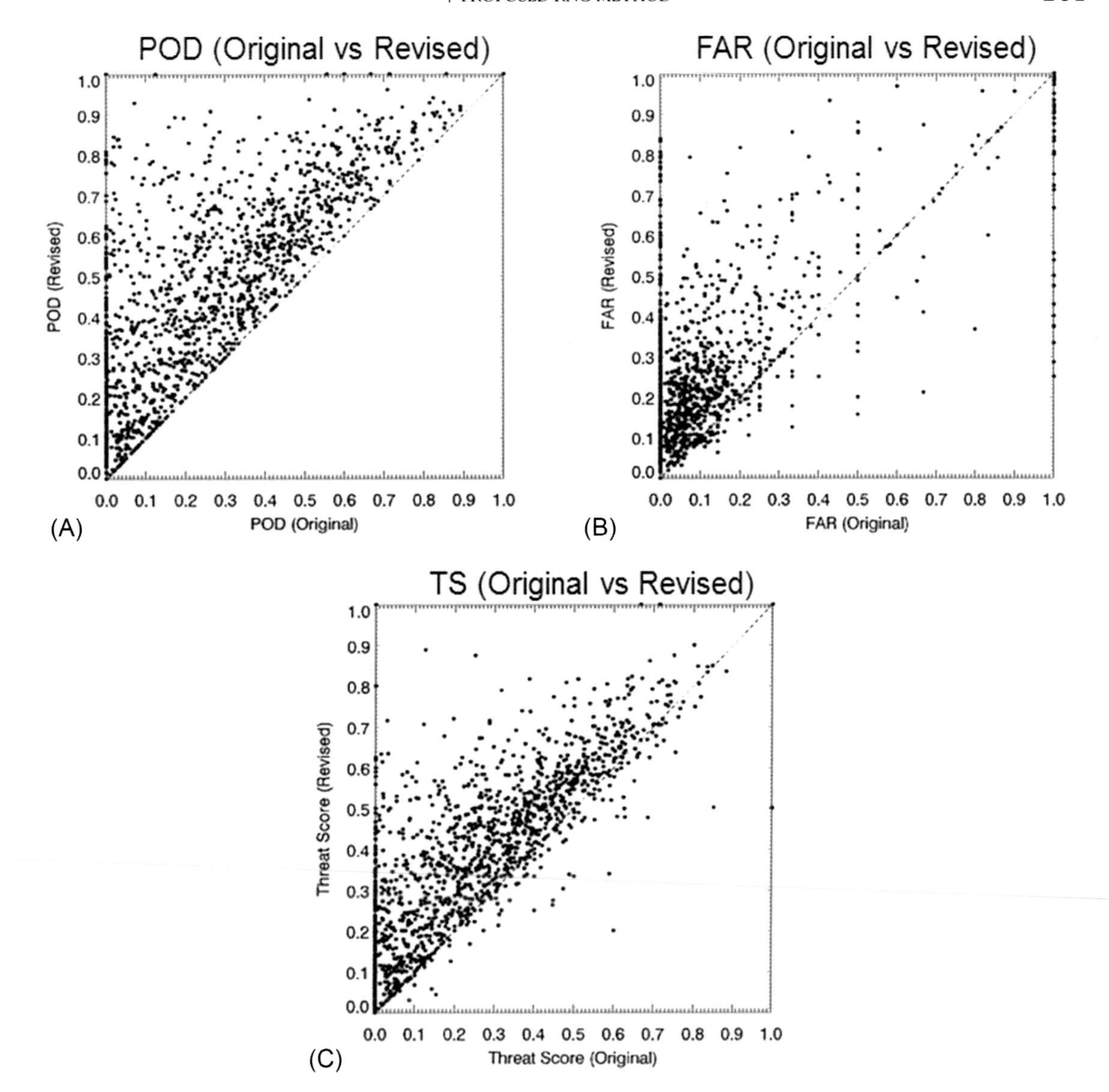

FIG. 12 Same as Fig. 3, except for results between the original and the revised RNC method.

TABLE 2 Changes of the Average Skill Scores From the Original Method to the Revised Method

	POD	FAR	TS	FB
MSPPS	0.483	0.300	0.400	0.689
Original	0.323	0.105	0.311	0.360
Revised	0.509	0.305	0.416	0.732

of the original, because of the increased detection of rain pixels, some cases show an improvement of the FAR. This shows that true detection (N_1) increases by more than the misclassification of pixels (N_3) using the revised method. Therefore, for about 84% of cases, the TS values of the revised method are improved; on average, the TS is improved from 0.311 to 0.416.

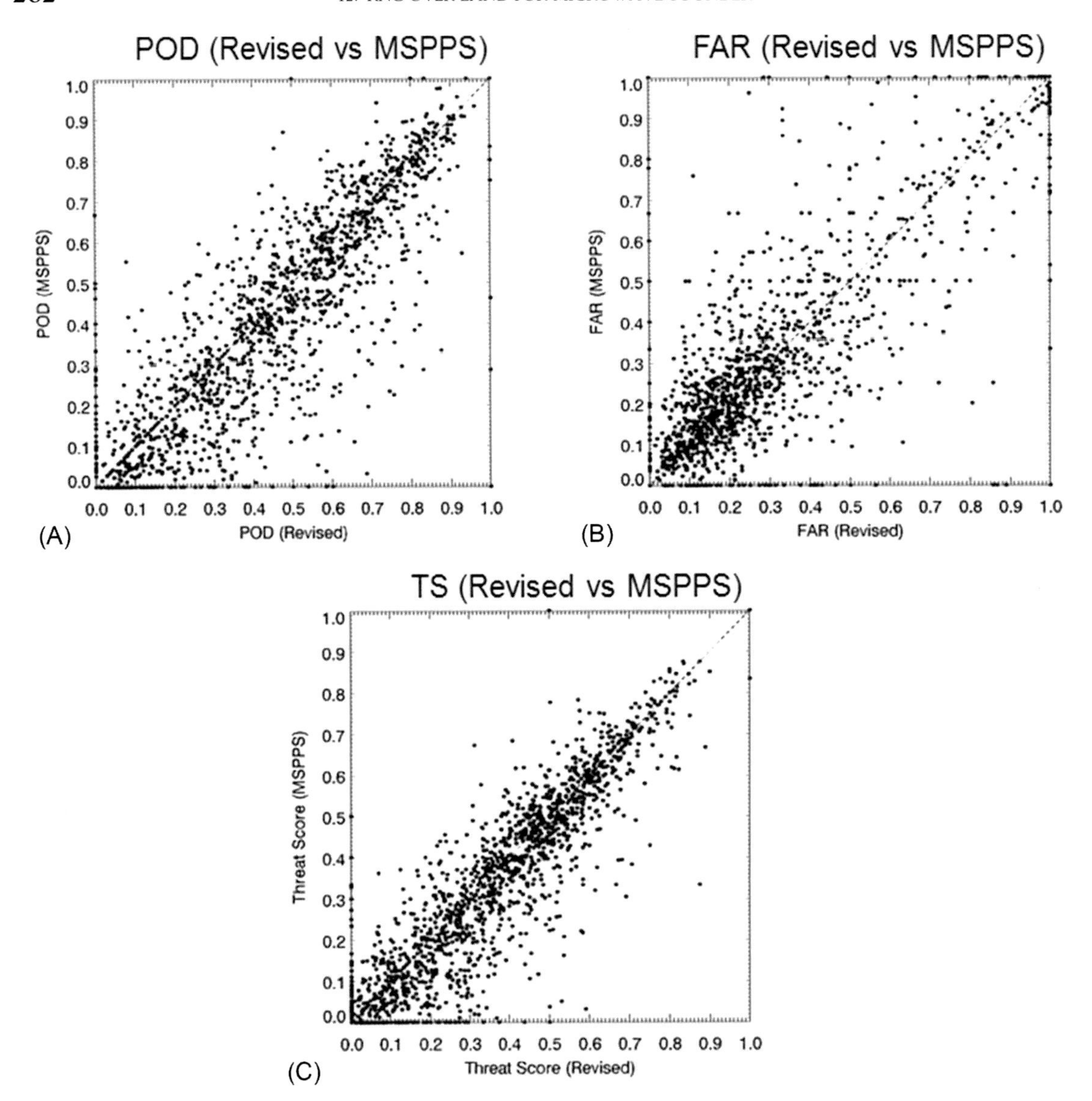

FIG. 13 Same as Figs. 3 and 12, except for results between the revised RNC method and the MSPPS.

Fig. 13 shows the difference in the skill scores of the revised method and the MSPPS in all matched-up cases. Because most of the shallow precipitation is missed by the original method, the POD and TS of the original method are considerably lower than those of the MSPPS, as shown in Fig. 3. On the other hand, because the detection of shallow precipitation is improved by the revised method, although the FAR of the revised method is similar to that of the MSPPS, the POD and TS of the revised method are improved and the average skill scores are slightly better than those of the MSPPS.

5 CONCLUSION

We proposed a new RNC method over land for the precipitation retrieval algorithm for MWSs. The proposed RNC method employed the SI from 89 and 150 GHz, as defined by Shige et al. (2009). To estimate the performance of the proposed RNC method, PR–AMSU matched-up cases were investigated.

Based on the PR–AMSU matched-up cases, the precipitation area was estimated using the scattering signatures of the 89- and 150-GHz channels. The precipitation area derived from the scattering signature at 150 GHz was larger than that derived from the scattering signature at 89 GHz in about 80% of the 2141 precipitation systems examined. On average, the precipitation area from the scattering signature at 150 GHz was about 47% larger than that obtained from the scattering signature of the 89-GHz channel. It was found that the use of the scattering signature at 150 GHz detected precipitation with a precipitation top height <4.5 km.

The performance of the original RNC method over land for the GSMaP algorithm was compared with PR observations. In the precipitation systems observed by the PR, while light rain regions were missed by the original method, areas of light rain that had storm heights of 3.52 km, on average, were detected by the revised method. Note that the rain area of the GSMaP_MWS was still smaller than that of the PR because of the failure to detect very shallow precipitation, such as the rain from cumulus clouds. This issue has been left for future work. Consequently, for 84% of all PR–AMSU matched-up cases, the TS value was improved and the average TS value improved from 0.311 to 0.416.

For comparison with the MSPPS, because most shallow precipitation was missed by the original method, the POD and TS of the GSMaP original method were considerably lower than those of the MSPPS, although the FAR of the original method was better. Conversely, because the detection of shallow precipitation was improved by the revised method, although the FAR of the revised method was similar to that of the MSPPS, the POD and TS of the revised method were improved and the average skill scores were slightly better than those of the MSPPS.

Although the revised method was developed for application in the precipitation retrieval algorithm of MWSs, the revised method is also applied to the retrieval algorithms of the new type of MWRs equipped with both imager and sounder channels, such as the Special Sensor Microwave Imager/Sounder (SSMIS) (Kunkee et al., 2008). As in this study, the space-borne precipitation radar is helpful for the development of the MWR algorithm. Therefore Dual-Frequency Precipitation Radar (DPR) with the Global Precipitation Measurement (GPM) Microwave Imager (GMI) onboard the GPM Core Observatory (Hou et al., 2014) will be promising for future works, particularly for the development of methods for precipitation at higher latitudes.

In this work, utilization of the 150-GHz channel was focused in order to detect areas of light precipitation that were not detected by the 89-Ghz channel. The TB 186 and TB 190 were used secondarily with the aim of screening out the desert surface and snow cover. This can be helpful for identifying a value of the 150-GHz channel in the AMSU retrieval. Nevertheless, an investigation of the TB190, as in previous papers such as Bennartz and Bauer (2003) and Chen and Staelin (2003), may be valuable, and a combination between the revised method and a method using the TB190 will be regarded as one for future tasks.

Acknowledgments

The authors would like to thank Dr. Kazumasa Aonashi (Meteorological Research Institute, Japan Meteorological Agency) and Ms. Misako Kachi and Dr. Riko Oki (Earth Observation Research Center, Japan Aerospace Exploration Agency) for their valuable comments. They also acknowledge the Cooperative Institute for Climate Studies server for providing the MSPPS Day-2 product.

References

Aonashi, K., Awaka, J., Hirose, M., Kozu, T., Kubota, T., Liu, G., Shige, S., Kida, S., Seto, S., Takahashi, N., Takayabu, Y.N., 2009. GSMaP passive microwave precipitation retrieval algorithm: algorithm description and validation. J. Meteor. Soc. Japan 87A, 119–136.

Bennartz, R., 2002. Precipitation analysis using the advanced microwave sounding unit in support of now casting applications. Meteorol. Appl. 9, 177–189.

Bennartz, R., Bauer, P., 2003. Sensitivity of microwave radiances at 85–183 GHz to precipitating ice particles. Radio Sci. 38 (4), 8075. https://doi.org/10.1029/2002RS002626.

Boukabara, S.-A., et al., 2011. MiRS: an all-weather 1DVAR satellite data assimilation and retrieval system. IEEE Trans. Geosci. Remote Sens. 49, 3249–3272. https://doi.org/10.1109/TGRS.2011.2158438.

Casella, D., Panegrossi, G., Sano, P., Milani, L., Petracca, M., Dietrich, S., 2015. A novel algorithm for detection of precipitation in tropical regions using PMW radiometers. Atmos. Meas. Tech. 8, 1217–1232. https://doi.org/10.5194/amt-8-1217-2015.

Chen, F.W., Staelin, D.H., 2003. AIRS/AMSU/HSB precipitation estimates. IEEE Trans. Geosci. Remote Sens. 41, 410–417.

Ferraro, R.R., Weng, F., Grody, N.C., Zhao, L., 2000. Precipitation characteristics over land from the NOAA-15 AMSU sensor. Geophys. Res. Lett. 27, 2669–2672.

Ferraro, R.R., Weng, F., Grody, N.C., Zhao, L., Meng, H., Kongoli, C., Pellegrino, P., Qiu, S., Dean, C., 2005. NOAA operational hydrological products derived from the advanced microwave sounding unit. IEEE Trans. Geosci. Remote Sens. 43 (5), 1036–1049.

Grody, N., Zhao, J., Ferraro, R., Weng, F., Boers, R., 2001. Determination of precipitable water and cloud liquid water over oceans from the NOAA 15 advanced microwave sounding unit. J. Geophys. Res. 106 (D3), 2943–2954.

Hou, A.Y., Kakar, R.K., Neeck, S., Azarbarzin, A.A., Kummerow, C.D., Kojima, M., Oki, R., Nakamura, K., Iguchi, T., 2014. The global precipitation measurement mission. Bull. Am. Meteorol. Soc. 95, 701–722. https://doi.org/10.1175/BAMS-D-13-00164.1.

Huffman, G.J., Adler, R.F., Bolvin, D.T., Gu, G., Nelkin, E.J., Bowman, K.P., Hong, Y., Stocker, E.F., Wolff, D.B., 2007. The TRMM multi-satellite precipitation analysis: quasi-global, multi-year, combined-sensor precipitation estimates at fine scale. J. Hydrometeorol. 8, 38–55.

Iguchi, T., Kozu, T., Meneghini, R., Awaka, J., Okamoto, K., 2000. Rainprofiling algorithm for the TRMM precipitation radar. J. Appl. Meteorol. 39 (12), 2038–2052.

Iguchi, T., Kozu, T., Kwiatkowski, J., Meneghini, R., Awaka, J., Okamoto, K., 2009. Theoretical basis of the evolution of the TRMM rain profiling algorithm. J. Meteor. Soc. Japan 87A, 1–30.

Joyce, R.J., Janowiak, J.E., Arkin, P.A., Xie, P., 2004. CMORPH: a method that produces global precipitation estimates from passive microwave and infrared data at high spatial and temporal resolution. J. Hydrometeorol. 5 (3), 487–503.

Kida, S., Shige, S., Kubota, T., Aonashi, K., Okamoto, K., 2009. Improvement of rain/no-rain classification methods for microwave radiometer observations over ocean using the 37-GHz emission signature. J. Meteor. Soc. Japan 87A, 165–181.

Kida, S., Shige, S., Manabe, T., 2010. Comparison of rain fractions over tropical and sub-tropical ocean obtained from precipitation retrieval algorithms for microwave sounders. J. Geophys. Res. 115. https://doi.org/10.1029/2010JD014279. D24101.

Kidd, C., Matsui, T., Chern, J., Mohr, K., Kummerow, C., Randel, D., 2016. Global precipitation estimates from cross-track passive microwave observations using a physically based retrieval scheme. J. Hydrometeorol. 17 (1), 383–400.

Kubota, T., Shige, S., Hashizume, H., Aonashi, K., Takahashi, N., Seto, S., Hirose, M., Takayabu, Y.N., Nakagawa, K., Iwanami, K., Ushio, T., Kachi, M., Okamoto, K., 2007. Global precipitation map using satelliteborne microwave radiometers by the gsmap project: production and validation. IEEE Trans. Geosci. Remote Sens. 45 (7), 2259–2275.

Kubota, T., Ushio, T., Shige, S., Kida, S., Kachi, M., Okamoto, K., 2009. Verification of high resolution satellite-based rainfall estimates around Japan using a gauge-calibrated ground radar dataset. J. Meteor. Soc. Japan 87A, 203–222.

Kummerow, C., Barnes, W., Kozu, T., Shiue, J., Simpson, J., 1998. The tropical rainfall measuring mission (TRMM) sensor package. J. Atmos. Ocean. Technol. 15, 808–816.

Kunkee, D.B., Poe, G.A., Boucher, D.J., Swadley, S.D., Hong, Y., Wessel, J.E., Uliana, E.A., 2008. Design and evaluation of the first special sensor microwave imager/sounder. IEEE Trans. Geosci. Remote Sens. 46, 863–883.

Kwon, E.-H., Sohn, B.-J., Chang, D.-E., Ahn, M.-H., Yang, S., 2008. Use of numerical forecasts for improving TMI rain retrievals over the mountainous area in Korea. J. Appl. Meteorol. Climatol. 47, 1995–2007. https://doi.org/10.1175/2007JAMC1857.1.

Laviola, S., Levizzani, V., 2011. The 183-WSL fast rain rate retrieval algorithm: part I: retrieval design. Atmos. Res. 99, 443–461.

Lin, X., Hou, A.Y., 2008. Evaluation of coincident passive microwave rainfall estimates using TRMM PR and ground measurements as references. J. Appl. Meteorol. Climatol. 47 (12), 3170–3187.

New, M., Todd, M., Hulme, M., Jones, P., 2001. Precipitation measurements and trends in the twentieth century. Int. J. Climatol. 21, 1899–1922.

Sano, P., Panegrossi, G., Casella, D., Di Paola, F., Milani, L., Mugnai, A., Petracca, M., Dietrich, S., 2015. The passive microwave neural network precipitation retrieval (pnpr) algorithm for amsu/mhs observations: description and application to european case studies. Atmos. Meas. Tech. 8, 837–857. https://doi.org/10.5194/amt-8-837-2015.

Seto, S., Kubota, T., Takahashi, N., Iguchi, T., Oki, T., 2008. Advanced Rain/No-Rain classification methods for microwave radiometer observations over land. J. Appl. Meteorol. Climatol. 47 (11), 3016–3029.

Shige, S., Kummerow, C.D., 2016. Precipitation-top heights of heavy orographic rainfall in the Asian monsoon region. J. Atmos. Sci. 73, 3009–3024. https://doi.org/10.1175/JAS-D-15-0271.1.

Shige, S., Yamamoto, T., Tsukiyama, T., Kida, S., Ashiwake, H., Kubota, T., Seto, S., Aonashi, K., Okamoto, K., 2009. The GSMaP precipitation retrieval algorithm for microwave sounders—part I: over-ocean algorithm. IEEE Trans. Geosci. Remote Sens. 47 (9), 3084–3097.

Shige, S., Kida, S., Yamamoto, T., Kubota, T., Aonashi, K., 2010. High-temporal global rainfall maps from satellite passive microwave radiometers, microwave and millimeter wave technologies. In: Mini, I. (Ed.), Modern UWB antennas and equipment. INTECH, pp. 301–312.

Shige, S., Yamamoto, M.K., Taniguchi, A., 2015. Improvement of TMI rain retrieval over the Indian subcontinent. In: Lakshmi, V. (Ed.), Remote Sensing of the Terrestrial Water Cycle, Geophysical Monograph. In: vol. 206. American Geophysical Union, pp. 27–42.

Surussavadee, C., Staelin, D.H., 2008. Global millimeter-wave precipitation retrievals trained with a cloud-resolving numerical weather prediction model, part 1: retrieval design. IEEE Trans. Geosci. Remote Sens. 46, 99–108. https://doi.org/10.1109/TGRS.2007.908302.

Taniguchi, A., Shige, S., Yamamoto, M.K., Mega, T., Kida, S., Kubota, T., Kachi, M., Ushio, T., Aonashi, K., 2013. Improvement of high-resolution satellite rainfall product for Typhoon Morakot (2009) over Taiwan. J. Hydrometeorol. 14, 1859–1871.

Vila, D., Ferraro, R., Joyce, R., 2007. Evaluation and improvement of AMSU precipitation retrievals. J. Geophys. Res. 112. https://doi.org/10.1029/2007JD008617. D20119.

Weng, F., Grody, N.C., 2000. Retrieval of ice cloud parameters using a microwave imaging radiometer. J. Atmos. Sci. 57 (8), 1069–1081.

Wilks, D.S., 2006. Statistical Methods in the Atmospheric Sciences, second ed. Academic Press. 467 p.

You, Y., Wang, N.Y., Ferraro, R., Meyers, P.A., 2016. Prototype precipitation retrieval algorithm over land for ATMS. J. Hydrometeorol. 17 (5), 1601–1621.

Zhao, L., Weng, F., 2002. Retrieval of ice cloud parameters using the advanced microwave sounding unit. J. Appl. Meteorol. 41 (4), 384–395.

CHAPTER

13

Remote Sensing of Precipitation from Airborne and Spaceborne Radar

Stephen J. Munchak

NASA Goddard Space Flight Center, Greenbelt, MD, United States

1 INTRODUCTION

The ability of weather radar to measure the location and intensity of precipitation was rapidly realized in the late 1940s following World War II. However, ground-based radars are limited in their ability to directly detect precipitation close to the ground far from the radar site due to ground clutter, refraction of the radar beam, and the curvature of the earth. Beam blockage by terrain also poses problems for radar coverage in mountainous areas. Coverage over oceans and other remote areas, where maintaining a ground radar would be difficult and costly, is also impractical. However, the precipitation that falls in these regions has important impacts on the global atmospheric circulations via latent heating (e.g., Hoskins and Karoly, 1981; Hartmann et al., 1984; Matthews et al., 2004) and can have a profound influence on weather patterns thousands of kilometers away. Likewise, knowledge of precipitation over land, particularly in the form of snow, is a crucial component of the mass balance equation for glaciers and ice sheets, which must be properly characterized for realistic climate simulations (Shepherd et al., 2012).

Airborne radar systems can provide high sensitivity and finely resolved vertical profiles to characterize precipitation microphysics for the benefit of model parameterizations and process understanding (e.g., Reinhart et al., 2014; Heymsfield et al., 2013; Rauber et al., 2017). However, in order to achieve true global coverage, it has been proposed from nearly the beginning of the space age to put a weather radar in space (Keigler and Krawitz, 1960). Efforts to do so began in earnest in the late 1970s and 1980s with the planning of the Tropical Rainfall Measuring Mission satellite (TRMM; Simpson et al., 1988; Okamoto et al., 1988) with its Ku-band Precipitation Radar (PR; Table 1), which was launched in 1997. TRMM was followed in 2014 by the Global Precipitation Measurement (GPM; Hou et al., 2014) mission with its Ku- and Ka-band Dual-Frequency Precipitation Radar (DPR; Table 1), which provides increased accuracy, sensitivity, and extension to higher latitudes. Both the TRMM PR and GPM DPR were intended not only to estimate precipitation directly from the radar data but also to construct a database of precipitation profiles to unify precipitation retrievals from passive microwave radiometers (Hou et al., 2014;

Remote Sensing of Aerosols, Clouds, and Precipitation
https://doi.org/10.1016/B978-0-12-810437-8.00013-X

© 2018 Elsevier Inc. All rights reserved.

TABLE 1 Key Parameters of Spaceborne Weather Radars Launched Prior to 2016

Name	Frequency	Half-Power Beamwidth and Nominal Ground Footprint	Nominal Altitude	Orbital Inclination	Vertical Resolution	Minimum Detectable Signal
TRMM PR	13.8 GHz	0.71° 4.3 km (1997–2001) 5.0 km (2001–14)	350 km (1997–2001) 402 km (2001–14)	35°	250 m	17 dBZ (1997–2001) 18 dBZ (2001–14)
CloudSat CPR	94 GHz	0.108° 1.4 × 1.7 km	705 km	98.2°	480 m	−30 dBZ (2006–11)
GPM KuPR	13.6 GHz	0.71° 5 km	407 km	65°	250 m	14 dBZ
GPM KaPR	35.55 GHz	0.71° 5 km	407 km	65°	250 m (MS) 500 m (HS)	18 dBZ (MS) 12 dBZ (HS)

The GPM DPR consists of two radars with matched beams: the KuPR and KaPR.

Kummerow et al., 2015), enabling more frequent coverage than is possible from a narrow swath on a single satellite. Meanwhile, in 2006, the CloudSat mission (Stephens et al., 2002) with a W-band Cloud Profiling Radar (CPR; Table 1), was launched into polar orbit, complementing TRMM and GPM by providing estimates of light precipitation at high latitudes (Behrangi et al., 2014).

In radar engineering and meteorology, it is common to refer to specific frequency ranges (bands) by letter designation. Table 2 lists bands commonly used for meteorological radars according to the IEEE Standard 521-2002. Some of these bands contain gas absorption lines, where atmospheric extinction can be orders of magnitude higher than the surrounding "window" regions. However, it has been proposed to use radars operating at two or more closely spaced frequencies near some of these bands to estimate vertical profiles of water vapor (e.g., Meneghini et al., 2005; Lebsock et al., 2015) by taking advantage of the differential attenuation.

TABLE 2 Band Designations, Frequency Ranges, and Significant Absorption Lines in Each Band

Designation	Frequency Range (GHz)	Significant Gas Absorption Lines
S	2–4	
C	4–8	
X	8–12	
Ku	12–18	
K	18–27	H_2O (22.235 GHz)
Ka	27–40	
V	40–75	O_2 (several lines 49–70 GHz)
W	75–110	
G	110–300	O_2 (118.75 GHz), H_2O (183.31 GHz)

In order to minimize size and weight, which strongly correspond to the cost of a satellite mission, it is necessary to use higher frequencies (Ku-, Ka-, or W-band) than are typical for ground radar systems. With increasing frequency, power and antenna size requirements for a desired sensitivity and horizontal resolution are reduced, but attenuation and multiple scattering, which can lead to ambiguity in converting reflectivity to precipitation rate, increase. Even at higher frequencies, the distance from low earth orbit results in ground footprints that are large relative to the scale of variability in most precipitation systems; this nonuniformity must also be considered in precipitation retrieval algorithms. This chapter is intended to provide an overview of the theoretical basis and some practical implementations of precipitation retrieval algorithms for nadir (or near-nadir) looking airborne and spaceborne weather radars at attenuating frequencies, without consideration of polarimetric quantities or Doppler velocity. While dual-polarimetric radars are widely used from ground-based platforms to identify preferentially oriented, nonspherical hydrometeors, at near-nadir incidence angles these measurements are of limited utility although the linear depolarization ratio measurements can be useful for identifying melting layers and nonspherical ice particles (Pazmany et al., 1994; Galloway et al., 1997). Doppler velocities are useful for inferring hydrometeor fall speeds and at multiple frequencies can be highly effective in discerning cloud liquid from rain (e.g., Kollias et al., 2007) as well as identification of ice particle habits (Kneifel et al., 2016). However, obtaining them from rapidly moving satellite platforms is a difficult engineering challenge that will first be attempted in the EarthCARE mission (Illingworth et al., 2015).

2 RADAR PRECIPITATION MEASUREMENT FUNDAMENTALS

The earliest attempts to measure rainfall with radar (Marshall et al., 1947) found that, in general, a power law relationship between radar reflectivity factor Z and rainfall rate R existed:

$$Z = aR^b \tag{1}$$

The coefficient a and exponent b of this power law were later provided by Marshall and Palmer (1948), whose values are still in wide use today. Despite this common usage, it was quickly recognized (e.g., Atlas and Chmela, 1957) that these parameters varied widely and seemed to be associated with synoptic conditions. It is now recognized (e.g., Brandes et al., 2006) that the power law of Marshall and Palmer (1948) is more representative of frontal stratiform rainfall, which is the predominant rainfall type in Ontario, Canada, where the radar and rainfall observations upon which this power law was based were taken. Convective and tropical rainfall, for example, is observed to have a smaller coefficient a (Tokay and Short, 1996). A more comprehensive review of varying power law relations is given by Battan (1973). For the purposes of this chapter, it is sufficient to recognize that the nonuniqueness of the Z-R relationship is a fundamental result of the general equations for radar reflectivity and rainfall rate:

$$Z = \int_{D_{min}}^{D_{max}} N(D) D^6 dD, \tag{2}$$

$$R = \frac{\pi}{6} \int_{D_{min}}^{D_{max}} D^3 N(D) V(D) dD, \tag{3}$$

where $v(D)$ is the drop fall speed. Owing to the fact that vertical air motions are small near the ground and that raindrops achieve terminal fall velocity within about 100 m (Section 10.3.6, Pruppacher and Klett, 1997), formulae relating terminal fall speed to drop size are often used. A simple power law such as $v(D) = 17.67D^{0.67}$ (where V is in m s^{-1} and D is in cm; Atlas and Ulbrich, 1977) is convenient for the calculation of Z-R power law coefficients by combining Eqs. (2), (3), especially when an analytic form of the drop size distribution $N(D)$ is assumed. Slightly more accurate power laws such as the one given by Beard (1976) account for different hydrodynamic regimes as drops grow in size, and this is the relationship used for all rain rate calculations in this chapter.

In Eq. (2), there is no dependence of Z on the radar wavelength. This is only valid when the particle size is much smaller than the wavelength. For larger sizes, the equivalent reflectivity factor Z_e is used instead:

$$Z_e = \frac{\lambda^4}{\pi^5 |K|^2} \int_{D_{\min}}^{D_{\max}} N(D) \sigma_b(D, \lambda) dD, \tag{4}$$

where K is a function of the complex index of refraction, λ is the radar wavelength in mm, σ_b is the backscattering cross section (in mm^2), and $N(D)$ is the number concentration of raindrops (in m^{-3}) per size interval, resulting in units of $mm^6 m^{-3}$ for Z_e. Note that Eqs. (3), (4) are also valid for frozen and melting particles. However, the definition of particle size (and fall speeds) becomes ambiguous for nonspherical particles, and even large raindrops exhibit some departures from sphericity. For the remainder of this chapter, the convention will be that D represents the diameter of an equal-mass homogeneous (solid or liquid) sphere, so that $N(D)$ is equivalent to a mass distribution, and $N(D)$ will be referred to as the particle size distribution (PSD).

The backscattering cross-section describes the amount of electromagnetic radiation that is scattered toward the source of incident radiation. For particles much smaller than the wavelength, the individual dipoles that comprise the particle can be treated as coherent scatterers, and the Rayleigh approximation holds:

$$\sigma_b = \frac{\pi^5 |K|^2 D^6}{\lambda^4} \tag{5}$$

and Eq. (4) becomes Eq. (2). As the size parameter $\pi D/\lambda$ approaches one, the Rayleigh approximation breaks down and Mie theory, which provides exact results for spheres, should be used. For particles that are rotationally symmetric about one axis (such as oblate or prolate spheroids, cylinders, or cones), σ_b can be calculated from the T-Matrix (Mishchenko and Travis, 1998). Finally, σ_b for arbitrarily shaped particles can be calculated with the discrete-dipole approximation (Draine and Flatau, 1994), a method often used for realistically shaped snowflakes (e.g., Liu, 2004; Kim, 2006; Petty and Huang, 2010; Kuo et al., 2016) and melting particles (Johnson et al., 2016). Plots of σ_b from all of these approximations can be found in Fig. 1 for spherical and oblate raindrops and spherical, cylindrical, and synthetically grown ice particles and aggregates.

Although the departures of reflectivity from Rayleigh theory are important, particularly for multifrequency radars, it is still the case that while Z is approximately proportional to the 6th moment of the DSD, rain rate is proportional to a much lower 3.67th moment. This is illustrated in Fig. 2, which shows the relative contribution of different drop sizes to reflectivity and rain rate for a typical exponential DSD. Thus the fundamental problem of radar meteorology is that multiple values of R can be associated with a single value of Z.

Aside from backscatter, another characteristic of precipitation particles that is critical for understanding radar measurements is the extinction cross-section σ_e. This quantity describes the amount of electromagnetic radiation absorbed and scattered by the particle, and like σ_b depends on the dielectric constant, particle size, and shape (for size parameters close to or greater than one). Fig. 3 shows the extinction efficiency for the same particle types in Fig. 1.

For particles with small size parameters, σ_e is given by:

$$\sigma_e = \frac{\pi^2 D^3}{\lambda} \Im\left(\frac{m^2 - 1}{m^2 + 2}\right), \tag{6}$$

where m is the complex index of refraction and $\Im(x)$ denotes the imaginary part of x. The bulk extinction coefficient, k_{ext}, can be calculated by integrating σ_e over the particle size distribution:

$$k_{\text{ext}} = \int_{D_{\min}}^{D_{\max}} \sigma_e N(D) dD, \tag{7}$$

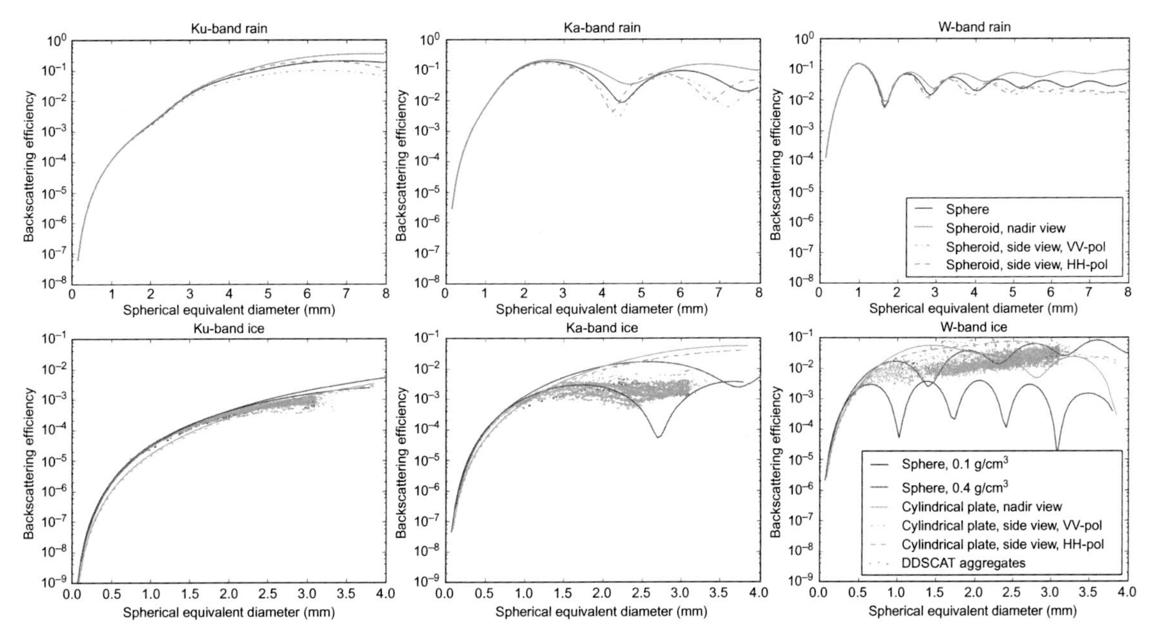

FIG. 1 Backscattering efficiencies calculated at Ku-, Ka-, and W-band frequencies for rain and ice particles. Spheroidal raindrops were modeled according to aspect ratio and canting angle distributions given in Beard et al. (2010). The cylindrical snow particles, which are an effective representation of hexagonal plates (Adams and Bettenhausen, 2012), were modeled with an aspect ratio $(D/h)=6$ and effective density of 0.6 g cm^{-3}. The DDSCAT particles (color indicates relative density) are from the database of Kuo et al. (2016).

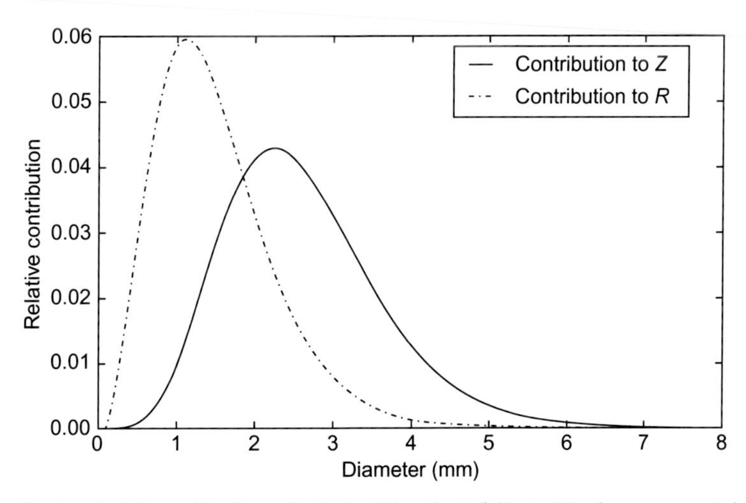

FIG. 2 Relative contribution (in 0.1-mm bins) to reflectivity (Z) and rainfall rate (R) of an exponential drop size distribution with a median volume-weighted diameter of 1.5 mm.

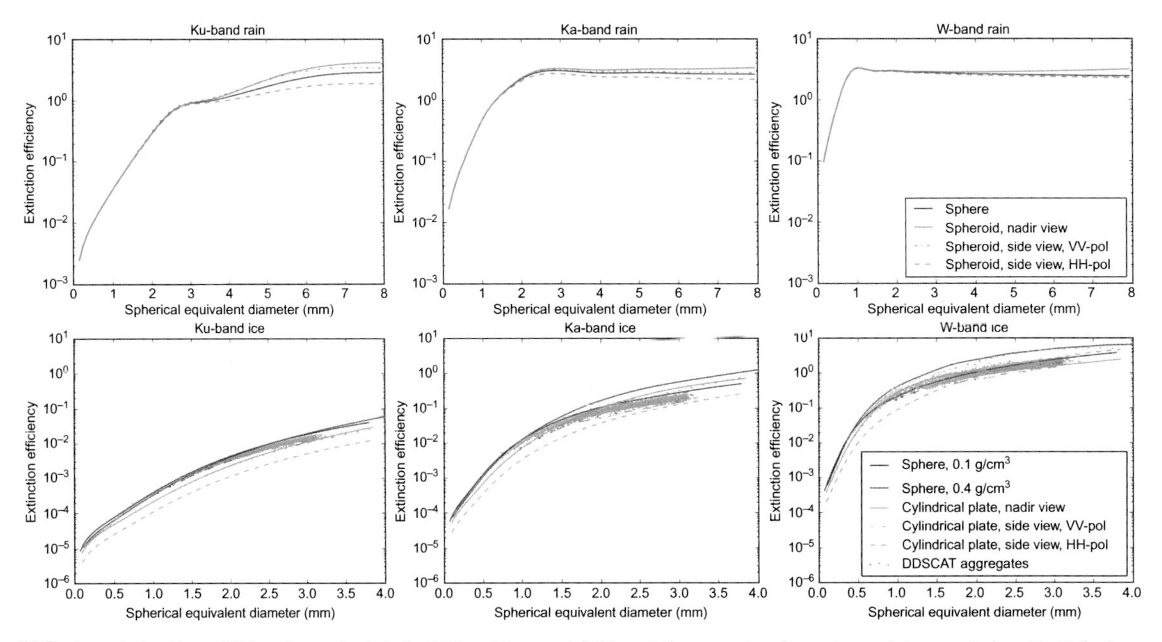

FIG. 3 Extinction efficiencies calculated at Ku-, Ka-, and W-band frequencies for rain and ice particles. Particle type description can be found in the Fig. 1 caption.

Note that in addition to precipitation particles, atmospheric gases such as oxygen and water vapor can have nonnegligible contributions to k_{ext} at some wavelengths common to airborne and spaceborne radars (Li et al., 2001; Tanelli et al., 2006; Ellis and Vivekanandan, 2010). Cloud water, which has a negligible contribution to radar reflectivity at the Ka-band and lower frequencies, nevertheless can contribute significantly to k_{ext} at these frequencies as well (Grecu and Olson, 2008). Note the independence of k_{ext} with respect to D_m in the limit of small D_m in Fig. 5. This follows from Eq. (6) and shows that in the limit of small particles, k_{ext} is directly proportional to the water/ice content. The effect of bulk extinction on the measured reflectivity can be calculated by integrating k_{ext} along the two-way radar propagation path:

$$Z_m(r) = e^{-0.2\ln(10)\int_0^r k_{ext}(s)ds} Z_e(r) \tag{8}$$

Note that while Z_e is an intrinsic property of the PSD at a given location, Z_m also depends on the integrated bulk extinction between the source of the radar signal and that location, creating another source of uncertainty when converting from Z to R. Therefore, understanding the vertical profile of the PSD, on which Z, k_{ext}, and R depend, is fundamental to the radar-precipitation profiling algorithms that are described in this chapter.

3 THE PARTICLE SIZE DISTRIBUTION

The previous section demonstrated that knowledge of the particle size distribution (PSD) is needed to convert Z to the physical integrated quantities such as precipitation rate R and water content W. It is often convenient to assume an analytical form of the PSD that describes the shape with a few (relative to a discrete bin representation) free parameters.

Although PSD models with as many as eight free parameters have been proposed (Kuo et al., 2004), most often the modified gamma distribution with three free parameters (Ulbrich, 1983) is used:

$$N(D) = N_0 D^{\mu} \exp(-\Lambda D) \tag{9}$$

The free parameters are often referred to as the intercept (N_0), slope (Λ), and shape (μ). These names describe the mathematical form of the distribution moreso than physical quantities, and it is difficult to impart any physical meaning to a value for any one of the parameters given in isolation. However, it is possible to recast these parameters in terms of physical quantities through the following relationships (Testud et al., 2001; Williams et al., 2014):

$$D_m = \frac{4+\mu}{\Lambda}, \tag{10}$$

$$N_w = N_0 D_m^{\mu} \frac{\Gamma(4+\mu)}{\Gamma(4)} \frac{256}{(4+\mu)^{4+\mu}}, \tag{11}$$

$$\sigma_m = \frac{D_m}{\sqrt{4+\mu}}, \tag{12}$$

where D_m is the mass-weighted mean diameter (also defined as the ratio of the 4th to 3rd moment of the PSD), N_w is the normalized intercept parameter defined such that it is equal to N_0 for an exponential ($\mu = 0$) PSD of the same water content and D_m (Bringi and Chandrasekar, 2001), and σ_m is the mass spectrum standard deviation. The median volume diameter, D_0, describes the particle size such that $\int_{D_{min}}^{D_0} D^3 dD = \int_{D_0}^{D_{max}} D^3 dD$, and for a gamma distribution $D_0 = \frac{3.67+\mu}{4+\mu} D_m$. Note that these expressions are only valid for a PSD where particle density is constant with size. For realistic frozen particles, this is often not the case, and formulae to convert PSD parameters from observed to solid-sphere-equivalent particle sizes can be found in Petty and Huang (2011).

Analysis of disdrometer observations has shown that neither the three free parameters in Eq. (9) nor the physical quantities in Eqs. (10)–(12) are statistically independent in rain (e.g., Haddad et al., 1996; Zhang et al., 2003; Munchak and Tokay, 2008; Williams et al., 2014) and relationships between the parameters can be formulated to reduce the degrees of freedom in Eq. (9). This is particularly useful for radar precipitation retrieval algorithms because it allows for a common basis from which to compute Z-R and Z-k_{ext} relationships, and, when joint probability distribution functions (pdfs) of PSD parameters are known, these a priori statistics can be used to constrain the retrieval of PSD parameters from radar profiles of reflectivity and other information.

To demonstrate the relationship between PSD parameters and radar reflectivity Z, Figs. 4 and 5 show reflectivity and extinction coefficients integrated over PSDs with D_m ranging from 0.1 to 3 mm and μ ranging from −1 to 3 for snow and following the σ_m-D_m relationships given by Williams et al. (2014) for rain. The integrated water content of all PSDs was normalized to $W = 1\ \text{g m}^{-3}$ to emphasize the importance of the shape of the PSD in terms of its mean and dispersion on the reflectivity and extinction. For these PSDs, the equivalent intercept N_w can be calculated from D_m and W (Testud et al., 2001):

$$N_w = \frac{256}{\pi \rho_w} \frac{W}{D_m^4}. \tag{13}$$

While analytic expressions relating Z, D_m, W, and R can be derived from the gamma distribution (Ulbrich and Atlas, 1998), statistical relationships can also be derived directly from disdrometer measurements without any assumptions of the PSD shape. In Fig. 6, scatter plots of k_{ext}, W, R, D_m, and N_w versus Z at Ku-band are shown for rain PSDs measured during the IFLoodS field experiment (Ryu et al., 2016). Most of these parameters (except N_w) have a strong correlation to reflectivity. As will be shown later in this chapter, it is particularly useful for radar-based

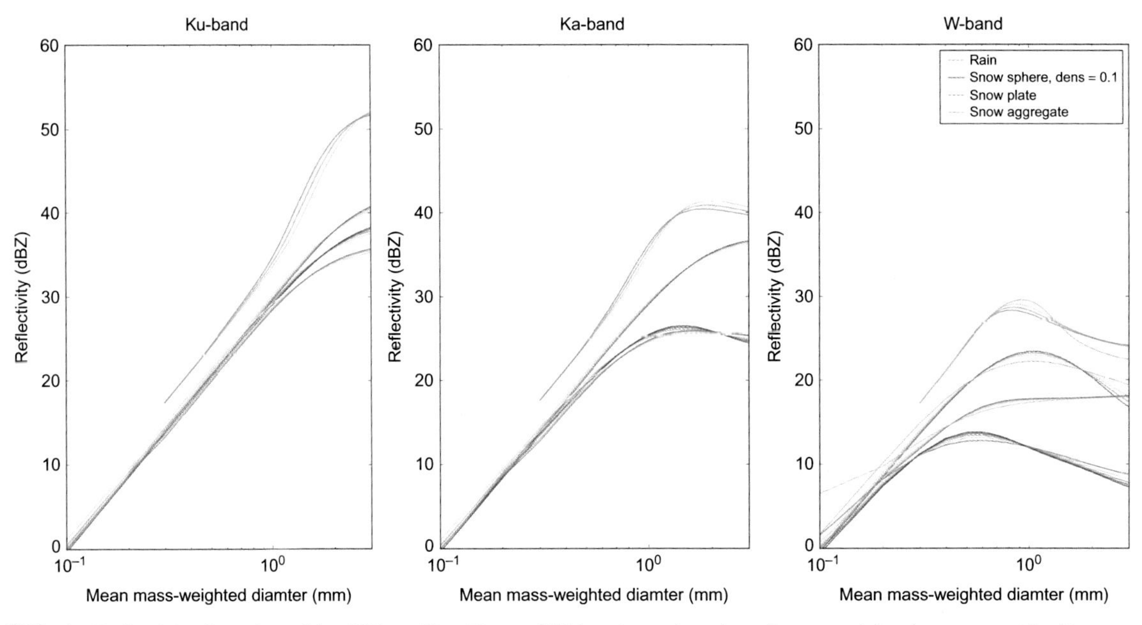

FIG. 4 Reflectivity for rain and ice PSDs at Ku-, Ka-, and W-band as a function of mass-weighted mean particle diameter Dm for various shape parameter assumptions. All PSDs contain 1 g m^{-3} of water content.

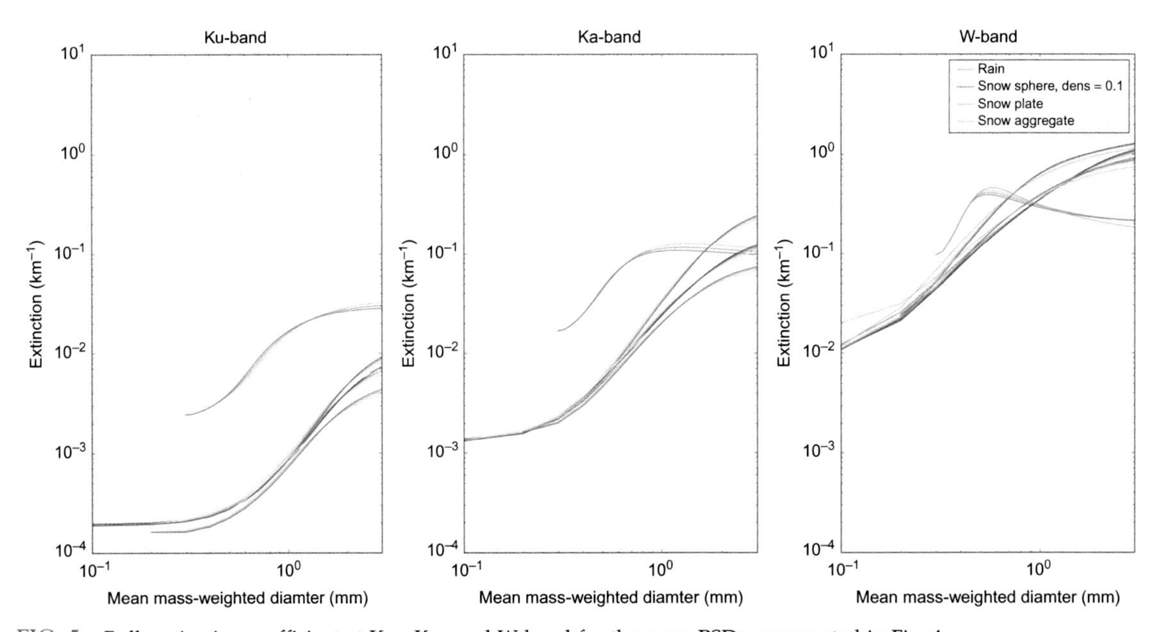

FIG. 5 Bulk extinction coefficient at Ku-, Ka-, and W-band for the same PSDs represented in Fig. 4.

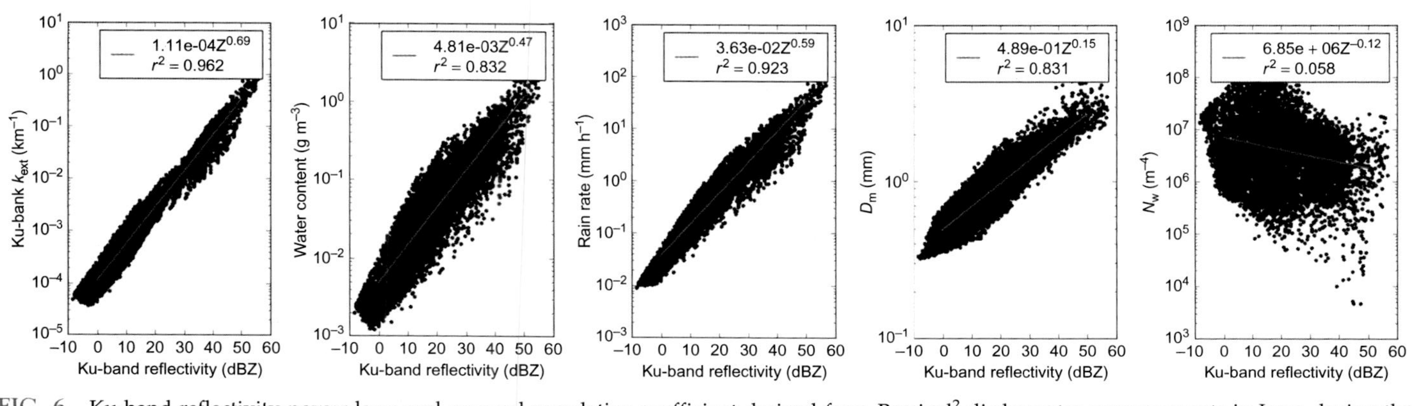

FIG. 6 Ku-band reflectivity power laws and squared correlation coefficient derived from Parsivel[2] disdrometer measurements in Iowa during the IFloodS field experiment for the integral parameters: attenuation coefficient at Ku-band (k_{ext}), liquid water content (W), rain rate (R), mean mass-weighted diameter (D_m), and normalized intercept parameter (N_w).

precipitation retrievals to modify a parameter that is uncorrelated to reflectivity, and measurements from IFLoodS and other field experiments suggest that N_w is a strong candidate for such a parameter. However, any integral parameter may be transformed to one that is uncorrelated with reflectivity by deriving a reflectivity power law relationship (such as Eq. 1) and converting to "normalized" quantities k', W', R', D_m', and N_w', e.g., (for R):

$$R' = \frac{R}{aZ^b} \tag{14}$$

The correlation plots and coefficients between the normalized parameters are shown in Fig. 7. Each panel shows the correlation between a pair of normalized parameters, and it is notable that the squared correlation coefficients are large, in many cases over 0.8. This has strong implications for radar retrievals because it implies that, for a given Z, if a normalized parameter such as W', D_m', Nw' (or just N_w) is known, then the other integral parameters, including attenuation, can be predicted to a high degree of accuracy.

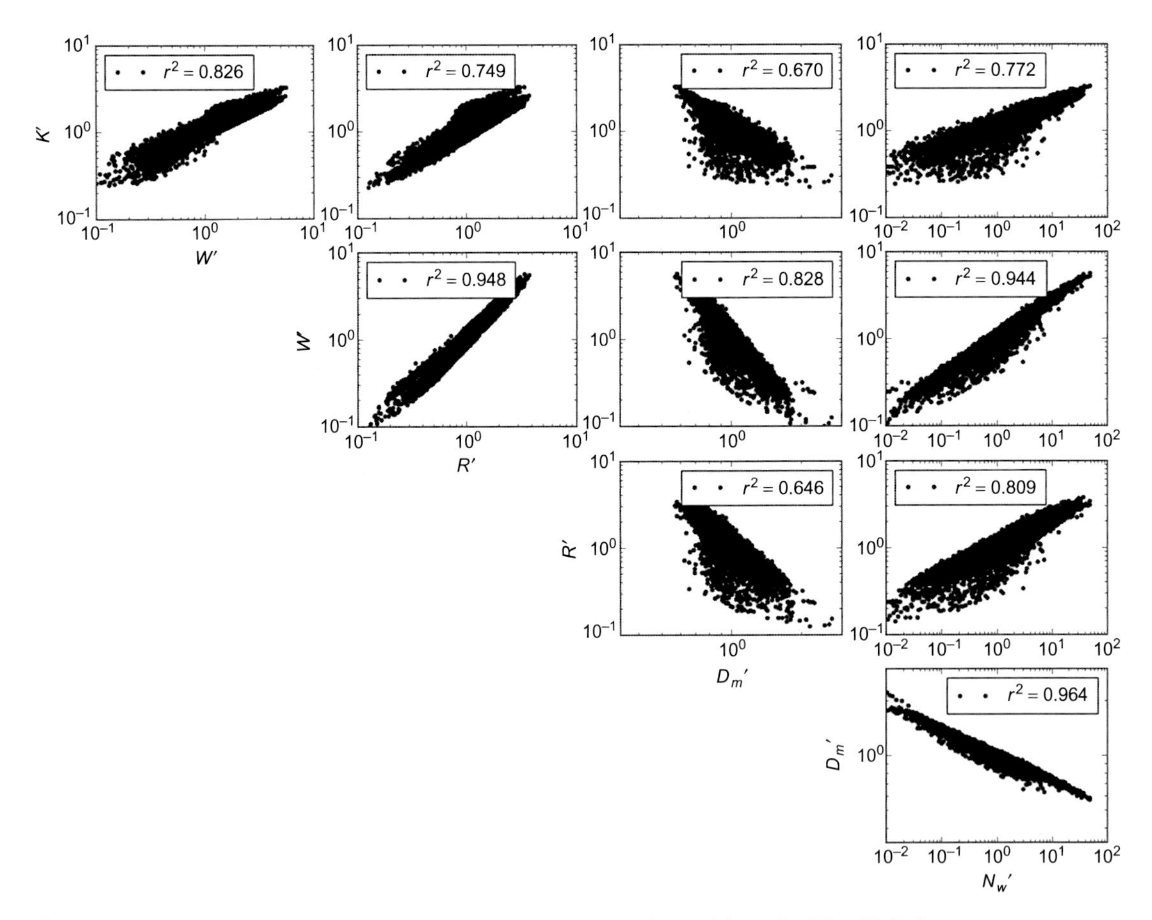

FIG. 7 Scatter diagrams for pairs of normalized integral quantities (Eq. 14) from the IFloodS disdrometer measurements. The squared logarithmic correlation coefficient is labeled in each panel.

4 SINGLE-FREQUENCY METHODS

Since many airborne and spaceborne radars operate at frequencies where attenuation by rain is significant, a correction for this attenuation must be made. A radar algorithm that corrects for attenuation was first described by Hitschfeld and Bordan (1954). If a Z-k_{ext} power law of the form

$$k_{ext} = \alpha Z_e^{\beta}, \tag{15}$$

and β is constant with respect to range, then the path-integrated attenuation (PIA), which is the integral in the exponential term of Eq. (8), can be solved for analytically:

$$\mathrm{PIA}(r) = \frac{-1}{0.1\ \ln(10)\beta} \ln\left(1 - 0.2\ \ln(10) \int_0^r \alpha(s) Z_m^{\beta}(s) ds\right) \tag{16}$$

However, this method is numerically unstable because small changes in the Z-k_{ext} relationship can lead to large changes in path-integrated attenuation (PIA) at far range gates. This can be inferred from Eq. (15) if the values of α, β, and Z_m are such that the argument of the logarithm is near or less than zero, in which case the PIA is undefined. An example, illustrated in Fig. 8, provides a physical interpretation of how linear increases in α lead to nonlinear increases in PIA. In physical terms, if the estimated attenuation is too large at a given range gate, then attenuation-corrected reflectivities will be overestimated in subsequent range gates. These overestimated reflectivities will produce an even larger attenuation correction in further range gates, which will result in reflectivity and attenuation estimates that can increase to unphysical values.

While attenuation is a potential source of error if not accurately corrected, it can also be an additional source of information. Eq. 8

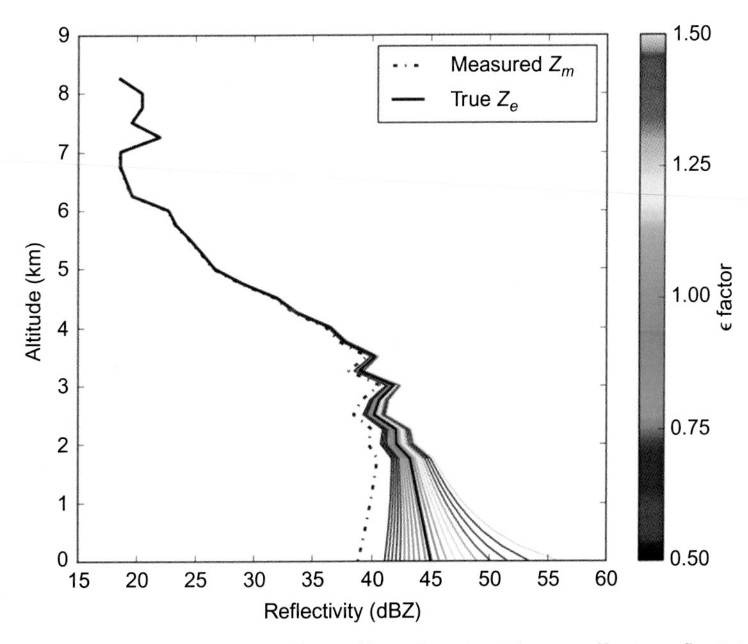

FIG. 8 An example of an attenuation-correcting radar profiling algorithm. The true effective reflectivity (solid) is attenuated to the measured signal (dashed). An attenuation correction is applied assuming Z-k relationships that are have been multiplied by factors ranging from 0.5 to 1.5.

includes a term describing the PIA. If a reflector of a known effective cross-section is placed at the end of the path, then the apparent decrease in this cross section from its known value is equal to the PIA. Fortuitously, a downward-looking radar has exactly such a reference cross-section in the earth's surface itself. This fact was recognized by Meneghini et al. (1983) and forms the fundamental basis for the surface reference technique (SRT).

The SRT is a method for obtaining an estimate of the path-integrated attenuation from a downward-looking radar that is independent of the reflectivity profile. Variations of this method have been used to obtain PIA estimates for use in the TRMM rain-profiling algorithm (Meneghini et al., 2000), CloudSat's rain profile product (Haynes et al., 2009; Mitrescu et al., 2010; Lebsock and L'Ecuyer, 2011), and GPM DPR products (Meneghini et al., 2015) as well as from airborne radars (Tanelli et al., 2006).

The SRT requires a reference value for the normalized surface backscatter cross-section (σ_0) that is unattenuated by precipitation. Then, the PIA can be calculated by taking the difference of the observed σ_0 (in dB units) from the reference:

$$\mathrm{PIA} = \sigma_{0,NR} - \sigma_{0,R}. \tag{17}$$

The key to this process is obtaining an accurate estimate of the unattenuated cross-section $\sigma_{0,NR}$. In practice, methods for obtaining this reference value include the along-track, across-track, temporal, and geophysical model function. The along-track reference is particularly useful because σ_0 varies strongly with incidence angle, thus $\sigma_{0,NR}$ can be estimated by averaging σ_0 in precipitation-free observations surrounding a precipitation observation at the same incidence angle. In the event that a precipitation feature happens to be long in the along-track direction but narrow in the cross-track direction, it may be preferable to use the cross-track reference. In this case, a quadratic function is fit to the precipitation-free σ_0 observations to obtain estimates of $\sigma_{0,NR}$ and PIA in the precipitation feature. The along-track and cross-track spatial methods are most applicable when the surface type does not vary from the region where $\sigma_{0,NR}$ estimates into the precipitation feature. This can be true over ocean surfaces, but over land or mixed water-land surfaces, $\sigma_{0,NR}$ can be highly variable, adding considerable uncertainty to the PIA estimate and often producing unphysically negative PIA values. In these cases, the temporal reference can be used. The temporal reference is a gridded average of $\sigma_{0,NR}$ and its standard deviation over multiple previous observations. In order to minimize the standard deviation of $\sigma_{0,NR}$, thus providing more accurate PIA estimates, it is necessary to reduce the grid size to a size approaching the radar footprint (Meneghini and Jones, 2011). However, this is a challenge for spaceborne radars without a repeating ground track because the number of samples in such a small grid cell at each incidence angle can be small and therefore not representative of changing surface conditions. Meneghini and Kim (2017) suggest optimally including neighboring grid cells in the $\sigma_{0,NR}$ calculation such that the standard deviation of $\sigma_{0,NR}$ is minimized.

An alternate method for calculation $\sigma_{0,NR}$ is the use of a geophysical model function (GMF). This method treats $\sigma_{0,NR}$ as a function of physical parameters. In general, σ_0 depends on the earth incidence angle, surface roughness, and dielectric constant. Over ocean surfaces, σ_0 can be largely explained by the surface wind speed and direction (relative azimuth to the radar look angle) and incidence angle (Li et al., 2002; Freilich and Vanhoff, 2003; Munchak et al., 2016). Over land surfaces, no such simple relationships exist but the behavior of σ_0 has been found to vary as a function of roughness, soil moisture, and vegetation coverage (Stephen et al., 2010; Puri et al., 2011). The advantage of using a GMF is that the representativeness of the spatial or temporal reference is not an issue; however, the GMF must be well calibrated. Also, a source for the physical parameters is required, which itself introduces some uncertainty even for a theoretically perfect GMF. Finally, there is the effect of precipitation on the surface itself (Seto and Iguchi, 2007), which by definition cannot be accounted for by the spatial or temporal references but could be included in a GMF.

Another innovative technique that is made possible with a dual-frequency radar is the dual-frequency SRT (dSRT; Meneghini et al., 2012). This technique determines the differential PIA at two frequencies by taking a difference of differences from Eq. (16). If there is a positive correlation between $\sigma_{0,NR}$ at the two frequencies in the reference dataset, which is often the case, then the error in the differential PIA will be reduced from the single-frequency error using the same reference data. Regardless of the source for the surface reference, it is critical that it can also provide an estimate of the standard deviation of $\sigma_{0,NR}$, which from Eq. (16) can be considered to be the standard deviation of the PIA. The larger the PIA is relative to this standard deviation, the more useful it is to constrain the precipitation PSD via the Z-k_{ext} relationship.

With an independent estimate of the PIA, it is possible to constrain the precipitation profiles to those that produce the same or similar PIA, given the uncertainty that comes from the SRT. The observed profile of Z_m, in conjunction with a vertical profile model of the Z-k_{ext} relationships, can be used to apportion the PIA vertically. As shown earlier in this chapter (Figs. 4 and 5), the Z-k_{ext} relationships depend on the PSD and precipitation phase, and so by modifiying the Z-k_{ext} relationship to produce a profile that matches the SRT PIA, the PSD is also implicitly modified. The vertical profile model is also used to extend the PSD into the range gates near the surface that are contaminated by ground clutter. This layer can be 1 km deep or more for a spaceborne radar depending on the beam width, pulse width, and incidence angle (Takahashi et al., 2016), and presents a difficulty in obtaining precipitation rates for shallow clouds either by failing to detect them entirely or by missing much of the growth of rain drops if it occurs within the clutter-affected gates (Shimizu et al., 2009). Sidelobe clutter presents a more severe problem in that it can contaminate range gates far from the surface at off-nadir angles. A thorough description of sidelobe clutter mitigation techniques for the GPM DPR is given by Kubota et al. (2016).

Returning to the nomenclature in Eq. (14), and assuming that the normalized PSD parameter is constant in the vertical (implying that all vertical variability is represented by Z), k' can be inserted into Eq. (16) as a multiplier to α, and a value for k' can be obtained that exactly matches the observed PIA. However, because there is always some uncertainty in the SRT PIA (which may even be negative in some cases), and very large or small values of k' may imply unphysical PSDs, it is valuable to derive a solution that considers both the error characteristics of the PIA and prior knowledge of the joint pdfs of PSD parameters (Fig. 7). With this knowledge, a maximum likelihood estimate can be found along with a pdf of solutions to the observed radar profile. Various PSD models and vertical profile models have been applied to TRMM (Iguchi et al., 2000, 2009), GPM (Seto and Iguchi, 2015), CloudSat (L'Ecuyer and Stephens, 2002; Mitrescu et al, 2010), and airborne radar (Amayenc et al., 1996; Grecu et al., 2011) data in some form of a probabilistic framework.

In a general form, probabilistic, or optimal estimation, retrieval frameworks seek to minimize an objective function of the form:

$$J = (y - f(x))^T S_y^{-1} (y - f(x)) + (x - x_a)^T S_a^{-1} (x - x_a), \quad (18)$$

where $\mathbf{y}$ is the observation vector, $\mathbf{x}$ is the retrieval parameter (or state) vector, f($\mathbf{x}$) is the forward model, $\mathbf{x_a}$ is the a priori state vector, $\mathbf{S_y}$ is the observation and model error covariance matrix, and $\mathbf{S_a}$ is the state error covariance matrix. Methods of finding the minimum of such objective functions include Gauss-Newton iterative minimization (Rodgers, 2000; Boukabara et al., 2011) and ensemble filter approaches (Evensen, 2006; Grecu et al., 2016). For the radar retrieval problem, $\mathbf{y}$ may be broadly defined as a set of observed reflectivites Z_m along a ray or even in three-dimensional space. Likewise, $\mathbf{x}$ can consist of multiple PSD parameters at each range gate. However, since single-frequency radar retrievals are underconstrained with respect to multiple PSD parameters, it is often more efficient to consider only a single column at a time and include to Z_m in the forward model, so that $\mathbf{y}$ contains only the PIA and $\mathbf{x}$ is reduced to a normalized parameter of the PSD (such as the k', D_m', W', R', or N_w' shown in Fig. 7). Implicit in this formulation is that the vertical behavior of the normalized PSD parameter is constant. Owing to the difficulty in making direct measurements of the PSD instantaneously over a vertical column, such an assumption is difficult to verify. Analyses of measurements made with multiparameter radars (Bringi et al., 2015) and profilers (Williams, 2016) suggest that the vertical decorrelation lengths of normalized PSD parameters are on the same order as the vertical extent of precipitation, implying that, absent any other measurements, the column-averaged normalized PSD parameter should be representative of the surface value.

In summary, the fundamental process of converting from a vertical profile of radar reflectivity to precipitation physical parameters is the correction of attenuation via a vertical profile model of the PSD and associated Z-k_{ext} relationships, and making adjustments to this model to achieve agreement with independent estimates of the PIA. The equations and assumptions presented so far are valid for uniform beam filling and no multiple scattering, which can greatly impact SRT PIA, Z-k_{ext}, and Z-R relationships. These issues will be discussed later in this chapter.

5 MULTIFREQUENCY METHODS

The frequency dependence of backscattering efficiency (when size parameter becomes similar to the wavelength) and extinction efficiency have motivated the development of radar systems operating at two or more frequencies (e.g., Sadowy et al., 2003; Li et al., 2008) or separate radars with matched beams (e.g., Chandrasekar et al., 2010; Battaglia et al., 2016) so that these differences can be used to further constrain the range-resolved parameters of the PSD. For any pair of frequencies υ_1 and υ_2, where $\upsilon_1 < \upsilon_2$, the dual-frequency ratio (DFR) is defined as:

$$\mathrm{DFR} = 10\log_{10}\left(\frac{Z_1}{Z_2}\right) = dBZ_1 - dBZ_2. \quad (19)$$

When a gamma distribution PSD is assumed Eq. (9), and the equation for effective reflectivity Eq. (4) is inserted into Eq. (19), the DFR has the convenient property of only depending on the shape and slope (or alternatively, D_m and σ_m) parameters of the gamma PSD. In principle, then, it is possible to retrieve D_m from the DFR (after attenuation correction has been performed) and an assumption regarding μ, such as a constant value (Liao et al., 2014), reflectivity-dependent value (Munchak and Tokay, 2008), μ-Λ relationship (Zhang et al., 2003), or σ_m-D_m relationship (Williams et al., 2014). At some frequency pairs, such as Ku-Ka, there is an ambiguity in that two values of D_m can be associated with the same DFR for a given μ (Fig. 9). In these cases, a probabilistic framework and additional observations (such as PIA) can be useful to determine which of these solutions is a better fit to all of the measurements. When the size parameter is small compared to both wavelengths, DFR will be close to zero and invariant with size, necessitating higher frequencies or differential attenuation to determine the PSD parameters.

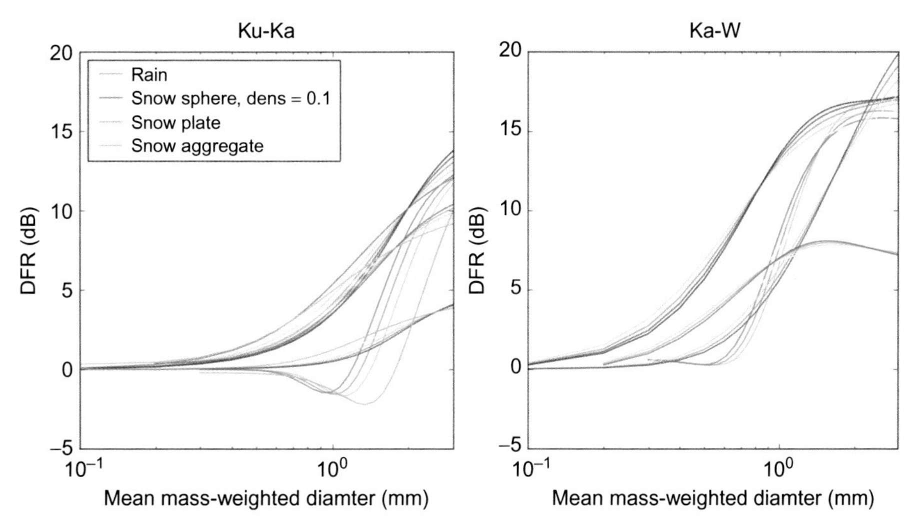

FIG. 9 Dual-frequency ratio vs D_m for Ku-Ka (left) and Ka-W (right) frequency pairs. Different values for μ (ice) or σ_m' (rain) are indicated by different shades of each color, as in Figs. 4 and 5.

The impact of adding a frequency on improving the accuracy of the retrieved rainfall rate can be examined with the IFloodS disdrometer dataset. Fig. 10 shows the output of a simple Bayesian retrieval using different frequencies individually and in combination. This framework predicts the rainfall rate based on a weighted average of all rainfall rates in the database, where the weights are based on the difference between observed reflectivity and reflectivity at n frequencies (divided by the measurement error σ_n) in the database:

$$R_i = \frac{\sum_j R_j w_j}{\sum_j w_j}, \tag{20}$$

where

$$w_j = \exp\left(-\sum_n \frac{\left(Z_{n,i} - Z_{n,j}\right)^2}{\sigma_n^2}\right). \tag{21}$$

Although the exact values in Fig. 10 should not be considered representative of all precipitation regimes, and represent the ideal case (perfect attenuation correction, no multiple scattering or nonuniform beam filling), the impact of additional frequencies is clear in the improved correlation and reduced relative error. It is also apparent that little additional information is provided by extra frequencies at very low rainfall rates, where backscattering follows the Rayleigh approximation for the Ku, Ka, and W bands.

Multiple-frequency measurements are also particularly useful in the melting layer and ice phase of precipitation. Le and Chandrasekar (2013) describe a method to identify the melting layer via the measured (nonattenuation-corrected) DFR. According to this method, the top of the melting layer is at a peak in the slope ($-d\mathrm{DFR}/dZ$), a DFR peak occurs within the melting layer due to Mie scattering by large aggregates, and as these collapse into raindrops a local minimum in the DFR signifies the bottom of the melting layer before differential attenuation begins to increase in the rain layer. Within the ice phase, multifrequency measurements are useful not only for determining mean particle size, but with three frequencies some indication of particle shape can also be discerned (Leinonen et al., 2012; Kulie et al., 2014; Kneifel et al., 2015). The basis for this can be seen when

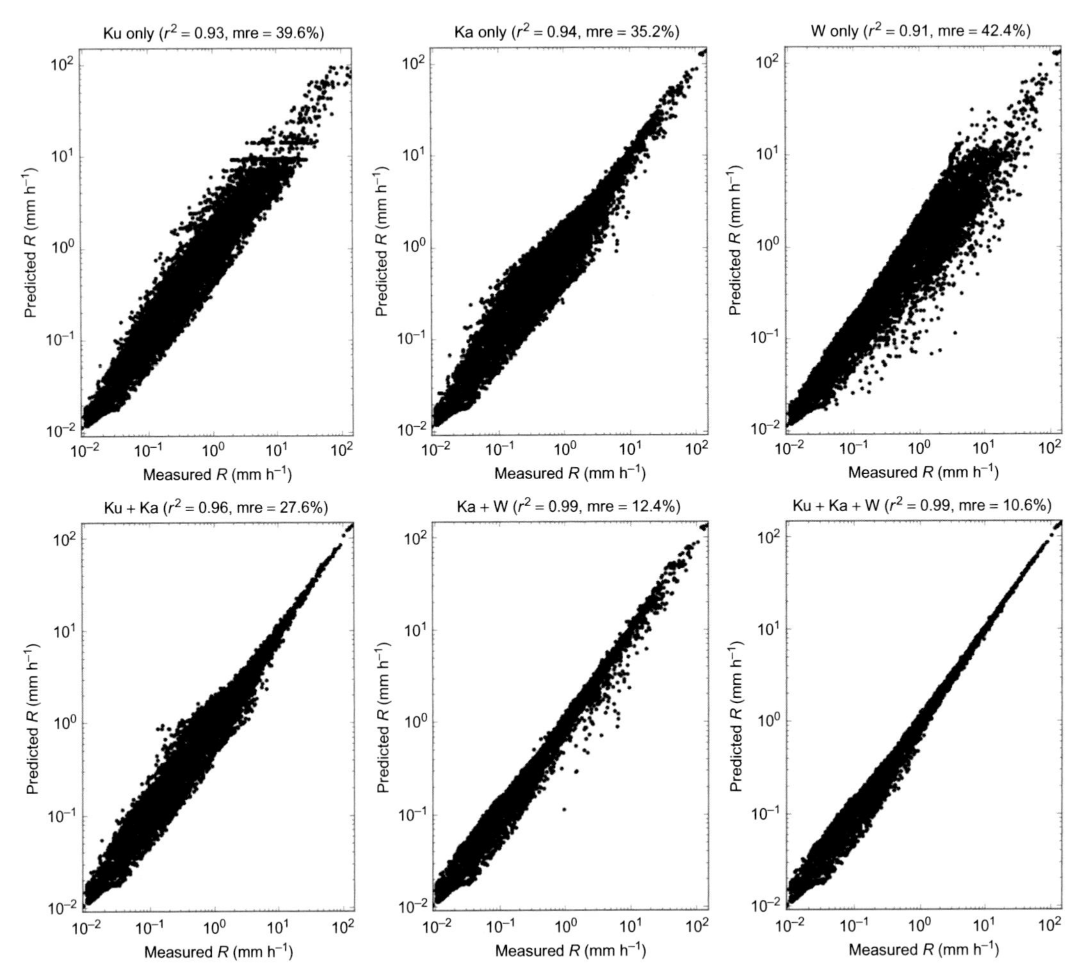

FIG. 10 Observed vs. predicted rainfall rate from Eq. (20) from the IFloodS PSD database for different combinations of Ku, Ka, and W-band reflectivity measurements, assuming a perfect attenuation correction is made and 1 dB measurement error. The squared correlation coefficient (r^2) and mean relative error (mre) are given for each combination.

the DFR curves in Fig. 9 are plotted against each other in Ku-Ka vs Ka-W space (Fig. 11). In these scattering models, for example, aggregates have much smaller Ka-W DFR than pristine plates. Leinonen and Szyrmer (2015) also found that rimed particles tend to have a lower Ka-W DFR for a given Ku-Ka DFR than nonrimed aggregates.

How can the ample information provided by radar measurements at multiple frequencies be incorporated into the probabilistic estimation framework Eq. (18)? Because actual reflectivity measurements are noisy, and attenuation in even moderate rainfall can be enough to overwhelm the PSD effect on DFR, strict bin-by-bin PSD retrievals and attenuation correction can be error-prone. One approach to mitigate these errors formulated by Grecu et al. (2011) is to assign only the higher-frequency reflectivities to the observation vector **y** and retrieve n-1

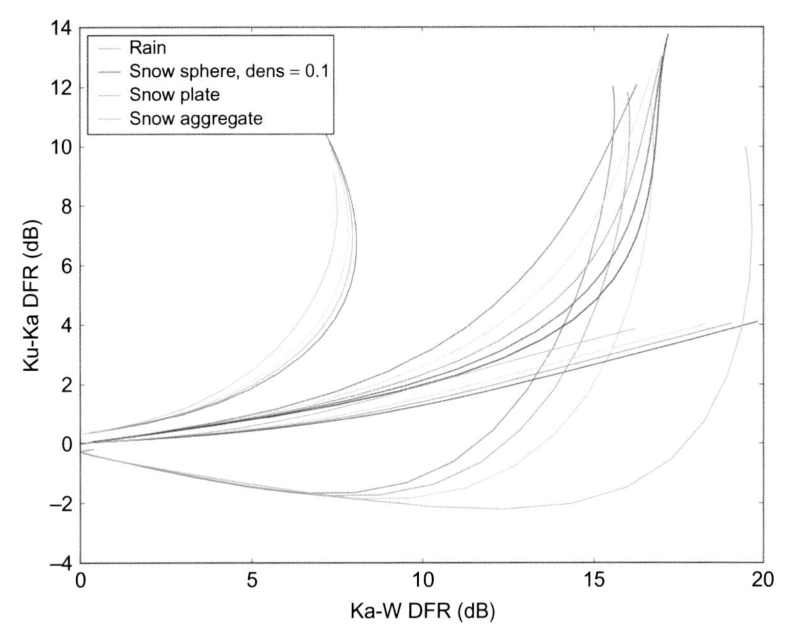

FIG. 11 Dual-frequency ratio at Ka and W bands vs dual-frequency ratio at Ku-Ka bands for the hydrometeor PSDs used in Figs. 4, 5, and 9.

range-resolved PSD parameters in **x**, where n is the number of radar frequencies. This range-resolved parameter should be independent of Z (N_w or a "normalized" PSD parameter such as $D_m{}'$ are good candidates) and should have a vertical constraint imposed, either via positive off-diagonal values in $\mathbf{S_a}$ or by retrieving this parameter at a reduced set of nodes within the profile and interpolating values between the nodes. Then, the forward model calculates the profile of Z_m at the higher frequency from a PSD that matches the lower-frequency reflectivity. PIA at multiple frequencies can also be included in the observation vector, or if the dSRT is used to reduce noise, the difference in PIA can be obtained from the forward model instead.

As with the single-frequency methods, under some circumstances multiple-frequency methods can be prone to error if the forward model does not account for nonuniform beam filling and multiple scattering. These phenomena will be discussed in subsequent sections.

6 EFFECTS OF NONUNIFORM BEAM FILLING

So far, it has been assumed that the PSD is uniform throughout a radar range gate, and thus the integral parameters such as Z, R, and k_{ext} do not depend on the horizontal resolution of the radar. Strictly speaking, it is questionable whether "uniform" PSDs exist in nature (Jameson and Kostinski, 2001), and analyses with networks of disdrometers reveal spatial variability on sub-100 m scales (e.g., Jameson et al., 2015). Practically, any nonuniformity will result in error when Z-R and Z-k_{ext} power laws are used to correct for attenuation and retrieve rainfall. As shown by Iguchi et al. (2000, 2009), even if everywhere in a volume, $R=aZ^b$ is strictly followed, if Z is not uniform within the volume then the volume mean value of R will be less than that implied by the volume mean value of Z:

$$\overline{R}=\overline{aZ_e^b}\leq a\overline{Z_e}^b, \tag{22}$$

because b is less than one, following Jensen's inequality. Likewise, because the exponent in the Z-k_{ext} relationships is also less than one, k_{ext} will also have a low bias relative to the true volume mean. This has further implications for the conversion of Z_m to Z_e and the interpretation of the SRT PIA. At any range gate, the measured reflectivity will be the average of the attenuated reflectivities, with the attenuation summed along each infinitesimally narrow pencil-beam direction (θ_i,ϕ_i), weighted by the antenna gain pattern;

$$Z_m(r) = \overline{Z_e(r,\theta_i,\phi_i)\mathrm{PIA}(r,\theta_i,\phi_i)} \tag{23}$$

The Chebyshev sum inequality for two ordered pairs, which states that if $a_1 \leq a_2 \leq \ldots \leq a_n$ and $b_1 \geq b_2 \geq \ldots \geq b_n$, then $\overline{ab} \leq \bar{a}\bar{b}$, can be used in this context if we let a represent $Z_e\,(r,\theta_i,\phi_i)$ and b represent the pencil-beam PIA (r,θ_i,ϕ_i). If Z_e and PIA are ordered accordingly within a radar footprint, then it follows that

$$Z_m \leq \left(\overline{Z_e}\right)\left(\overline{\mathrm{PIA}}\right), \tag{24}$$

Thus the effective attenuation, that is, Z_m/Z_e, is less then or equal to the SRT PIA (assuming a uniform surface backscatter cross-section within the radar field-of-view) if the Chebyshev condition is met.

To further illustrate the effects of nonuniform beam filling (NUBF), two scenarios will be considered: one in which each range gate is composed of varying proportions of empty space and a uniform PSD, and one in which the shape of the PSD (D_m, μ) is constant but the concentration follows a lognormal distribution within each range gate. An analysis of high-resolution airborne data convolved to lower-resolution satellite footprints by Tanelli et al. (2012) found that these two scenarios fit the vast majority of observations.

For the first scenario, the effect of NUBF on the measurements and retrievals are illustrated in Fig. 12 for the sample profile shown in Fig. 8. The fraction of the beam containing precipitation was varied from 10% to 100% while scaling the PSD such that the column-averaged Z_e and R are equal to the uniform case. The measured reflectivity near the surface increases from the uniform case as the precipitation fraction decreases as long as the vertical autocorrelation is not high. This is because the PIA along each pencil beam is more evenly distributed than Z_e at any given level. When the profiles become highly vertically correlated, the attenuation is able to reach higher values and $\overline{Z_m}$ is reduced from the uniform case. Regardless of the precipitation fraction or vertical correlation structure, the SRT PIA is always reduced from the uniform case. This can be inferred from the beam-integrated equation for the normalized surface backscatter cross section:

$$\sigma_{0,R} = \iint \sigma_{0,NR}(r,\theta)e^{-0.1\ln(10)\mathrm{PIA}(r,\theta)}drd\theta. \tag{25}$$

If $\sigma_{0,NR}$ is constant, then the reduction in $\sigma_{0,R}$ (in dB units) is equal to

$$\mathrm{PIA_{SRT}} = 10\log_{10}\left(\overline{e^{-0.1\ln(10)\mathrm{PIA}(r,\theta)}}\right) \leq \overline{\mathrm{PIA}(r,\theta)} \tag{26}$$

by Jensen's inequality. The reduction is most severe as the precipitation fraction decreases and vertical correlation increases, effectively exposing more unattenuated surface, which dominates the measured σ_0.

When a Hitchfeld-Bordan attenuation correction is applied to these profiles under the assumption of uniform beam filling and no adjustment to the intrinsic Z-k and Z-R relationships to match the PIA, the errors in Z_e and R near the surface correspond to the difference in Z_m near the surface compared to the uniform case. When the SRT PIA is used to adjust the Z-k and Z-R relationships, Z_e and R are biased low due to the low bias of the SRT PIA when NUBF occurs.

The same set of plots for the second scenario, a filled beam with lognormally varying Z_e, is shown in Fig. 13. The behavior with respect to the standard deviation of dBZ_e is similar to the

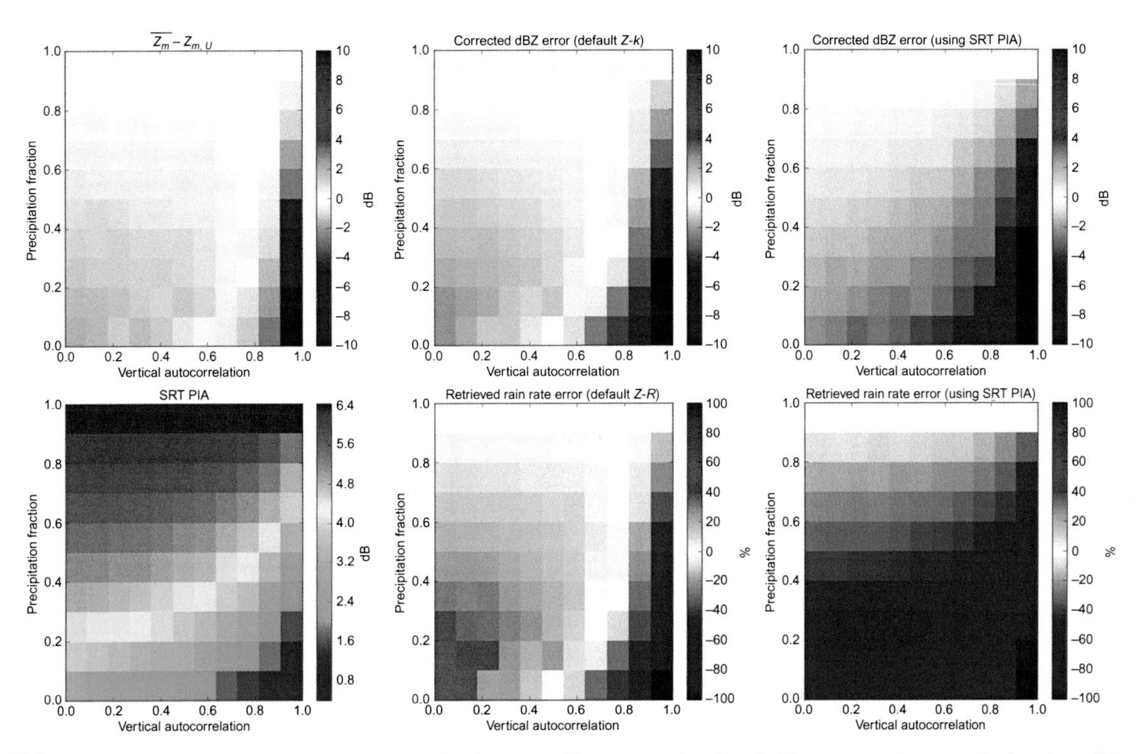

FIG. 12 Nonuniform beam filling scenarios for the Z_e profile presented in Fig. 8. These scenarios vary the fraction of the beam containing precipitation at each level on the vertical axis and the autocorrelation of precipitation from one vertical level to the next on the horizontal axis. In these scenarios, the precipitation is uniform where present and scaled such that the average nonattenuated reflectivity is constant at each level. The difference in measured reflectivity from a uniform beam, SRT PIA, and errors in corrected reflectivity and retrieved rainfall rate near the surface (using default relationships and the SRT PIA to adjust the PSD) are shown.

behavior with respect to the precipitation fraction in the first scenario, although reversed in sign (with larger dBZ_e variability corresponding to smaller precipitation fractions). The behavior with respect to vertical correlation structure is also similar, except that the reduction in reflectivity and SRT PIA occurs only when the precipitation structure is almost perfectly aligned with the radar look vector. When the standard deviation of dBZ_e is only a few dB, the errors from the H-B algorithm are small when the SRT PIA is used, but increase for large variability of dBZ_e and very large or small vertical autocorrelation. The bias of the unmodified Hitschfeld-Bordan retrieval is closer to zero in these examples, but it must be remembered that variability in the *Z-k* relationship was not considered when generating the synthetic profiles, which would add additional random error to the Hitchfeld-Bordan results.

The cases presented in Figs. 12 and 13 are not meant to be a universal guide to NUBF biases, since such a guide would require analysis of several different profile shapes, patterns of nonuniformity, and PSD perturbations. These examples do serve to illustrate the mechanics of NUBF and its dependence on the structure of the horizontal and vertical variability. Although a low bias in the SRT PIA relative to the effective PIA was always present in these examples,

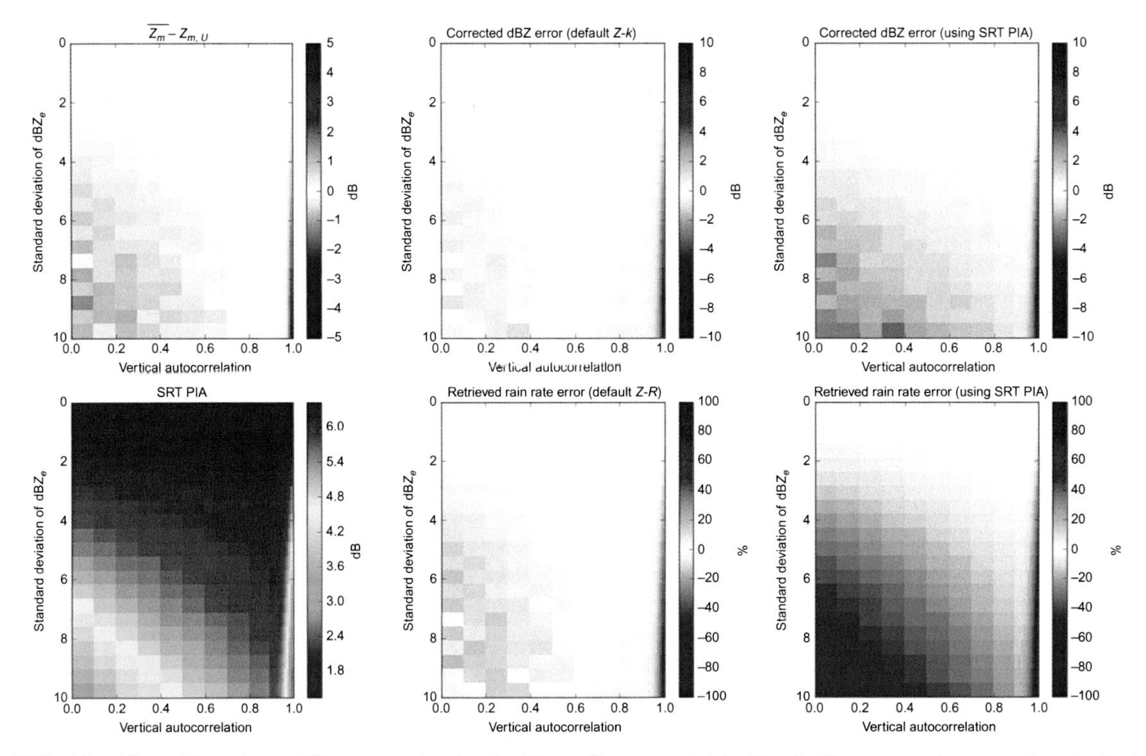

FIG. 13 Nonuniform beam filling scenarios for the Z_e profile presented in Fig. 8. These scenarios vary the standard deviation of *dBZe* at each level (scaled such that the average nonattenuated reflectivity is unmodified from the original profile) on the vertical axis and the autocorrelation of precipitation from one vertical level to the next on the horizontal axis. The difference in measured reflectivity from a uniform beam, SRT PIA, and errors in corrected reflectivity and retrieved rainfall rate near the surface (using default relationships and the SRT PIA to adjust the PSD) are shown.

one can imagine a scenario where this is not the case. If a beam is slanted with respect to a precipitation shaft, then it is possible to violate the Chebyshev condition, meaning the near-surface reflectivity can become anticorrelated with the pencil-beam PIA. Given this complex behavior, how can precipitation retrieval algorithms accurately account for NUBF under all circumstances where it may occur?

In the TRMM PR radar-profiling algorithm (Iguchi et al., 2009), an analytical solution to the NUBF problem is derived under the assumption that k_{ext} follows a gamma distribution within each range gate and that the nonuniformity is range-independent (perfectly correlated in the beam direction). This requires an estimate of the coefficient of variation, σ_n. Similarly, the GPM combined radar-radiometer algorithm subdivides a beam into components with lognormally varying reflectivities, but this method requires an estimate of the variability as well. Universal estimates can be used, but these can introduce regional and seasonal biases since the scale of variability of precipitation is often storm-dependent (Tokay and Bashor, 2010). A desirable alternative is to use some local context, such as the convective/stratiform classification of the profile and the horizontal variability of SRT PIA estimates at the large scale, which can be used to inform the estimate of the local variability

(Kozu and Iguchi, 1999). Oversampling of the radar beam can produce finer-scale estimates of the SRT PIA that may be useful for this purpose as well (Takahashi et al., 2016). Since the SRT PIA is sensitive to NUBF, the probabilistic framework Eq. (18) can include a parameter describing the NUBF variability, which can aid in fitting SRT PIA and multifrequency radar measurements that cannot be easily explained with uniform beam filling.

7 MULTIPLE SCATTERING

Another phenomenon that can hinder the straightforward interpretation of radar reflectivity profiles is multiple scattering (MS). As the name suggests, this occurs when a sufficient amount of energy from the radar pulse undergoes multiple scattering events before returning to the receiver (Fig. 14). Because the range of the scatterer to the radar is resolved via the time delay, echoes associated with MS enhance the reflectivity at range gates further from the radar than where the MS occurs (one indicator of MS is echo resolved below the surface). While MS is ubiquitous for radars with a finite beamwidth, it becomes important for the interpretation of radar measurements when the multiply scattered signal becomes comparable to the single-scattered signal.

An extensive review of MS in the context of meteorological radars is given by Battaglia et al. (2010), and only a short summary is presented here focusing on (1) causes of MS, (2) simulations of MS, and (3) mitigation of MS in precipitation profiling algorithms. For the purposes of this section, "photon" refers to an infinitesimally small parcel of energy transmitted by the radar. Three properties of the radar system and scatterer (hydrometeors) are important in determining if MS will be significant:

1. Radar fields-of-view of comparable or larger size relative to the photon mean free path, which is inversely proportional to k_{ext}.
2. High single-scatter albedo, which increases the probability of a scattering event instead of an absorption event each time a photon interacts with the scatterer.
3. Phase functions that are peaked in the forward direction, keeping photons within the radar beam.

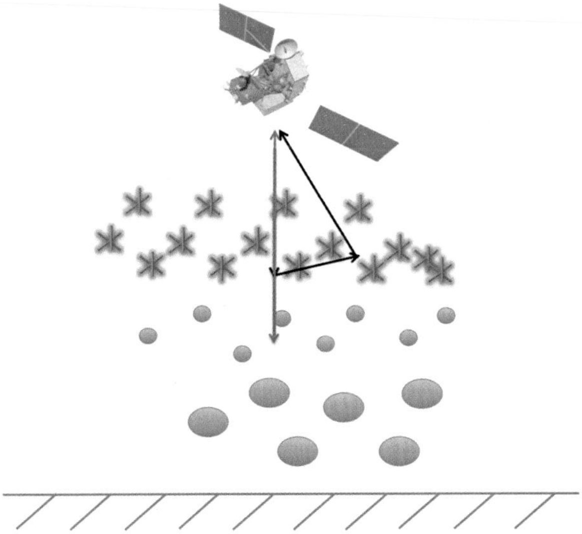

FIG. 14 Schematic depiction of the multiple scattering process. The path of a multiple-scattered photon (black) is interpreted as an echo from lower levels of the cloud following a single scattering event (red).

Analytical expressions quantifying MS have been derived by Kobayashi et al. (2005, 2007) for continuous wave and by Ito et al. (2007) for a time-dependent (pulse) wave. Hogan and Battaglia (2008) developed a computationally efficient time-dependent two-stream approximation for plane-parallel atmospheres. Although computationally expensive, Monte Carlo simulations are well suited to examining the MS problem by sampling the probabilities of scattering and absorption events within a three-dimensional scattering medium (Battaglia et al., 2005, 2006).

To illustrate the effect of different radar parameters (frequency, beamwidth) under a large dynamic range of hydrometeor concentrations, Monte Carlo MS radar simulations were performed for a tropical mesoscale convective system simulated by the Weather Research and Forecasting model with two-moment microphysics (Morrison et al., 2009) representing cloud water, cloud ice, snow, graupel, and rain (Fig. 15).

Observations were simulated from a flight line that passed directly over a deep convective core as well as a broad stratiform anvil overspreading some shallow cumulus cells at Ku-, Ka-, and W-band from an altitude of 20 km (Fig. 16) and 400 km (Fig. 17) to represent high-altitude airborne and satellite observation, respectively. Both sets of simulations at all frequencies were assuming a 1-degree half-power beamwidth, which corresponds to a ground footprint size of 350 m for the airborne simulation and 7 km for the satellite simulation. While this generally underestimates the beamwidth of airborne radars and overestimates the beamwidth of satellite-based radars presently in operation (Table 1), this is helpful for providing lower and upper bounds of expected MS effects from both vantage points. From the airborne perspective, MS enhancement is mostly limited to regions of extreme attenuation in the convective core between 50 and 100 km along track. Magnitudes vary from a few dB at Ku-band, 10–20 dB at Ka-band, to 30–40 dB at W-band. The stratiform region between 100 and 150 km, which contains significant amounts of ice aloft and little to no absorbing cloud water (resulting in high single-scatter albedo), also exhibits a few dB of MS enhancement at Ka-band and significant MS (>10 dB) at W-band. From the satellite perspective, the wider beam width results in greater MS effects that reach to higher levels of the cloud than the airborne simulation. Significant echo that appears to be originating below the surface is also evident.

For the retrieval framework Eq. (18), MS can be represented with a forward model that considers it, such as the Monte Carlo methods or the Hogan and Battaglia (2008) plane-parallel model. Since it is evident from Figs. 16 and 17 that MS does not significantly affect many profiles, it can be useful to first identify MS effects before invoking a computationally expensive forward model. Battaglia et al. (2014) argue that the differential slope (dZ_m/dh) at a pair of frequencies should equal the differential attenuation plus the change in DFR that is mediated by changes in the PSD (Fig. 9) with respect to height, which can be seen by combining Eqs. (8), (19):

$$\mathrm{DFR}_m(r) = 10\log_{10}\left(\frac{Z_{e,1}(r)}{Z_{e,2}(r)}\right) - 2\int_0^r k_{\mathrm{ext},1}(s) - k_{\mathrm{ext},2}(s)ds. \quad (27)$$

Differentiating with respect to r,

$$\frac{d\mathrm{DFR}_m}{dr} = \frac{d(dBZ_{e,1} - dBZ_{e,2})}{dr} + 2(k_{ext,2}(r) - k_{ext,1}(r)). \quad (28)$$

Because Z_e is often nearly constant with height, particularly below the melting layer, and attenuation increases predictably with frequency (Eq. 6 and Fig. 5), MS effects that reduce the differential slope to values that cannot be

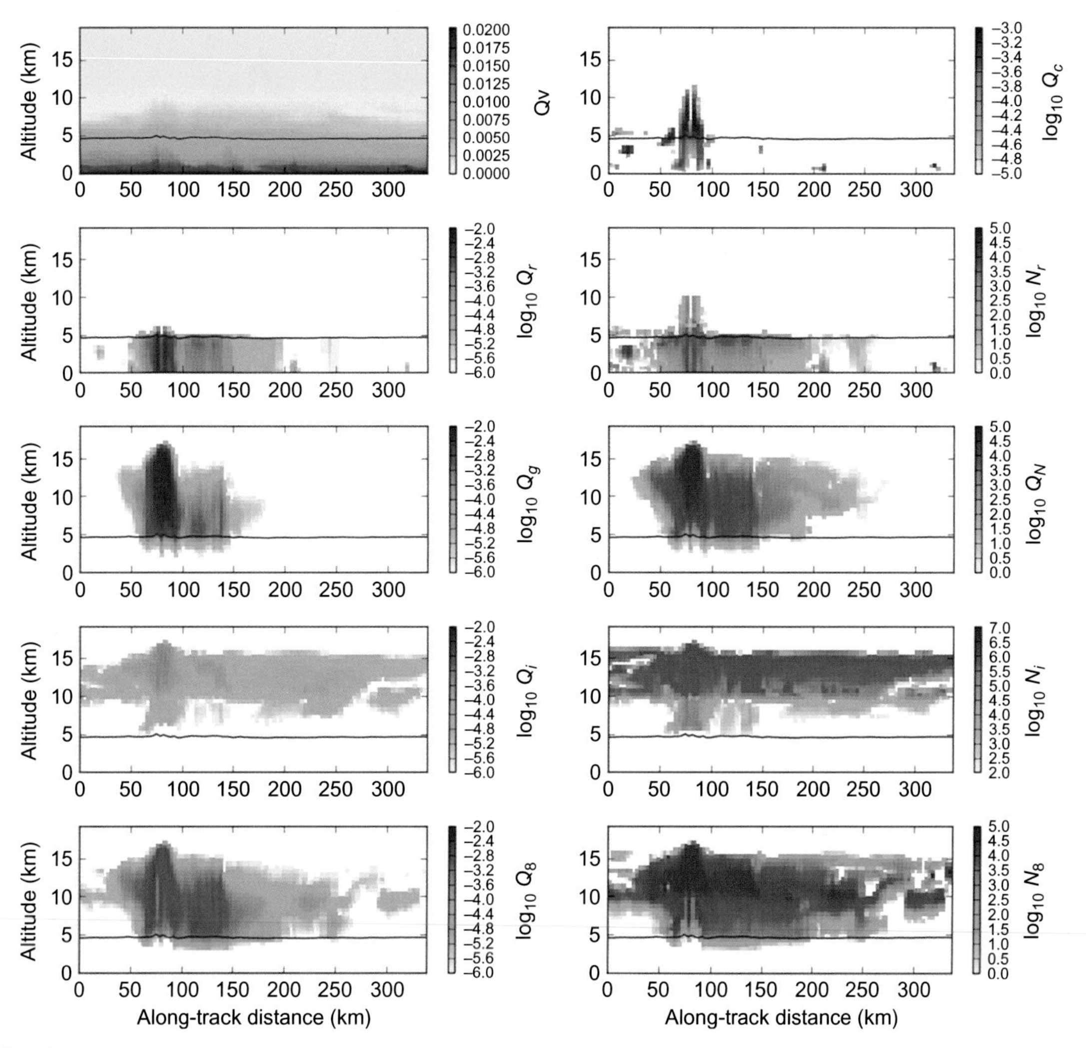

FIG. 15 Cross-sections of water vapor mixing ratio (Q_v), cloud water mixing ratio (Q_c), rain mixing ratio (Q_r) and number concentration (N_r), graupel mixing ratio (Q_g) and number concentration (N_g), cloud ice mixing ratio (Q_i) and number concentration (N_i), and snow mixing ratio (Q_s) and number concentration (N_s) along a synthetic flight line over a WRF simulation of a tropical mesoscale convective system. The 0°C isotherm is indicated by the blue line.

explained by any physical PSD can be a clear sign of multiple scattering that can "trigger" an MS-enabled forward model.

When the SRT PIA is used, MS effects must also be considered. If MS is weak compared to the surface return, it may safely be ignored. If MS is dominant, then there will be no peak in the surface return at the expected range gates (e.g., W-band in Fig. 17), and SRT algorithms should be able to identify these cases and fail to provide a meaningful PIA value. However, when the MS effect is of comparable magnitude to the surface return, the result will be an enhancement of $\sigma_{0,R}$ that must be accounted for to avoid a low bias in the SRT PIA (Haynes et al., 2009).

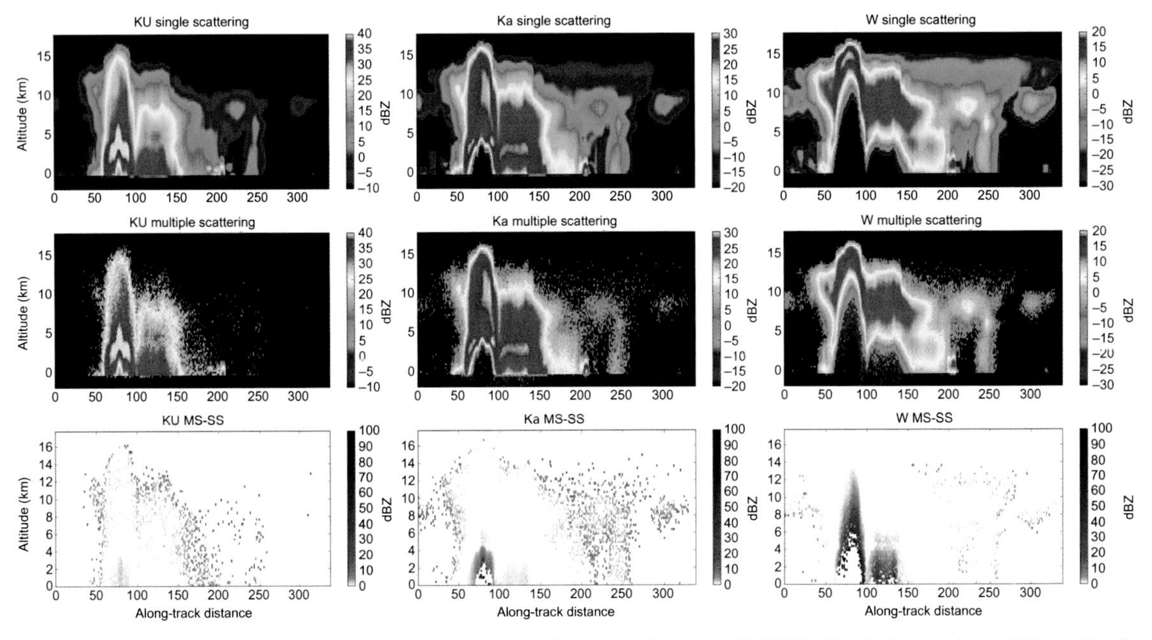

FIG. 16 Nadir-viewing high-altitude airborne radar simulations of the tropical MCS in Fig. 15 for a radar at 20-km altitude with a 1-degree half-power Gaussian beamwidth. The top row shows single-scattering simulations at Ku-, Ka-, and W-band, the middle row shows Monte Carlo multiple-scattering simulations, and the bottom row shows the difference (multiple-scattering enhancement) at each frequency.

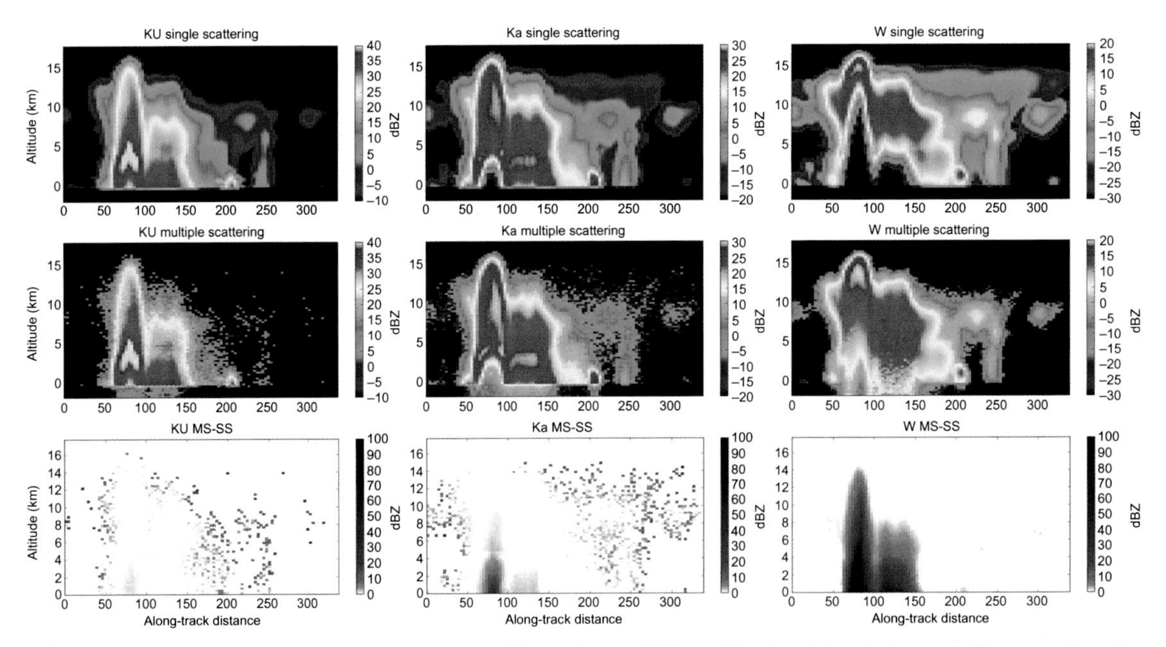

FIG. 17 Same as Fig. 16 for a nadir-viewing satellite radar at 400 km altitude with a 1-degree half-power Gaussian beamwidth.

8 RADAR-RADIOMETER METHODS

Many times, airborne and spaceborne radar measurements are taken simultaneously with passive microwave measurements of the same scene. Passive microwave measurements can provide independent estimates of precipitation profiles (Evans et al., 1995; Kummerow et al., 1996) and, from a satellite perspective, can provide greatly enhanced coverage relative to radars (Hou et al., 2014). The placement of radar and radiometer on the same platform provides an opportunity for radiometer precipitation retrievals to be built from radar profiles (Kummerow et al., 2011, 2015). The radiometer measurements can also be used more directly to provide additional constraints to the radar profile (Haddad et al., 1997; Grecu et al., 2004, 2016; Munchak and Kummerow, 2011). This section will provide a brief overview of the types of constraints that are provided by common frequencies on passive microwave instruments and how these can be integrated into the probabilistic retrieval framework Eq. (18).

The upwelling radiance measured by a passive microwave radiometer (commonly expressed in units of brightness temperature (*Tb*)) is a complex function of processes along the radiometer line-of-sight described by the vector radiative transfer equation:

$$\frac{d\vec{I}(\hat{n},\lambda)}{ds} = -\mathbf{K}\vec{I} + \vec{a}\,B(\lambda,T) + \int_{4\pi} \mathbf{Z}(\hat{n},\hat{n}',\lambda)\,\vec{I}\,d\hat{n}' \tag{29}$$

which, in order of the terms on the right-hand side, states that the radiance is reduced by extinction (absorption plus scattering), increased by emission (proportional to absorption), and increased by scattering into the path. In precipitation at microwave frequencies, none of these terms can be ignored and the sensitivity of *Tb* to the PSD profile depends on whether absorption/emission (1st and 2nd term) or scattering (1st and 3rd term) is dominant and on the surface type, which determines the background against which the precipitation "signal" is superimposed (Munchak and Skofronick-Jackson, 2013). From Figs. 4 and 5 it can be inferred that absorption is more important in the liquid phase, particularly at small size parameters, whereas scattering becomes dominant in the ice phase and at higher size parameters. The surface type plays a role because water surfaces are highly reflective, polarized, and, to a good approximation, reflections can be treated as specular (Meissner and Wentz, 2012). Thus, at low frequencies, where absorption dominates, the Tb increases with the liquid water path (first row of Fig. 18). Over land, which is less reflective, the emission from the surface is closer to the physical temperature of the surface and so the additional emission by hydrometeors in the atmosphere has little effect on *Tb* (second row of Fig. 18). However, at higher frequencies, particularly when ice is present, extinction by scattering reduces the *Tb* significantly (lower right panels of Fig. 18 and top left of Fig. 19). Several radiative transfer models exist that can solve Eq. (27) under assumptions of a plane-parallel atmosphere (e.g., Wiscombe, 1977; Evans and Stephens, 1991; Heidinger et al., 2006) or for three-dimensional atmospheres (Evans, 1998; Davis et al., 2005).

In the framework of a radar retrieval, these models can be used to convert PSD profiles derived from single- or multi-frequency measurements to Tb at any frequency for which the hydrometeor scattering properties, absorption by atmospheric gases, and surface emissivity are known. The problem then becomes one of how to represent the radiometer measurements in **y** and f(**x**) in Eq. (18). If the radiometer line-of-sight and beamwidth are identical to the radar, the *Tb* in f(**x**) can be computed directly from the radar profile if three-dimensional effects are ignored. However, this is rarely the case, as passive microwave radiometers often have wider fields-of-view and are offset at an oblique angle (Fig. 19) in order to take advantage of

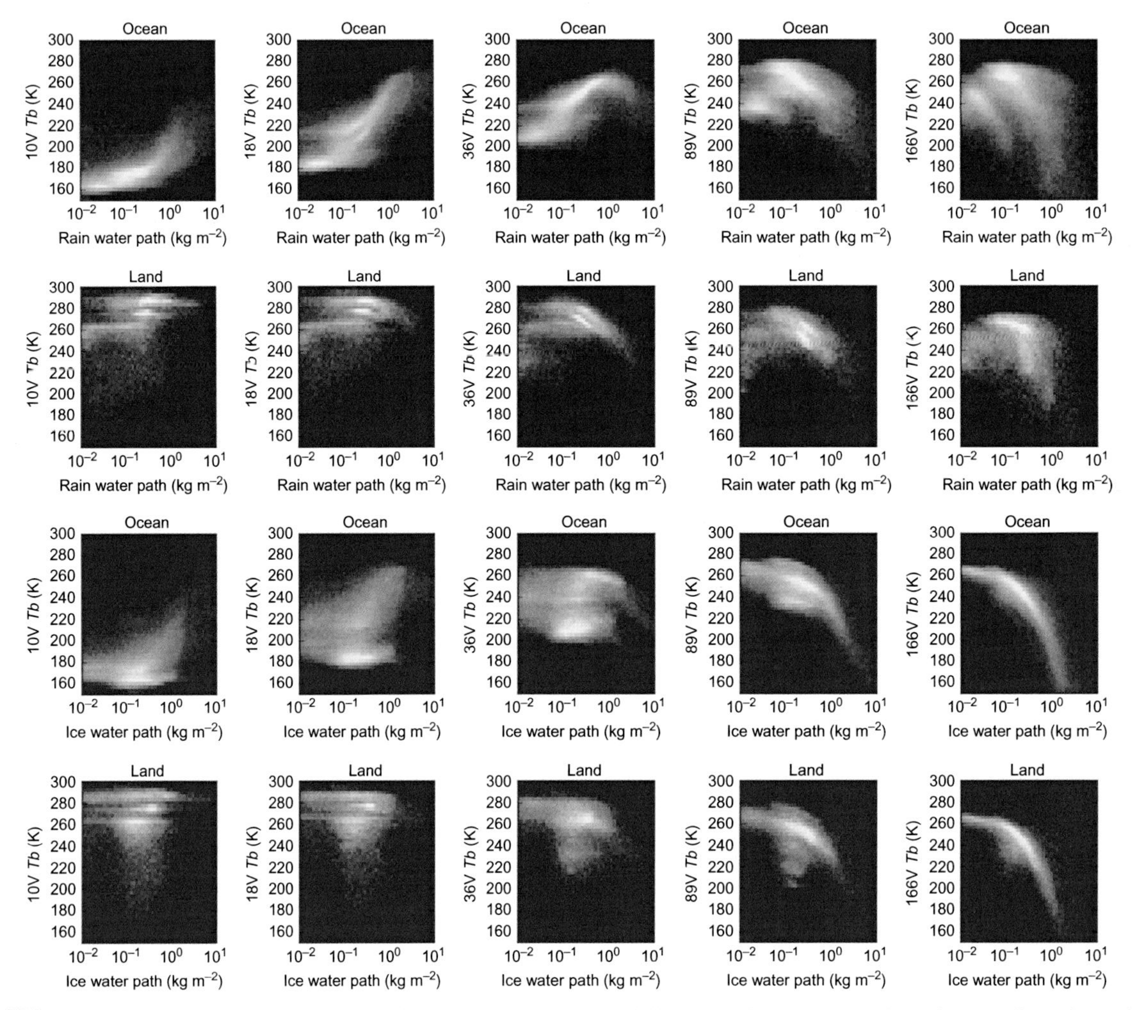

FIG. 18 Two-dimensional histograms of radar-derived integrated liquid and ice water path and vertically polarized observed brightness temperature from one day of GPM data at 10.65, 18.7, 36.64, 89, and 166 GHz, separately for land and water surfaces.

polarization differences that arise from surface and precipitation scattering (e.g., Prigent et al., 2005; Wang et al., 2013; Adams and Bettenhausen, 2012; Gong and Wu, 2017). One solution to this dilemma, and the one that is used by the GPM combined radar-radiometer algorithm (Grecu et al., 2016), is deconvolution, which weights *Tbs* in an oversampled field in such a way as to achieve a common resolution while conserving the observed *Tb* (Petty and Bennartz, 2016) over the scene. Another option is to expand **y** and **x** to present observations and state variables describing multiple radar columns so that the slant path and native resolution radiometer observations can be more accurately modeled (Munchak and Kummerow, 2011). However, since finding the minimum of Eq. (18) involves iterative matrix multiplication and inversion (Rodgers, 2000), which scale as the third power of the length of **y** and **x**, this

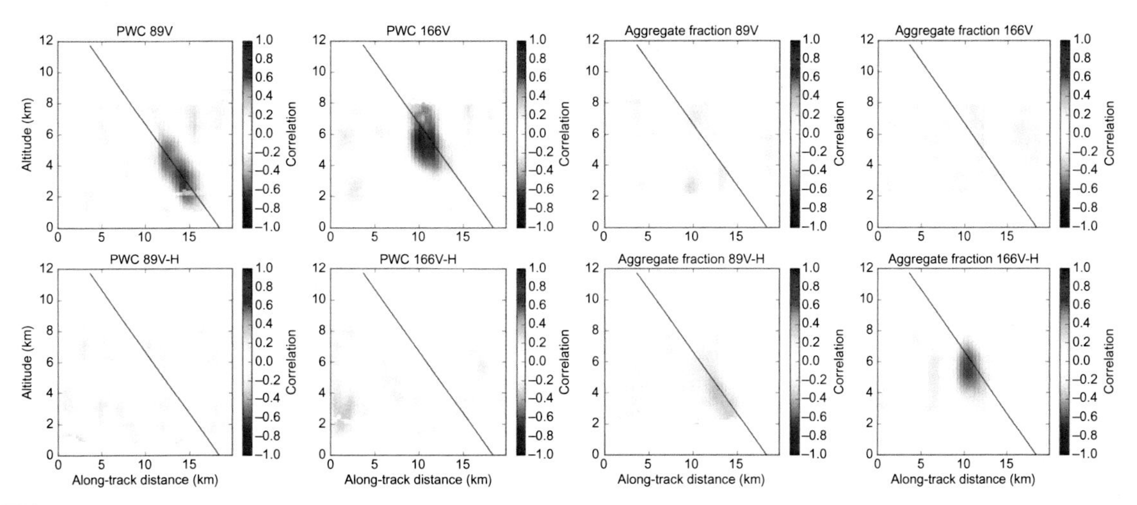

FIG. 19 Correlations between precipitation water content (PWC; left) or aggregate/pristine fraction (right) and 89/166 GHz brightness temperatures (top) and polarization difference (bottom) along a 53-degree slant path through an approximately uniform stratiform precipitation field. These correlations were calculated from an ensemble of retrievals all consistent with measured Ku-band reflectivity profiles during the OLYMPEX field experiment on December 3, 2015.

can be computationally expensive unless there are only a few state variables (e.g., D_m' at a few vertical levels) per radar column. Despite these challenges, the additional information provided by combining passive and active measurements of precipitation can better constrain surface properties such as wind over the ocean (Munchak et al., 2016) and precipitation profiles, particularly when ice is involved (Olson et al., 2016) than multifrequency radar profiles alone.

9 SUMMARY

The placement of weather radar on airborne and satellite platforms has proven to be of great utility for precipitation monitoring and the understanding of microphysical processes from subkilometer to global scales. While these platforms provide a unique vantage point from ground-based radars, they also present some challenges and limitations. The primary challenges for these platforms are limitations on weight, antenna size, and power consumption, all of which can be mitigated to some extent by the use of higher frequencies (Ku through W-band) than are typical for ground systems (S- or C-band). While these higher frequencies can provide sufficient sensitivity and angular resolution, they are much more prone to attenuation and multiple scattering, which respectively decrease and increase the measured reflectivity from that which is intrinsic to the precipitation PSD. A third phenomenon, nonuniform beam filling, can also cause departures in reflectivity from the uniform case and, like multiple scattering, its severity increases with the beamwidth.

In order to accurately retrieve the profile of precipitation PSD parameters from reflectivity profiles at one or more frequencies, optionally augmented by independent estimates of attenuation and/or passive microwave radiances, an optimal estimation framework has been described which minimizes a cost function consisting of normalized differences between observed and simulated measurements plus differences between the retrieved and a priori state vector. This framework is most robust when used with a forward model capable of simulating all aspects of the measurements that are relevant to the instrument, depending on frequency and beamwidth (e.g., polarization, multiple scattering, nonuniform beam filling), and characterization of the PSD in the state vector using parameters that are uncorrelated to reflectivity.

As the capabilities of airborne and satellite-based radars continue to increase in the future, this framework can be used to understand new measurements as well as to critically evaluate the forward model. As an example of this iterative process, the inability of homogenous spherical models of ice to represent multifrequency radar measurements led to advances in particle-scattering models for realistically shaped crystals and crystal aggregates, and as these models become widely available they should be considered the standard for use in retrievals. It is by this iterative process that retrieval uncertainty becomes better defined and is reduced by the addition of multiple, colocated measurements.

Acknowledgments

Appreciation is extended to Ali Tokay (NASA GSFC/ University of Maryland, Baltimore County) for providing the Parsivel[2] data from IFloodS, Kwo-sen Kuo (NASA GSFC/ Unviersity of Maryland, College Park) for the single scattering database of pristine and aggregate snow crystals, Ian Adams (Naval Research Laboratory) for the Monte Carlo multiple scattering model, and the APR3 team at NASA JPL for providing the data used in Fig. 19. Appreciation is also extended to Dr. Robert Meneghini (NASA GSFC) and an anonymous reviewer for providing comments that improved the clarity and accuracy of this chapter.

References

Adams, I.S., Bettenhausen, M.H., 2012. The scattering properties of horizontally aligned snow crystals and crystal approximations at millimeter wavelengths. Radio Sci. 47 (5).

Amayenc, P., Diguet, J.P., Marzoug, M., Tani, T., 1996. A class of single- and dual-frequency algorithms for rain-rate profiling from a spaceborne radar. Part II: tests from airborne radar measurements. J. Atmos. Oceanic Technol. 13 (1), 142–164.

Atlas, D., Chmela, A.C., 1957. Physical-synoptic variations of raindrop size parameters. In: Proc. Sixth Weather Radar Conf, pp. 21–29.

Atlas, D., Ulbrich, C.W., 1977. Path- and area-integrated rainfall measurement by microwave attenuation in the 1–3 cm band. J. Appl. Meteorol. 16 (12), 1322–1331.

Battaglia, A., Ajewole, M.O., Simmer, C., 2005. Multiple scattering effects due to hydrometeors on precipitation radar systems. Geophys. Res. Lett. 32. L19801, https://doi.org/10.1029/2005GL023810.

Battaglia, A., Ajewole, M., Simmer, C., 2006. Evaluation of radar multiple-scattering effects from a GPM perspective part I: model description and validation. J. Appl. Meteorol. Climatol. 45, 1634–1647. https://doi.org/10.1175/JAM2424.1.

Battaglia, A., Tanelli, S., Kobayashi, S., Zrnic, D., Hogan, R.J., Simmer, C., 2010. Multiple-scattering in radar systems: a review. J. Quant. Spectrosc. Radiat. Transf. 111 (6), 917–947.

Battaglia, A., Tanelli, S., Heymsfield, G.M., Tian, L., 2014. The dual wavelength ratio knee: a signature of multiple scattering in airborne Ku–Ka observations. J. Appl. Meteorol. Climatol. 53 (7), 1790–1808.

Battaglia, A., Mroz, K., Lang, T., Tridon, F., Tanelli, S., Tian, L., Heymsfield, G.M., 2016. Using a multiwavelength suite of microwave instruments to investigate the microphysical structure of deep convective cores. J. Geophys. Res.-Atmos. 121 (16), 9356–9381.

Battan, L.J., 1973. Radar Observation of the Atmosphere. The University of Chicago Press, Chicago, IL, USA. 324 pp.

Beard, K.V., 1976. Terminal velocity and shape of cloud and precipitation drops aloft. J. Atmos. Sci. 33, 851–864. https://doi.org/10.1175/1520-0469(1976)033<0851:TVASOC>2.0.CO;2.

Beard, K.V., Bringi, V.N., Thurai, M., 2010. A new understanding of raindrop shape. Atmos. Res. 97 (4), 396–415.

Behrangi, A., Stephens, G., Adler, R., Huffman, G., Lambrigtsen, B., Lebsock, M., 2014. An update on the oceanic precipitation rate and its zonal distribution in light of advanced observations from space. J. Clim. 27, 3957–3965. https://doi.org/10.1175/JCLI-D-13-00679.1.

Boukabara, S.A., Garrett, K., Chen, W., Iturbide-Sanchez, F., Grassotti, C., Kongoli, C., Ferraro, R., 2011. MiRS: An all-weather 1DVAR satellite data assimilation and retrieval system. IEEE Trans. Geosci. Remote Sens. 49 (9), 3249–3272.

Brandes, E., Zhang, G., Sun, J., 2006. On the influence of assumed drop size distribution form on radar-retrieved thunderstorm microphysics. J. Appl. Meteorol. Climatol. 45, 259–268. https://doi.org/10.1175/JAM2335.1.

Bringi, V.N., Chandrasekar, V., 2001. Polarimetric Doppler Weather Radar: Principles and Applications. Cambridge University Press., Cambridge, UK. 636 pp.

Bringi, V., Tolstoy, L., Thurai, M., Petersen, W., 2015. Estimation of spatial correlation of drop size distribution parameters and rain rate using NASA'S s-band polarimetric radar and 2D video disdrometer network: two case studies from MC3E. J. Hydrometeor. 16, 1207–1221. https://doi.org/10.1175/JHM-D-14-0204.1.

Chandrasekar, V., Schwaller, M., Vega, M., Carswell, J., Mishra, K.V., Meneghini, R., Nguyen, C., 2010. Scientific and engineering overview of the NASA dual-frequency dual-polarized Doppler radar (D3R) system for GPM ground validation. In: Geoscience and Remote Sensing Symposium (IGARSS), 2010 IEEE International (pp. 1308-1311). IEEE.

Davis, C., Emde, C., Harwood, R., 2005. A 3-D polarized reversed Monte Carlo radiative transfer model for millimeter and submillimeter passive remote sensing in cloudy atmospheres. IEEE Trans. Geosci. Remote Sens. 43 (5), 1096–1101.

Draine, B.T., Flatau, P.J., 1994. Discrete dipole approximation for scattering calculations. J. Opt. Soc. Am. A 11, 1491–1499.

Ellis, S.M., Vivekanandan, J., 2010. Water vapor estimates using simultaneous dual-wavelength radar observations. Radio Sci. 45 (5).

Evans, K.F., 1998. The spherical harmonics discrete ordinate method for three-dimensional atmospheric radiative transfer. J. Atmos. Sci. 55 (3), 429–446.

Evans, K.F., Stephens, G.L., 1991. A new polarized atmospheric radiative transfer model. J. Quant. Spectrosc. Radiat. Transf. 46, 412–423.

Evans, K.F., Turk, J., Wong, T., Stephens, G.L., 1995. A Bayesian approach to microwave precipitation profile retrieval. J. Appl. Meteorol. 34 (1), 260–279.

Evensen, G., 2006. Data Assimilation: The Ensemble Kalman Filter. Springer, Berlin. 280 pp.

Freilich, M.H., Vanhoff, B.A., 2003. The relationship between winds, surface roughness, and radar backscatter at low incidence angles from TRMM precipitation radar measurements. J. Atmos. Ocean. Technol. 20, 549–562.

Galloway, J., Pazmany, A., Mead, J., McIntosh, R.E., Leon, D., French, J., Vali, G., 1997. Detection of ice hydrometeor alignment using an airborne W-band polarimetric radar. J. Atmos. Ocean. Technol. 14 (1), 3–12.

Gong, J., Wu, D.L., 2017. Microphysical properties of frozen particles inferred from global precipitation measurement (GPM) microwave imager (GMI) polarimetric measurements. Atmos. Chem. Phys. 17, 2741–2757. https://doi.org/10.5194/acp-17-2741-2017.

Grecu, M., Olson, W., 2008. Precipitating snow retrievals from combined airborne cloud radar and millimeter-wave radiometer observations. J. Appl. Meteorol. Climatol. 47, 1634–1650. https://doi.org/10.1175/2007JAMC1728.1.

Grecu, M., Olson, W.S., Anagnostou, E.N., 2004. Retrieval of precipitation profiles from multiresolution, multifrequency active and passive microwave observations. J. Appl. Meteorol. 43, 562–575.

Grecu, M., Olson, W., Munchak, S., Ringerud, S., Liao, L., Haddad, Z., Kelley, B., McLaughlin, S., 2016. The GPM combined algorithm. J. Atmos. Ocean. Technol. 33, 2225–2245. https://doi.org/10.1175/JTECH-D-16-0019.1.

Grecu, M., Tian, L., Olson, W.S., Tanelli, S., 2011. A robust dual-frequency radar profiling algorithm. J. Appl. Meteorol. Climatol. 50 (7), 1543–1557.

Haddad, Z.S., Durden, S.L., Im, E., 1996. Parameterizing the raindrop size distribution. J. Appl. Meteorol. 35, 3–13.

Haddad, Z.S., Smith, E.A., Kummerow, C.D., Iguchi, T., Farrar, M.R., Durden, S.L., Alves, M., Olson, W.S., 1997. The TRMM 'day-1' radar/radiometer combined rain-profiling algorithm. J. Meteorol. Soc. Jpn. 75, 799–809.

Hartmann, D., Hendon, H., Houze, R., 1984. Some Implications of the mesoscale circulations in tropical cloud clusters for large-scale dynamics and climate. J. Atmos. Sci. 41, 113–121.

Haynes, J.M., L'Ecuyer, T.S., Stephens, G.L., Miller, S.D., Mitrescu, C., Wood, N.B., Tanelli, S., 2009. Rainfall retrieval over the ocean with spaceborne W-band radar. J. Geophys. Res.-Atmos. 114 (D8).

Heidinger, A.K., O'Dell, C., Bennartz, R., Greenwald, T., 2006. The successive-order-of-interaction radiative transfer model. Part I: model development. J. Appl. Meteorol. Climatol. 45 (10), 1388–1402.

Heymsfield, G., Tian, L., Li, L., McLinden, M., Cervantes, J., 2013. Airborne radar observations of severe hailstorms: implications for future spaceborne radar. J. Appl. Meteorol. Climatol. 52, 1851–1867. https://doi.org/10.1175/JAMC-D-12-0144.1.

Hitschfeld, W., Bordan, J., 1954. Errors inherent in the radar measurement of rainfall at attenuating wavelengths. J. Meteor. 11, 58–67.

Hogan, R., Battaglia, A., 2008. Fast lidar and radar multiple-scattering models part II: wide-angle scattering using the time-dependent two-stream approximation. J. Atmos. Sci. 65, 3636–3651. https://doi.org/10.1175/2008JAS2643.1.

Hoskins, B., Karoly, D., 1981. The steady linear response of a spherical atmosphere to thermal and orographic forcing. J. Atmos. Sci. 38, 1179–1196.

Hou, A., Kakar, R., Neeck, S., Azarbarzin, A., Kummerow, C., Kojima, M., Oki, R., Nakamura, K., Iguchi, T., 2014. The global precipitation measurement mission. Bull. Am. Meteorol. Soc. 95, 701–722. https://doi.org/10.1175/BAMS-D-13-00164.1.

Iguchi, T., Kozu, T., Meneghini, R., Awaka, J., Okamoto, K.I., 2000. Rain-profiling algorithm for the TRMM precipitation radar. J. Appl. Meteorol. 39 (12), 2038–2052.

Iguchi, T., Kwiatkowski, J., Meneghini, R., Awaka, J., Okamoto, K.I., 2009. Uncertainties in the rain profiling algorithm for the TRMM precipitation radar. J. Meteorol. Soc. Jpn. Ser. II 87, 1–30.

Illingworth, A.J., Barker, H.W., Beljaars, A., Ceccaldi, M., Chepfer, H., Clerbaux, N., Fukuda, S., 2015. The Earth-CARE satellite: the next step forward in global measurements of clouds, aerosols, precipitation, and radiation. Bull. Am. Meteorol. Soc. 96 (8), 1311–1332.

Ito, S., Kobayashi, S., Oguchi, T., 2007. Multiple scattering formulation of pulsed beam waves in hydrometeors and its application to millimeter wave weather radar. IEEE Geosci. Remote Sens. Lett. 4, 13–17.

Jameson, A., Kostinski, A., 2001. What is a raindrop size distribution? Bull. Am. Meteorol. Soc. 82, 1169–1177.

Jameson, A., Larsen, M., Kostinski, A., 2015. On the variability of drop size distributions over areas. J. Atmos. Sci. 72, 1386–1397. https://doi.org/10.1175/JAS-D-14-0258.1.

Johnson, B.T., Olson, W.S., Skofronick-Jackson, G., 2016. The microwave properties of simulated melting precipitation particles: sensitivity to initial melting. Atmos. Meas. Tech. 9 (1), 9–21.

Keigler, J.E., Krawitz, L., 1960. Weather radar observation from an earth satellite. J. Geophys. Res. 65, 2793–2808.

Kim, M.J., 2006. Single scattering parameters of randomly oriented snow particles at microwave frequencies. J. Geophys. Res. Atmos. 111 (D14).

Kneifel, S., von Lerber, A., Tiira, J., Moisseev, D., Kollias, P., Leinonen, J., 2015. Observed relations between snowfall microphysics and triple-frequency radar measurements. J. Geophys. Res. Atmos. 120, 6034–6055. https://doi.org/10.1002/2015JD023156.

Kneifel, S., Kollias, P., Battaglia, A., Leinonen, J., Maahn, M., Kalesse, H., Tridon, F., 2016. First observations of triple-frequency radar Doppler spectra in snowfall: interpretation and applications. Geophys. Res. Lett. 43 (5), 2225–2233.

Kobayashi, S., Oguchi, T., Tanelli, S., Im, E., 2005. Second-order multiple-scattering theory associated with back-scattering enhancement for a millimeter wavelength weather radar with a finite beam width. Radio Sci. 42. RS6015, https://doi.org/10.1029/2004RS003219.

Kobayashi, S., Oguchi, T., Tanelli, S., Im, E., 2007. Backscattering enhancement on spheroid-shaped hydrometeors: considerations in water and ice particles of uniform size and Marshall-Palmer distributed rains. Radio Sci. 42. RS2001, https://doi.org/10.1029/2006RS003503.

Kollias, P., Miller, M.A., Luke, E.P., Johnson, K.L., Clothiaux, E.E., Moran, K.P., Widener, K.B., Albrecht, B.A., 2007. The atmospheric radiation measurement program cloud profiling radars: second-generation sampling strategies, processing, and cloud data products. J. Atmos. Ocean. Technol. 24, 1199–1214.

Kozu, T., Iguchi, T., 1999. Nonuniform beamfilling correction for spaceborne radar rainfall measurement: implications from TOGA COARE radar data analysis. J. Atmos. Ocean. Technol. 16 (11), 1722–1735.

Kubota, T., Iguchi, T., Kojima, M., Liao, L., Masaki, T., Hanado, H., Oki, R., 2016. A statistical method for reducing sidelobe clutter for the ku-band precipitation radar on board the GPM core observatory. J. Atmos. Ocean. Technol. 33 (7), 1413–1428.

Kulie, M., Hiley, M., Bennartz, R., Kneifel, S., Tanelli, S., 2014. Triple-frequency radar reflectivity signatures of snow: observations and comparisons with theoretical ice particle scattering models. J. Appl. Meteorol. Climatol. 53, 1080–1098. https://doi.org/10.1175/JAMC-D-13-066.1.

Kummerow, C., Olson, W.S., Giglio, L., 1996. A simplified scheme for obtaining precipitation and vertical hydrometeor profiles from passive microwave sensors. IEEE Trans. Geosci. Remote Sens. 34 (5), 1213–1232.

Kummerow, C.D., Ringerud, S., Crook, J., Randel, D., Berg, W., 2011. An observationally generated a priori database for microwave rainfall retrievals. J. Atmos. Ocean. Technol. 28, 113–130. https://doi.org/10.1175/2010JTECHA1468.1.

Kummerow, C., Randel, D., Kulie, M., Wang, N., Ferraro, R., Joseph Munchak, S., Petkovic, V., 2015. The evolution of the goddard profiling algorithm to a fully parametric scheme. J. Atmos. Ocean. Technol. 32, 2265–2280. https://doi.org/10.1175/JTECH-D-15-0039.1.

Kuo, K., Smith, E., Haddad, Z., Im, E., Iguchi, T., Mugnai, A., 2004. Mathematical–physical framework for retrieval of rain dsd properties from dual-frequency ku–ka-band satellite radar. J. Atmos. Sci. 61, 2349–2369.

Kuo, K., Olson, W., Johnson, B., Grecu, M., Tian, L., Clune, T., van Aartsen, B., Heymsfield, A., Liao, L., Meneghini, R., 2016. The microwave radiative properties of falling snow derived from nonspherical ice particle models. part I: an extensive database of simulated pristine crystals and aggregate particles, and their scattering properties. J. Appl. Meteorol. Climatol. 55, 691–708. https://doi.org/10.1175/JAMC-D-15-0130.1.

L'Ecuyer, T.S., Stephens, G.L., 2002. An estimation-based precipitation retrieval algorithm for attenuating radars. J. Appl. Meteor. 41, 272–285. https://doi.org/10.1175/1520-0450(2002)041<0272:AEBPRA>2.0.CO;2.

Le, M., Chandrasekar, V., 2013. Hydrometeor profile characterization method for dual-frequency precipitation radar onboard the GPM. IEEE Trans. Geosci. Remote Sens. 51 (6), 3648–3658.

Lebsock, M.D., L'Ecuyer, T.S., 2011. The retrieval of warm rain from CloudSat. J. Geophys. Res.-Atmos. 116 (D20).

Lebsock, M.D., Suzuki, K., Millán, L.F., Kalmus, P.M., 2015. The feasibility of water vapor sounding of the cloudy boundary layer using a differential absorption radar technique. Atmos. Meas. Tech. 8, 3631–3645. https://doi.org/10.5194/amt-8-3631-2015.

Leinonen, J., Szyrmer, W., 2015. Radar signatures of snowflake riming: a modeling study. Earth Space Sci. 2 (8), 346–358.

Leinonen, J., Kneifel, S., Moisseev, D., Tyynelä, J., Tanelli, S., Nousiainen, T., 2012. Evidence of nonspheroidal behavior in millimeter-wavelength radar observations of snowfall. J. Geophys. Res. 117. D18205, https://doi.org/10.1029/2012JD017680.

Li, L., Sekelsky, S., Reising, S., Swift, C., Durden, S., Sadowy, G., Dinardo, S., Li, F., Huffman, A., Stephens, G., Babb, D., Rosenberger, H., 2001. Retrieval of atmospheric attenuation using combined ground-based and airborne 95-GHz cloud radar measurements. J. Atmos. Ocean. Technol. 18, 1345–1353. https://doi.org/10.1175/1520-0426(2001)018<1345:ROAAUC>2.0.CO;2.

Li, L., Im, E., Durden, S.L., Haddad, Z.S., 2002. A surface wind model–based method to estimate rain-induced radar path attenuation over ocean. J. Atmos. Ocean. Technol. 19, 658–672.

Li, L., Heymsfield, G., Carswell, J., Schaubert, D., Creticos, J., Vega, M., 2008. High-altitude imaging wind and rain airborne radar (HIWRAP). In: Geoscience and Remote Sensing Symposium, 2008. IGARSS 2008. IEEE International (Vol. 3, pp. III-354). IEEE.

Liao, L., Meneghini, R., Tokay, A., 2014. Uncertainties of GPM DPR rain estimates caused by dsd parameterizations. J. Appl. Meteorol. Climatol. 53, 2524–2537. https://doi.org/10.1175/JAMC-D-14-0003.1.

Liu, G., 2004. Approximation of single scattering properties of ice and snow particles for high microwave frequencies. J. Atmos. Sci. 61, 2441–2456. https://doi.org/10.1175/1520-0469(2004)061<2441:AOSSPO>2.0.CO;2.

Marshall, J.S., Palmer, W.M.K., 1948. The distribution of raindrops with size. J. Meteorol. 5 (4), 165–166.

Marshall, J., Langille, R., Palmer, W., 1947. Measurement of rainfall by radar. J. Meteorol. 4, 186–192.

Matthews, A.J., Hoskins, B.J., Masutani, M., 2004. The global response to tropical heating in the Madden–Julian oscillation during the northern winter. Q. J. R. Meteorol. Soc. 130, 1991–2011. https://doi.org/10.1256/qj.02.123.

Meissner, T., Wentz, F.J., 2012. The emissivity of the ocean surface between 6 and 90 GHz over a large range of wind speeds and earth incidence angles. IEEE Trans. Geosci. Remote Sens. 50 (8), 3004–3026.

Meneghini, R., Jones, J.A., 2011. Standard deviation of spatially averaged surface cross section data from the TRMM precipitation radar. IEEE Geosci. Remote Sens. Lett. 8 (2), 293–297.

Meneghini, R., Kim, H., 2017. Minimizing the standard deviation of spatially averaged surface cross-sectional data from the dual-frequency precipitation radar. IEEE Trans. Geosci. Remote Sens. 55 (3), 1709–1716.

Meneghini, R., Eckerman, J., Atlas, D., 1983. Determination of rain rate from a spaceborne radar using measurements of total attenuation. IEEE Trans. Geosci. Remote Sens. 21, 34–43.

Meneghini, R., Iguchi, T., Kozu, T., Liao, L., Okamoto, K., Jones, J., Kwiatkowski, J., 2000. Use of the surface reference technique for path attenuation estimates from the TRMM precipitation radar. J. Appl. Meteorol. 39, 2053–2070.

Meneghini, R., Liao, L., Tian, L., 2005. A feasibility study for simultaneous estimates of water vapor and precipitation parameters using a three-frequency radar. J. Appl. Meteorol. 44, 1511–1525. https://doi.org/10.1175/JAM2302.1.

Meneghini, R., Liao, L., Tanelli, S., Durden, S.L., 2012. Assessment of the performance of a dual-frequency surface reference technique over ocean. IEEE Trans. Geosci. Remote Sens. 50, 2968–2977. https://doi.org/10.1109/TGRS.2011.2180727.

Meneghini, R., Kim, H., Liao, L., Jones, J., Kwiatkowski, J., 2015. An initial assessment of the surface reference technique applied to data from the dual-frequency precipitation radar (DPR) on the GPM satellite. J. Atmos. Ocean. Technol. 32, 2281–2296. https://doi.org/10.1175/JTECH-D-15-0044.1.

Mishchenko, M.I., Travis, L.D., 1998. Capabilities and limitations of a current FORTRAN implementation of the T-matrix method for randomly oriented, rotationally symmetric scatterers. J. Quant. Spectrosc. Radiat. Transf. 60 (3), 309–324.

Mitrescu, C., L'Ecuyer, T., Haynes, J., Miller, S., Turk, J., 2010. CloudSat precipitation profiling algorithm—model description. J. Appl. Meteorol. Climatol. 49, 991–1003. https://doi.org/10.1175/2009JAMC2181.1 DOI:10.1175%2F2009JAMC2181.1#External link, opens new window#_blank.

Morrison, H., Thompson, G., Tatarskii, V., 2009. Impact of cloud microphysics on the development of trailing stratiform precipitation in a simulated squall line: comparison of one-and two-moment schemes. Mon. Weather Rev. 137 (3), 991–1007.

Munchak, S., Kummerow, C., 2011. A modular optimal estimation method for combined radar–radiometer precipitation profiling. J. Appl. Meteorol. Climatol. 50, 433–448. https://doi.org/10.1175/2010JAMC2535.1.

Munchak, S.J., Skofronick-Jackson, G., 2013. Evaluation of precipitation detection over various surfaces from passive microwave imagers and sounders. Atmos. Res. 131, 81–94.

Munchak, S.J., Tokay, A., 2008. Retrieval of raindrop size distribution from simulated dual-frequency radar measurements. J. Appl. Meteorol. Climatol. 47, 223–239. https://doi.org/10.1175/2007JAMC1524.1.

Munchak, S., Meneghini, R., Grecu, M., Olson, W., 2016. A consistent treatment of microwave emissivity and radar backscatter for retrieval of precipitation over water surfaces. J. Atmos. Ocean. Technol. 33, 215–229. https://doi.org/10.1175/JTECH-D-15-0069.1.

Okamoto, K., Awaka, J., Nakamura, K., Ihara, T., Manabe, T., Kozu, T., 1988. A feasibility study of rain radar for the tropical rainfall measuring mission. J. Commun. Res. Lab. 35, 109–208.

Olson, W.S., Tian, L., Grecu, M., Kuo, K.S., Johnson, B.T., Heymsfield, A.J., Meneghini, R., 2016. The microwave radiative properties of falling snow derived from nonspherical ice particle models. part II: initial testing using radar radiometer and in situ observations. J. Appl. Meteorol. Climatol. 55 (3), 709–722.

Pazmany, A.L., McIntosh, R.E., Kelly, R.D., Vali, G., 1994. An airborne 95 GHz dual-polarized radar for cloud studies. IEEE Trans. Geosci. Remote Sens. 32 (4), 731–739.

Petty, G.W., Bennartz, R., 2016. Field-of-view characteristics and resolution matching for the global precipitation measurement (GPM) microwave imager (GMI). Atmos. Meas. Tech. https://doi.org/10.5194/amt-2016-275.

Petty, G., Huang, W., 2010. Microwave backscatter and extinction by soft ice spheres and complex snow aggregates. J. Atmos. Sci. 67, 769–787. https://doi.org/10.1175/2009JAS3146.1.

Petty, G., Huang, W., 2011. The modified gamma size distribution applied to inhomogeneous and nonspherical particles: key relationships and conversions. J. Atmos. Sci. 68, 1460–1473. https://doi.org/10.1175/2011JAS3645.1.

Prigent, C., Defer, E., Pardo, J.R., Pearl, C., Rossow, W.B., Pinty, J.-P., 2005. Relations of polarized scattering signatures observed by the TRMM microwave instrument with electrical processes in cloud systems. Geophys. Res. Lett. 32. L04810, https://doi.org/10.1029/2004GL022225.

Pruppacher, H.R., Klett, J.D., 1997. Microphysics of Clouds and Precipitation, second rev. and enl. ed. Kluwer Academic Publishers, Dordrecht. 954 pp.

Puri, S., Stephen, H., Ahmad, S., 2011. Relating TRMM precipitation radar land surface backscatter response to soil moisture in the southern United States. J. Hydrol. 402 (1), 115–125.

Rauber, R.M., Ellis, S.M., Vivekanandan, J., Stith, J., Lee, W., McFarquhar, G.M., Jewett, B.F., Janiszeski, A., 2017. Finescale structure of a snowstorm over the Northeastern United States: a first look at high-resolution HIAPER Cloud Radar Observations. Bull. Amer. Meteor. Soc. 98, 253–269. https://doi.org/10.1175/BAMS-D-15-00180.1.

Reinhart, B., Fuelberg, H., Blakeslee, R., Mach, D., Heymsfield, A., Bansemer, A., Durden, S., Tanelli, S., Heymsfield, G., Lambrigtsen, B., 2014. Understanding the relationships between lightning, cloud microphysics, and airborne radar-derived storm structure during hurricane Karl (2010). Mon. Weather Rev. 142, 590–605. https://doi.org/10.1175/MWR-D-13-00008.1.

Rodgers, C.D., 2000. Inverse Methods for Atmospheric Sounding: Theory and Practice. World Scientific, Singapore. 238 pp.

Ryu, Y., Smith, J., Baeck, M., Cunha, L., Bou-Zeid, E., Krajewski, W., 2016. The regional water cycle and heavy spring rainfall in iowa: observational and modeling analyses from the IFloodS campaign. J. Hydrometeor. 17, 2763–2784. https://doi.org/10.1175/JHM-D-15-0174.1.

Sadowy, G.A., Berkun, A.C., Chun, W., Im, E., Durden, S.L., 2003. Development of an advanced airborne precipitation radar. (Technical Feature). Microw. J. 46 (1), 84–93.

Seto, S., Iguchi, T., 2007. Rainfall-induced changes in actual surface backscattering cross sections and effects on rain estimates by spaceborne precipitation radar. J. Atmos. Ocean. Technol. 24, 1693–1709. https://doi.org/10.1175/JTECH2088.1.

Seto, S., Iguchi, T., 2015. Intercomparison of attenuation correction methods for the GPM dual-frequency precipitation radar. J. Atmos. Oceanic Technol. 32, 915–926. https://doi.org/10.1175/JTECH-D-14-00065.1.

Shepherd, A., Ivins, E.R., Geruo, A., Barletta, V.R., Bentley, M.J., Bettadpur, S., Horwath, M., 2012. A reconciled estimate of ice-sheet mass balance. Science 338 (6111), 1183–1189.

Shimizu, S., Tagawa, T., Iguchi, T., Hirose, M., 2009. Evaluation of the effects of the orbit boost of the TRMM satellite on PR rain estimates. J. Meteorol. Soc. Jpn. Ser. II 87, 83–92.

Simpson, J., Adler, R.F., North, G.R., 1988. A proposed tropical rainfall measuring mission (TRMM) satellite. Bull. Am. Meteorol. Soc. 69 (3), 278–295.

Stephen, H., Ahmad, S., Piechota, T.C., Tang, C., 2010. Relating surface backscatter response from TRMM precipitation radar to soil moisture: results over a semi-arid region. Hydrol. Earth Syst. Sci. 14 (2), 193–204.

Stephens, G.L., Vane, D.G., Boain, R.J., Mace, G.G., Sassen, K., Wang, Z., Miller, S.D., 2002. The CloudSat mission and the A-Train: a new dimension of space-based observations of clouds and precipitation. Bull. Am. Meteorol. Soc. 83 (12), 1771–1790.

Takahashi, N., Hanado, H., Nakamura, K., Kanemaru, K., Nakagawa, K., Iguchi, T., Yoshida, N., 2016. Overview of the end-of-mission observation experiments of precipitation radar onboard the tropical rainfall measuring mission satellite. IEEE Trans. Geosci. Remote Sens. 54 (6), 3450–3459.

Tanelli, S., Durden, S.L., Im, E., 2006. Simultaneous measurements of ku- and ka-band sea surface cross sections by an airborne Radar. IEEE Geosci. Remote Sens. Lett. 3 (3), 359–363. https://doi.org/10.1109/LGRS.2006.872929.

Tanelli, S., Sacco, G.F., Durden, S.L., Haddad, Z.S., 2012. Impact of non-uniform beam filling on spaceborne cloud and precipitation radar retrieval algorithms. SPIE Asian-Pacific Remote Sensing, International Society for Optics and Photonics, Kyoto, Japan (pp. 852308–852308).

Testud, J., Oury, S., Black, R., Amayenc, P., Dou, X., 2001. The concept of "Normalized" distribution to describe raindrop spectra: a tool for cloud physics and cloud remote sensing. J. Appl. Meteorol. 40, 1118–1140.

Tokay, A., Bashor, P.G., 2010. An experimental study of small-scale variability of raindrop size distribution. J. Appl. Meteorol. Climatol. 49 (11), 2348–2365.

Tokay, A., Short, D.A., 1996. Evidence from tropical raindrop spectra of the origin of rain from stratiform versus convective clouds. J. Appl. Meteorol. 35 (3), 355–371.

Ulbrich, C.W., 1983. Natural variations in the analytical form of the raindrop size distribution. J. Clim. Appl. Meteorol. 22, 1764–1775.

Ulbrich, C., Atlas, D., 1998. Rainfall microphysics and radar properties: analysis methods for drop size spectra. J. Appl. Meteorol. 37, 912–923.

Wang, J.R., Skofronick-Jackson, G.M., Schwaller, M.R., Johnson, C.M., Monosmith, W.B., Zhang, Z., 2013. Observations of storm signatures by the recently modified conical scanning millimeter-wave imaging radiometer. IEEE Trans. Geosci. Remote Sens. 51 (1), 411–424. https://doi.org/10.1109/TGRS.2012.2200690.

Williams, C., 2016. Reflectivity and liquid water content vertical decomposition diagrams to diagnose vertical evolution of raindrop size distributions. J. Atmos. Ocean. Technol. 33, 579–595. https://doi.org/10.1175/JTECH-D-15-0208.1.

Williams, C., Bringi, V., Carey, L., Chandrasekar, V., Gatlin, P., Haddad, Z., Meneghini, R., Joseph Munchak, S., Nesbitt, S., Petersen, W., Tanelli, S., Tokay, A., Wilson, A., Wolff, D., 2014. Describing the shape of raindrop size distributions using uncorrelated raindrop mass spectrum parameters. J. Appl. Meteorol. Climatol. 53, 1282–1296. https://doi.org/10.1175/JAMC-D-13-076.1.

Wiscombe, W., 1977. The delta-*M* method: rapid yet accurate radiative flux calculations for strongly asymmetric phase functions. J. Atmos. Sci. 34, 1408–1422.

Zhang, G., Vivekanandan, J., Brandes, E.A., Meneghini, R., Kozu, T., 2003. The shape–slope relation in observed gamma raindrop size distributions: statistical error or useful information? J. Atmos. Ocean. Technol. 20, 1106–1119.

Further Reading

Kojima, M., Coauthors, 2012. Dual-frequency precipitation radar (DPR) development on the global precipitation measurement (GPM) core observatory. In: Shimoda, H., et al. (Eds.), Earth Observing Missions and Sensors: Development, Implementation, and Characterization II. International Society for Optical Engineering (SPIE Proceedings, Vol. 8528), 85281A. https://doi.org/10.1117/12.976823.

CHAPTER

14

Status of High-Resolution Multisatellite Precipitation Products Across India

Satya Prakash, Ashis K. Mitra†, Rakesh M. Gairola‡, Hamid Norouzi*, Damodara S. Pai§*

*City University of New York, Brooklyn, NY, United States

†National Centre for Medium Range Weather Forecasting, Noida, India

‡Space Applications Centre, Ahmedabad, India

§India Meteorological Department, Pune, India

1 INTRODUCTION

Precipitation is one of the crucial atmospheric variables in the global hydrological cycle, and plays a key role in the Earth's water and energy budget. Accurate estimate of precipitation at various spatiotemporal scales is critical for a number of applications in hydrology, water resources and food security, hydroelectric power sectors, meteorology, and climatology. A reliable precipitation estimate is also important for the validation of numerical weather prediction model outputs for its integration in practical applications and further advancement in the models. Following the advancement in remote sensing techniques, more than a dozen satellite-based precipitation products are available to users (Sorooshian et al., 2011; Kucera et al., 2013). However, comprehensive error characteristics of each product are crucial for its integration in any specific application (Turk et al., 2008; Tian and Peters-Lidard, 2010; AghaKouchak et al., 2012; Shah and Mishra, 2016). India is a unique subcontinent to evaluate any global or quasiglobal satellite-based precipitation product due to its complex topography and highly variable monsoon rainfall (Fig. 1) spanning from June to September. Additionally, there is a fairly good network of rain gauge observations across the country (Rajeevan and Bhate, 2009; Pai et al., 2014).

Geostationary satellites carry visible and infrared sensors suitable for precipitation estimation, and provide consistent cloud top characteristics of the specified regions of the globe. Such observations are very important in the tropical regions to monitor the monsoon weather systems and extreme events. After the launch of Kalpana-1 satellite by the Indian Space Research Organisation (ISRO) in late 2002, three infrared-based rainfall products, based on distinct algorithms, were developed and are operational at the India Meteorological Department (IMD) and at the Meteorological and

Remote Sensing of Aerosols, Clouds, and Precipitation
https://doi.org/10.1016/B978-0-12-810437-8.00014-1

© 2018 Elsevier Inc. All rights reserved.

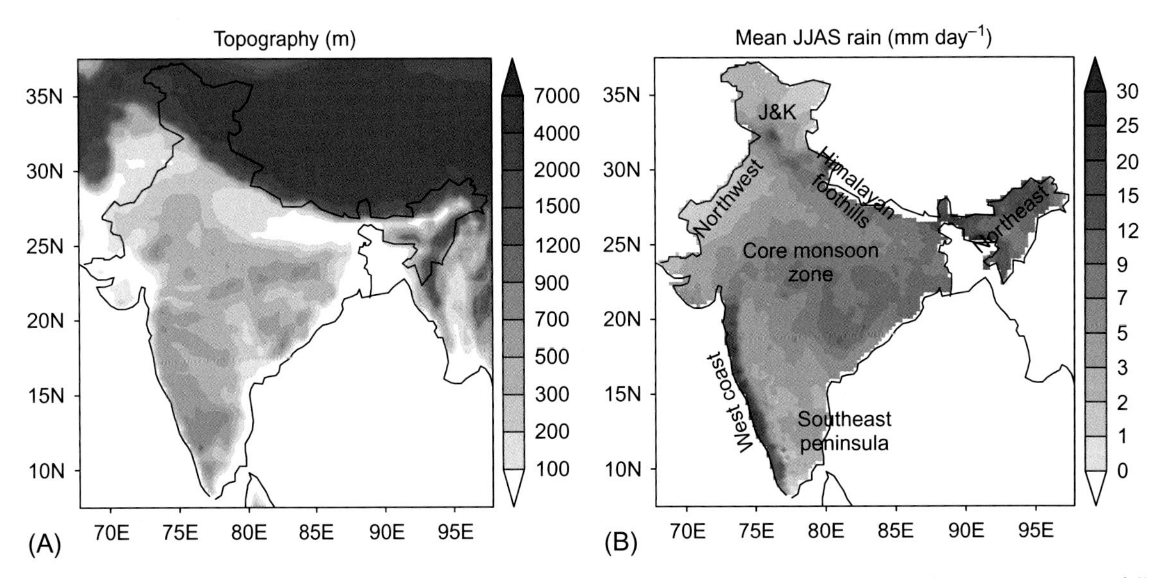

FIG. 1 (A) Spatial distribution of topography over India and surrounding regions, (B) mean southwest monsoon rainfall over India derived from IMD gauge-based observations for 1971–2014. Some important subregions within the study area are also indicated.

Oceanographic Satellite Data Archival Centre (MOSDAC), and are used for a number of applications (Prakash et al., 2010, 2011; Mahesh et al., 2014; Patel et al., 2015; Bushair et al., 2016). These three algorithms, after suitable fine-tunings and modifications, are also implemented by the INSAT-3D satellite (e.g., Fig. 2), launched in July 2013 (Gairola et al., 2014; Varma et al., 2015). But, these infrared-based precipitation estimates have considerable biases because they are based on the indirect relationship between the cloud top temperature and surface rainfall. Alternatively, microwave sensors onboard the low-Earth orbiting satellites provide more accurate estimates of precipitation due to direct interactions with hydrometeors. However, microwave-based estimates have rather coarser spatiotemporal resolution. Nevertheless, passive microwave retrievals of precipitation have deficiencies in light rainfall and snowfall detection (Behrangi et al., 2014). Hence, the synergism of infrared and microwave precipitation estimates, resulting in multisatellite precipitation estimates, is a viable option for more accurate precipitation at finer spatial and temporal scales (Sorooshian et al., 2000; Xie et al., 2002; Joyce et al., 2004; Huffman et al., 2010). The multisatellite precipitation estimates benefit from the relative merits of both types of space borne sensors.

With the launch of the Tropical Rainfall Measuring Mission (TRMM) satellite with the first space borne precipitation radar in late 1997, a number of high-resolution multisatellite precipitation products (MSPPs) were developed for near real-time and research applications. These MSPPs combine precipitation estimates from available space borne infrared and microwave sensors. The TRMM Multisatellite Precipitation Analysis (TMPA; Huffman et al., 2010), available every 3-h at 0.25° latitude/longitude resolution, is one of most widely used high-resolution MSPPs among them. This MSPP shows relatively lower bias and rather smaller errors in the tropics as compared to other contemporary MSPPs (Prakash et al., 2014; Liu, 2015; Maggioni et al., 2016). The TRMM satellite, a precursor for the Global Precipitation Measurement (GPM), was decommissioned in June 2015 and provided more than 17 years of unprecedented tropical and subtropical precipitation

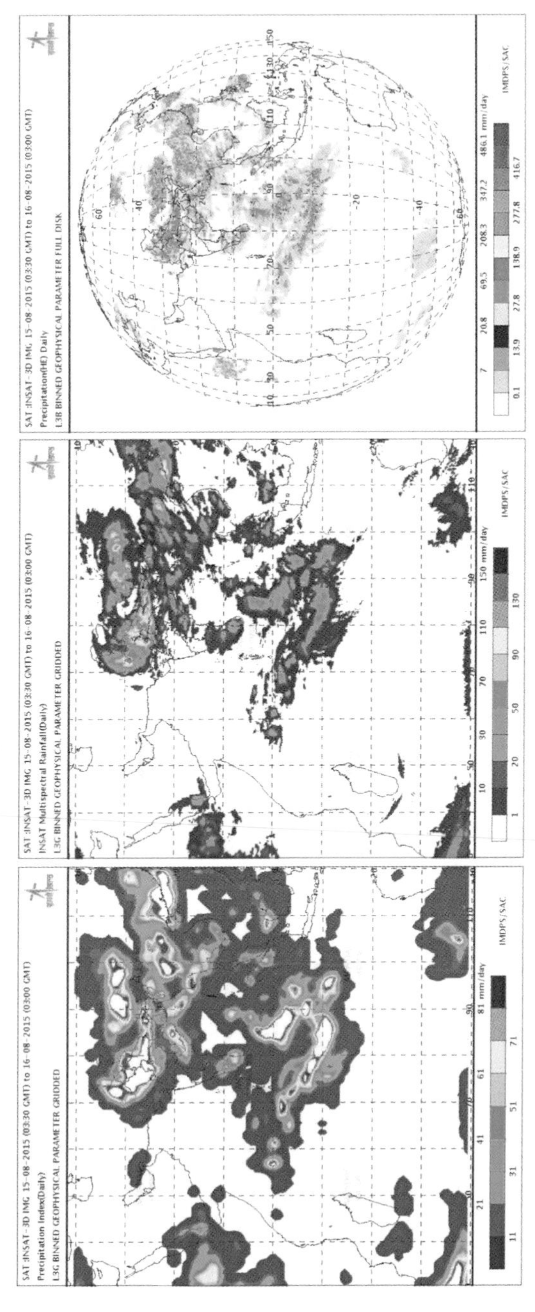

FIG. 2 An example of daily precipitation derived from three distinct operational algorithms using the INSAT-3D satellite. *Courtesy: http://www.mosdac.gov.in.*

estimates (Houze et al., 2015). On February 28, 2014, the National Aeronautics and Space Administration (NASA) and the Japan Aerospace Exploration Agency (JAXA) jointly deployed the GPM Core Observatory as a successor to the TRMM satellite (Hou et al., 2014). The GPM-based MSPP, known as Integrated Multi-satellitE Retrievals for GPM (IMERG), was released thereafter (Huffman et al., 2014). It is available at finer spatial (0.1° latitude/longitude) and temporal (half-hourly) resolutions as compared to the TMPA. The IMERG product based on Day-1 algorithm was recently evaluated at global and regional scales for a shorter record of time (Guo et al., 2016; Liu, 2016; Prakash et al., 2016a,b; Tang et al., 2016), which showed an overall better performance than TMPA estimates.

This chapter highlights the recent evaluations of different high-resolution MSPPs (both TRMM- and GPM-based) over India, especially for the southwest monsoon season that yields about two-thirds of the annual rainfall across the country. The potential of the combined use of local rain gauges and high-resolution MSPPs for near real-time applications is also discussed.

2 A REVIEW OF RECENT EVALUATIONS OF HIGH-RESOLUTION MSPPs ACROSS INDIA

For a wider applicability of the available high-resolution MSPPs, several studies were made to evaluate or compare these estimates at global and regional scales. It should also be noted that MSPPs are used to undergo intermittent revisions and consequently, newer versions of the products get released. In this section, studies related to the evaluation of recent versions of TRMM-era high-resolution MSPPs over India for the southwest monsoon period are highlighted. Several high-resolution MSPPs such as TMPA-3B42, Climate Prediction Center (CPC) Morphing technique (CMORPH; Joyce et al., 2004), Precipitation Estimation from Remotely Sensed Information using Artificial Neural Networks (PERSIANN; Sorooshian et al., 2000), Naval Research Laboratory (NRL)-blended (Turk and Miller, 2005), Global Satellite Mapping of Precipitation (GSMaP) moving vector with Kalman filter method (GSMaP_MVK; Kubota et al., 2009), CPC-Rainfall Estimation Algorithm (CPC-RFE; Xie et al., 2002), etc. based on partially different data sets and distinct algorithms, were comprehensively evaluated against gauge-based observations over the country (Prakash et al., 2014, 2015a,b,c,d, 2016c; Bharti and Singh, 2015; Rana et al., 2015; Sunilkumar et al., 2015; Shah and Mishra, 2016). Among these MSPPs, the TMPA-3B42 research product and CPC-RFE use gauge analysis for bias correction over the land. These studies recognized that even though all the MSPPs are capable of identifying the broad-scale monsoon features, they have biases and errors. In general, the TMPA-3B42 version 7 research product is superior to others over the country due to a relatively smaller bias and less error. Additionally, the version 7 of the TMPA-3B42 research product showed appreciably better performance than its predecessor, version 6 (Prakash et al., 2015d; Rana et al., 2015). The research product of TMPA-3B42 V7 is also better than near real-time product across the country due to robust calibration and inclusion of gauge analysis for bias correction in the research product (Prakash et al., 2014, 2016c). Prakash et al. (2015a) highlighted the latest assessment studies of the TMPA-3B42 V7 product over India and the surrounding oceanic regions. Larger biases in MSPPs were seen over the orographic regions of the Western Ghats and the Himalayan foothills due to varied topography along with coupled atmospheric-oceanic monsoon interactions (Houze, 2012; Prakash et al., 2014; Bharti and Singh, 2015). The underestimation of intermittent shallow precipitation along the west coast by all the MSPPs was also evident (Sunilkumar et al., 2015). The seasonal variations in the error characteristics of TMPA-3B42 products across the country were also investigated (Prakash et al., 2015b). Larger systematic error, primarily due to bias, in both near real-time and research TMPA products were found during the premonsoon season (March to May).

After the launch of the GPM Core Observatory, two high-resolution global MSPPs—IMERG (Huffman et al., 2014) and GSMaP version 6 (Ushio et al., 2013) were released. These two GPM-based MSPPs products were developed using different algorithms. Both products use gauge analysis from distinct sources for bias correction in their respective research version products. It should also be noted that GSMaP MSPPs are available for both TRMM (e.g., GSMaP_MVK) and GPM era (e.g., GSMaP version 6). The capability of the IMERG product in heavy rainfall detection was recently assessed against gauge-based observations over India (Prakash et al., 2016b). A preliminary analysis for the southwest monsoon season of 2014 showed a notable improvement in IMERG over TMPA-3B42 for heavy rainfall detection. An assessment of both the GPM-based MSPPs (IMERG and GSMaP version 6) against gauge-based observations for the southwest monsoon season of 2014 showed that the MSPPs have not only improved in resolutions, but also biases were reduced as compared to TMPA-3B42 estimates. The missed and false precipitation biases were also noticeably reduced in the GPM-based MSPPs (Prakash et al., 2016a). Even though the errors in the GPM-based MSPPs were reduced overall, the MSPPs still have rather larger uncertainty over the orographic regions.

3 EVALUATION OF FIVE HIGH-RESOLUTION MSPPs FOR THE INDIAN MONSOON 2014

In this section, the latest versions of three widely used TRMM-era MSPPs (TMPA-3B42 version 7, CMORPH version 1.0, and PERSIANN) and two GPM-based MSPPs (IMERG and GSMaP version 6) are evaluated against gauge-based observations over India. The assessment is done for the southwest monsoon season of 2014 at a daily scale. Since the gauge-based data set is available at daily scale ending at 0300 Universal Time Constant (UTC) and at 0.25° latitude/longitude resolution, all the MSPPs were re-projected at the same spatial resolution and accumulated to daily scale ending at 0300 UTC. Fig. 3 presents the mean daily rainfall from all these precipitation products over the Northern India for September 3–6, 2014. During this period, heavy rainfall occurred over Jammu & Kashmir that led to catastrophic flooding. It can be seen that rainfall features from IMERG is in better agreement with gauge observations. Heavy rainfall is also captured well from the IMERG product, which is underestimated in other MSPPs. The underestimation of rainfall is larger in PERSIANN. This case study convincingly shows the improvement in IMERG over TMPA in precipitation estimation.

Two subregions within India were chosen for further analysis of the daily monsoon rainfall in 2014. One of them is Core Monsoon Zone (CMZ) and another is the west coast (indicated in Fig. 1B). Both the subregions get a higher mean rainfall during the monsoon. The rainfall over CMZ is crucial for active/break spells of the monsoon over India (Rajeevan et al., 2010), which is vital for agricultural practices. The west coast gets heavy rainfall during the monsoon due to low-level monsoon jet and varied topography. Satellite-based rainfall estimates have larger errors over this region. It should also be noted that both subregions have fairly good networks of gauge observations (Pai et al., 2014). Fig. 4 illustrates the time-series of daily monsoon rainfall for both subregions within India. All MSPPs are able to capture the daily variations in monsoon rainfall adequately over the CMZ, whereas there are considerable differences among them over the west coast. CMORPH and PERSIANN highly underestimate monsoon rainfall over the west coast.

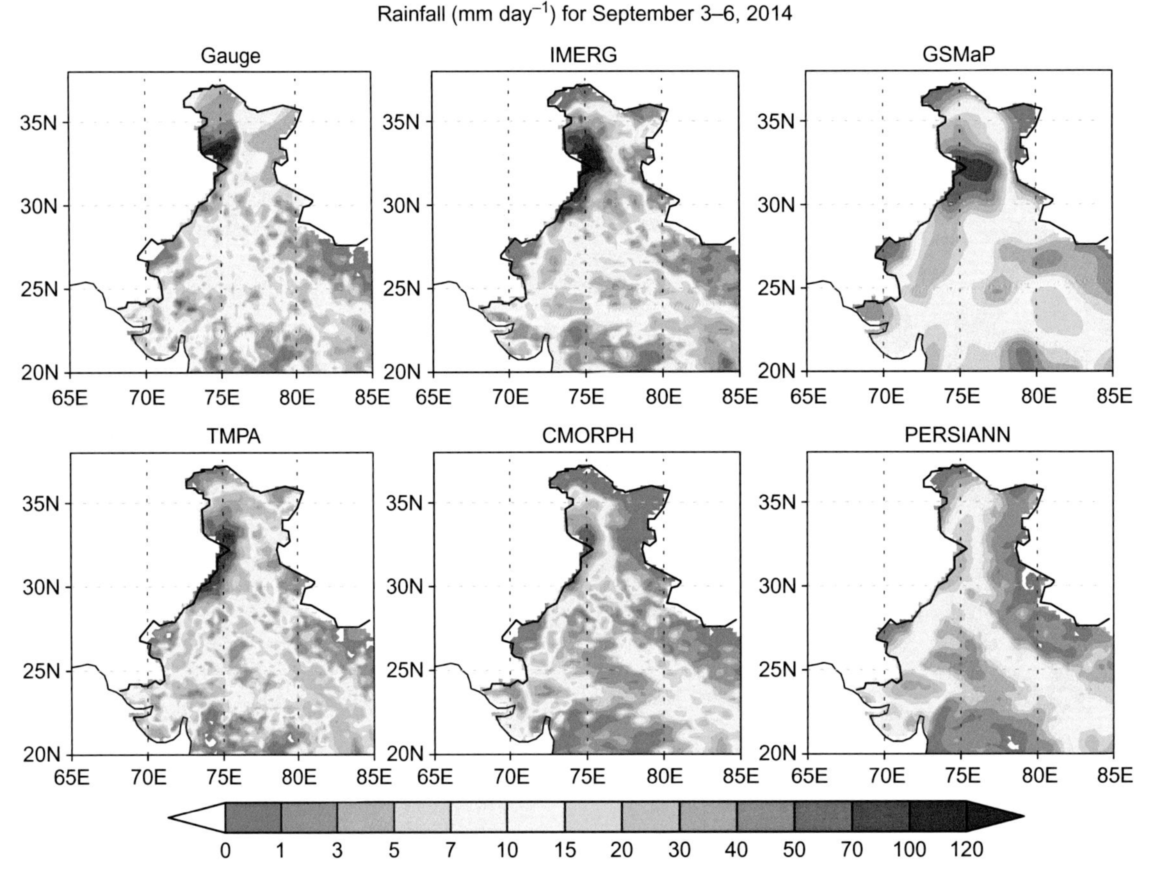

FIG. 3 Mean daily rainfall over the northern India for September 3–6, 2014 from gauge-based observations and five multi-satellite products.

The use of gauge analysis in IMERG, GSMaP, and TMPA products might be one of the reasons of their better performance over the CMORPH and PERSIANN over the orographic regions of the west coast.

Fig. 5 shows the box plots of daily rainfall averaged over both subregions. IMERG shows better performance than other MSPPs over the CMZ. CMORPH and PERSIANN show rather lower mean seasonal monsoon rainfall over both subregions. IMERG shows larger maximum precipitation values than gauge observations over the west coast. This may lead to eventual overestimation of heavy rainfall over this orographic region. CMORPH and PERSIANN have the largest biases over this region. Over the west coast, GSMaP performs marginally better than IMERG. Fig. 6 shows the Taylor diagram (Taylor, 2001), representing correlation, root-mean-square error and standard deviation of different precipitation estimates. IMERG shows marginal improvement over TMPA over both the regions, whereas GSMaP shows higher correlation and lower error over the west coast. These results again support that GPM-based MSPPs are improved over the TRMM-era MSPPs. However, error is still larger over the orographic regions. The cumulative distribution

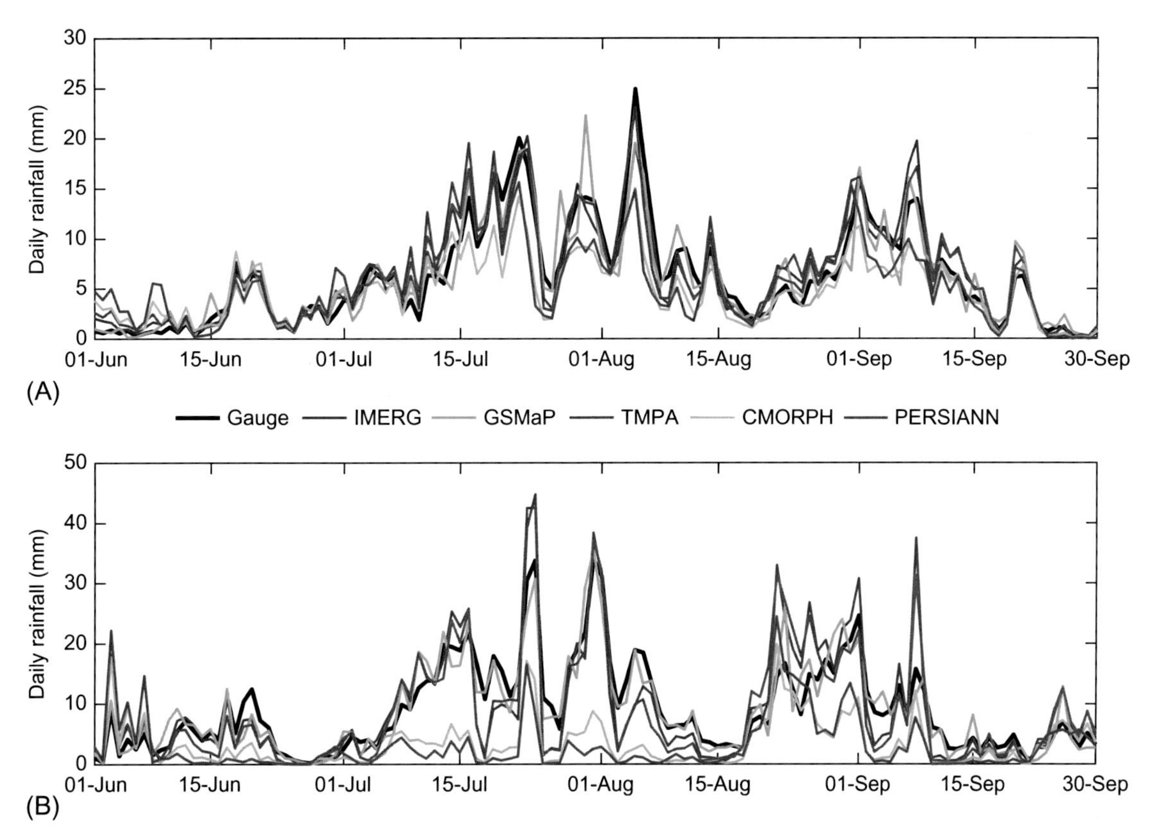

FIG. 4 Time-series of daily rainfall averaged over (A) core monsoon zone and (B) west coast for the southwest monsoon season of 2014.

function (Fig. 7) shows that TMPA substantially underestimates light rainfall over both subregions, which is notablely improved in GPM-based MSPPs. It should be noted that the results presented here are based on only one monsoon season. An extensive evaluation over a longer time-period is essential for a robust conclusion.

4 OPERATIONAL MERGED SATELLITE-GAUGE RAINFALL PRODUCT IN INDIA

As discussed in the earlier sections, MSPPs benefit from the relative merits of infrared and microwave satellite-based sensors, but have biases. The use of local rain gauge observations would essentially reduce the bias of the MSSPs and can be used for several hydrometeorological applications (Krishnamurti et al., 2009; Mitra et al., 2009; Gairola et al., 2015). Mitra et al. (2009) developed a near real-time merged TMPA and gauge product for the Indian monsoon region using the successive correction method. This daily gridded rainfall at 1° latitude/longitude resolution was found to be superior than other available data sets and represents broad-scale monsoon features more realistically (Mitra et al., 2013). Further, this algorithm was applied with a larger number of gauge observations

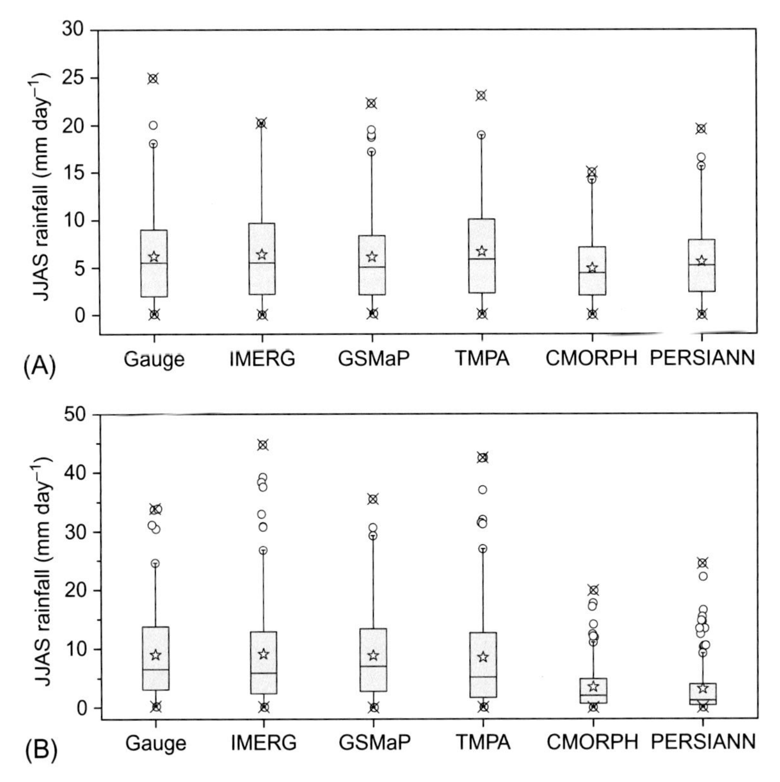

FIG. 5 Box plots of daily rainfall averaged over (A) core monsoon zone and (B) west coast for the southwest monsoon season of 2014. The cross symbols indicate minimum and maximum daily rainfall, the open circles represent outliers, and the star symbols show the mean rainfall.

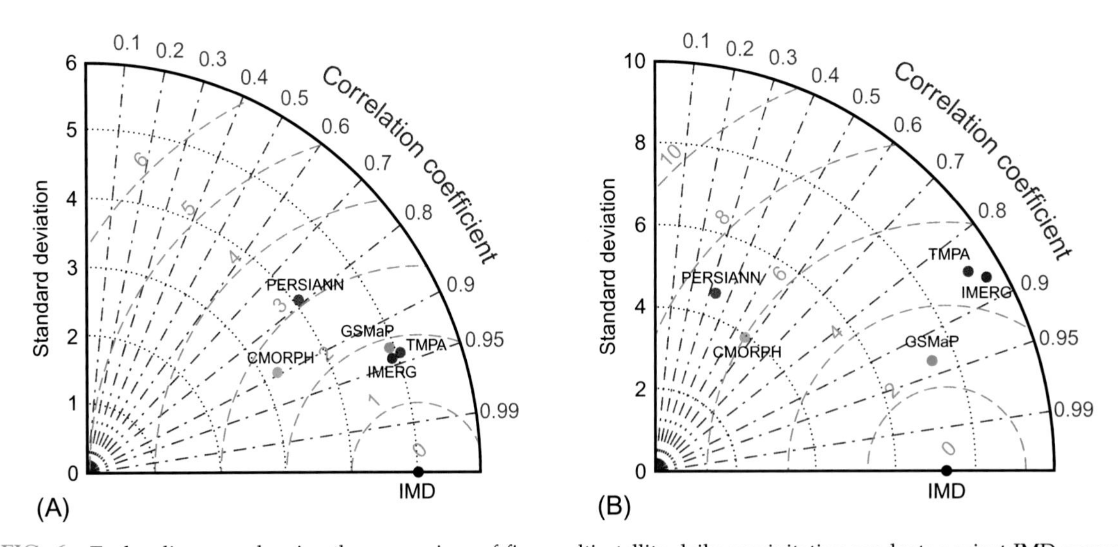

FIG. 6 Taylor diagrams showing the comparison of five multisatellite daily precipitation products against IMD gauge-based observations averaged over (A) core monsoon zone and (B) west coast for the southwest monsoon season of 2014.

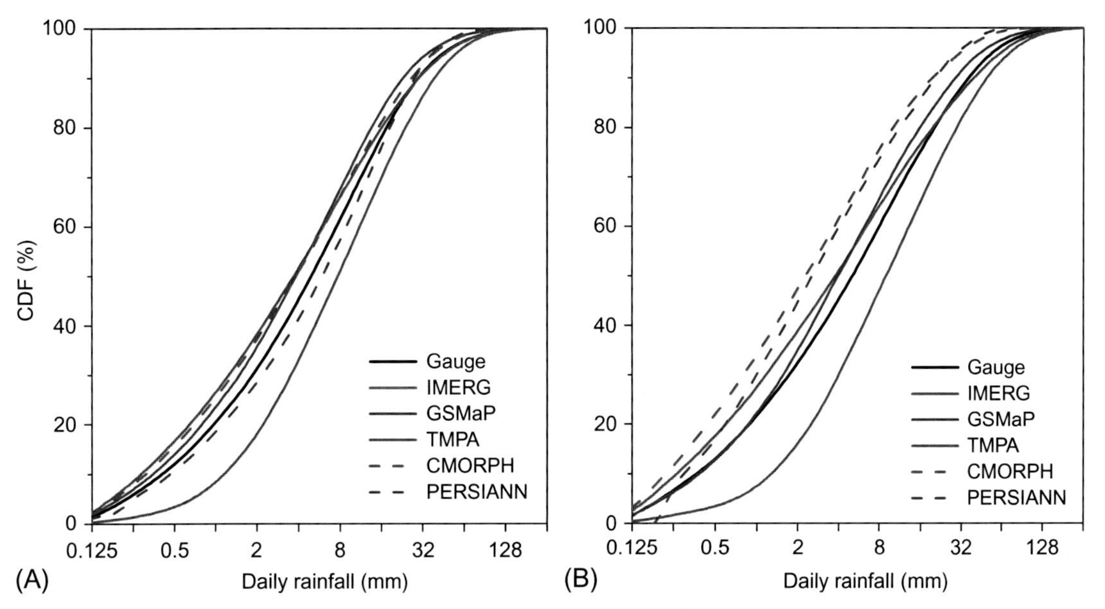

FIG. 7 Cumulative distribution functions of daily precipitation products over (A) core monsoon zone and (B) west coast for the southwest monsoon season of 2014.

and consequently, daily rainfall estimates at 0.5° latitude/longitude resolution was produced in near real-time at operational basis (Mitra et al., 2014). This data set was widely used for numerical model outputs verification and other hydrological applications. One example of merged TMPA and gauge daily rainfall estimate for July 23–29, 2014 is shown in Fig. 8, which was operational at IMD Pune and at the National Centre for Medium Range Weather Forecasting (NCMRWF) until late 2014. The northwestward propagation of monsoon rainfall is seen very well in this illustration. The weekly mean rainfall is also shown for this period.

After the release of the IMERG near real-time precipitation product, an attempt was made to combine available gauge data with it for finer resolution merged product. Consequently, a new merged IMERG and gauge daily rainfall product at 0.25° latitude/longitude resolution was made operational at NCMRWF in late 2015, which is available with a latency period of 1 day. There are more than 1500 gauge observations available in near real-time. This new merged rainfall product is better than its predecessor in terms of improved quality and finer spatial resolution. This rainfall product is freely available to users. Fig. 9 shows the daily merged IMERG and gauge rainfall estimate for May 16–22, 2016, just before the onset of the southwest monsoon over India. During this period, a cyclonic storm "Roanu" developed in the southeastern Bay of Bengal and moved along the eastern coast of India (close to coast) through Bangladesh. The associated daily rainfall clearly shows heavy rainfall over these regions. Thus, this merged rainfall product is supposed to be optimal for near real-time applications and to monitor the progress of monsoon.

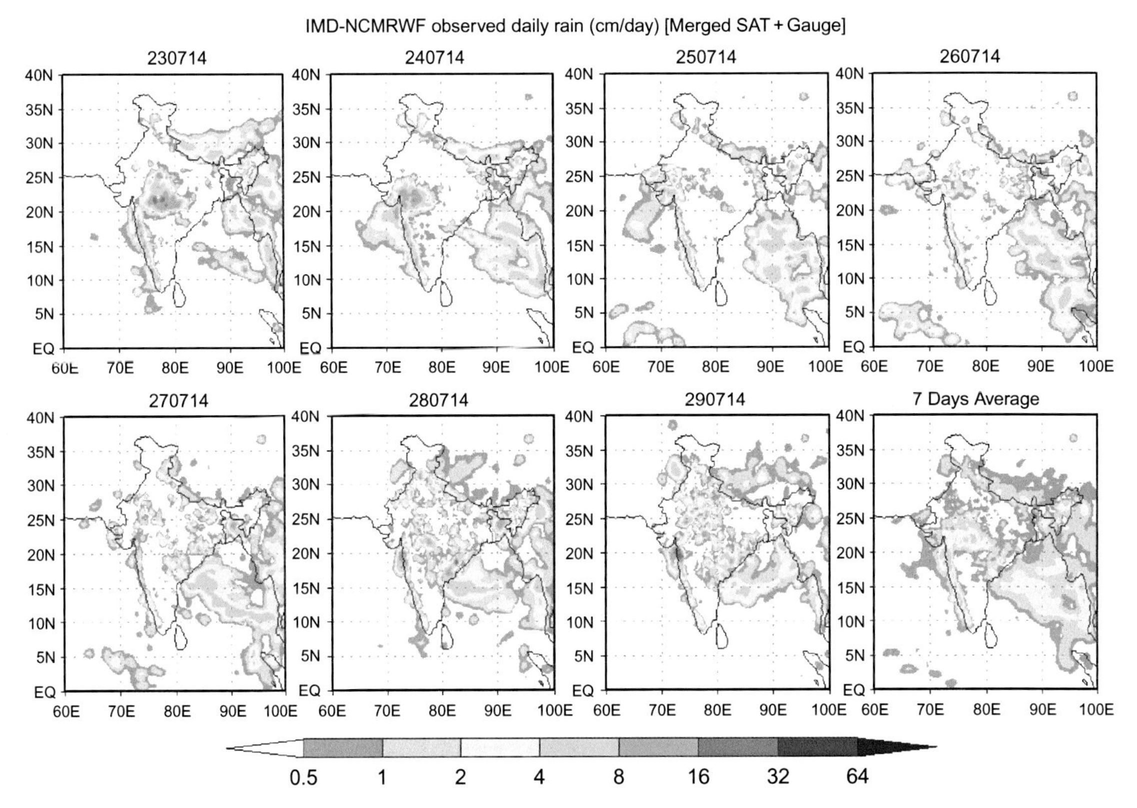

FIG. 8 Daily and weekly rainfall from operational near real-time merged satellite-gauge (TMPA and gauge) over the Indian region for July 23–29, 2014.

5 SUMMARY AND CONCLUSION

Reliable estimate of precipitation is crucial for several applications ranging from hydrometeorology to climatology. After the launch of the TRMM satellite, precipitation estimation techniques got a rapid boost and several high-resolution MSPPs were developed to study the tropical and subtropical precipitation characteristics. However, there are uncertainties in the MSPPs due to high spatiotemporal variability of precipitation and some limitations of retrieval techniques. A comprehensive error estimate is essential to use any MSPP in any specific application and for further advancement in retrieval algorithms. India is an ideal test-bed to evaluate any MSPP due to an appreciably good network of rain gauges. Additionally, India is a subcontinent having varied topographic structure as well as large variability of the monsoon rainfall.

In recent years, several studies were performed to characterize the errors in the different TRMM-era high-resolution global or quasiglobal MSPPs over the Indian subcontinent. In this chapter, these recent evaluations of high-resolution MSPPs over India for the southwest monsoon period were reviewed. In general, the TMPA-3B42 product is proven to be superior to the other TRMM-era MSPPs. Two finer

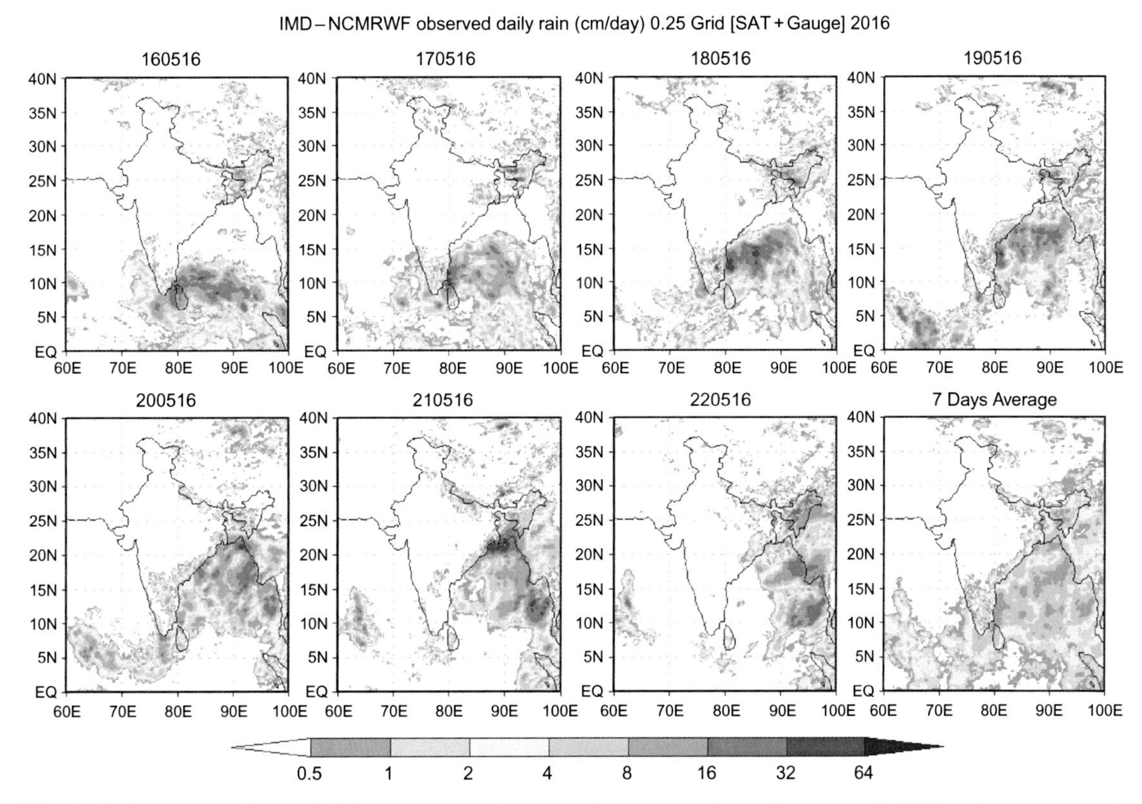

FIG. 9 Daily and weekly rainfall from operational near real-time merged satellite-gauge (IMERG and gauge) over the Indian region for May 16–22, 2016.

resolution MSPPs, IMERG and GSMaP version 6, were released after the launch of the GPM Core Observatory. Both GPM-based MSPPs were also compared to TMPA-3B42 and gauge-based observations across India. An analysis was done for the two major subregions within India for the monsoon season of 2014. Two GPM-based and three TRMM-era MSPPs were evaluated which showed that GPM-based estimates improved over the TRMM-based MSPPs. Nonetheless, the GPM-based MSPPs still have larger errors over the orographic regions. A more comprehensive evaluation of GPM-based multisatellite precipitation estimates for longer periods is further required for their widest usage and applications in various sectors. Moreover, the MSPPs used to undergo some revisions and new versions of products get released eventually. Hence, continuous evaluation of the updated MSPPs is essential. The evaluation of these MSPPs at their native resolutions using station data is also very important for the assessment of high-resolution regional models and hydrological applications.

The use of local rain gauges with high-resolution MSPPs provides optimal estimate of precipitation for near real-time applications and model verification. Such an operational merged satellite-gauge precipitation estimate exists in India for the monsoon region, which was recently upgraded with the IMERG estimate. A brief description of this merged precipitation data set with some examples was also presented. Furthermore, the additional use of automatic rain gauges and weather radars would essentially benefit the merged rainfall product over India.

Acknowledgments

The respective sources of precipitation products and images used in this study are thankfully acknowledged. The authors would also like to express appreciation to Drs. G. J. Huffman (NASA), T. N. Krishnamurti (FSU), A. AghaKouchak (UCI), and Z. Liu (GMU) for helpful discussions.

References

AghaKouchak, A., Mehran, A., Norouzi, H., Behrangi, A., 2012. Systematic and random error components in satellite precipitation data sets. Geophys. Res. Lett. 39, L09406. https://doi.org/10.1029/2012GL051592.

Behrangi, A., Tian, Y., Lambrigtsen, B.H., Stephens, G.L., 2014. What does CloudSat reveal about global land precipitation detection by other spaceborne sensors? Water Resour. Res. 50, 4893–4905. https://doi.org/10.1002/2013WR014566.

Bharti, V., Singh, C., 2015. Evaluation of error in TRMM 3B42V7 precipitation estimates over the Himalayan region. J. Geophys. Res. Atmos. 120, 12458–12473. https://doi.org/10.1002/2015JD023779.

Bushair, M.T., Prakash, S., Patel, S., Gairola, R.M., 2016. Assessment of Kalpana-1 rainfall product over Indian meteorological sub-divisions during the summer monsoon season. J. Indian Soc. Remote Sens. 44 (1), 67–76. https://doi.org/10.1007/s12524-015-0465-1.

Gairola, R.M., Prakash, S., Bushair, M.T., Pal, P.K., 2014. Rainfall estimation from Kalpana-1 satellite data over Indian land and oceanic regions. Curr. Sci. 107 (8), 1275–1282.

Gairola, R.M., Prakash, S., Pal, P.K., 2015. Improved rainfall estimation over the Indian monsoon region by synergistic use of Kalpana-1 and rain gauge data. Atmosfera 28 (1), 51–61.

Guo, H., Chen, S., Bao, A., Behrangi, A., Hong, Y., Ndayisaba, F., Hu, J., Stepanian, P.M., 2016. Early assessment of integrated multi-satellite retrievals for global precipitation measurement over China. Atmos. Res. 176–177, 121–133. https://doi.org/10.1016/j.atmosres.2016.02.020.

Hou, A.Y., Kakar, R.K., Neeck, S., Azarbarzin, A.A., Kummerow, C.D., Kojima, M., Oki, R., Nakamura, K., Iguchi, T., 2014. The Global Precipitation Measurement mission. Bull. Am. Meteorol. Soc. 95 (5), 701–722. https://doi.org/10.1175/BAMS-D-13-00164.1.

Houze Jr., R.A., 2012. Orographic effects on precipitating clouds. Rev. Geophys. 50, RG1001. https://doi.org/10.1029/2011RG000365.

Houze Jr., R.A., Rasmussen, K.L., Zuluaga, M.D., Brodzik, S.R., 2015. The variable nature of convection in the tropics and subtropics: a legacy of 16 years of the Tropical Rainfall Measuring Mission satellite. Rev. Geophys. 53, 994–1021. https://doi.org/10.1002/2015RG000488.

Huffman, G.J., Adler, R.F., Bolvin, D.T., Nelkin, E.J., 2010. The TRMM multi-satellite precipitation analysis (TMPA). In: Hossain, F., Gebremichael, M. (Eds.), Satellite Applications for Surface Hydrology. Springer, pp. 3–22. https://doi.org/10.1007/978-90-481-2915-7_1.

Huffman, G.J., Bolvin, D. T., Braithwaite, D., Hsu, K., Joyce, R., Xic, P., 2014. NASA Global Precipitation Measurement (GPM) Integrated Multi-satellite Retrievals for GPM (IMERG), Algorithm Theoretical Basis Document (ATBD), Version 4.4, NASA, 30 p. http://pmm.nasa.gov/sites/default/files/document_files/IMERG_ATBD_V4.4.pdf.

Joyce, R.J., Janowiak, J.E., Arkin, P.A., Xie, P., 2004. CMORPH: a method that produces global precipitation estimates from passive microwave and infrared data at high spatial and temporal resolution. J. Hydrometeorol. 5, 487–503.

Krishnamurti, T.N., Mishra, A.K., Simon, A., Yatagai, A., 2009. Use of a dense rain-gauge network over India for improving blended TRMM products and downscaled weather models. J. Meteorol. Soc. Jpn. 87A, 393–412. https://doi.org/10.2151/jmsj.87A.393.

Kubota, T., Ushio, T., Shige, S., Kida, S., Kachi, M., Okamoto, K., 2009. Verification of high resolution satellite-based rainfall estimates around Japan using gauge-calibrated ground radar dataset. J. Meteorol. Soc. Jpn. 87A, 203–222. https://doi.org/10.2151/jmsj.87A.203.

Kucera, P.A., Ebert, E.E., Turk, F.J., Levizzani, V., Kirschbaum, D., Tapiador, F.J., Loew, A., Borsche, M., 2013. Precipitation from space: advancing Earth system science. Bull. Am. Meteorol. Soc. 94 (3), 365–375. https://doi.org/10.1175/BAMS-D-11-00171.1.

Liu, Z., 2015. Comparison of versions 6 and 7 3-hourly TRMM multi-satellite precipitation analysis (TMPA) research products. Atmos. Res. 163, 91–101. https://doi.org/10.1016/j.atmosres.2014.12.015.

Liu, Z., 2016. Comparison of Integrated Multisatellite Retrievals for GPM (IMERG) and TRMM Multisatellite Precipitation Analysis (TMPA) monthly precipitation products: initial results. J. Hydrometeorol. 17, 777–790. https://doi.org/10.1175/JHM-D-15-0068-1.

Maggioni, V., Meyers, P.C., Robinson, M.D., 2016. A review of merged high-resolution satellite precipitation product accuracy during the Tropical Rainfall Measuring Mission (TRMM) era. J. Hydrometeorol. 17, 1101–1117. https://doi.org/10.1175/JHM-D-15-0190-1.

Mahesh, C., Prakash, S., Gairola, R.M., Shah, S., Pal, P.K., 2014. Meteorological sub-divisional scale rainfall monitoring using Kalpana-1 VHRR measurements. Geogr. Res. 52 (3), 328–336. https://doi.org/10.1111/1745-5871.12068.

Mitra, A.K., Bohra, A.K., Rajeevan, M.N., Krishnamurti, T.N., 2009. Daily Indian precipitation analyses formed from a merge of rain-gauge with TRMM TMPA satellite derived rainfall estimates. J. Meteorol. Soc. Jpn. 87A, 265–279.

Mitra, A.K., Momin, I.M., Rajagopal, E.N., Basu, S., Rajeevan, M.N., Krishnamurti, T.N., 2013. Gridded daily Indian monsoon rainfall for 14 seasons: merged TRMM and IMD gauge analyzed values. J. Earth Sys. Sci. 122 (5), 1173–1182.

Mitra, A.K., Prakash, S., Momin, I.M., Pai, D.S., Srivastava, A.K., 2014. Daily merged satellite gauge real-time rainfall dataset for Indian region. Vayumandal 40, 33–43.

Pai, D.S., Sridhar, L., Rajeevan, M., Sreejith, O.P., Satbhai, N.S., Mukhopadhyay, B., 2014. Development of a new high spatial resolution ($0.25° \times 0.25°$) long period (1901–2010) daily gridded rainfall data set over India and its comparison with existing data sets over the region. Mausam 65 (1), 1–18.

Patel, S., Prakash, S., Bhatt, B., 2015. An assessment of Kalpana-1 rainfall product for drought monitoring over India at meteorological sub-division scale. Water Int. 40 (4), 689–702. https://doi.org/10.1080/02508060.2015.1072784.

Prakash, S., Mahesh, C., Gairola, R.M., Pal, P.K., 2010. Estimation of Indian summer monsoon rainfall using Kalpana-1 VHRR data and its validation using rain gauge and GPCP data. Meteorol. Atmos. Phys. 110 (1-2), 45–57. https://doi.org/10.1007/s00703-010-0106-8.

Prakash, S., Mahesh, C., Gairola, R.M., 2011. Large-scale precipitation estimation using Kalpana-1 IR measurements and its validation using GPCP and GPCC data. Theor. Appl. Climatol. 106 (3-4), 283–293. https://doi.org/10.1007/s00704-011-0435-7.

Prakash, S., Sathiyamoorthy, V., Mahesh, C., Gairola, R.M., 2014. An evaluation of high-resolution multisatellite rainfall products over the Indian monsoon region. Int. J. Remote Sens. 35 (9), 3018–3035. https://doi.org/10.1080/01431161.2014.894661.

Prakash, S., Mitra, A.K., Momin, I.M., Gairola, R.M., Pai, D.S., Rajagopal, E.N., Basu, S., 2015a. A review of recent evaluations of TRMM Multisatellite Precipitation Analysis (TMPA) research products against ground-based observations over Indian land and oceanic regions. Mausam 66 (3), 355–366.

Prakash, S., Mitra, A.K., AghaKouchak, A., Pai, D.S., 2015b. Error characterization of TRMM Multisatellite Precipitation Analysis (TMPA-3B42) products over India for different seasons. J. Hydrol. 529, 1302–1312. https://doi.org/10.1016/j.jhydrol.2015.08.062.

Prakash, S., Mitra, A.K., Pai, D.S., 2015c. Comparing two high-resolution gauge-adjusted multisatellite rainfall products over India for the southwest monsoon period. Meteorol. Appl. 22 (3), 679–688. https://doi.org/10.1002/met.1502.

Prakash, S., Mitra, A.K., Momin, I.M., Pai, D.S., Rajagopal, E.N., Basu, S., 2015d. Comparison of TMPA-3B42 versions 6 and 7 precipitation products with gauge-based data over India for the southwest monsoon period. J. Hydrometeorol. 16 (1), 346–362. https://doi.org/10.1175/JHM-D-14-0024.1.

Prakash, S., Mitra, A.K., AghaKouchak, A., Liu, Z., Norouzi, H., Pai, D.S., 2016a. A preliminary assessment of GPM-based multi-satellite precipitation estimates over a monsoon dominated region. J. Hydrol. https://doi.org/10.1016/j.jhydrol.2016.01.029.

Prakash, S., Mitra, A.K., Pai, D.S., AghaKouchak, A., 2016b. From TRMM to GPM: how well can heavy rainfall be detected from space? Adv. Water Res. 88, 1–7. https://doi.org/10.1016/j.advwatres.2015.11.008.

Prakash, S., Mitra, A.K., Rajagopal, E.N., Pai, D.S., 2016c. Assessment of TRMM-based TMPA-3B42 and GSMaP precipitation products over India for the peak southwest monsoon season. Int. J. Climatol. 36 (4), 1614–1631. https://doi.org/10.1002/joc.4446.

Rajeevan, M., Bhate, J., 2009. A high resolution daily gridded rainfall dataset (1971–2005) for mesoscale meteorological studies. Curr. Sci. 96 (4), 558–562.

Rajeevan, M., Gadgil, S., Bhate, J., 2010. Active and break spells of the Indian summer monsoon. J. Earth Sys. Sci. 119, 229–247.

Rana, S., McGregor, J., Renwick, J., 2015. Precipitation seasonality over the Indian subcontinent: an evaluation of gauge, reanalyses, and satellite retrievals. J. Hydrometeorol. 16 (2), 631–651. https://doi.org/10.1175/JHM-D-14-0106.1.

Shah, H.L., Mishra, V., 2016. Uncertainty and bias in satellite-based precipitation estimates over India subcontinental basins: implications for real-time streamflow simulation and flood prediction. J. Hydrometeorol. 17 (2), 615–636. https://doi.org/10.1175/JHM-D-15-0115.1.

Sorooshian, S., Hsu, K.-L., Gao, X., Gupta, H.V., Imam, B., Braithwaite, D., 2000. Evolution of the PERSIANN system satellite-based estimates of tropical rainfall. Bull. Am. Meteorol. Soc. 81, 2035–2046.

Sorooshian, S., AghaKouchak, A., Arkin, P., Eylander, J., Foufoula-Georgiou, E., Harmon, R., Hendrickx, J.M.H., Imam, B., Kuligowski, R., Skahill, B., Skofronick-Jackson, G., 2011. Advanced concepts on remote sensing of precipitation at multiple scales. Bull. Am. Meteorol. Soc. 92, 1353–1357. https://doi.org/10.1175/2011BAMS3158.1.

Sunilkumar, K., Rao, T.N., Saikranthi, K., Rao, M.P., 2015. Comprehensive evaluation of multisatellite precipitation estimates over India using gridded rainfall data. J. Geophys. Res. Atmos. 120, 8987–9005. https://doi.org/10.1002/2015JD023437.

Tang, G., Ma, Y., Long, D., Zhong, L., Hong, Y., 2016. Evaluation of GPM Day-1 IMERG and TMPA Version-7 legacy products over Mainland China at multiple spatiotemporal scales. J. Hydrol. 533, 152–167. https://doi.org/10.106/j.jhydol.2015.12.008.

Taylor, K.E., 2001. Summarizing multiple aspects of model performance in a single diagram. J. Geophys. Res. 106 (D7), 7183–7192.

Tian, Y., Peters-Lidard, C.D., 2010. A global map of uncertainties in satellite-based precipitation measurements. Geophys. Res. Lett. 37, L24407. https://doi.org/10.1029/2010GL046008.

Turk, F.J., Miller, S.D., 2005. Toward improved characterization of remotely sensed precipitation regimes with MODIS/AMSR-E blended data techniques. IEEE Trans. Geosci. Remote Sens. 43 (5), 1059–1069. https://doi.org/10.1109/TGRS.2004.841627.

Turk, F.J., Arkin, P., Ebert, E.E., Sapiano, M.R.P., 2008. Evaluating high-resolution precipitation products. Bull. Am. Meteorol. Soc. 89, 1911–1916.

Ushio, T., Tashima, T., Kubota, T., Kachi, M., 2013. Gauge adjusted Global Satellite Mapping of Precipitation (GSMaP_Gauge). In: Proc. 29th ISTS, 2013-n-48.

Varma, A.K., Gairola, R. M., Goyal, S., 2015. Hydro-estimator: modification and validation. Technical report, SAC/EPSA/AOSG/SR/03/2015, p. 27.

Xie, P., Yarosh, Y., Love, T., Janowiak, J.E., Arkin, P.A., 2002. A real-time daily precipitation analysis over South Asia. In: 16th Conf. of Hydrology, Amer. Meteorol. Soc., Orlando, FL.

CHAPTER

15

Real-Time Wind Velocity Retrieval in the Precipitation System Using High-Resolution Operational Multi-radar Network

Haonan Chen, Venkatachalam Chandrasekar

Colorado State University, Fort Collins, CO, United States

1 INTRODUCTION

High-winds, especially the localized high-impact wind phenomena such as microbursts and tornadoes, are among the most destructive natural disasters, often leading to substantial loss of life and property. Such events are more common in the United States due to the unique climatology and geography of the continent (Perkins, 2002). The United States experiences more than 1200 tornadoes and more than 80 deaths and 1500 injuries associated with tornadoes every year (http://www.spc.noaa.gov). Monitoring tornadoes, especially the full vector winds in real-time, is a challenging task due to the limitations of traditional weather sensing instruments. In this chapter, we discuss the application of a high-resolution operational Doppler radar network with regard to vector wind fields retrieval within precipitation systems.

The multiple-Doppler radar technique has been used to retrieve wind velocity information since the late 1960s (e.g., Browning and Wexler, 1968; Armijo, 1969; Miller and Strauch, 1974; Gal-Chen, 1982; Chong et al., 1983). A number of countries have deployed operational radar networks for weather monitoring and forecasting, such as the U.S. Weather Surveillance Radar-1988 Doppler (WSR-88DP) network, also known as the Next-Generation Radar (NEXRAD) network. The operational radar networks typically consist of long-range microwave (e.g., S or C band) radars. For example, the WSR-88DP network is comprised of about 160 S-band (wavelength ~10 cm) radar sites that are operated according to a set of predefined scan strategies. The WSR-88DP network can provide general weather information across the United States with an update rate of ~5–6 min. However, the capacity of the WSR-88DP radar is

Remote Sensing of Aerosols, Clouds, and Precipitation
https://doi.org/10.1016/B978-0-12-810437-8.00015-3

© 2018 Elsevier Inc. All rights reserved.

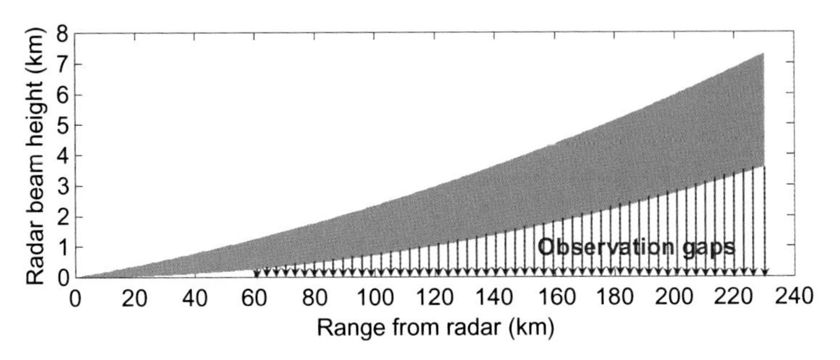

FIG. 1 The lowest (0.5 degrees) beam height of the WSR-88DP volume coverage pattern (VCP12) scanning mode, as a function of range from radar. The beam height is calculated based on the 4/3 Earth radius model.

limited due to its coverage deficiency. Fig. 1 shows the lowest (0.5 degrees) radar beam height as a function of range from radar for the WSR-88DP scanning strategy of Volume Coverage Pattern 12 (VCP12). The beam height is calculated based on the 4/3 Earth radius model (Schelleng et al., 1933). At the WSR-88DP's maximum coverage range of 230 km, the beam center is about 5.4 km above ground level (AGL) due to the effect of Earth's curvature. Compounding the terrain blockage, more than 70% of the atmosphere below 1-km altitude AGL cannot be observed (see also Fig. 2), where many hazardous weather events occur such as tornadoes and flash floods. In addition, the WSR-88DP radar coverage is nonoverlapping (at very high altitudes, if any), and the average spacing between radars is about 230 km in the eastern United States and 345 km in the western United States. This nonoverlapping deployment makes it difficult to conduct vector wind retrievals. On the other hand, tornadoes and microbursts can form within a few minutes, and can be relatively localized and fast-moving. Even the single-Doppler radar observations from WSR-88DP cannot monitor these phenomena effectively due to the coarse temporal resolution caused by its slow scanning rate.

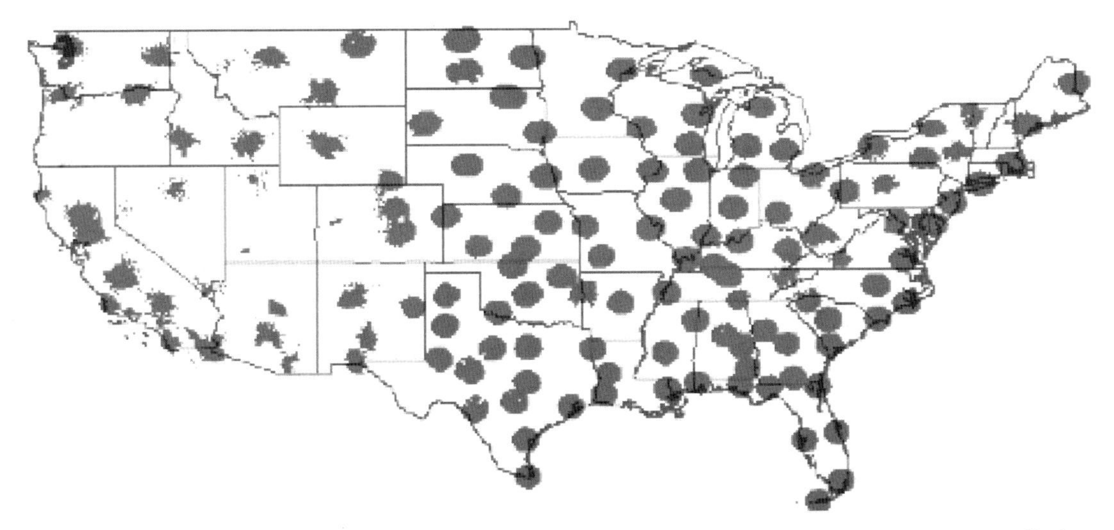

FIG. 2 WSR-88DP radar coverage at 1-km AGL over the continental United States. The coverage map is essentially derived based on the center beam height of 0.5-degree elevation. *Courtesy of NOAA/NWS/OST.*

An additional data source is indispensable in order to produce reliable and efficient wind products with the current operational radar network.

In order to overcome the coverage limitations of the WSR-88DP radar network and enhance the spatial-temporal resolution of radar observations, the National Science Foundation Engineering Research Center (NSF-ERC) for CASA introduced an alternative sensing paradigm through the deployment of a dense network of shorter-range, low-power, high-resolution dual-polarization X-band radars (wavelength ~3 cm) (McLaughlin et al., 2009; Chandrasekar et al., in press). With the enhanced radar observations, especially in the lower troposphere (1–3 km AGL), the new sensing paradigm-termed distributed collaborative adaptive sensing (DCAS)-can significantly improve the forecast and responses to hazardous weather events. Fig. 3 shows the simplified architecture of a DCAS system, which includes distributed high-resolution X-band Doppler radars; algorithms that dynamically process the collected data, detect ongoing weather features, and manage system resource allocations; and interfaces that enable end-users to interact with the system. Compared to the scanning strategy adopted by NWS radars, the DCAS paradigm rapidly reconfigures each radar node in response to the changing atmospheric conditions and end-user needs. The first such research network of this kind, termed the Integrated Project 1 (IP1), consisted of four radar nodes. It was deployed in the "tornado alley" over southwestern Oklahoma for the study of precipitation and hazardous winds (McLaughlin et al., 2009; Chandrasekar et al., 2012). Since Spring 2012, CASA, in collaboration with the NWS and North Central Texas Council of Governments (NCTCOG), has been operating the first urban operational radar network in the Dallas-Fort Worth (DFW) metroplex. The main goal of this network is urban weather hazard detection and mitigation (Chandrasekar et al., 2013). The network topology of CASA radars allows for high-resolution observation of the lower troposphere while providing large areas of overlapping coverage. At the overlapping regions, multiple-Doppler analyses can be conducted to retrieve the vector wind velocity and wind patterns. Along with other real-time operational

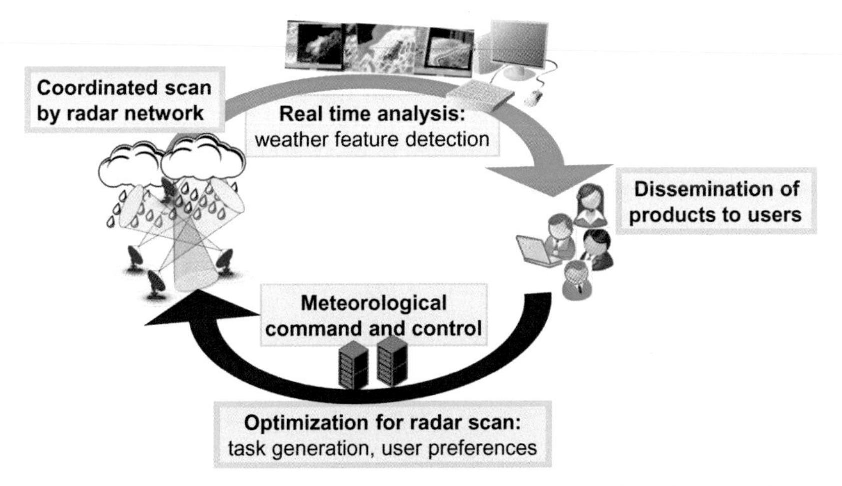

FIG. 3 Simplified architecture of a distributed collaborative adaptive sensing (DCAS) system. The real-time data and products are disseminated to various end-users, including NWS weather forecasters, emergency managers, and CASA's researchers themselves. The radar scanning strategy is adapted to end-users' feedback.

CASA products such as quantitative precipitation estimation (QPE) and hail detection, the high-resolution radar network serves as a critical emergency weather warning tool.

This chapter presents the principles and applications of the multiple-Doppler weather radar technique with an emphasis on the implementation at the high-resolution operational CASA radar networks. In Section 2, an overview of Doppler radar observations as well as the fundamental concept of multiple-Doppler retrieval is described. With this introduction, the real-time multiple-Doppler wind retrieval system designed for high-resolution CASA radar networks is detailed in Section 3. Sample wind retrieval products in the presence of tornadoes, high-winds, and downbursts are provided in Section 4. Section 5 summarizes the main findings and suggests directions for future operational multiple-Doppler radar analysis.

2 MULTIPLE-DOPPLER METHODOLOGY FOR WIND RETRIEVAL

2.1 Radar Background

Radar is the acronym for "RAdio Detection And Ranging." It is an object-detection system that uses radio waves to determine the range, angle, and/or velocity of targets. Essentially, radars operate by sending electromagnetic waves toward targets to determine their properties based on the return signal. Meteorological targets such as thunderstorms are composed of large numbers of hydrometeors extending over a large space. Modern pulse radars treat these as distributed targets within a sample volume that is defined by the radar's beamwidth and sample range spacing (Fig. 4).

The Doppler frequency shift is given by:

$$f_d = -\frac{2V_R}{\lambda} \tag{1}$$

where f_d is the Doppler frequency; λ is the radar wavelength; and V_R is the target radial velocity, which is the velocity component along the pointing axis of the radar (Bringi and Chandrasekar, 2001). For a constant wavelength, the Doppler frequency shift is solely dependent on the velocity. Measuring the mean frequency shifts in backscatter from moving particles yields the mean radial velocity measured by radar:

$$V_R = -\frac{\lambda f_d}{2} \tag{2}$$

In practice, pulse Doppler weather radars measure the Doppler shift by detecting the change in phase shift of the return signal over sample time. The maximum unambiguous velocity can be described as:

$$V_{\max} = \frac{\lambda}{2} \times f_{\max} = \frac{\lambda}{4T_S} \tag{3}$$

where T_S is the pulse repetition time (PRT) (Bringi and Chandrasekar, 2001). Obviously, there are limitations on the maximum velocity that can be resolved unambiguously. Either increasing the wavelength or reducing the PRT- or both-should be done in order to increase the maximum unambiguous velocity. However, care must be taken when trying to achieve

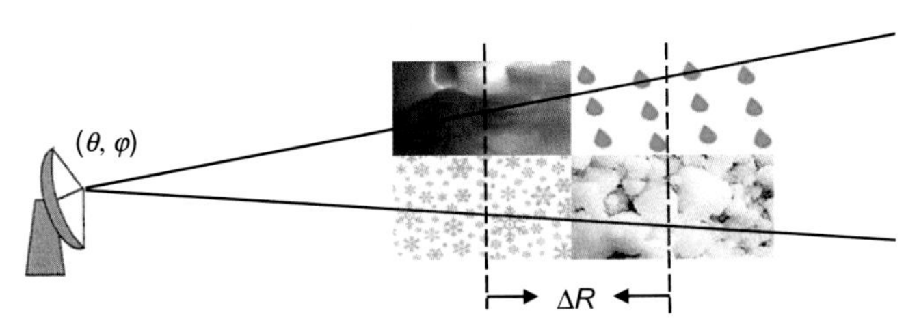

FIG. 4 Doppler weather radar sensing distributed targets within a sample volume. The sample volume size is determined by the radar's horizontal and vertical beamwidths θ and φ, and range spacing ΔR.

higher V_{max} due to the range velocity ambiguity. In addition, it should be noted that although a pulse Doppler radar by itself is a powerful tool for weather monitoring and remote sensing, it cannot extrapolate the entire kinematics of weather systems because it can only measure radial velocities. Its true strength lies in working together as a multiple-Doppler radar network, where different components of radial velocities observed by different radars can determine true air motion vectors. The high-resolution CASA dual-polarization Doppler radar networks are such systems. For the given radar system/frequency (i.e., X-band for CASA), the PRT is more adjustable. In CASA, modern pulsing schemes with advanced signal processing were developed to overcome the range velocity ambiguity (Bharadwaj et al., 2010).

2.2 Principle of Multi-Doppler Retrieval

The essence of multiple-Doppler wind retrieval from a radar network is to get the two- or three-dimensional velocity components in Cartesian coordinates from the nonorthogonal radial velocities measured by the radar nodes. Fig. 5 shows conceptual diagrams of the multiple-Doppler retrieval problem. In the Cartesian coordinate system, the velocity of a particle at (x,y,z) within a thunderstorm can be expressed by a triplet $(u,v,w+w_f)$, where $u,v,$ and w are the corresponding velocity components in eastward, northward, and vertical, respectively. w_f is the terminal velocity of falling particles. In applications, the true vertical air motion w should be separated from the Doppler measurement $(w+w_f)$. The projections of the particle's motion onto radars' line of sight are:

$$
\begin{aligned}
V_R^1 &= u\sin\Phi_1\cos\theta_1 + v\cos\Phi_1\cos\theta_1 \\
&\quad + (w+w_f)\sin\theta_1 \\
&\vdots \\
V_R^m &= u\sin\Phi_m\cos\theta_m + v\cos\Phi_m\cos\theta_m \\
&\quad + (w+w_f)\sin\theta_m
\end{aligned}
\tag{4}
$$

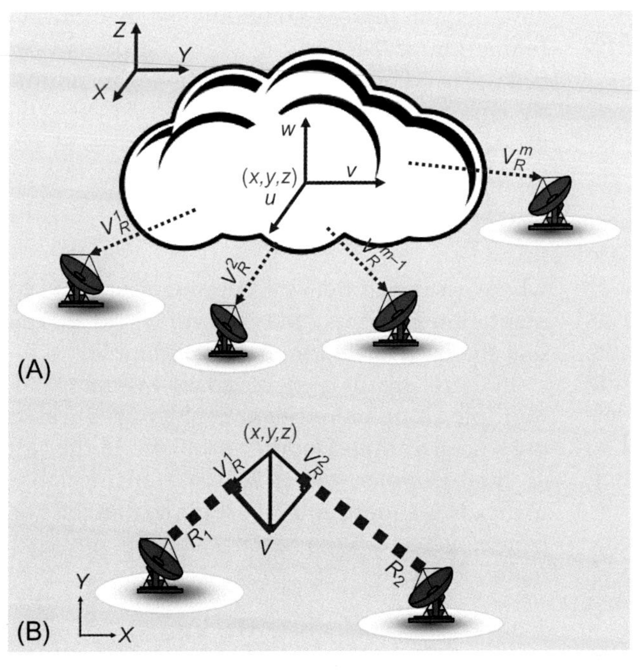

FIG. 5 (A) Conceptual diagram showing the multiple-Doppler retrieval problem (i.e., retrieving three-dimensional wind velocity components from radar-measured radial velocities in a network environment). (B) diagram showing the projection of particle velocity onto radar-measured radial velocity.

where V_R^m is the radial velocity measured by radar node m; and Φ_m and θ_m are the azimuth and elevation angles of the radial beam of radar node m, respectively. The azimuth angle is measured clockwise from 0 degrees north.

Taking into account the geometric relation in Cartesian coordinates, Eq. (4) can also be expressed as:

$$V_R^m = \frac{1}{r_m}\left[u(x - x_m) + v(y - y_m) + (w + w_f)(z - z_m)\right] \tag{5}$$

for radar node m at (x_m, y_m, z_m) with slant range $r_m = \sqrt{(x - x_m)^2 + (y - y_m)^2 + (z - z_m)^2}$. In addition, the advection effects caused by storm motion need to be taken into account in order to adjust for the difference between multiple-Doppler synthesis time t and the data sample time $t + \Delta t$. Previous studies in Gal-Chen (1982) conclude that instead of V_R^m it is more accurate to advect the product $V_R^m r_m$. Therefore, for a thunderstorm moving with velocity component (U, V), Eq. (5) should be replaced with:

$$\frac{\left[V_R^m r_m\right]_{t+\Delta t}}{\left[r_m\right]_t} = \left[\frac{u(x - x_m + U\Delta t)}{\left[r_m\right]_t} + \frac{v(y - y_m + V\Delta t)}{\left[r_m\right]_t} + \frac{(w + w_f)(z - z_m)}{\left[r_m\right]_t}\right]_t \tag{6}$$

That is, radar-measured radial velocities are first multiplied by slant ranges from the radar at sample time $t + \Delta t$. Then, the product is advected at the storm motion to new locations, where it is divided by slant ranges at the Doppler synthesis time t. Accordingly, the coefficients of u and v in Eq. (6) are modified to account for the change in the radar-pointing direction.

Putting the radial velocities into vector $\mathbf{V_R} = \left[\frac{\left[V_R^1 r_1\right]_{t+\Delta t}}{\left[r_1\right]_t} \; \cdots \; \frac{\left[V_R^m r_m\right]_{t+\Delta t}}{\left[r_m\right]_t}\right]^T$, and using the following matrix form:

$$\mathbf{H} = \begin{bmatrix} \frac{(x - x_1 + U\Delta t)}{[r_1]_t} & \frac{(y - y_1 + V\Delta t)}{[r_1]_t} & \frac{(z - z_1)}{[r_1]_t} \\ \vdots & \vdots & \vdots \\ \frac{(x - x_m + U\Delta t)}{[r_m]_t} & \frac{(y - y_m + V\Delta t)}{[r_m]_t} & \frac{(z - z_m)}{[r_m]_t} \end{bmatrix},$$

a linear system can be formed as follows:

$$\mathbf{V_R} = \mathbf{H}\left[u \;\; v \;\; w + w_f\right]^T \tag{7}$$

In practice, the terms in matrix $\mathbf{H}$ are known, which are mainly determined by the geometry of radar measurements. The three-dimensional wind velocity components can then be attained using the generalized least square method:

$$\left[u \;\; v \;\; w + w_f\right]^T = \left(\mathbf{H}^T\mathbf{H}\right)^{-1}\mathbf{H}^T\mathbf{V_R} \tag{8}$$

The horizontal wind components u and v can be retrieved directly from the solution in Eq. (8), provided that at least two radars are available. But the retrieval of the vertical wind component is rather challenging due to the particle falling and variation of vertical air pressure. For CASA radars, low elevation scans are generally performed, which produce radial velocities that have small vertical components. In addition, estimation of the vertical wind component can be improved by incorporating the mass continuity equation into the system. In this case, the mass continuity equation can be expressed as:

$$\frac{\partial(\rho u)}{\partial x} + \frac{\partial(\rho v)}{\partial y} + \frac{\partial(\rho w)}{\partial z} = 0 \tag{9}$$

where ρ is the air density. For operational implementation in CASA radar networks, we assume that the local variation of ρ is negligible.

If there are only two radars available, the multiple-Doppler retrieval problem reduces to the so-called dual-Doppler problem. In the case of dual-Doppler retrieval, the transformation matrix $\mathbf{H}$ is square and invertible so that the horizontal velocity components can be first estimated from:

$$\left[u' \;\; v'\right]^T = \mathbf{H}^{-1}\mathbf{V_R} \tag{10}$$

The estimates in solution (Eq. 10) can be further improved by substituting the following form into the mass continuity equation in Eq. (9):

$$[u \;\; v]^T = [u' \;\; v']^T + (w + w_f)\boldsymbol{\varepsilon} \tag{11}$$

Similar to the transformation matrix **H**, **ε** here is dependent on the geometry of radar layout and measurement.

2.3 Error Analysis

The linear system in Eq. (7) is essentially an inversion problem that converts observations from a nonorthogonal vector space to an orthogonal vector space. During this inversion process, the variances of the retrieved wind velocity components generally exceed the measurement errors in the radial velocities. With the least square solution given by Eq. (8), the variance of retrieved wind products can be assessed as:

$$\mathrm{Cov}\left[u \;\; v \;\; w + w_f\right]^T = \left(\mathbf{H}^T\mathbf{H}\right)^{-1}\mathbf{H}^T E\left[\mathbf{V_R}\mathbf{V_R^T}\right] \mathbf{H}\left(\mathbf{H}^T\mathbf{H}\right)^{-T} \tag{12}$$

Assuming the errors involved in the radar velocity measurements are independent and equal, the variance can be written in a normalized form:

$$\frac{\mathrm{tr}\left(\mathrm{Cov}\left[u \;\; v \;\; w + w_f\right]^T\right)}{3\mathrm{Var}\left(V_R^m\right)} = \mathrm{tr}\left(\mathbf{H}^T\mathbf{H}\right)^{-T} \tag{13}$$

Recalling that the transformation matrix **H** only depends on the advection and geometry of radar measurement, the variance in Eq. (13) can be used to evaluate the impact of geometry on the multiple-Doppler wind retrieval analysis (Davies-Jones, 1979). Taking the initial estimates of horizontal wind components in Eq. (10) as an example, if we get rid of the known items such as the retrieval time and location, the normalized variance in dual-Doppler retrieval can be obtained as Eq. (14) after modest algebraic manipulation of Eqs. (4)–(6):

$$\frac{\sigma_{u'}^2 + \sigma_{v'}^2}{2\sigma_{V_R}^2} = \frac{1}{\sin^2(\Phi_1 - \Phi_2)} \frac{\cos^2\theta_1 + \cos^2\theta_2}{2\cos^2\theta_1 \cos^2\theta_2} \tag{14}$$

Eq. (10) is selected because the initial estimates are the fundamental variables driving the solution of the mass continuity equation. The first term in Eq. (14) is related to the angle subtended by the two Doppler radar beams on the horizontal plane. Literature suggests that this angle of 30 degrees or greater is adequate for Doppler wind retrieval (Davies-Jones, 1979). The second term is related to the elevation angles of this radar pair. The retrieval errors become larger as the elevation angle increases.

In applications, an upper bound needs to be imposed on Eq. (13) to ensure the accuracy of the retrieved wind components. To achieve this, a subset of radar coverage can be found for multiple-Doppler wind retrieval, especially when multiple-Doppler pairs exist. This principle is commonly used for Doppler radar network design.

Synchronization of observations from multiple radars is another error source for multiple-Doppler retrieval. To meet this requirement, the high-resolution CASA operational radar network is designed with a small "heartbeat" for a volume-scan. More details about the multiple-Doppler system design for CASA radar networks, including the best Doppler pair selection and data synchronization, are presented in Section 3.

3 REAL-TIME MULTIPLE-DOPPLER RETRIEVAL SYSTEM FOR CASA RADAR NETWORKS

3.1 CASA Radar Networks

CASA, established in 2003, is a multisector partnership among academia, industry, and government dedicated to developing next-generation weather sensing networks. Today's weather forecast and warning systems utilize data from high-power, long-range radars that have limited ability to observe the lower part of the atmosphere due to the Earth's curvature, leading to undersampling of meteorological conditions in the lower troposphere where most weather activities occur. Compounding the problem is the coarse resolution that these radars provide, with sample volumes extending to many cubic kilometers as range increases. CASA tries to overcome these limitations by deploying dense networks of X-band Doppler radars that operate at a shorter-range. These small radars are generally deployed just a few tens of kilometers apart. Overall these innovative networks will save lives and property by detecting the region of the lower atmosphere currently below conventional radar coverage.

The networked approach (see Fig. 3) employed by CASA adaptively operates the radars within a dynamic information technology infrastructure, dictating the radars to scan areas of interest according to the changing weather conditions and end-user needs (Chandrasekar et al., 2013, in press). Up-to-the-minute radar observations and products are transmitted to the organizations that make critical decisions about the weather. In addition, the small CASA radars can be readily deployed on small towers with small land footprints or existing infrastructure elements such as rooftops and communications towers.

CASA has implemented scalable research and operational testbeds in Oklahoma, Puerto Rico, and most recently, the DFW metroplex, to sense infrasound signatures of tornadoes, severe thunderstorms, and other weather hazards. In this chapter, radar observations from Oklahoma (i.e., the IP1 testbed) and the DFW urban testbeds are used extensively for demonstration purposes. In the following, a brief overview of the CASA Oklahoma research testbed and operational DFW urban network is provided. For more details, the reader is referred to McLaughlin et al. (2009), Junyent et al. (2010), and Chandrasekar et al. (2012) about the IP1 testbed, and Chandrasekar et al. (2013), Chen and Chandrasekar (2015), and Chandrasekar et al. (in press) about the DFW network.

3.1.1 CASA IP-1 Testbed

The CASA IP1 testbed is the first radar network developed by CASA (McLaughlin et al., 2009). It serves as a prototype of the DCAS system for high spatiotemporal resolution sensing of severe storms in the lower atmosphere. The test bed, located approximately 45 km southwest of Oklahoma City, Oklahoma, consists of four mechanically scanning, automated, short-range X-band Doppler radars covering an area of about 7000 km^2. Fig. 6 illustrates the coverage map of the CASA IP1 radars. These radars, called KCYR, KLWE, KRSP, and KSAO, are located in the towns of Cyril, Lawton, Rush Springs, and Chickasha, respectively. Each radar node is approximately 30 km away from the next unit. The circles in Fig. 6 correspond to a 40-km range from the radars. The range resolution of IP1 radars is 75 m. The location of the testbed was chosen based on its climatological and meteorological properties. Being in tornado alley, this testbed has about a 77% chance of experiencing at least one tornado each year, and severe storms are almost 100% guaranteed every year. This area receives an average of four tornado warnings and 53 thunderstorm warnings per year (http://www.spc.noaa.gov).

3.1.2 Operational CASA DFW Urban Network

The DFW region is one of the largest inland metropolitan areas in the United States. It experiences a wide range of natural weather hazards, including urban flash flooding, tornadoes, and hail. Monitoring the rapidly changing meteorological conditions in such a region in a timely manner is necessary for emergency management and decision-making. Therefore, it is an ideal location to demonstrate the CASA DCAS concept for urban applications. The DFW radar network is CASA's first urban operational testbed. Centered in the DFW testbed is a

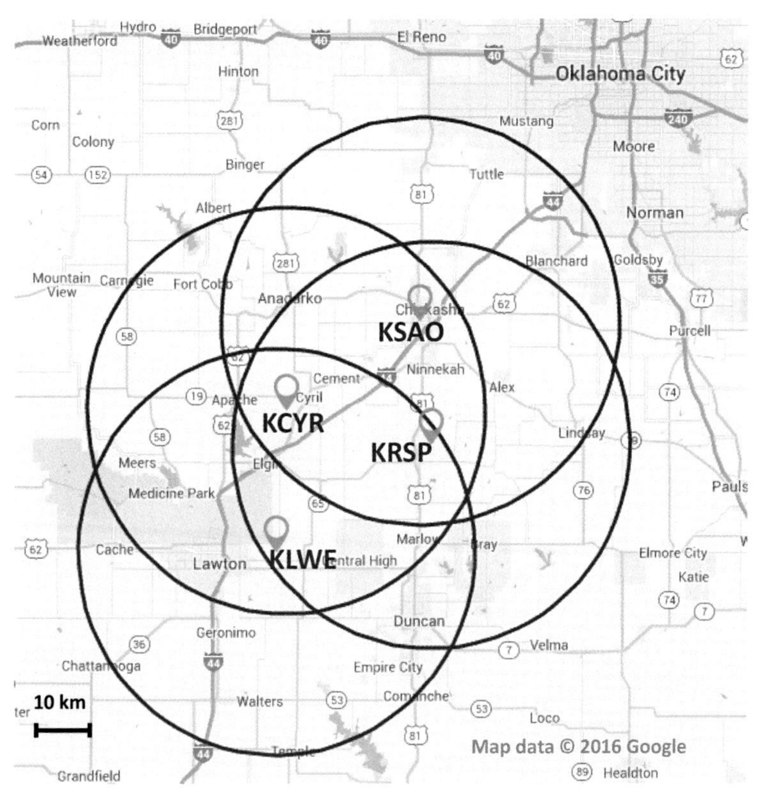

FIG. 6 Map of the CASA IP1 test bed in southwestern Oklahoma showing the Doppler radar sites and 40-km coverage range rings.

network of eight boundary-layer observing, dual-polarization, X-band Doppler radars. Fig. 7 shows the geographical layout of the eight X-band radars as well as the S-band WSR-88DP radar deployed in Fort Worth (i.e., the KFWS radar). The letter symbols such as "XMDL" correspond to the naming of various radars. The smaller *blue circles* in Fig. 7 correspond to the 40-km coverage range rings of the X-band radars and the larger *red circle* corresponds to the 100-km range distance from the KFWS radar. This urban remote sensing network covers 12 out of the 16 counties in the DFW area, providing coverage to most of the 6.5 million people in this region (Chandrasekar et al., in press). The major objectives of this operational urban radar network are to: (1) develop high-resolution three-dimensional mapping of the boundary-layer atmospheric conditions, and monitor severe weather hazards including high-winds, tornadoes, hail, and flash floods; (2) create neighborhood-scale impact-based warnings and forecasts for a wide range of public and private sector decision-makers that lead to benefits for public safety and the economy; and (3) demonstrate the added value of CASA X-band radar networks to the existing and future NWS sensors and products, and assess optimal combinations of multiscale observing systems (Chandrasekar et al., 2013; Chen and Chandrasekar, 2015).

The systems deployed in this urban network are based on new technologies developed within the CASA project. In addition, the

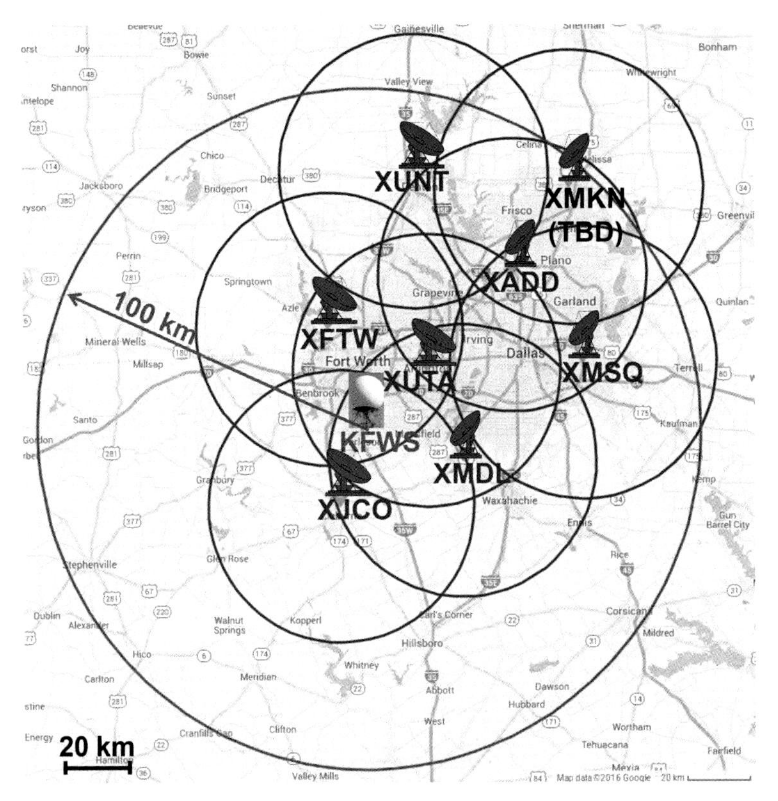

FIG. 7 The layout of S-band KFWS WSR-88DP radar (100-km range ring in *red*) and DFW dual-polarization X-band Doppler radars (40 km range rings in *blue*).

existing in situ and remote sensors, such as WSR-88DP, Terminal Doppler Weather Radar (TDWR), and rain gauges, are used for creating new products and for validation purposes.

3.1.3 Network Design Consideration for Multiple-Doppler Retrieval

The radar layout is designed to optimize multiple-Doppler retrievals by maximizing coverage overlap and compensating for minimum beam-crossing angle blindspots. Here, the beam-crossing angle refers to the angle of intersection between beams of two radars. A small beam-crossing angle corresponds to two radars scanning a region between them that is close along the axis connecting the two radars. In such a case, the two radars would measure Doppler velocities that are approximately equal and opposite, giving very little or no orthogonal component to triangulate the true vector velocity components. Taking the CASA IP1 testbed as an example, Fig. 8A shows the mosaic cross-beam angles at 1-km altitude. For most of the overlapping regions, it can be seen that the cross-beam angle is higher than 30 degrees, which is sufficient for multiple-Doppler wind retrieval.

As shown in Figs. 6 and 7, extensive overlap exists in the CASA radar networks, where multiple-Doppler analysis can be conducted. In many regions of the radar network, multiple-Doppler pairs exist, which provides an option to use another Doppler pair that has highly orthogonal beam-crossing angles to

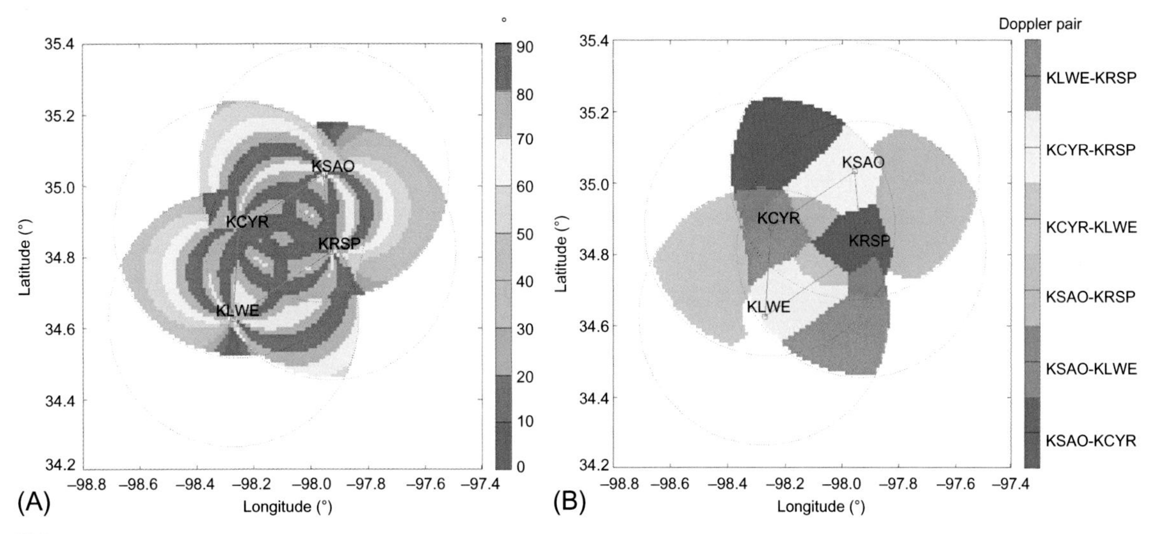

FIG. 8 Radar locations and network coverage at CASA Oklahoma testbed: (A) the mosaic cross-beam angles at 1-km altitude; (B) the optimized best dual-Doppler pairs (Wang et al., 2008).

compensate for the region where the blindspot occurred. In the CASA network design, an optimization procedure is implemented to find the best Doppler pairs and map them to the target volumes. The optimization is based on the geometry-related error criterion in Eq. (14), orthogonality of the beam-crossing angle, and other constraints such as the compromise between elevation angle and range of a given resolution volume. Again, taking the Oklahoma testbed as an example, the optimal dual-Doppler pairs at 5 km altitude are illustrated in Fig. 8B. It should be noted that the dual-Doppler optimization is less preferred when more than two radars are available. That is, multiple-Doppler wind retrieval will be conducted with three or more radars. We also want to note that in an operational environment, the Doppler wind retrieval performance greatly relies on the networked radar scan strategy. The computational resource in operational CASA radar networks is highly competitive such that optimal and adaptive scan strategies should be implemented. The networked scan strategy is achieved through regulating a few key processes, including streaming data from individual radar nodes to the radar operations center, identifying meteorological features from the radar data, and determining each radar's future scan task based on detected features. In addition, given the radar locations in the operational CASA network, the normalized geometry-based vector wind retrieval errors in Eq. (14) are precomputed over a three-dimensional coverage grid. The error table along with other constraints such as the detected storm location is used to determine the suitable Doppler pairs. Fig. 9 shows a sample operational scan display and scan domain relative to the whole coverage domain afforded by CASA radar nodes in Oklahoma for the thunderstorm event on May 24, 2011. The observed reflectivity field is depicted with colors, whereas the radar coverage is depicted by the *white-shaded* areas. The left figure shows observations at 2225 UTC, and the figure to the right illustrates adaptive scanning after 15 min (i.e., 2240 UTC). This event, which will be detailed in Section 4.1, is characterized with tornadic storms advecting toward the northeast.

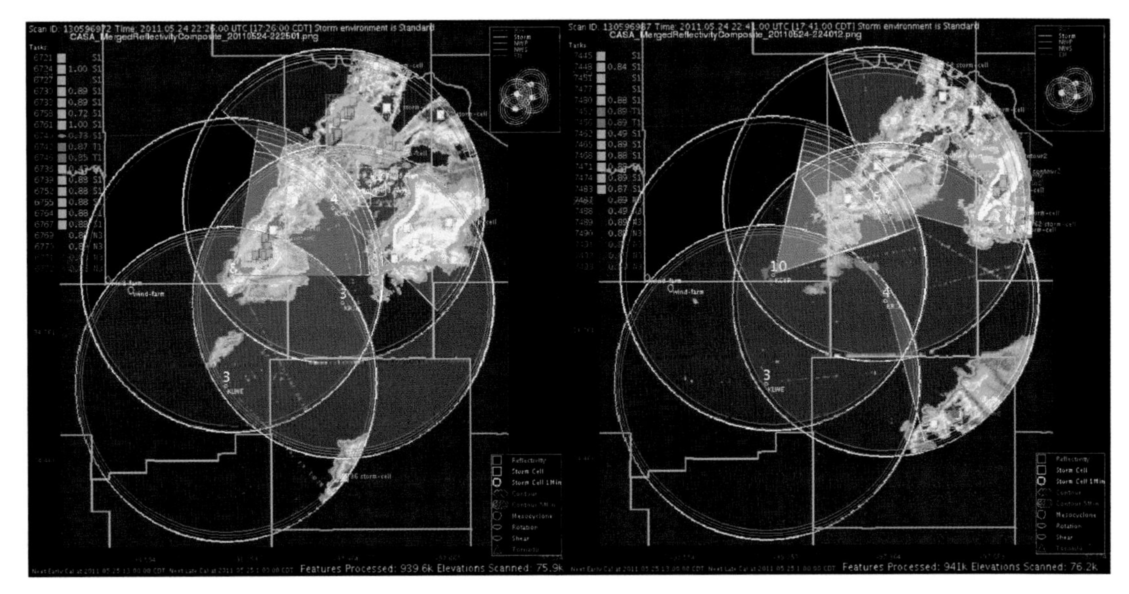

FIG. 9 Example operational adaptive scan display. Left figure shows scan information and observations at 2225 UTC, May 24, 2011, whereas figure on right illustrates the adaptive scan information after 15 min (i.e., 2240 UTC). The observed reflectivity field is depicted with *colors*. Radar coverage is depicted by the *white-shaded* areas. The storm was advecting toward the northeast.

In CASA, the volume-scan heartbeat is configured as about 1 min, with radars initiating new volume scans usually within 15 s. With this fast scanning scheme for small X-band radars and the relatively slow motion of precipitation systems, the adaptive CASA radar network can create a high-frequency real-time "snapshot" of the weather structure. The higher-resolution CASA radar observations consequently result in higher-resolution Doppler retrievals and thus better detection of severe weather features.

3.2 CASA Multiple-Doppler Retrieval System

Fig. 10 illustrates the framework of a real-time multiple-Doppler wind retrieval system designed for operational CASA radar networks. Externally, the retrieval system appears as a single functional block that takes in radar data sweeps and outputs two-dimensional and/or three-dimensional wind products. It shares computing and data resources with other simultaneous product generation systems running in the radar operations center, including but not limited to the hail detection system (Chandrasekar et al., in press) and QPE system (Chen and Chandrasekar, 2015). All the CASA product generation systems are autonomous real-time processes. Once executed, each system creates an independent processing branch until finished. It then awaits the execution order of the next DCAS loop iteration. To meet the real-time requirement, each execution falls within the radar network's scanning heartbeat time (i.e., 1 min). Internally, the multiple-Doppler process is broken down into four major sections, including real-time radar data transmission, data aggregation, the retrieval core system itself, and real-time product distribution. These components must interact seamlessly under precise timing to ensure a robust function of the system. Herein, the data transmission and aggregation

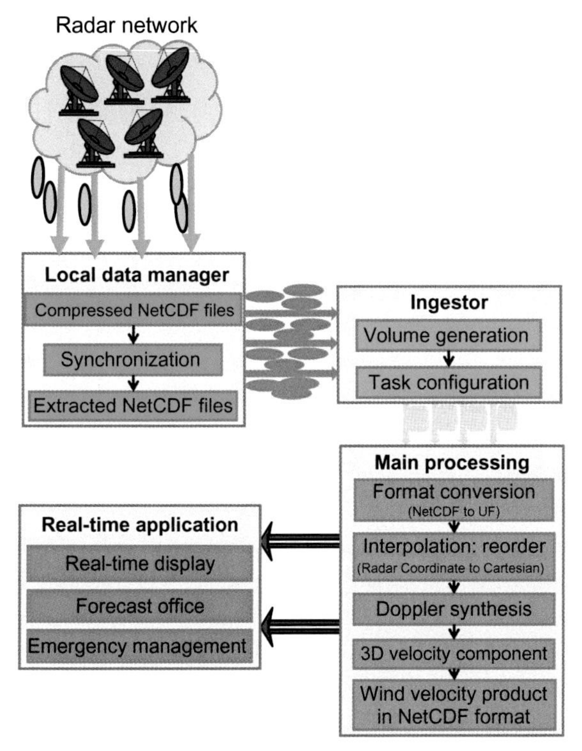

FIG. 10 Framework of the real-time multiple-Doppler wind retrieval system for operational high-resolution CASA radar network.

can be considered as preprocessing, which prepares volume data for the main multiple-Doppler processing. The main processing revolves around two components, data interpolation and Doppler synthesis. These functions can be realized by legacy software such as *REORDER* for interpolation (Oye and Case, 1995) and *CEDRIC* for Doppler synthesis (Miller and Anderson, 1991), which have been widely used with proven proficiency in terms of both accuracy and speed. Therefore, the *REORDER* and *CEDRIC* duo has been implemented in CASA radar networks. Mapping radar volume data from its radial coordinate to Cartesian space through interpolation provides a common coordinate system for overlapping scans in the testbed, which is a necessary geometric condition for performing multiple-Doppler retrieval. As outlined in Section 2, the gridded volume data from each radar is synthesized into a single composite grid with resolved three-dimensional velocity components.

This section provides an in-depth description of the key components and their interactions in the overall multiple-Doppler process flow.

3.2.1 *Data Transmission and Aggregation*

The data preprocessing is controlled by the Local Data Manager (LDM), which is a software system for efficient and reliable distribution of arbitrary but finite-sized data via the Internet. It operates on a client-server model with the data source being the servers and the data sinks being clients. In CASA, the LDM is regarded as an event-driven process in the sense that it distributes the radar data to the radar operations center as soon as it is generated. The process is achieved through the Internet and LDM publish-subscribe protocols. In particular, each CASA radar hosts an LDM server, and the LDM residing in the radar operations center is technically a client downloading data from each radars' LDM servers. Other LDM clients can also be set up on any arbitrary Internet-connected device to stream data from CASA radars.

In CASA radar networks, the real-time radar data variables streamed from individual radar nodes are separate plan position indicator (PPI) or range height indicator (RHI) sweep files in radial coordinates centered at each radar, which are incompatible with the multiple-Doppler retrieval core system. In addition, latency issues may occur during data transmission, which will cause the radars' sweeps to arrive out of order. To this end, an *Ingestor* program is implemented and embedded into the LDM to operate on the received data. The key function of *Ingestor* is to decompress the incoming data, extract their scanning information, and synchronize them to respective radar and volumes. Technically it consists of three separate processes that share an interprocess communication channel. Firstly, *Ingestor* parses the incoming file name containing the radar name and data sampling time, and enforces the correct grouping and sequence of the radar data sweeps. Secondly, the data sweep files are decompressed and scan information from each data sweep is extracted, including elevation angle and scan time. Thirdly, *Ingestor* organizes the sweeps into respective volumes for each radar, and entire volumes of data from multiple radars serve as input to the main processing of the CASA multiple-Doppler system.

3.2.2 Radar Data Resampling

As outlined in Section 2.2, resampling radar volume data from polar coordinates onto Cartesian space is a prerequisite of the final Doppler synthesis and retrieval. For the operational CASA radar network, the interpolation is performed as distance-weighted averaging of range gates to various grid points based on radius of influence criteria. Fig. 11 illustrates the geometric relations of the interpolation scheme. For distance weighting to a certain grid point, each range gate in the input rays is assigned its Cartesian coordinate in (x, y, z). Then its distance to the grid point is compared to the grid point's radius of influence R. Radar range gates within the radius of influence are used to determine the weighted value of that point. For the CASA multiple-Doppler system, we use a conventional Cressman weighting scheme (Cressman, 1959), for which the nth range gate's weight becomes

$$W_n = \frac{R^2 - R_n^2}{R^2 + R_n^2} \quad n = 1,2,3,\ldots,N \tag{15}$$

where the radius of influence R is given by $R^2 = dX^2 + dY^2 + dZ^2$. The dX, dY, and dZ values that define R are user-specified parameters. For identical dX, dY, and dZ, the weighting volume literally becomes a sphere of influence. The final data value at grid point k is then obtained as:

$$a_k = \frac{\sum_{n=1}^{N} W_n a_n}{\sum_{n=1}^{N} W_n} \tag{16}$$

It should be noted that a variety of weighting schemes are available from the literature, including uniform, closest point, and exponential. In addition, the radius of influence can also be defined by components of azimuth, elevation, and range, in which case the user specifies ΔAzimuth, ΔElevation, and ΔRange. From the specified azimuth, elevation, and range information, dX, dY, and dZ can be calculated. This scheme has the effect of increasing the radius of influence for grid points farther away from the radar. It can help interpolate the otherwise missing grid point for scans where there are spatial gaps in between, especially at far ranges from the radar. But it may also cause blurring

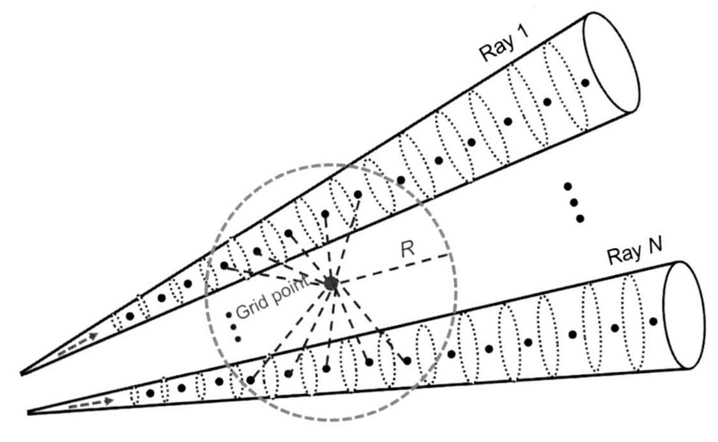

FIG. 11 Data interpolation via range-weighted averaging within a radius of influence.

of weather features. Therefore, a constant radius of influence is applied since the high-resolution X-band radars are designed and operated to use a combination of beamwidth and scanning elevations that leave no gaps within their volume scans.

For real-time applications, resolutions on all three axes are set to 0.5 km. This value is chosen based on two primary concerns: computational complexity and data resolution. Both concerns are tightly related to the speed limitation of data interpolation for multiple-Doppler retrieval. The computational complexity is not a serious constraint because it can be readily addressed by upgrading hardware to match the real-time requirements. However, doubling the resolution in cubic relations will multiply the number of grid points by eight, scaling interpolation times by eight as well. As with any brute-force method, refining the resolution from 500 m down to CASA radar's radial range resolution (less than 100 m) is unfeasible and unnecessary.

3.2.3 *Doppler Synthesis and Retrieval*

The core Doppler retrieval function is applied on the gridded/interpolated radar data. From Section 3.2.2, each radar volume is interpolated to a common Cartesian grid representing the entire testbed. For Doppler synthesis, the gridded data needs to be superimposed to properly create the regions of dual- or multi-Doppler overlap. Fig. 12 shows a diagram of the superposition of CASA radars in the Oklahoma testbed. This superposition inherently imposes resolution requirements on the data grids. Although the synthesis algorithm can handle data grids of mixed resolutions, 0.5-km resolution for three directions is kept for the CASA operational radar network. As aforementioned, this is a compromise between covering the entire testbed and keeping a sustainable number of grid points to ensure real-time performance.

In order to implement the Doppler retrieval methodology described in Section 2.2, appropriate approximation and discretization are applied when necessary. For example, reflectivity thresholding is applied to the gridded radar data to ensure product quality. The core multiple-Doppler retrieval for CASA is realized by *CEDRIC* (Miller and Anderson, 1991), which includes a few key steps: (1) taking Cartesian coordinated datasets that contain radial velocity fields, and setting appropriate coordinate boundaries for the Cartesian grids; (2) conducting Doppler synthesis to get (u, v, w) by resolving the least square error solution to the system equations outlined in Section 2.2; and (3) calculating horizontal divergences of the solutions to (u, v), which form the terms of the mass continuity relation in Eq. (9). The divergence terms are integrated to solve the mass continuity equation in discrete form. All the key processes are

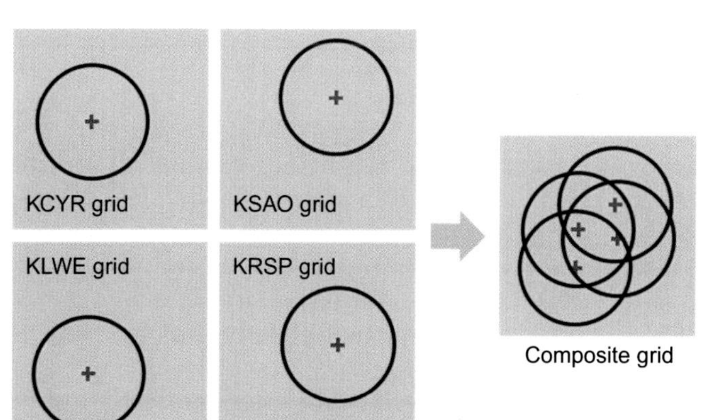

FIG. 12 Grid synthesis via superposition for *CEDRIC* analysis.

performed by the *SYNTHES* operation in the *CEDRIC* package. For more details about the usage of *CEDRIC* and the command structure of *SYNTHES*, interested readers are referred to Miller and Anderson (1991).

3.2.4 *Advection*

As mentioned in Section 2.2, storm advection should be taken into account for real-time operational multiple-Doppler retrievals. The CASA Doppler synthesis system incorporates two stages of storm advection correction. The first stage, which is described in Section 2.2, accounts for nonsimultaneous observations inherent in the sweeping motion of radar scans and across radar networks. The spatial displacement of a fast-moving storm within CASA's 1 min heartbeat is corrected using Eq. (6). The second stage of advection correction performed by the CASA multiple-Doppler retrieval system is the removal of mean storm motion altogether from the velocity products. This is done to isolate the internal air dynamics of the storm from the moving frame of reference of the storm itself. This can reveal features such as vortexes that are otherwise obscured by the storm movement. It is a reflection of the fact that even if there is no spin in the absolute air motion relative to ground, an observer within the storm may still feel a spin. Mathematically, this correction removes the constant components from areas of convergence or divergence, leaving the pure derivative. In an operational CASA multiple-Doppler system, subtracting the mean storm motion is performed in *CEDRIC* operations. The storm motion in both stages is obtained from nearby sounding stations. Particularly, the air motion at a midlevel atmosphere altitude corresponding to 700 mbar is used for mean storm motion. However, we want to note that since the sounding data is collected at a single point, errors may be introduced when assuming a constant air motion for the whole radar network domain. In general, larger, slower-moving, and longer-lasting precipitation systems tend to be more accurately portrayed by the sounding data. The detailed advection methodology is beyond the scope of this chapter. One can find more information from Gal-Chen (1982).

4 OBSERVATIONS, RESULTS, AND VALIDATION

This section presents CASA radar network observations and wind retrieval products in the presence of tornadoes, high-winds, and downbursts, which are characterized by different airflow structures. We intend to select a number of interesting events from both the Oklahoma and DFW testbeds. The case studies presented here were demonstrated by ground reports from NWS and social media.

4.1 EF4 Tornado in CASA IP1 Testbed (May 24, 2011)

The severe weather event on May 24, 2011, was preceded by several days of calm weather as a midlatitude cyclone moved eastward through the south central United States. The cyclone first developed on May 21 over the central states, with a low-pressure center over Wyoming and South Dakota. By May 23, the low-pressure center was located over western Oklahoma and northern Texas. The system continued its eastward motion over Oklahoma's unstable air, creating the potential for extreme weather on Tuesday, May 24. Several supercell thunderstorms developed over western and central Oklahoma in the evening, producing two EF4 tornadoes and many other less-intense tornadoes throughout south central Oklahoma. A tornado warning was issued at 2142 UTC for northwestern Grady County, and another at 2150 UTC for northern Grady County. The first EF4 tornado touched down at 2206 UTC in Grady County, 16 min after the last warning was issued. Another tornado warning was issued at 2210 UTC for southern Grady County and western McClain County, then another at 2223 UTC for northwestern Cleveland County, northwestern Grady County, and northwestern McClain County. Three minutes later, the second EF4 tornado touched down in Grady County (NOAA, NWS, 2011). The two EF4 tornadoes were well captured by the operational CASA IP1 radar network. Fig. 13 shows sample reflectivity observations from two CASA radars (i.e., KSAO and KRSP radar) at 2238 UTC. The two tornadoes are marked with white ellipses "A" and "B" in Fig. 13. In particular, tornado A was observed by both KRSP and KSAO radar for most of the time of evolution, which provides us an opportunity to conduct multiple-Doppler analysis. Therefore, tornado A is selected as one of the cases to be investigated in this chapter. This tornado reached the ground around 2226 UTC, traveled 23 miles, and lasted about 39 min. The peak wind was about 306 km/h. Fig. 14 shows the path of this EF4 tornado and its relative location to the CASA KSAO radar. The multiple-Doppler retrieval system was operating without any incident during this high-impact event. Fig. 15 illustrates sample wind vector products and reflectivity observations for an area of 16 km by 16 km in a 2-min span (2240–2241 UTC). This area was essentially where the tornado hook was during the time span. The latitude/longitude of the origin (0 km, 0 km) in Fig. 15 is (34.8276°N, −98.1007°W) (not shown), which is approximately the center of the CASA IP1 testbed.

The tornado moved continually to the northeast, until a short movement to the right from 2237 to 2241 UTC (see *blue rectangular box* in Fig. 14). This small turn was crucial for the real-time tornado forecast, especially because the operational WSR-88DP radar (KTLX radar in this case) did not have the spatial or temporal resolution to monitor such a small change. The CASA radar network, however, was ideal for this situation because of its high spatial and temporal resolution, which allowed emergency managers to more effectively direct citizens and save lives and property.

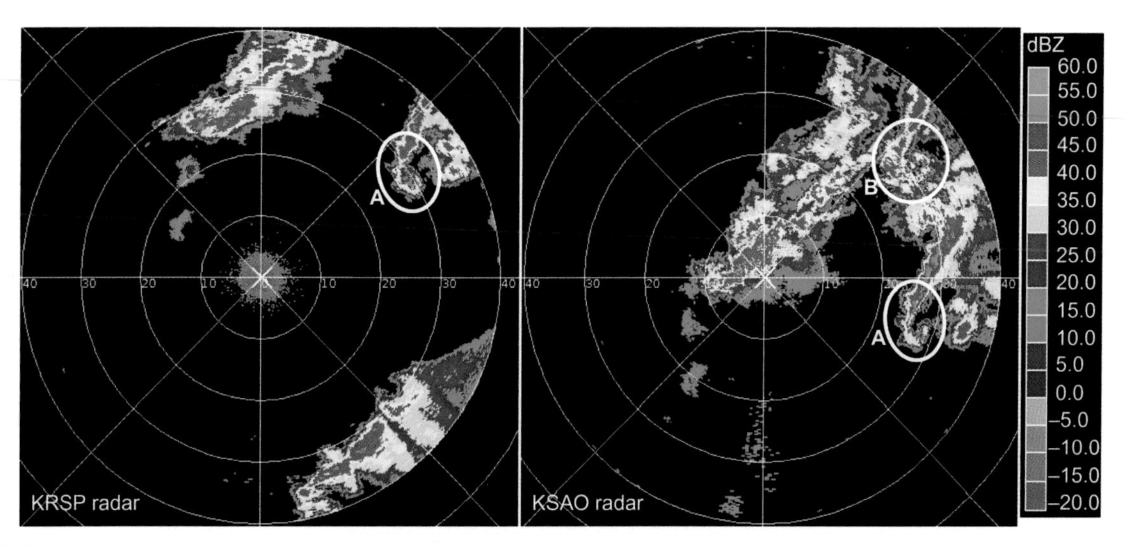

FIG. 13 Sample reflectivity observations from two CASA IP1 radars during the May 24, 2011, tornado event at 2238 UTC: (left) KRSP radar, (right) KSAO radar. The data are based on 2-degree elevation scans. The *white ellipses* denote tornado hook echoes. At this time frame, tornado "A" was observed by both KRSP and KSAO radars.

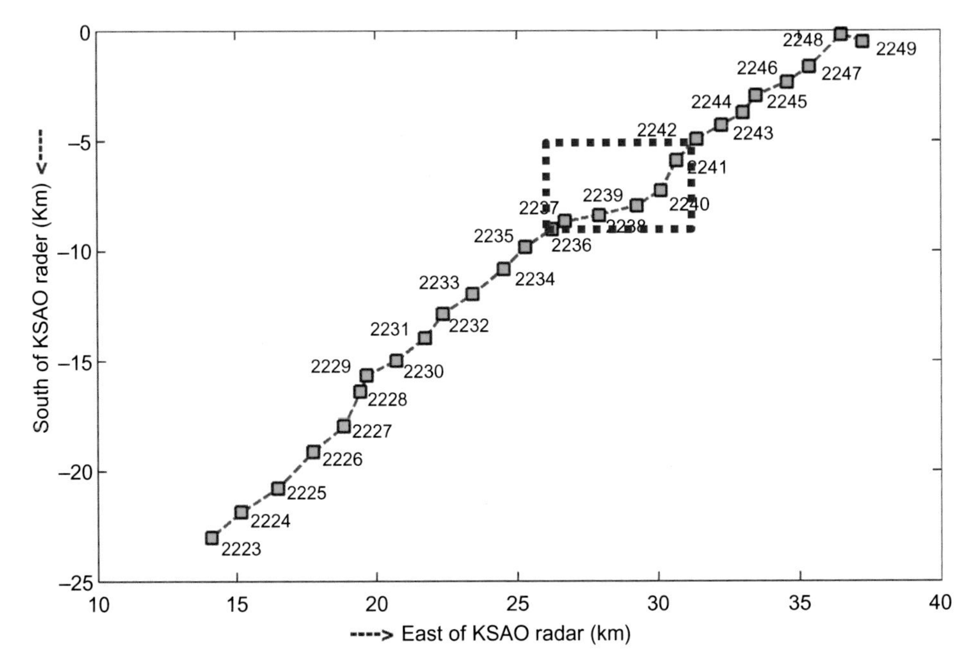

FIG. 14 Path of the May 24, 2011, EF4 tornado and its relative location to a CASA Doppler radar (i.e., KSAO radar). The tornado moved continually to the northeast, until a short movement to the right from 2237 to 2241 UTC. This small turn was crucial during real-time forecasting, especially because the operational NWS radar (KTLX radar in this case) does not have the spatial or temporal resolution to be able to sense this small movement.

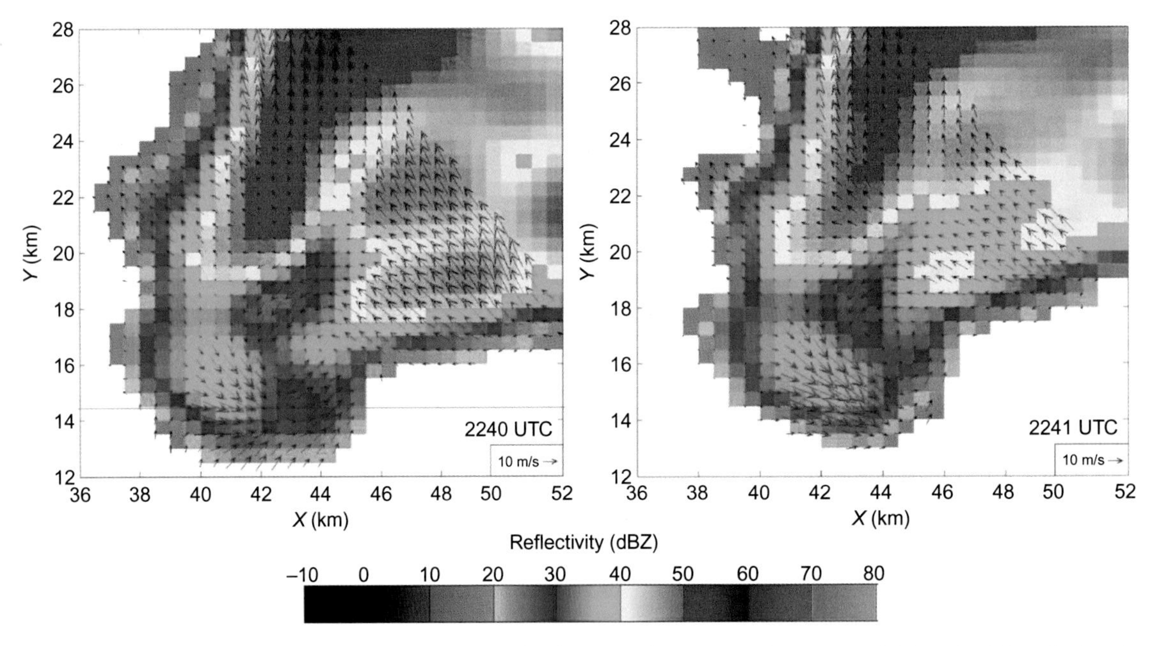

FIG. 15 Sample multiple-Doppler wind products and reflectivity observations for an area of 16 km by 16 km during the May 24, 2011, tornado event. This area is essentially where the tornado hook is. The lat/lon of origin (0 km, 0 km) is (34.8276°N, −98.1007°W), which is approximately the center of CASA IP1 testbed.

4.2 EF0 Tornado in the CASA DFW Network (May 8, 2014)

On May 8, 2014, a large-scale lift ahead of an upper-level shortwave, combined with ample instability and adequate moisture, evolved in the North Texas region. Severe thunderstorms were observed moving through this area. Scattered convection developed in the afternoon, and a linear mesoscale convective system had formed by the late afternoon. An EF0 tornado was reported in Cockrell Hill (32.757N, −96.889W) in Dallas County around 2014 UTC on May 8, 2014. Fig. 16 shows a screenshot of the NWS tornado report for this event. The tornado path length is about 800 m, and the path width is about 137 m, according to the NWS report. The tornado only lasted 2 min (2014–2015 UTC). Although there were no fatalities or injuries, the damaging downburst winds damaged numerous trees and brought down power lines across areas in and around Dallas. The most significant damage occurred in Dallas County due to damaging wind gusts with estimated peak winds of about 130 km/h. The EF0 tornado caused damage to a warehouse building in Cockrell Hill. Several windows were blown out of the warehouse. The building suffered roof damage as well as the tornado moved from the southwest to the northeast.

```
PUBLIC INFORMATION STATEMENT
NATIONAL WEATHER SERVICE FORT WORTH TX
411 PM CDT SAT MAY 10 2014

...NWS DAMAGE SURVEY FOR 05/08/14 TORNADO AND THUNDERSTORM WIND EVENT
UPDATE...

.OVERVIEW...SEVERE THUNDERSTORMS MOVED THROUGH THE NORTH TEXAS REGION ON
THURSDAY MAY 8TH. SURVEY CREWS FOUND EVIDENCE OF STRAIGHT LINE WIND DAMAGE IN
PARTS OF DALLAS AND JOHNSON COUNTIES IN THE DAMAGE SURVEYS CONDUCTED ON
FRIDAY. AN ADDITIONAL CREW WENT BACK OUT THIS MORNING AND FOUND DEFINITIVE
EVIDENCE OF A TORNADO IN COCKRELL HILL...IN THE AREA SOUTH OF I-30 AND NORTH
OF U.S. 180...AND EAST OF NORTH COCKRELL HILL ROAD...AND WEST OF NORTH
WESTMORLAND ROAD.

.TORNADO #1 COCKRELL HILL...

RATING:                   EF-0
ESTIMATED PEAK WIND:      80 MPH
PATH LENGTH /STATUTE/:    0.5 MILES
PATH WIDTH /MAXIMUM/:     150 YARDS
FATALITIES:               0
INJURIES:                 0

START DATE:               MAY 08 2014
START TIME:               314 PM CDT
START LOCATION:           COCKRELL HILL
START LAT/LON:            32.757 / -96.889

END DATE:                 MAY 08 2014
END TIME:                 315 PM CDT
END LOCATION:             COCKRELL HILL
END LAT/LON:              32.760 / -96.882
```

Available at http://www.srh.noaa.gov/data/warn_archive/FWD/PNS/0510_211208.txt

FIG. 16 NWS tornado report for the EF0 tornado event in Cockrell Hill on May 8, 2014.

For the entire event, the CASA multi-Doppler retrieval system was continuously operating, monitoring the event. Fig. 17 shows the reflectivity observations and multi-Doppler retrieval results during this EF0 tornado event. At 2014 UTC, the retrieved maximum velocity is about 70 mph, and it became 74 mph at 2015 UTC. The estimated vorticity and vortex are also shown in Fig. 17, from which we can clearly see the vortex evolution and tornado movement in a 2-min span. The vortex locations (*white crosses* in Fig. 17B and D) agree well with the location where the tornado was reported.

4.3 High-Winds in the CASA DFW Network (October 2, 2014)

On October 2, 2014, severe thunderstorms packing winds of up to 200 km/h tore through the DFW area. The severe storm began to develop shortly before 1800 UTC on October 2, 2014, when a severe thunderstorm watch was issued for most of the regions in North Texas. The storms developed near Jack, Wise, and Parker counties before moving east. The severe thunderstorm warning was effective until 2200 UTC for Dallas County. This fast-moving storm left widespread damage and power outages as winds

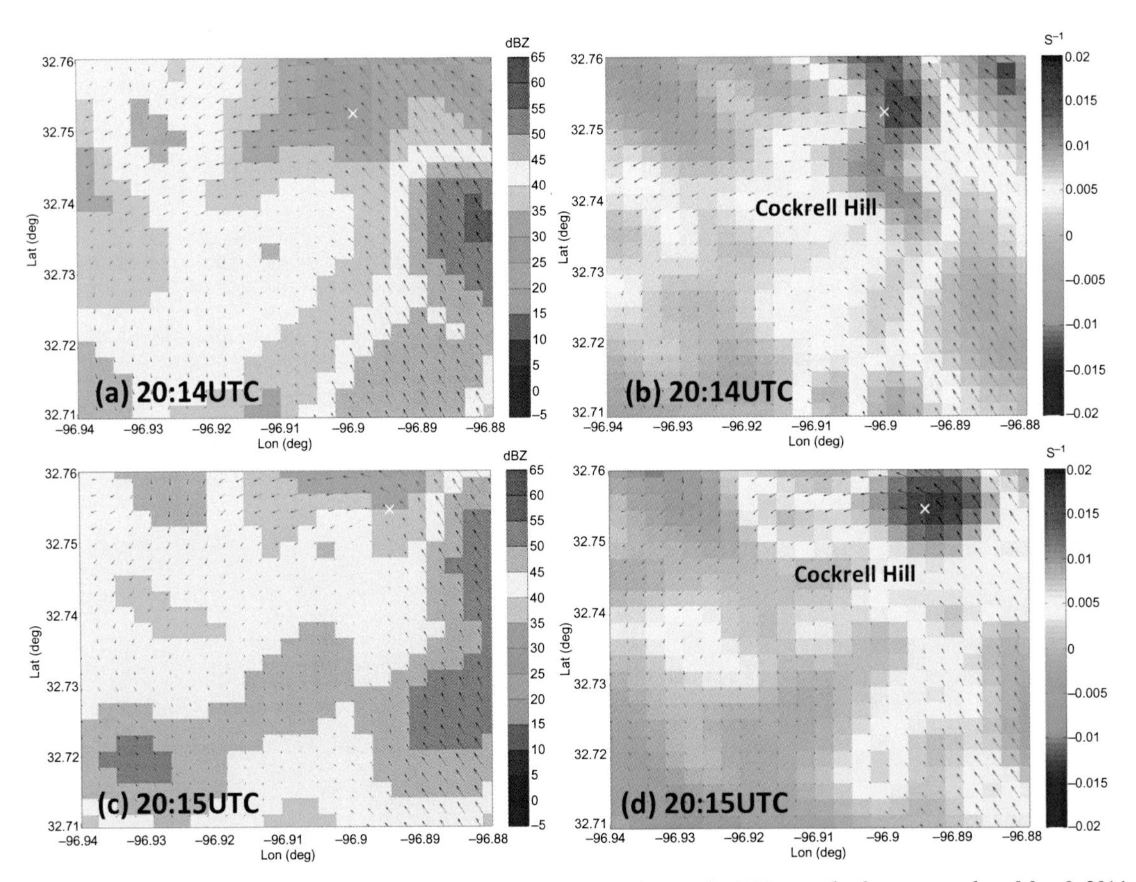

FIG. 17 Reflectivity observations and multiple-Doppler products during the EF0 tornado that occurred on May 8, 2014: (A) and (C) are the observed reflectivity overlaid with wind directions at 2014 UTC and 2015 UTC, respectively; whereas (B) and (D) illustrate the vorticity field and vortex (marked with *white crosses*) at 2014 UTC and 2015 UTC, respectively. The maximum velocity at 2014 UTC is about 113 km/h, and 120 km/h at 2015 UTC.

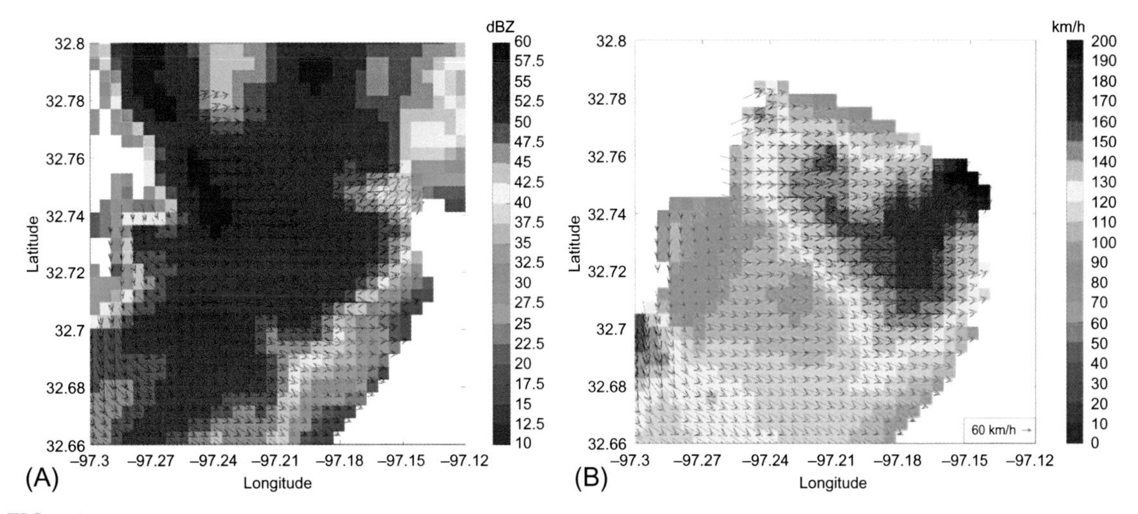

FIG. 18 Real-time multiple-Doppler retrieval products during the high-wind event in the DFW Metroplex on October 2, 2014, at 2053 UTC: (A) Composite reflectivity overlaid with estimated wind directions. (B) wind directions overlaid with wind speed. The maximum wind speed is about 200 km/h.

downed utility poles and tree limbs. Many flights were canceled at DFW International Airport.

It was concluded that the significant damage was not caused by rain (less than 10 mm of rain was observed at the DFW airport), but instead by straight-line winds. The multiple-Doppler retrieval system for the DFW urban radar network was operating during this high-wind event. Fig. 18 shows sample real-time Doppler wind retrieval products at 2053 UTC, when the peak wind was reported. The retrieved wind speed and directions are also indicated in Fig. 18. It can be seen that the peak wind speed reached about 200 km/h at the location near (−97.15°W, 32.75°N). The estimated peak wind and corresponding location agree fairly well with the ground report (NOAA, SPC, 2014), which again demonstrates the excellent performance of the designed multi-Doppler wind retrieval system.

4.4 Downbursts Over DFW International Airport (August 12, 2016)

August 12, 2016 was officially the hottest day of the year. DFW International Airport set a record by reaching 107°F, the highest in 4 years according to the NWS Fort Worth office. After temperatures at DFW Airport hit the record high at 1400 local time, they dropped more than 20 degrees to 85°F 2 h later, when a fast-moving thunderstorm passed through. As the cold front interacted with a low-pressure system moving from Louisiana, a severe thunderstorm warning was briefly issued for Fort Worth and Benbrook in the afternoon, and another severe thunderstorm warning was issued later for northeast Tarrant County and Dallas County. According to the NWS report, the storm caused some damage in Irving, downing utility poles, uprooting trees, and ripping part of the roof off an apartment complex. All inbound flights into DFW Airport were being held at their departing airport until 1715 (2215 UTC) to let the storms pass. Departures were delayed by 30–45 min. Particularly, several microbursts developed across Tarrant County and Dallas County.

The multiple-Doppler retrieval system of the CASA DFW operational radar network was able to capture the fine details of the evolution of downbursts during this event. Fig. 19 illustrates the reflectivity observations as well as the retrieved wind products with a focus on DFW International Airport (marked with *black*

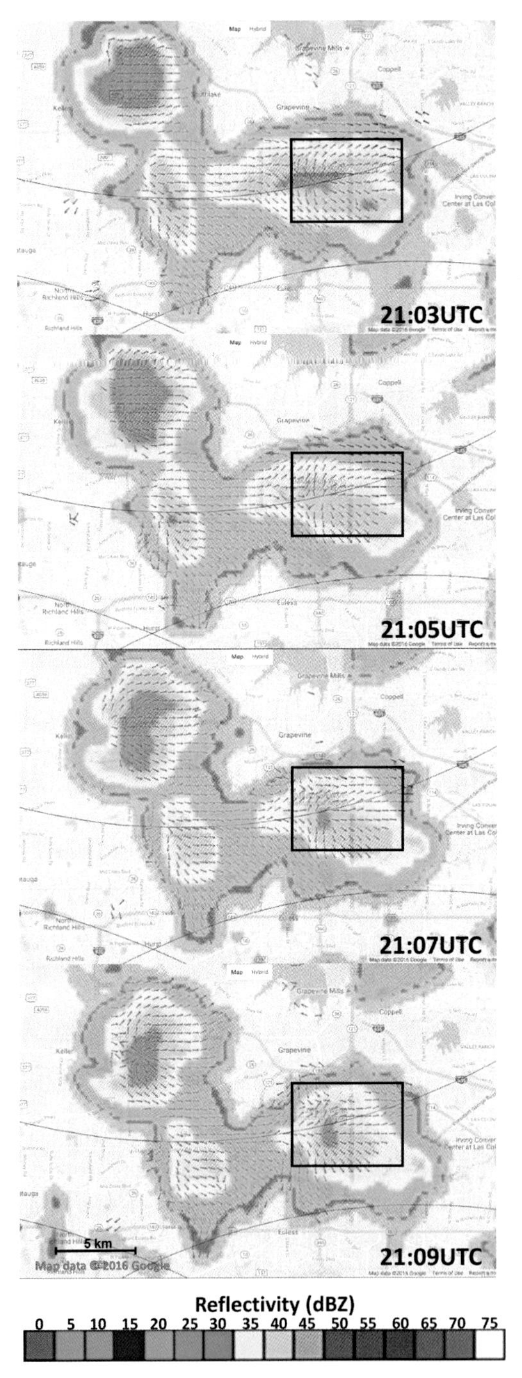

FIG. 19 Evolution of downbursts over the DFW International Airport region (marked with a *black rectangle*) on August 12, 2016. The downbursts formed and dissipated in a few minutes, moving across the airport. The wind directions are shown by *arrows in black*, and the background *color* represents the radar reflectivity.

rectangles), where a downburst occurred. The wind directions are represented by *arrows in black*. At 1603 (2103 UTC), the downburst was right in the DFW International Airport region, and it moved across the airport to the northwest in a few minutes. The CASA wind products were sent to the Federal Aviation Administration (FAA) air traffic management evaluators and dispatch instructors at airlines to help air traffic management.

4.5 Validation

Although the selected events presented in this chapter were verified by ground reports from NWS, especially at the extreme locations such as where the tornado or downburst occurred, the quantitative validation of real-time multiple-Doppler wind products is extremely difficult. In the research environment, simulated radar observations are often used to evaluate the consistency of the retrieved winds. In such an environment, a reference airflow derived from either an analytical wind field or cloud model outputs is typically sampled to generate simulated radar observations.

However, the exact real-time wind information cannot be observed by in situ instruments in practice, especially at the resolution of each grid pixel (500 m × 500 m for the CASA radar network), which obviously make it almost impossible to validate the real-time multiple-Doppler wind retrievals. Some attempts have been made on this ground. For example, Bousquet et al. (2016) used radial velocity measurements collected by an airborne radar to cross-compare with the multiple-Doppler winds produced by a radar network in southern France. Nevertheless, we want to note that for operational applications, such high-resolution validation is neither feasible nor necessary. By nature, the ground-based Doppler radars are able to provide desirable velocity information at broad spatial coverage. On the other hand, the vector wind products based on Doppler radar observations can be used to evaluate numerical model-based wind fields (Beck et al., 2014).

5 SUMMARY

Air motion retrieval has been pursued since the birth of Doppler radars. Traditional single-Doppler radar only measures the wind velocity component toward or away from the radar-pointing direction. Combining radial velocity data from several Doppler radars, the three-dimensional wind structure can be rendered. A large number of previous studies have been devoted to the multiple-Doppler radar analysis, including optimal radar network design and solutions to the mass continuity equation. Although the mathematical principles behind multiple-Doppler retrieval are relatively straightforward, implementing it in real-time in an operational radar network has never become true. One of the major limitations is that current operational radar networks are often incapable of providing dense observations, especially at upper levels. Besides the high sensitivity to errors and computational limitations for real-time applications, these fundamental coverage and resolution limitations greatly impede the application of current operational radar networks as a severe wind warning tool. Additional information such as sounding data has to be included to obtain accurate wind retrieval. To this end, research-based mobile Doppler radar systems have been developed to enhance the sensing of high-impact wind phenomena. Although the research-based products have greatly improved our understanding of severe weather events, they are very limited for operational deployment.

CASA has introduced an innovative sensing paradigm through deploying a dense network of short-range high-resolution dual-polarization X-band radars. Compared to the static scanning strategy adopted by state-of-the-art operational radars, CASA implements an electronic scan strategy that can optimize the scanning of each radar node in response to the changing weather features and end-user feedback. With the enhanced radar observations, especially in the lower troposphere, better monitoring and forecast of severe weather can be achieved in terms of QPE and forecast, hydrometeor classification, flash flood forecast, and three-dimensional wind retrieval. This chapter detailed the high-resolution real-time multiple-Doppler retrieval system designed for the operational CASA radar networks. The dynamic adaptive scan strategy for Doppler synthesis was described. Sample observations and products in the presence of tornadoes, high-winds, and microbursts in CASA's Oklahoma testbed and the DFW urban network were presented. The excellent performance of the real-time operational CASA wind products was demonstrated by ground reports from NWS and social media. In addition, the products are distributed to the NWS forecast office, the emergency management department, and other user groups for issuing warnings and other real-time applications.

Acknowledgments

This work was primarily supported by the U.S. National Science Foundation (NSF) Hazards SEES program. The support from NOAA/NWS and the North Central Texas Council of Governments (NCTCOG) is also acknowledged. The authors are grateful for the comments from key CASA team members, namely, Dr. Francesc Junyent from Colorado State University, Brenda Philips, Apoorva Bajaj, and Eric Lyons from the University of Massachusetts-Amherst.

References

Armijo, L., 1969. A theory for the determination of wind and precipitation velocities with Doppler radars. J. Atmos. Sci. 26, 570–573.

Beck, J., Nuret, M., Bousquet, O., 2014. Model wind field forecast verification using multiple-Doppler syntheses from a National Radar Network. Weather Forecast. 29, 331–348.

Bharadwaj, N., Chandrasekar, V., Junyent, F., 2010. Signal processing system for the CASA Integrated Project I radars. J. Atmos. Ocean. Technol. 27, 1440–1460.

Bousquet, O., Delanoë, J., Bielli, S., 2016. Evaluation of 3D wind observations inferred from the analysis of airborne and ground-based radars during HyMeX SOP-1. Q. J. R. Meteorol. Soc. 142, 86–94.

Bringi, V.N., Chandrasekar, V., 2001. Polarimetric Doppler Weather Radar: Principles and Applications. Cambridge University Press, Cambridge, United Kingdom.

Browning, K.A., Wexler, R., 1968. The determination of kinematic properties of a wind field using Doppler radar. J. Appl. Meteorol. 7, 105–113.

Chandrasekar, V., Wang, Y., Chen, H., 2012. The CASA quantitative precipitation estimation system: a five year validation study. Nat. Hazards Earth Syst. Sci. 12, 2811–2820.

Chandrasekar, V., Chen, H., Philips, B., Seo, D.J., et al., 2013. The CASA Dallas Fort Worth Remote Sensing Network ICT for Urban Disaster Mitigation. EGU General Assembly, Vienna.

Chandrasekar V., Chen H. and Philips B.J., Principles of high-resolution radar network for hazard mitigation and disaster management in an urban environment, J. Meteorol. Soc. Jpn. in press.

Chen, H., Chandrasekar, V., 2015. The quantitative precipitation estimation system for Dallas–Fort Worth (DFW) urban remote sensing network. J. Hydrol. 531, 259–271.

Chong, M., Testud, J., Roux, F., 1983. Three-dimensional wind field analysis from dual-Doppler radar data. Part II: minimizing the error due to temporal variation. J. Clim. Appl. Meteorol. 22, 1216–1226.

Cressman, G.P., 1959. An operational objective analysis system. Mon. Weather Rev. 87, 367–374.

Davies-Jones, R.P., 1979. Dual-Doppler Radar coverage area as a function of measurement accuracy and spatial resolution. J. Appl. Meteorol. 18, 1229–1233.

Gal-Chen, T., 1982. Errors in fixed and moving frame of references: applications for conventional and Doppler radar analysis. J. Atmos. Sci. 39, 2279–2300.

Junyent, F., Chandrasekar, V., McLaughlin, D., Insanic, E., Bharadwaj, N., 2010. The CASA Integrated Project 1 networked radar system. J. Atmos. Ocean. Technol. 27, 61–78.

McLaughlin, D., et al., 2009. Short-wavelength technology and the potential for distributed networks of small radar systems. Bull. Am. Meteorol. Soc. 90, 1797–1817.

Miller, L.J., Anderson, W., 1991. Multiple-Doppler Radar Wind Synthesis in CEDRIC. National Center for Atmospheric Research, Mesoscale and Microscale Meterology Division, Boulder, Colorado, USA.

Miller, L.J., Strauch, R.G., 1974. A dual Doppler radar method for the determination of wind velocities within precipitating weather systems. Remote Sens. Environ. 3, 219–235.

NOAA, NWS, 2011. Information for the Tornado Outbreak of May 24, 2011. National Oceanic and Atmospheric

Administration, National Weather Service Weather Forecast Office, Norman, OK. Available from: http://www.crh.noaa.gov/oun/?n=events-20110524. (Cited 2011).

NOAA, SPC, 2014. 20141002's Storm Reports. National Oceanic and Atmospheric Administration Storm Prediction Center, Norman, OK. Available from: http://www.spc.noaa.gov/exper/archive/event.php?date=20141002. (Cited 2014).

Oye, D., Case, M., 1995. REORDER: A Program for Gridding Radar Data-Installation and Use Manual for the UNIX Version. National Center for Atmospheric Research, Atmospheric Technology Division, Boulder, Colorado, USA.

Perkins, S., 2002. Tornado alley, USA: new map defines nation's twister risk. Sci. News 161, 296–298.

Schelleng, J.C., Burrows, C.R., Ferrel, E.B., 1933. Ultra-short wave propagation. Proc. IRE 21, 427–463.

Wang, Y., Chandrasekar, V., Dolan, B., 2008. Development of scan strategy for dual Doppler retrieval in a networked radar system. In: Proceedings of International Geoscience and Remote Sensing Symposium 2008.

Index

Note: Page numbers followed by *f* indicate figures, *t* indicate tables, and *b* indicate boxes.